Indian Clinical
Neurosurgery

Volume II
Malignant Brain Tumors

Indian Clinical Neurosurgery

Volume II
Malignant Brain Tumors

Editor
Anil K Singh MS, MCh.
Professor & Head, Department of Neurosurgery
G.B. Pant Hospital, New Delhi

Associate Editors
Sanjiv Sinha
Daljit Singh
Vikas Gupta
Ashish Goyal

CBS PUBLISHERS & DISTRIBUTORS
4596/1-A, 11 Darya Ganj, New Delhi -110 002 (India)
CBS Homepage : http://www.cbspd.com

ISBN: 81-239-0801-6

First Edition 2002

© Editors & Publisher
All right reserved. No part of this publication may be reproduced, stored in a retrieval system, or transmitted in any form, or by any means-electronic, mechanical, photo-copying, recording or otherwise without the prior permission of the editors and publishers.

Published and Distributed by:
Satish K. Jain for
CBS Publishers & Distributors
4596/1A, 11, Darya Ganj, New Delhi-110 002

Production Director : Vinod K. Jain

The editors and publishers have made efforts to ascertain the authenticity of the material provided by the contributors and to trace the copyright holders for borrowed material. If any omissions have occurred and publishers will be pleased to make the necessary arrangements at the first opportunity.

Laser type setting by:
NS Computers, New Delhi (India)

Printed at:
Asia Printographs, Shahdara, Delhi

Foreword

I am glad to learn that the second volume of the Indian Clinics of Neurological Surgery is being brought out by Dr. A K Singh and his team. This volume is based on the presentations at the 2nd International Clinic of Neurological Surgery held at G B Pant Hospital in September 2001, and which focussed on the management of malignant brain tumors.

These clinics are a welcome addition to the continuing medical education programmes run by different societies and institutions. The unique feature of these clinics is their emphasis on selected areas of neurosurgical practice – a format which permits in depth discussion of the chosen problems. The second clinic discussed the problems of managing these hopeless tumors of brain in a holistic manner—from pathology to the latest modalities of treatment. The chapter on meta-analysis is an eye opener—have the advanced armamentarium at our disposal really made all that much difference to the eventual outcome? It also emphasizes the need for more Class II and if possible Class I evidence regarding the best modality of treatment of these lesions. I hope the younger generation will pick up the gauntlet.

The chapters are well written presenting the different perspectives of various clinicians involved in the care of these lesions and the references are adequate. This book should be of immense practical use to all those who take it upon themselves to manage malignant lesions of the brain—the proverbial last frontier of clinical medicine. One very good thing about these books, though containing all the relevant and uptodate information, is that they are affordable by most Indians.

I hope the editors will continue in their efforts to bring many more such volumes devoted to different aspects of neurosurgical practice.

PROF. S N BHAGWATI
Director, Neurosurgery
Bombay Hospital
Mumbai

Acknowledgement

The Editors are grateful to all the contributors who, despite their hectic schedules, were able to send in their manuscripts in time, and then also to return the same to us after incorporating our suggestions in the very short time that we were able to give them.

This book would not have been possible but for the persistent efforts of Mr. S K Jain, Mr. B R Sharma and Mr. K B Sharma of CBS Publishers & Distributors. Had they relented even an inch, we would have taken them for the entire rope. Their constant suggestions and solutions are largely responsible for making the material presentable. We thank them.

We also acknowledge the efforts of our senior residents, who not only relieved us of some of our clinical duties, but also actively contributed to the writing of chapters by collecting the material. Mr. Sree Kumar, our web administrator, has been very helpful in obtaining the latest information from the net. Mr. Chadha and Mr. Deepak have helped in typing, and then retyping, the manuscripts of the contributions from our department.

Lastly, and most importantly, we are keenly aware of the sacrifices made by our families over the past 2 months so that we could concentrate on writing and editing this book so as to meet the publishers' deadline. May God bless them.

Room No. 523, Academic Block,	**Editors**
G B Pant Hospital,	*A K Singh*
New Delhi 110 002	*S Sinha*
	D Singh
	V Gupta
	A Goyal

Preface

Gliomas constitute a significant segment of the clinical load of an average practising neurosurgeon, and by extension of practising neuropathologists and neuroradiologists as well. They remain a challenge to all neuroscientists even though recent years have seen some promising breakthroughs in our understanding of the molecular basis of these tumors. There are newer modalities of treatment which offer better response rates, though none has, as yet, made a statistically significant difference in a clinical setting.

The second Indian Clinic of Neurological Surgery, held at GB Pant Hospital, New Delhi, from 24th to 26th September, 2001, was devoted to discussing various aspects of pathology, presentation, diagnosis, and treatment of gliomas. Professor P N Tandon, Emeritus Professor of Neurosurgery at AIIMS, New Delhi, who has devoted a lifetime to the study of gliomas and who continues to be actively associated with research in their biological behaviour, gave the keynote address providing a bird's eye view of the evolution of management of these lesions. The participating faculty represented experiences from across the breadth and length of this country. Unfortunately the Canadian faculty was unable to attend due to the September 11 terrorist attacks on USA.

This, the second volume of Indian Clinical Neurosurgery, like the first volume, is based on the presentations made during the Clinic. The articles presented here endeavour to review various aspects of management of gliomas. The presence of conflicting views reflects not only the varying experiences of different practitioners but also the imperfect state of our art and science. As we tread further along the road, in future, some of these differences are likely to be resolved. However, anyone desirous of resolving these differences will have to first acquaint himself with what has been done already. This volume makes an attempt to provide that information.

We, the editors, are glad that the first volume found acceptance amongst students and practitioners. We hope that this volume will also be found useful by those for whom it is meant, i.e. the students and the busy practitioners who do not have the time to scour all the current literature on the subject.

15th December 2001
New Delhi

Editor
A K Singh

Associate Editors
S Sinha, D Singh
V Gupta, A Goyal

List of Contributors

ABRAHAM, MARY, MD — Prof., Dept. of Anaesthesia, GB Pant Hospital, New Delhi, India

AHUJA, NIRAJ, MD — Assistant Professor, Dept. of Psychiatry, GB Pant Hospital, New Delhi, India

ARORA, KAVITA, MD — Sr. Res. Associate, Dept. of Psychiatry, GB Pant Hospital & MAMC, New Delhi, India

BAHADUR, AK — Prof., Dept. of Radiation Oncology, MAMC, New Delhi, India

BAMPOE, JOSEPH, MB, ChB, FRCS(I) — UHN, Univ. of Toronto, Canada

BANKATA, SUDHANSHU — Consultant, Radiology, MR Centre, New Delhi, India

BATRA, ARUN — INMAS, Delhi, India

BEHARI, SANJAY — Assistant Professor, Dept. of Neurosurgery, SGPGIMS, Lucknow, India

BERNSTEIN, MARK, MD, FRCS(C) — UHN, Univ. of Toronto, Canada

CHANDRA, PS — Assistant Prof., Dept. of Neurosurgery, AIIMS, New Delhi, India

CHANDRAMOULI, BA — Prof., Dept. of Neurosurgery, NIMHANS, Bangalore, India

CHAUDHURI, SWAPNA — Dept. of Physiology, Univ. College of Medicine, Kolkata. India

CHAUDHURI, S — Dept. of Neurosurgery, Bangur Inst. of Neurology, Kolkata, India

CHOWDHURY, DEBASHISH, MD, DM — Assistant Professor, Dept. of Neurology, GB Pant Hospital, Delhi University, New Delhi, India

DASH, HARI HARA — Prof. & Head, Dept. of Neuro-anesthesia, AIIMS, New Delhi, India

DEOPUJARI, CE, MS, MCh. MSC — Bombay Hospital, Inst. of Medical Sciences, Mumbai, India

DUA, SANJEEV, MS, MCh — Head, Dept. of Neurosurgery, GTB Hospital, New Delhi, India

DUA, VIKRAM, MS, MCh — Sr. Resident, Dept. of Neurosurgery, GB Pant Hospital, New Delhi, India

DUBEY, SUDHIR — Dept. of Neurosurgery, NIMHANS, Bangalore, India

GANAPATHY, K, MS (NEURO), MNAMS (NEURO), FACS, FICS, Ph.D, Sr. Consultant Neurosurgeon, Apollo Hospital, Chennai, India

GANJOO, P — Associate Professor, Dept. of Anaesthesia, GB Pant Hospital, New Delhi, India

GOYAL, ASHISH, MS, MCh — Sr. Research Associate, Dept. of Neurosurgery, GB Pant Hospital, New Delhi, India

GUHA, ABHIJIT, MD, FRCSC, FACS — Prof., Dept. of Neurosurgery, UHN, Univ. of Toronto, Canada

GULATI, PRAVEEN, MD — Consultant, Radiology, MR Centre, New Delhi, India

GUPTA, VIKAS, MS, MCh — Sr. Research Associate, Dept. of Neurosurgery, GB Pant Hospital, New Delhi, India

HALDAR, S, MD — Dept. of Radiation Oncology, Rajiv Gandhi Cancer Inst., New Delhi, India

HUKKU, S, MD — Dept. of Radiation Oncology, Rajiv Gandhi Cancer Inst., New Delhi, India

HUSSAIN, M, MS, MCh — Prof. & Head, Dept. of Neurosurgery, KG Medical College, Lucknow, India

JAIN, S — Addl. Prof., Dept. of Radiology, SGPGIMS, Lucknow, India

JAIN, VK — Prof. & Head, Dept. of Neurosurgery, SGPGIMS, Lucknow, India

JILOHA, RC, MD — Prof. & Head, Dept. of Psychiatry, GB Pant Hospital & MAMC, New Delhi, India

KHOSLA, VK, MS, MCh — Prof. & Head, Dept. of Neurosurgery, PGIMER, Chandigarh, India

KHUSHU, S — INMAS, Delhi, India

KRISHNAN, PRASAD — Dept. of Neurosurgery, SGPGIMS, Lucknow, India

KULKARNI, VAIJAYANTEE — Dept. of Neurosurgery, CMC, Vellore, India

KUMAR, A — Dept. of Neurosurgery, GTB Hospital, New Delhi, India

KUMAR, SHALEEN — Dept. of Radiology, SGPGIMS, Lucknow, India

KUMAR, SUSHIL, MS, MCh — Dean, MAMC, New Delhi, India

LAPERRIERE, NORMAND J, MD — Univ. of Toronto, Ontario, Canada

MAHADEVAN, A — Dept. of Neuropathology, NIMHANS, Bangalore, India

MAHAJAN, HARSH, MD — Dept. of Nuclear Medicine and Bone Densitometry, Sir Ganga Ram Hospital, New Delhi, India

MAHAPATRA, AK — Prof., Dept. of Neurosurgery, AIIMS, New Delhi, India

MEHTA, VS, MS, MCh — Prof. & Head, Neurosurgery, AIIMS, New Delhi, India

MISHRA, AJAY — Sr. Res. Associate, Dept. of Neurosurgery, GTB Hospital, New Delhi, India

MISHRA, JAYANT, MCh — Consultant Neurosurgeon, VIMHANS, New Delhi, India

MOHANTY, BIBEKANANDA — Assistant Professor, Dept. of Neuro-anaesthesia, AIIMS, New Delhi, India

MOORTHY, RANJITH K — Dept. of Neurosurgery, CMC, Vellore, India

MUKHOPADHYAY, SANJAY — Dept. of Pathology, AIIMS, New Delhi, India

PANIGRAHI, MANAS — Dept. of Neurosurgery, NIMS, Hyderabad, India

PANKAJ, PROMILA — Dept. of Nuclear Medicine and Bone Densitometry, Sir Ganga Ram Hospital, New Delhi, India

PRAHARAJ, SS — Dept. of Neurosurgery, NIMHANS, Bangalore, India

RAJSHEKHAR, VEDANTAM — Prof., Dept. of Neurosurgery, CMC, Vellore, India

RATH, GK, MD — Prof., Dept. of Radiation Oncology, AIIMS, New Delhi, India

RAVI, V — Dept. of Neuropathology, NIMHANS, Bangalore, India

REDDI, RAJASHEKHAR — Sr. Resident, Dept. of Neurology, Sir Ganga Ram Hospital, New Delhi, India

REDDY, AK, MS, MCh — Prof. & Head, Deptt. of Neurosurgery, NIMS, Hyderabad, India

SARKAR, CHITRA — Prof. & Head, Dept. of Neuro Pathology, AIIMS, New Delhi, India

SARKAR, S — Dept. of Haematology, School of Tropical Medicine, Kolkata, India

SASTRY, KVR — Prof. & Head, Dept. of Neurosurgery, NIMHANS, Bangalore, India

SAWLANI, V — Asst. Prof., Dept. of Radiology and Radiotherapy, SGPGIMS, Lucknow, India

SEN, ISHITA — Dept. of Nuclear Medicine and Bone Densitometry, Sir Ganga Ram Hospital, New Delhi, India

SETHI, NITIN K — Sr. Resident, Dept. of Medicine, Sir Ganga Ram Hospital, New Delhi, India

SETHI, PK — Chairman, Dept. of Neurology, Sir Ganga Ram Hospital, New Delhi, India

SHANKAR A — Dept. of Neurosurgery, CMC, Vellore, India

SHANKAR, SK — Dept. of Neurosurgery, NIMHANS, Bangalore, India

SHARMA, AJAY, MS, MCh — Prof., Dept. of Neurosurgery, GB Pant Hospital, Delhi University, New Delhi, India

SHARMA, DN, MD — Dept. of Radiation Oncology, AIIMS, New Delhi, India

SHARMA, MEHAR CHAND — Dept. of Neuro Pathology, AIIMS, New Delhi, India

SINGH, AK, MS, MCh — Prof. & Head, Dept. of Neurosurgery, GB Pant Hospital, Delhi University, New Delhi, India

SINGH, DALJIT, MS, MCh — Associate Prof., Dept. of Neurosurgery, GB Pant Hospital, Delhi University, New Delhi, India

SINGH PRAKASH — Consultant, Dept. of Neurosurgery, RR Centre, Army Medical Corps, New Delhi, India

SINHA, SANJIV, MS, MCh — Associate Professor, Dept. of Neurosurgery, GB Pant Hospital, Delhi University, New Delhi, India

SRIDHAR, K, MS, DNB — Consultant Neurosurgeon, VHS Hospital, Chennai, India

SUBIMAL ROY — Consultant Pathologist, Sir Ganga Ram Hospital, New Delhi

TANDON, MONICA, MD — Assistant Professor, Dept. of Anaesthesia, GB Pant Hospital, New Delhi, India

TANDON, PN, MS, FRCS, FNA, DSC — Chairman, National Brain Research Centre, New Delhi, India

TATKE, MEDHA MD, Ph.D. — Prof., Dept. of Pathology, GB Pant Hospital, New Delhi, India

TRIPATHI, RP, MD, MAMS, FICRI — Head, NMR Research Centre, INMAS, Delhi, India

TRIPATHY, P — Dept. of Neurosurgery, Nizam's Institute of Medical Sciences, Hyderabad, India

VANI SANTOSH — Associate Prof., Dept. of Pathology, NIMHANS, Bangalore, India

VATSAL, DK, MS, MCh — Sr. Research Associate, Dept. of Neurosurgery, K G Medical College, Lucknow, India

VENNIYOOR, AJ — Dept. of Neurosurgery, RR Centre, Army Medical Corps, New Delhi, India

ZADEH, GELARE, MD — UHN, Univ. of Toronto, Canada

Contents

Foreword — v

Preface — vii

List of Contributors — ix

I. Gliomas – General Overview

1. Glioma Therapy: Past, Present & Future
   – PN Tandon — 1

2. Epidemiology and Classification of Malignant Brain Tumours
   – Subimal Roy, Chitra Sarkar, Mehar Chand Sharma — 11

3. Tumor Markers: An Overview
   – Chitra Sarkar, Mehar Chand Sharma, Sanjay Mukhopadhyay — 35

4. Role of Radiotherapy in Malignant Gliomas
   – GK Rath, DN Sharma — 56

5. Stereotactic Radiosurgery for Malignant Gliomas
   – K Ganapathy — 68

6. Stereotactic Biopsy: Techniques, Indications, Limitations and Complications
   – Vaijayantee Kulkarni, Vedantam Rajshekhar — 80

7. Brain Tumour Experimental Trial Designs
   – Normand J Laperriere — 90

II. Principles of Surgery

8. Planning and Executing a Classical Approach to Supratentorial Gliomas
   – AK Singh, Vikas Gupta, Vikram Dua — 97

9. What is Adequate surgery for Glioma? The rationale for cytoreduction
   – SS Praharaj, Sudhir Dubey, KVR Sastry — 108

10. Surface vs Deep seated lesion: different or same surgical strategy
    – A K Reddy, Pradipta Tripathy, Manas Panigrahi — 117

11. Surgery for Eloquent Area Glioma
    – S Sinha, Daljit Singh, Ashish Goyal — 126

12. Stereotactic Biopsy and Radiotherapy for Supratentorial Eloquent Region Gliomas
 — A Shankar, Vedantam Rajshekhar ... 132

13. Surgical Strategies for Intraventricular Gliomas
 — CE Deopujari ... 137

14. Is Microsurgery required in every case of Intraparenchymal Glioma?
 — K Sridhar ... 143

15. Imaging in Glioma
 — Praveen Gulati, Sudhanshu Bankata ... 149

16. Post-operative imaging—Which and when?
 — V Sawlani, S Jain, Shaleen Kumar ... 160

17. Closure Techniques and Controversies
 — Sushil Kumar, Vikas Gupta, Daljit Singh ... 167

18. Awake Craniotomy (Conscious sedation) for mapping of eloquent areas of the brain in Neurosurgical practice
 — Daljit Singh, S Sinha, A Goyal ... 173

III. Low Grade Gliomas

19. Low Grade Gliomas–Case for delayed surgery
 — VK Jain, Prasad Krishnan, Sanjay Behari ... 180

20. Low Grade Gliomas: Early Radiotherapy
 — Normand J Laperriere ... 190

21. Delayed Radiation Therapy in Low Grade Gliomas
 — AK Bahadur ... 195

22. A Meta-Analysis of low grade Astrocytomas
 — Sanjeev Dua, Ajay Mishra, A Kumar ... 198

IV. Brain Stem Gliomas

23. Surgical Approaches to Brain Stem Gliomas
 — VS Mehta, PS Chandra ... 217

24. Brain Stem Gliomas: Stereotactic Biopsy and Radiotherapy
 — Ranjith K Moorthy, Vedantam Rajshekhar ... 230

25. Brain Stem Gliomas: Radiotherapy without Biopsy
 — Normand J Laperriere ... 238

26. Perioperative Management of Patients with Brainstem Gliomas
 — Bibekananda Mohanty, Hari Hara Dash ... 242

V. Advances in the management of Gliomas

27. Biological Therapies of Human Astrocytomas — Gelah Zadeh, Abhijit Guha ... 256

28. Immunomodulation of Gliomas — Swapna Chaudhuri, S Sarkar, S Chaudhuri ... 280

29. Cortical Mapping and Functional MRI: Value in surgical planning — RP Tripathi, Arun Batra, S Khushu ... 307

30. Role of Rapid Histopathological and Cytological Diagnostic Techniques — Medha Tatke ... 315

31. Malignant Gliomas: Radiosensitizers — Normand J Laperriere ... 324

VI. Metastasis

32. Intracranial Metastasis – A Dilemma: to Operate or not to Operate — Prakash Singh, AJ Venniyoor ... 330

33. Surgical Management of Intracranial Metastasis — Joseph Bampoe, Mark Bernstein ... 341

34. Stereotactic Radiosurgery for Brain Secondaries — K Ganapathy ... 350

VII. Management of Posterior Fossa Malignant Tumors

35. Surgical Management of Medulloblastomas — BA Chandramouli ... 358

36. Posterior Fossa Ependymomas — M Hussain, D K Vatsal ... 366

37. Controversies in Management of Malignant Posterior Fossa Tumours — AK Mahapatra ... 375

38. Anaesthetic Considerations in Paediatric Posterior Fossa Surgery — Mary Abraham, Monica S Tandon, Pragati Ganjoo ... 387

39. Malignant Tumors of the Posterior Fossa – Radiotherapy and Chemotherapy — S Hukku, S Haldar ... 404

VIII. Pineal Masses: Surgical Management

40. How I do it – Pineal Tumor — VK Khosla ... 409

41. Surgery for Pineal Region Tumours in Children
 – Ajay Sharma ... 417

IX. Medical Management of Gliomas

42. Medical Management of Gliomas and Terminal Care
 – Debashish Chowdhury ... 424

43. Neuropsychiatric Aspects of Brain Tumors: A Focus on Gliomas
 – Niraj Ahuja, Kavita Arora, RC Jiloha ... 437

X. Recurrent Gliomas

44. Tumor Markers in Neuro-Oncology – Diagnostic and Prognostic Significance in Recurrent Gliomas
 – Vani Santosh, SK Shankar, S Dubey, BA Chandramouli ... 451

45. Recurrence versus Radionecrosis: An imaging challenge
 – Ishita Sen, Promila Pankaj, Harsh Mahajan ... 462

46. Reoperation for Recurrent Gliomas—Rationale and Techniques
 – AK Singh, Vikas Gupta, S Sinha ... 466

47. Stereotactic Irradiation for Recurrent Gliomas
 – K Ganapathy ... 476

48. Gamma Knife Radiosurgery of Recurrent Gliomas and Metastasis
 – Jayant Mishra ... 481

XI. Neuro AIDS

49. Epidemiology and Pathology of CNS Tumors in AIDS
 – Vani Santosh, A Mahadevan, V Ravi ... 489

50. Management of Central Nervous System AIDS Tumours
 – PK Sethi, Rajashekhar Reddi, Nitin K Sethi ... 500

1
Glioma Therapy: Past, Present and Future

PN Tandon

"The surgery for brain tumors may be likened, without being trivial, to a form of major sport which is played against an invisible but utterly relentless antagonist quick to take advantage of every misplay and faulty move". Harvey Cushing

INTRODUCTION

After nearly 120 years since Godlee[1] performed the first operation for a glioma in 1884, there have been many advances in diagnostic techniques, anaesthetic and pre and post operative management strategies, surgical refinements and adjuvant therapies (Radiotherapy, Chemotherapy, Immunotherapy and Genetherapy). All these have won us many battles, but winning the war is still not in sight. No doubt a glioma can now be diagnosed when it is no more than a centimeter, it can be "completely" (at least macroscopically) excised without inflicting any damage to the surrounding brain and with negligible operative mortality. We know a great deal about its molecular biology and pathology and have means to target oncolytic radiation, drugs, or genes to the tumour, sparing the surrounding brain. Notwithstanding all these advances, it is an undisputed fact that a majority, if not all, victims of atleast malignant gliomas, die of their tumour unconquered. Table 1 reflects the current state of our knowledge.

Most of the discussion that follows is primarily concerned with malignant gliomas (anaplastic astrocytomas and glioblastomas). No doubt much of this is also applicable to the low grade tumours as well.

Diagnosis: One of the greatest advances in the management of gliomas has been in respect to diagnosis. The current generation of neurosurgeons cannot visualize the problems associated with using the invasive diagnostic procedures, pneumo/ventriculography or angiography with their limited information, difficulties in interpretation and inherent discomfort and risks to the patient. CT and MR have no doubt revolutionized this field. It is now possible to visualize even very small lesions, as well as those located in depth. The extent of the tumour, the degree of associated changes like necrosis, cystic degenerative and oedema are clearly seen. MR spectroscopy and PET scan can reveal the metabolic characteristics of the tumours. Nevertheless, the problem of unequivocally determining

Table 1 : *Gliomas*

Christened by VIRCHOW in 1863
Initially classified by BAILEY & CUSHING 1926
First Reported Surgery BENNETT & GODLEE 1884

ETIOLOGY	:	UNKNOWN
PATHOGENESIS	:	ILL UNDERSTOOD
PATHOLOGY	:	UNABLE TO PREDICT BIOLOGY
IMMUNOLOGY	:	INTRIGUING
MOLECULAR BIOLOGY	:	INCOMPLETE INFORMATION
DIAGNOSIS	:	GENERALLY DELAYED
TREATMENT	:	LACKS CONSENSUS
PROGNOSIS	:	DISMAL
RESEARCH EFFORTS	:	UNCRITICAL

NO CURE IN SIGHT

---

the histological nature of the lesion and more so the histological grade of the tumour persists.[2,3] In some respects MRI is an improvement on CT, and if MR spectroscopy is added one can have better idea about the nature and grade of the lesion. Three dimensional images combined with functional MR (fMR) can be utilized to delineate the relationship of the tumour to eloquent regions of the brain. Collectively all this information adds to the level and degree of diagnostic capability. For an individual patient, however, biopsy and histological examination remain the gold standard.

Surgery: Surgery remains the mainstay of the treatment. Thankfully the external decompression of early years, which prolonged the survival marginally but at the cost of an ugly, painful swelling at the site of operation, which hardly relieved the suffering of the patient but distressed the family, is no more practiced. McKenzie's refinement of internal decompression was no doubt a step in the right direction but carried a mortality of nearly 50% till as late as 1940s. Many surgeons advocated free hand needle biopsy, only adding to the mortality rate. Advances in anaesthetic and surgical techniques helped to reduce this mortality by half over the next two decades. Introduction of dexamethasone in the pre and post operative period, first recommended by Galicich et al,[4] proved to be a distinct landmark in reducing the operative mortality dramatically. Currently the operative mortality is no more than 1-2 per cent. Simultaneously there has been reduction in post-operative morbidity and improvement in the quality of survival. But there have been only marginal gains in the duration of survival. There has been a progressively increasing consensus that "maximal" or "radical" tumour removal, without risking any damage to the surrounding brain is superior to only biopsy or partial resection.[5-10]

It may be interesting to quote two studies to illustrate that till a decade ago there were conflicting views expressed regarding the extent of surgery. Winger et al claimed, "The present series of unselected patients demonstrated *for the first time* (italics added by present author) that extent of resection is a significant prognostic factor independent of tumor location and all other parameters". They concluded that patients with gross resection lived longer than those with partial resection. Lobectomy did not improve survival time in the sub-total resection group. In contrast to this Nazzara and Neuwelt et al[11] equally vehemently challenged this. Following a review of last 50 years of publications they claimed, "This analysis shows that there is *little justification for dogmatic statements concerning the relationship between increasing patients survival times and aggressive surgical management* in adult with supratentorial intermediate and high grade astrocytomas if patients receive post-operative radiotherapy". (Italics added by present author).

Personally the present author favours the practice of gross tumour removal since it is now generally accepted that surgery plus radiotherapy offer the longest survival times. The extent of cyto-reduction directly influences the effectiveness of radiotherapy.[5, 12, 13]

While surgery was generally accepted as the first line of treatment for gliomas, the present author was surprised to encounter a peculiar resistance on the part of colleagues especially from the UK, as revealed at the time of a symposium organized to commemorate the centenary of Godlee's operation. In any case even among those who practiced surgical decompression there were many skeptics who denied this treatment to patients with dominant hemisphere tumours.[14] Among several others, we have demonstrated the scope of surgery for tumours in such locations.[5, 12, 13, 15] Thus not only 95 per cent of patients with dominant hemispherical gliomas with no speech defect did not suffer any deficit, approximately 65% with pre-operative dysphasia became normal or had significant improvement in their speech. This also applied to other pre-operative deficits like hemiparesis/plegia.[13]

Tumours crossing the midline or involving the basal ganglia were also regarded by many as a contradiction for surgery. In a series of 200 patients of supratentorial gliomas who had survived surgery for one year or more there were 40 whose tumour had crossed the midline, 11 with involvement of the basal ganglia and six with both these features. Radical tumour removal was carried out on all of them without any increase in mortality and morbidity.[13] With the availability of functional MRI, per-operative ultrasonography and use of CUSA and lasers, there has been further improvement in these results.

Radiotherapy: On the basis of a long term follow-up of a large series of cases the Montreal School had already established the benefits of post-operative radiotherapy in 1960s.[16] Elvidge[17,18] had provided a detailed study on the usefulness of post-operative radiotherapy in a large series of astrocytic tumours followed for many years. Jelsma and Bucy[19] found that optimum treatment for glioblastomas was obtained by extensive tumour resection and radiotherapy. This has been repeatedly confirmed by the American Brain Tumor Study Group[20] and multicentric Scandinavian study[21] and many others. Thus post-operative radiotherapy came to be accepted as the standard practice irrespective of the nature of surgery - biopsy, partial or total removal. The conventional whole head irradiation used earlier has been replaced by generous megavoltage therapy local fields,[9] using linear accelerator or neutron beam, hyper-fractionation and accelerated fractionation techniques and a number of different regimes have had their advocates. A variety of radiosensitizers like bromouridine (BUDR), metronidazole and misonidazole have been tried but none has found general acceptance. In selected cases, interstitial brachytherapy using radioisotopes like I^{125} or I^{192} [22] provided dramatic results. However, a consensus on the best regime still evades since none of these techniques have significantly improved the long term survival. The role of radiosurgery or conformal radiotherapy using "gamma knife" or linear accelerator based "X" knife is still to be established. This may prove to be useful for deep-seated tumours or small residual or recurrence.

Chemotherapy: A variety of chemotherapeutic agents—semustine (CCNU), carmustine (BCNU, AONU), bleomycine, vincristine, procarbazine, methotrexate—have been tried either singly or in a variety of combinations, following surgery and radiotherapy, or in the interval between the two, both for the primary tumour or reserved for recurrence. Even intracarotid route or use of agents to break down blood-brain barrier for better access to the tumour failed to provide additional gains.[23] Inspite of a series of well controlled trials the overall results are not encouraging. As late as 1980, Walker et al[20] concluded, "on the basis of the results of this study, it appears best to use radiotherapy *(implied post-operative)* in the treatment of malignant gliomas and to continue the search for different chemotherapeutic regimens to combine with radiotherapy." A few years later, Hildebrand[24] observed that various chemotherapeutic agents are able to produce objective remission, but do not prolong mean (or median) survival time, nor the time of recurrence. He, however, could not find any evidence to suggest that any combination of drugs was superior to single agent therapy.

Recently Temozolomide, an oral, second generation alkylating agent has been claimed to exhibit a broad spectrum of anti-tumour activity with predictable pharmacokinetics and limited toxicity profile. This had primarily been used for patients with failed radiotherapy with or without chemotherapy with nitrosourea. Significant progression-free survival benefit at 6 months was observed in both glioblastoma and anaplastic astrocytomas.[25] A very high cost and lack of long-term survival data are atleast two factors preventing its being accepted as a component of routine therapeutic armamentarium.

It may be mentioned that BCNU, CCNU, methotrexate and vincristine in various permutation and combination are currently being recommended for routine use after surgery and radiotherapy for medulloblastoma.[26,27] A subgroup of aggressive oligodendrogliomas has been shown to be highly chemosensitive.[28]

In summary it appears that chemotherapy as an adjuvant has a place in overall management of gliomas in selected cases and primarily for recurrences and tumours unresponsive to conventional regimens.

Immunotherapy: Feasibility of targeting antibodies tagged to radio-isotope as a therapeutic strategy was first suggested by Mahaley and Day.[29] Isotope encephalometry widely practiced as a diagnostic procedure at that time had demonstrated preferential accumulation of the isotope in the tumour. Hopes generated by initial reports by Trouillas,[30] Young and Kaplan[31] have not proved to be lasting. The major limitation has been failure to identify tumour specific antigens even though claims have been made to the contrary from time to time. In recent years, attempts have been made to utilize biological response modifiers like human recombinant interleukin-2 (rIL2) to enhance the preferential entry of chemotherapeutic agents into the tumour. Attempts are being made to use genetically engineered cells, injected directly in the brain, to secrete IL2.[32]

Future Directions: Nearly fifteen years ago the author had concluded, "There is no doubt that improved methods of diagnosis and technically high standard of surgery are essential to provide relief to our patients with "masses" (tumours), but the basic understanding of their biology at molecular level is essential to meet their challenges appropriately". A few years later the gains due to improved methods of diagnosis and technically high standards of surgery made possible by operating microscope, laser and CUSA were already a reality. But so was the prediction that "the ultimate solution is unlikely to come from 'Sharper knife' or more powerful source of radiation, but from a better understanding of the biology of these tumours".[13]

The recent advances in molecular biology, genetics and biotechnology have provided valuable information on a variety of chromosomal abnormalities, cell surface molecules (antigens and receptors), pathways of signal transduction (including the role of a number of oncogenes, their mutations, over expression and inactivation and protein products), various growth factors, (EGFR, PDGF) including the angiogenic factor (Table 2). The role these play in cell multiplication on one hand and apoptosis on the other is getting to be known. Fresh hopes have been aroused with the completion of the human genome sequencing. A number of gene therapy protocols have been developed.

Table 2 : *Genetic abnormalities in CNS Tumours*

Glial Precursor
- P53 mutation/ chromosome 17p loss
- PDGF/PDGFR over expression
- Chromosome 22q loss

Astrocytoma
- Deregulation of the P'6 pathway: RB or P16 inactivation
- cDK4 gene amplification
- Chromosome 19q loss
- Expression of invasion associated molecules and angiogenic factor

Anaplastic Astrocytoma
- Chromosome I0p and I0q loss
- Expression of invasion - associated molecules and angiogenic factor

Glioblastoma

(Source: Pardos & Levin Seminars in Oncology 27, 1-10, 2000)

But as of today except for marginal gains no major breakthrough is in sight.

This brings me to the current research efforts in respect to gene therapy which offers some hope for future.

GENE THERAPY:

Recently, novel approaches have been proposed for treatment of malignant gliomas, based on gene transfer by chemical, viral or cellular vectors. The various strategies used in gene therapy include tumour sensitization to chemotherapy, stimulation of anti-tumour immune response, inhibition of tumoural neovascularization or suppressor gene replacement in tumour cells.[33] Surgical approaches for this purpose have been diverse–local free hand injection, stereotactic injection into the tumour, local administration into CSF. These may have been supplemented by techniques to overcome Blood-Brain-Barrier. Many different vectors have been designed to transfer the transgene. The most commonly used strategy consisted of retrovirus – induced transfer of Herpes Simplex Thymidine Kinase (tk) gene to tumour followed by Ganciclovire.[34] Grafting retrovirus packaging cell lines to selectively deliver "killer" or "suppression" gene using BAG vector bearing *E.coli LacZ* gene was tried by Short et al.[35] Genetically engineered fibroblasts modified to secrete IL-2 delivered intracerebrally were used by Glick et al,[32] Sobol et al,[36] Glick et al,[37] Whittle,[38] Mineta et al.[39] Gene transfer to enhance cytotoxic efforts of eyclophosphamide using P450, 2BI gene has also been tried by Wei et al.[40] Adeno-virus transduction of gene to over-express wild type P53 to enhance radiosensitivity was studied by Broaddus et al 1999. The most exciting recently developed therapeutic approaches for gliomas involve combination of immunotherapy and neural stem cells.[41] Neural stem cells can be engineered to produce therapeutic molecules and have the potential to travel along the white matter like neoplastic cells.[42] Experimental evidence indicates that while differentiating progenitor cells may release a factor with an anti-proliferative effect.

The technique is still experimental. This has not prevented a few surgeons from trying it for treatment of patients with glioblastoma or recurrent malignant tumour. Palu et al[43] from Padova, Italy, claimed to have assessed retrovirus mediated gene therapy for the first time in humans suffering from glioblastoma. They produced evidence of transgene activity in the treated tumour. Rampling et al[44] used this approach for the treatment of 9 patients with recurrent malignant gliomas, 4 of whom were alive and well 14-24 months later. No controlled large series trials have so far been reported.

Based on the animal experiments a number of problems associated with gene therapy need to be resolved before it could be recommended for human trials. These include: (i) inefficient transfection of host cells by viral vectors; (ii) restricted delivery across blood-brain and brain-tumour barrier; (iii) inability to reach tumour cells infiltrating brain parenchyma at some distance from the main tumour; (iv) transient gene expression; (v) toxicity of viral proteins, (vi) immune response to transfected protein.

FUTURE PERSPECTIVE:

The above account illustrates that many novel research ideas are currently being persued primarily in the labs and some even in the clinics. The advances up to date have no doubt improved the prognosis, the duration and quality of survival but a cure is nowhere in sight. Any attempt to venture a prediction for future will be foolhardy as history of science provides enough example of futility and failability of such predictions. Some of the most revolutionary discoveries or inventions were never predicted by the leaders of these disciplines. But once discovered they proved to be so simple and obvious. The unraveling of mystery of the atom or the gene fall in this category. Anyhow let me make a few guesses. Surgery will remain an important constituent of the therapeutic armamentarium in foreseeable future. It will become more precise and safe with routine use of computerized three dimensional imaging, per-operative ultrasonography, and selective use of automations now under development. Improvements in radiation therapy and chemotherapeutic agents, in various permutations and combinations are likely to enhance the period and quality of survival. Application of atleast some gene therapy techniques may provide further benefits. However, I for one, am not convinced that all these will ever succeed in providing the much sought after cure. One thing that appears certain to me, notwithstanding the limited gains achieved so far, is that the final answer will emerge from more in depth studies on biology of these tumours. I have a feeling that glioma research is not receiving the attention and investment urgently required. A vigorous rethinking and concerted research efforts are urgently called for and not the prevailing spirit of helplessness.

REFERENCES:

1. Bennett AH, Godlee RJ: Excision of a tumour from the brain. *Lancet* 1884;2:1090.
2. Choksey M.S. Valentine A, Shawdon A, Free CEL, Lindsay KW: Computed tomography in the diagnosis of malignant brain tumours: do all patients require biopsy *J Neurol Neurosurg Psychiat* 1989;54:821-25.
3. Wu RH, Lang ZJ, Du CC: Pathological studies of CT findings in supratentorial astrocytomas. Correlation between low density lesion and changes in fine structure: *Chinese Med J* 1991; 104:685-92.
4. Galicich JH, French LA. Melby JC: Use of dexamethasone in treatment of cerebral oedema associated with brain tumours. *Lancet* 1961;81:46-48.
5. Tandon PN, Agarwal SP, Mahapatra AK, Roy S: "Radical" surgical decompression of supratentorial gliomas. Do the results justify operation? In Walker MD, Thomas DGT (eds) Biology of Brain Tumours, Martin Nijhoff, Boston. 1986;277-86.
6. Ciric I, Ammirati M. Vick N, Mikhael M: Supratentorial gliomas: Surgical reconsiderations and immediate post-operative results: Gross total resection versus partial resections. *Neurosurgery.* 1987;21:21-26.
7. Fadul c, Wood J, Thaler H, Galidich J: Morbidity and mortality of craniotomy for excision of supratentorial gliomas. *C Neurology* 1988; 38: 1374-79.
8. Winger MS, MacDonald DR, Cairncross JG: Supratentorial anaplastic gliomas in adults: The prognostic importance of extent of resection and prior low-grade glioma. *J Neurosurg* 1989;71:487-93.
9. Shibamoto Y, Yamashita J, Takahashi M, Yamasaki T, Kikuchitt, Abe M: Supratentorial malignant gliomas: an analysis of radiation therapy in 178 cases. *Radiother Oncol* 1990;18:9-17.
10. Bricolo A, Turazzi S, Cristofori L, Gerosa M: Experience in "radical" surgery of supratentorial gliomas in adults. *J Neurol Scien* 1990;34: 297-98.
11. Neuwelt EA, Nazzaro JM, Gumerlock MK: Is there a role for biopsy in the treatment of supratentorial high grade glioma? *Clin Neurosury* 1990;73:384-407.
12. Tandon PN: Classification of brain tumors: Controversies surgical view point: In Neurooncology; Proceedings of the National Seminar. National Institute of Mental. Health and Neurosciences, Bangalore. 1981;245-52.
13. Tandon PN: Supratentorial gliomas: The unrelenting challenge. *Neurol India* 1994;42:131-46.
14. Garfield JS: Gliomas, In, Northfield's Surgery of the Nervous system ed JD Miller 1987.

15. Tandon PN, Mahapatra AK, Khosia A: Operations on gliomas involving speech, centre. *Acta Neurochirurg* 1993;56:67-71.
16. BouchardJ: Radiation Therapy of Tumours and Diseases of the Nervous system. Kimpton, London, 1966.
17. Elvidge AR, Martinez-Coll A: Long-term follow-up of 106 ~casesofasrocytoma 1928-1939. *J Neurosurg* 1956;13:230-43.
18. ElvidgeAR: Long-term survival in astrocytoma series. *J Neurosurg* 1968;28:399.
19. Jelsma R, Bucy PC: Glioblastoma mutiforme. Its treatment and some factors effecting survival. *Arch Neurol* 1969;20:161-71.
20. Walker MD, Green SB, Byar DP, Alexander Jr E: Randomized comparisons of, radiotherapy and nitrosoureas for treatment of malignant gliomas after surgery. *New Engi J Med* 1980;303:1323-29.
21. Kristiansen K, Hagen S, Kollevold T, et al: Combined modality therapy of operated~astrocytoma grad III, and IV. Confirmation of the value of post- operative irradiation and lack of potentiation of bleomycin on survival. *Cancer* 1981;47:649-52.
22. Kelly PJ: Computer-assisted streotaxis: new approach for the management of intracranial intra-axial tumors. *Neurology* 1986;37:535-41.
23. Stewart DJ, Garhovac Z, Hugenhbitz H, Russel N: Combined intera- arterial and systemic chemotherapy for intracerbral tumours. *Neurosurgery* 1987;21:207-14.
24. Hildebrand J: Combined modality treatment in malignant supratentorial gliomas. In Tumours of the Brain, (Ed) Bleehan NW.
25. Springer Verlag 1986; 209-Alfred Yung WK: Temozolomide in malignant gliomas. *Sem Oncol* 2000; 27:27-34.
26. Berry MP, Jenkin RDT, Keen CW, et al: Radiation treatment of medulloblastoma-A 21 year review. *J Neurosurg* 1981;55: 43-47.
27. McIntoshN, Chen M, Sartain PA, et al: Adjuvant chemotherapy for medulloblastoma. *Cancer* 1985;56:1316-18.
28. Cairncross JC, MacDonald DR, Ramsay DA: Aggressive oligodendro gliomas: a chemosensitive tumour. *Neurosurgery* 1992;31:78-81.
29. Mahaley S. Jr, Day ED: Immunological studies of human gliomas. *J Neurosurg* 1965;23:363-66.
30. Trouillas P: Immunologie et immunotherapy des tumeurs c'er'brales Etat actuel, Rev. Neurol (Paris) 1973;128:23-25.
31. Young HP, Kaplan A: Immunotherpy of human gliomas, In: Handbook offcCancer Immunology, Immunotherapy (Ed) Waters H, Garland ST, PM Press 1978; 357.
32. Glick RP, Lichtor T, Mogharbel A, Taylor CA, Cohen EP: Intracerebral versus subcutaneous immunization with allogeneic fibroblasts genetically engineered to secrete interleukin-2 in the treatment of central nervous system glioma and melanoma. *Neurosurg* 1997;41:898-907.
33. Couraud PO, Quinonero J, Tchelingerian JL, Vignais L: Novel gene therapeutic approaches for brain tumours. *Neurpath Appi Neurobiol* 1996;22:405-33.
34. Wildner 0: In *situ* use of suicide gene for therpy of brain tumours. *Ann Med* 1999;31:421-29.
35. Short MP, Choi Be, Lee JK, et al: Gene delivery to glioma cells in rat brain by grafting of a retro virus packaging cell line. *J Neurosci Res* 1990;27:427-39.
36. Sobol RE, Fakhrai H, Shawler D, et al: Interleukin -2 Gene therapy in a patient with qlioblastoma. *Gene theory* 1995;2:164-67.
37. Glick RP, Lichtor T, de Zoeten E, Deshmukh P, Cohen EP: Prolongation of survival of mice with glioma treated with semiallogeneic fibroblasts secreting interleukin-2. *Neurosurg* 1999;45:867-74.
38. Whittle IR: Editorial, Gene therapy for brain tumours. *Brit J Neurosurg* 1995;9:717-20.
39. Mineta T, Rabkin SD, Yazaki T, Hunter WD, Martuza RL: Attenuated " multi–mutated herpes simplex virus–1 for treatment of malignant gliomas. *Nat Med* 1995;1:938-43.
40. Wei MX, Tamiya T, Chase M: Experimental tumor therapy in mice using the cyclophosphamide activating cytochrome P450 2 BI gene. *Gene Ther* 1994;5:969-78.

41. Noble M: Can neural stem cells be used as therapeutic vehicles in the treatment of brain tumours. *Nat Med* 2000;6:369-70.
42. Benedetti S, Pirola B, Polio B, et al: Gene therapy of experimental brain tumors using: neural progenitor cells? *Nat Med* 2000;6:447-50.
43. Palu G, Cavaggioni A, Cavi P, et at: Gene therapy of glioblastama multiforme via combined expression of suicide and cytokine genes: a pilot study in humans. *Gene Ther* 1999;6:330-37.
44. Rampling R, Cruickshank G, Papanastassiou V, et al: Toxicity evaluation of replication–competent herpes simplex virus (ICP 34.5 null mutant 1716) in patients with recurrent malignant glioma. *Gene Ther* 2000;7:859-66.

2
Epidemiology and Classification of Malignant Brain Tumors

S Roy, C Sarkar, MC Sharma

Malignant tumors of the brain constitute 2% of all malignant tumors in man. 75% of these are primary and 25% are metastatic. Of the primary malignant brain tumors, 60% are neuroectodermal in origin – 50% being gliomas and 10% being medulloblastomas. Of all the glial tumors in the brain, astrocytic tumors comprise 30%, oligodendrogliomas 5-10% and ependymomas 5-10%. There is a male preponderance observed in all groups of malignant brain tumors.[1]

The common tumors which metastasize to the brain include carcinomas from lung, breast, kidney and colon. The other metastatic tumors include melanomas, sarcomas, germ cell tumors and leukemia/lymphomas.[1]

(I) EPIDEMIOLOGY OF BRAIN TUMORS

The epidemiological study of brain tumors has been limited by a number of factors unique to these neoplasms. Firstly, it is necessary to separate primary from metastatic brain tumors. Secondly, the histopathology of primary brain tumors is complex and a general consensus on their classification has emerged only in the last decade. Thirdly, brain tumors are rare tumors and hence obtaining adequate number of cases for studies is logistically difficult, often requiring collaboration across clinical and regional groups.

(A) Incidence rates in USA

In USA, the Central Brain Tumor Registry of the United States (CBTRUS) has centralized data from population based registries.[2, 3] Incidence rates for all primary brain tumors from 10 regions of USA as compiled by CBTRUS is shown in Table 1.[2-4] Rates are based on a four year period (1990-1993) and are rate standardized to the 1970 USA population to allow for comparison with other incidence reports.[3] In USA, the most common brain tumors reported are meningiomas, glioblastomas and astrocytomas.

Table 1 : *Numbers and annual incidence rates / 100,000 person-years of brain tumors by histological grouping.*

DIAGNOSIS	TOTAL NO.	RATE (%)
Pilocytic astrocytoma	246	0.3
Low grade diffuse astrocytoma	187	0.2
Anaplastic astrocytoma	565	0.5
Glioblastoma	2968	**2.6**
Astrocytoma NOS / Malignant glioma NOS	1387	**1.2**
Oligodendrogliomas	350	0.3
Anaplastic oligodendrogliomas	68	0.1
Mixed glioma	126	0.1
Ependymoma / anaplastic ependymomas / variants	321	0
Choroid plexus tumor	45	0.0
Glioneuronal tumors	143	0.1
Pineal tumors	31	0.1
Medulloblastomas	237	0.2
Nerve sheath tumors (benign and malignant)	992	0.9
Meningiomas	3442	**2.8**
Lymphomas	476	0.4
Pituitary tumors	1329	1.1

NOS : Not otherwise specified

(B) Incidence Rates in Other Countries

A few studies based on assessing rates by country of origin have been conducted **Table 2**. These studies have the advantage of comparable definitions and reporting procedures.[5] Brain tumor rates have been reported to be higher among Europeans and Israelies and lower among Africans and Asians living in Israel.[4] Eastern Europeans have been reported to have higher brain tumor rates than other Los Angeles County residents.[4] Women born in Eastern Europe and the Middle East have been reported to have higher incidence rates of meningiomas, while women born in Southern Europe had higher rates of glioma[4] **Table 2**.

Table 2: *International variation : Incidence of malignant brain/nervous system tumors (age 35-64 years) for selected high- and low-rate regions.*#

REGISTRY	MALES	FEMALES
High rate regions		
Sweden	16.9	16.9
Denmark	16.5	15.9
Norway	15.2	15.5
Israel : all Jews	14.3	15.3
UK : South Thames	3.4	10.1
Poland : Warsaw	12.5	8.1
US : Bay area whites	11.8	7.5
Low rate regions		
Slovenia	6.6	4.6
China : Tianjin	6.4	6.4
US : Puerto Rico	6.4	3.3
Belarus	5.9	3.7
Hong Kong	5.1	4.5
Japan : Osaka	3.4	2.4
India : Bombay	3.3	2.4

* Selected regions were those reported more than 100 male cases
\# Table compiled from IARC data for 1983-87 (Parin et al 1992 [Ref.4]
  [Modified from Davis FG and Preston-Hartin S (Ref.3)]

INCIDENCE RATES IN INDIA

The data collected from Neurosurgical centers in India is not uniform. However some of the statistics available is enlisted below :

GBPH, New Delhi (1992 to 2000)

Total number of gliomas	: 1588
Astrocytomas Grade I	: 189 (11.9%)
Astrocytomas Grade II	: 187 (11.8%)
Astrocytomas Grade III	: 206 (13%)
Astrocytomas Grade IV	: 259 (16.3%)
Oligdendeogliomas	: 103 (6.5%)
Mixed Gliomas	: 92 (5.8%)
Ependymomas	: 52 (3.3%)
Medulloblastomas	: 203 (12.8%)

PNET	;	69 (4.3%)
Choroid Plexus Carcinoma	:	7 (0.44%)
Others	:	21 (1.32%)

PGIMER, Chandigarh

Total number of intracranial tumors (3 years)	:	873
Gliomas	:	351 (40%)
Non-glial tumors	:	522 (60%)

SGPGIMS, Lucknow

Total number of intracranial tumors (1997-2001)	:	942
Astrocytic tumors (Grades I, II & III)	:	598 (63%)
Glioblastoma (Grade IV)	:	349 (37%)

Nizam's Institute of Medical Sciences, Hyderabad

Total number of gliomas (1997-2000)	:	228
Astrocytomas	:	138 (60.5%)
Glioblastoma	:	63 (27%)
Oligodendroglioma	:	16 (7%)
Ependymoma	:	7 (3%)

AIIMS, New Delhi

Total number of primary intracranial neoplasms (1989-98) :		5833
Total number of astrocytic tumors	:	1729 (30%)
Pilocytic astrocytoma (Grade I)	:	188 (11%)
Low grade diffuse astrocytoma (Grade II)	:	605 (35%)
Anaplastic astrocytoma (Grade III)	:	234 (13.5%)
Glioblastoma (Grade IV)	:	702 (40.5%)
Oligodendroglioma	:	56 (2%)
Anaplastic oligodendrogliomas	:	3
Oligoastrocytoma	:	185 (3.2%)
Anaplastic oligoastrocytoma	:	78 (1.3%)
Ependymoma	:	47 (0.8%)
Anaplastic ependymoma	:	31 (0.5%)
Medulloblastoma	:	181 (3.1%)
Lymphoma	:	80 (1.4%)

(II) CLASSIFICATION OF BRAIN TUMORS

The problem entailed in the classification of brain tumors has long provoked a good deal of controversy. By and large, however, most classifications have reflected a general acceptance of histologic tumor type that are based on identification of reasonably precise cytologic features linking neoplastic elements to normal cell types found in the mature and developing central nervous system.

Most modern classifications are based upon that of Bailey and Cushing.[6] These authors studied the embryogenesis of the various cellular components of the CNS. They then attempted to classify the tumors observed in terms of the different morphologic stages through which these cells pass in ontogenesis, using metallic techniques to supplement routine staining methods. Having enumerated 20 cell types, they proposed a classification based on a pyramidal scheme of cytogenesis, including 14 main groups as follows:

1. Medulloepithelioma
2. Medulloblastomas
3. Pineoblastoma
4. Pinealoma
5. Ependymoblastoma
6. Ependymoma
7. Neuroepithelioma
8. Spongioblastoma : (a) multiforme, (b) unipolare
9. Astrocytoma
10. Astrocytoma : (a) Protoplasmaticum, (b) Fibrillare
11. Oligodendroglioma
12. Neuroblastoma
13. Ganglioneuroma
14. Papilloma choroideum

While most CNS tumors recognized today can be identified in that list, current terminologies reflect a number of significant modifications in the definition of some. The International Classification of Central Nervous System tumors drafted under the auspices of the World Health Organization (WHO) in 1979[7] represented the first consensus achieved after many years of study and discussion by portents of different schools. It is, in its essential lines derived from the classification of Bailey and Cushing. While, as with any other classification scheme, criticisms of detail were levelled against it, its main contribution was to reconcile the terminologies and concepts of sometimes widely differing view points and thus achieve, by compromise, a reasonable measure of uniformity on the most commonly recognized tumor entities and their most frequent variants.

The first WHO classification underwent several revisions and modifications as newer concepts in tumor biology and pathogenesis emerged. The most recent WHO classification is the one propounded in 2000[8] and this is showing in **Table 3**.

Table 3 : *Revised WHO Classification (2000)*

TUMORS OF THE NEUROEPITHELIAL TISSUE

Astrocytic Tumors
Diffuse astrocytoma (Grade II)
Fibrillary astrocytoma
Protoplasmic astrocytoma
Gemistocytic astrocytoma
Anaplastic Astrocytoma (Grade III)
Glioblastoma (Grade IV)
    Giant cell glioblastoma
    Gliosarcoma
Pilocytic Astrocytoma (Grade I)
Pleomorphic Xanthoastrocytoma (Grade II)
Subependymal Giant Cell Astrocytoma (Grade I)

Oligodendroglial Tumors
Oligodendroglioma (Grade II)
Anaplastic oligodendrogliomas (Grade III)

Mixed Gliomas
Oligoastrocytoma (Grade II)
Anaplastic oligoastrocytoma (Grade III)

Ependymal Tumors
Ependymoma (Grade II)
    Cellular
    Papillary
    Clear cell
    Tanycytic
Anaplastic ependymoma (Grade III)
Myxopapillary ependymoma (Grade I)
Subependymoma (Grade I)

Choroid Plexus Tumors
Choroid plexus papillomas (Grade I)
Choroid plexus carcinoma

Glial Tumors Of Uncertain Origin
Astroblastoma (Grade not assigned)
Gliomatosis cerebri (Grade III)
Chordoid glioma of the third ventricle (Grade II)

Neuronal And Mixed Neuronal-Glial Tumors
Gangliocytoma (Grade I)
Dysplastic gangliocytoma of cerebellum (Lhermitte-Duclos)
Desmoplastic infantile astrocytoma / Ganglioglioma (Grade I)
Dysembryoplastic neuroepithelial tumor (DNT) (Grade I)
Ganglioglioma (Grade I or II)
Anaplastic Ganglioglioma (Grade III)
Central neurocytoma (Grade III)
Cerebellar liponeurocytoma (Grade I or III)
Paraganglioma of the filum terminale (Grade I)

Neuroblastic Tumors
Olfactory neuroblastoma (Esthesioneuroblastoma)
Olfactory neuroepithelioma
Neuroblastomas of the adrenal gland and sympathetic nervous system

Pineal Parenchymal Tumors
Pineocytoma (Grade II)
Pineoblastoma (Grade IV)
Pineal parenchymal tumor of intermediate differentiation

Embryonal Tumors
Medulloepithelioma
Ependymoblastoma
Medulloblastoma (Grade IV)
- Desmoplastic medulloblastoma
- Large cell medulloblastoma
- Medullomyoblastoma
- Melanotic medulloblastoma

Supratentorial primitve neuroectodermal tumor (PNET) (Grade IV)
- Neuroblastoma
- Ganglioneuroblastoma

Atypical teratoid/rhabdoid tumor (Grade IV)

TUMOURS OF PERIPHERAL NERVES

Schwannoma (Grade I)
(Neurilemmoma, Neurinoma)
- Cellular

Plexiform
Melanotic
Neurofibroma (Grade I)
Plexiform
Perineurioma (Grade I)
Intraneural perineurioma
Soft tissue perineurioma
Malignant Peripheral Nerve Sheath Tumor (MPNST) (Grade III or IV)
Epithelioid
MPNST with divergent mesenchymal and/or epithelial differentiation
Melanotic
Melantoic psammomatous

TUMORS OF THE MENINGES

Tumours Of The Meningothelial Cells

Meningioma (Grade I)
    Meningothelial (Grade I)
    Fibrous (fibroblastic) (Grade I)
    Transitional (mixed) (Grade I)
    Psammomatous (Grade I)
    Angiomatous (Grade I)
    Microcystic (Grade I)
    Secretory (Grade I)
    Lymphoplasmacyte – rich (Grade I)
    Metaplastic (Grade I)
    Clear cell (Grade II)
    Chordoid (Grade II)
Atypical (Grade II)
Papillary (Grade III)
Rhabdoid (Grade III)
Anaplastic meningiomas (Grade III)

Mesenchymal, Non-Meningothelial Tumors

Lipoma
Angiolipoma
Hibernoma
Liposarcoma (intracranial)
Solitary fibrous tumour
Fibrosarcoma
Malignant fibrous histiocytoma
Leiomyoma
Leiomyosarcoma

Rhabdomyoma
Rhabdomyosarcoma
Chondroma
Chondrosarcoma
Osteoma
Osteosarcoma
Osteochondroma
Haemangioma
Epithelioid Haemangio-endothelioma
Haemangiopericytoma (Grade II or III)
Angiosarcoma
Kaposi sarcoma

Primary Melanocytic Lesions
Diffuse melanocytosis
Melanocytoma
Malignant melanoma
Meningeal melanomatosis

Tumors of the Uncertain Histogenesis
Haemangioblastoma

LYMPHOMAS AND HAEMOPOIETIC NEOPLASMS
Malignant lymphomas
Plasmacytoma
Granulocytic sarcoma

GERM CELL TUMORS
Germinoma
Embryonal carcinoma
Yolk sac tumour
Choriocarcinoma
Teratoma
    Mature
    Immature
    Teratoma with malignant transformation
Mixed germ cell tumours

TUMOURS OF THE SELLAR REGION
Craniopharyngioma
    Adamantinomatous
    Papillary
    Granular cell tumour

METASTATIC TUMORS

(III) CLASSIFICATION AND GRADING OF GLIOMAS

The name 'glioma' has been used as a generic term that encompasses all the tumors of central neuroepithelial origin including those composed exclusively or predominantly of undifferentiated precursor cells, of neuroblastic cells in various stages of differentiation and of mixed populations exhibiting different directions of cellular maturation.[9]

Any scheme of the histologic grading has two main goals : (a) the tumor grade must predict clinical behaviour and (b) the grading criteria must be sufficiently objective and defined to minimize variation among observers and to maximize reproducibility.[9]

Classification of gliomas by histological typing based on this histogenesis of the tumor cells and grading based on the degree of anaplasia has been one of the most common parameters used by most workers for prognostic assessment and choice of treatment in gliomas. However, this remains till today a highly controversial and confusing subject both to neuropathologists and to clinicians. A pathologist often justifies the use of different diagnostic terms for tumors showing slight morphological variations, even though there are no differences in treatment or survival.[10] And "while differences of opinion among neuropathologists can be debated knowledgeably among this group of experts, such arguments foster confusion and hamper communication between pathologists and clinicians".[10,11]

Of the numerous grading systems in use for astrocytoma, most utilize three or four grades of malignancy. In addition, a key assumption of any valid grading scheme of malignancy is that representative histologic material has been obtained and evaluated.[9] Histologic grading of diffuse infiltrating gliomas may present important technical problems. Since anaplastic features may be focal, the prognostic value attached to the grading of small or stereotactic needle biopsies is totally dependent on adequate and representative sampling of the neoplasm. Not infrequently, the peripheral portion of a glioma has the classic appearance of a well-differentiated or low grade astrocytoma, but the deeper parts which are less accessible to resection or biopsy may already have developed the histologic character of glioblastoma.[9]

Gliomas have been classified in multiple ways since the 19th century. Historically, successive classifications have emphasized one or another histologic point, hoping to provide information simultaneously on the cellular genesis of the tumor, the etiology of the tumor, degree of malignancy and the prognosis. The problem has been further complicated by the fact that gliomas are heterogenous tumors, with a considerable number of cell types being involved in the neoplastic process and variations occurring not only between the different groups of gliomas but often within the same tumor.[12]

A brief outcome of the various classification and grading schemes for gliomas is given below :

BAILEY AND CUSHING (1926)

The first widely used classification of glial neoplasms was that proposed by **Bailey and Cushing in 1926.**[6] These authors studied the embryogenesis of various cellular components of the central nervous system (CNS). They then attempted to classify the tumors in terms of the different morphological stages through which the cells pass during ontogenesis. Accordingly, the glial tumors were subdivided into :

(i) Astrocytoma − (a) Protoplasmaticum and (b) Fibrillare

(ii) Astroblastoma

(iii) Spongioblastoma − (a) Unipolare and (b) Multipolare

(iv) Oligodendroblastoma

(v) Oligodendroglioma

(vi) Ependymoma

(vii) Ependymoblastoma

The main drawback of this classification was that it totally overlooked the role played by anaplasia in the composition and evolution of many gliomas.[12,13-15]

KERNOHAN ET AL (1949)

In 1949, **Kernohan et al**[16] introduced a system of grading gliomas adhering to the concept that gliomas may arise from a pre-existing adult cell type still capable of proliferation by a process of anaplastic transformation or dedifferentiation. They divided the gliomas into five main groups namely :

1. Astrocytoma
2. Ependymoma
3. Oligodendroglioma
4. Neuroastrocytoma
5. Medulloblastoma

Kernohan et al[16] then introduced a system of grading all these gliomas through one to four in ascending order of malignancy based on 8 features namely (i) cellularity (ii) pleomorphism (iii) giant cells (iv) vascularity (v) endothelial proliferation (vi) mitoses (vii) necrosis and (viii) infiltration zone.

The histological criteria for the Kernohan grading as enunciated by Ringertz[17] were as follows :

1. **Grade 1**

   All cells appear as on normal astrocytes, no pleomorphism,

Cellularity : Not always increased over normal brain

No mitoses

Vascularity as in normal brain

Slight proliferation of the vessel walls

Broad infiltration zone

2. **Grade 2**

   Major proportion of cells are like normal astrocytes

   Slight to moderate pleomorphism of a minor proportion of cells

   Cellularity as in grade 1

   No mitoses

   Vascularity as in grade 1

   Slight proliferation of the vessel walls more frequent than grade 1

   Broad infiltration zone

3. **Grade 3**

   50-75% cells are like normal astrocytes

   Moderate pleomorphism of other cells

   Occasional polymorphonucleated or multinucleated giant cells

   Cellularity : 1-1/2 times that of normal brain

   Mitoses averaging in 1 in every other high-power field (HPF)

   Vascularity increased

   Proliferation of the vessel walls is frequent and often pronounced

   Necrosis present

   Narrowing infiltration zone

4. **Grade 4**

   Few normal astrocytes

   Marked anaplastic transformation with bizarre cell forms with abundant polymorphonucleated and multinucleated giant cells.

   Cellularity 3 times that of normal brain

   Mitoses frequent: 4-5 in every HPF

   Vascularity high

   Marked proliferation of vessel walls

   Necrosis frequent and extensive

   Narrow infiltration zone

Kernohan and Sayre[18] claimed that there was a good correlation of their grading with survival.

RINGERTZ (1950)

Ringertz[17] objected to Kernohan's 4 step grading system which he regarded as exaggerated since he could not find any prognostic difference between grades 3 and 4. The average postoperative survival in grade 3 patients was 11.5 months and grade 4 cases 6.6 months. Similarly the 3 year survival rate in patients of grade 3 was 14.3% as against 3.8% in grade 4. Consequently, he recommended a 3 step grading in which he regarded grade 2 cases as intermediate forms between grade 1 (benign) and grade 3 (glioblastoma). The histological criteria used for his grading were as follows :

1. **Astrocytoma**

    All or nearly all the cells appear as normal astrocytes

    No or very slight pleomorphism; variation in size of nuclei in 25%

    Cellularity : 1-2 times normal brain in 75% of cases

    No mitosis

    Vascularity as in normal brain

    Slight proliferation in vessel walls in 10% cases

2. **Intermediate type**

    Major proportion of cells appear as normal astrocytes, the rest being moderately pleomorphic. A rather pronounced variation in nuclear size in 33% cases. Occasional multinucleated cell.

    Cellulartiy: 1-3 times that of normal brain.

    Mitosis in most cases; but never more than 1 in every third HPF

    Vascularity as astrocytomas

    Slight proliferation of vessel walls in 33% of cases; moderate in 10%

    Necrosis absent or very slight.

    Broad infiltration zone

3. **Glioblastoma**

    Pleomorphic changes, often marked in 50% cases with abundant multinucleated cells. In a number of cases however, a more monomorphous but malignant cell type.

    Cellularity: 3-4 times that of normal brain in 10% the rest up to 5 times.

    Mitoses: in 20% one in about every other HPF; in the rest more frequent

    Vascularity mostly considerable incrased

Proliferation of the vessel walls in almost every cases, often very strong
Necrosis: Frequent and widespread
Narrow infiltration zone

Presence of frequent and widespread necrosis was his main criteria for assigning a tumor to grade 3. Using this system, Ringertz[17] reported a good correlation with survival. The average postoperative survival for the 3 grades was 63 months, 32.2 months and 11 months respectively and similarly the 3 years survival rate was 57.5%, 21.8% and 2.9% respectively. He proposed a similar 3-tier grading system for oligodendrogliomas and ependymomas also.

CRITICISMS OF KERNOHAN AND RINGERTZ GRADING SYSTEMS

Zulch[19] found a significant difference in survival using Kernohan's 4-tier system namely 5 years or more for grade 1, 3 to 5 years for grade 2, 2-3 years for grade 3 and 6 to 15 months for grade 4 patients. However, several other authors have failed to demonstrate a significant difference in survival between astrocytomas of Kernohan's grade 1 and 2[20-23] and also between grades 3 and 4.[22-26] Further, this system was found to be highly subjective, difficult to apply and its reproducibility low mainly because of lack of firm criteria for separation of the grades.[7,12,23,27]

In addition to the subjectivity, over the years the other objections and skepticisms which have been raised concerning the efficacy of grading gliomas on surgical specimens are as follows.[7,12,23,27]

1. The biopsy material may not be representative of the tumor as a whole. Anaplasia is so often a localized development that the prognostic value to be attached to the grading of biopsies causes serious limitations. Sometimes the superficial portion of a glioma has the classic appearance of an astrocytoma but the deep and inaccessible parts have already developed the histological characters of glioblastoma.[7,12]

2. The grading systems lump together histologically distinct astrocytic tumor types with distinctly different prognostic significance.[7,12] Thus juvenile pilocytic astrocytomas which have a favourable prognosis both in children and adults[12,28] do not figure in Kernohan's or Ringertz's classification. Gemistocytic astrocytomas which often behave more aggressively[12,29] are also not recognized as a separate entity by them. Recognition of histological subtypes is also important because some gliomas like subependymal giant cell astrocytoma and pleomorphic xanthoastrocytomas can appear alarmingly pleomorphic histologically but are often biologically indolent.[30-32] However, in Kernohan's and Ringertz's systems all these entities are grouped under one head 'Astrocytomas'. Indiscriminate lumping of histological subtypes complicates the evaluation of therapeutic regimens in low grade astrocytoma series,[23] as illustrated by reports showing adjuvant radiotherapy to be either of benefit[20] or to be ineffective.[27] Similar observations

have been reported regarding the efficacy of radical surgical[20, 21, 33] and other showing no effect.[22]

3. The histological criteria for grading the degree of malignancy have been considered uniformly for all the histological types of tumor, though this is not the case.

   Thus, oligodendrogliomas may show increased mitoses but have a benign course.[12, 34, 35]

   Similarly, ependymomas are generally slowly growing tumors.[12, 36] However, the prognosis in malignant ependymomas is unpredictable and no correlation has been found between histopathological features and length of postoperative survival in these tumors.[12, 37]

4. Grading makes it extremely difficult to place tumors with mixed cell populations into an already pre-determined tumor category.[7]

Nelson et al[25, 38] and Burger et al[26] followed the 3-tier system of classification of astrocytic tumors into astrocytomas, anaplastic astrocytoma (or astrocytoma with anaplastic features) and glioblastoma similar to that proposed by Ringertz[17] with presence of necrosis being the main feature to categorize a tumor as glioblastoma. Using this classification, Nelson et al[25, 38] reported that the prognosis was uniformly poor in patients of glioblastoma in whom the 18 months survival was seen in only 15% cases and the median survival was 8 months. However, the prognosis in anaplastic astrocytoma was uncertain. Although the median survival was 28 months, approximately 38% of patients in this age group died within 18 months following randomization. Later, Burger et al[26] also found highly significant differences in the age, duration of preoperative symptoms and postoperative randomized survival between the two groups.

FIRST WHO CLASSIFICATION (ZULCH, 1979)

The first World Health Organization (WHO) histological typing of tumors of the central nervous system was developed by Zulch in 1979.[7] It was in its essential lines derived from the classification of Bailey and Cushing (1926), in that the typing was based on the hypothetic histogenesis of the tumor cells, together with the predominant cell type in the tumor.

The 1979 WHO histological typing of gliomas[7] was as follows :

1. Astrocytic tumors
   (a) Astrocytoma
       Variants – fibrillary, protoplasmic, gemistocytic
   (b) Anaplastic astrocytoma
   (c) Subependymal giant cell astrocytoma

(d) Pilocytic astrocytoma
(e) Astroblastoma
2. Oligodendroglial tumors
   (a) Oligodendroglioma
   (b) Anaplastic oligodendroglioma
   (c) Mixed oligoastrocytoma
3. Ependymal tumors
   (a) Ependymoma
      Variants – Myoxpapillary, papillary, subependymoma
   (b) Anaplastic ependymoma
4. Poorly differentiated and embryonal tumors
   (a) Glioblastoma
      Variants – GBM with sarcomatous component, giant cell glioblastoma
   (b) Medulloepithelioma
   (c) Primitive polar spongioblastoma

Grading was not included in the classification but was mentioned in the discussion of individual tumor types where each tumor had been assigned a grade. Thus, pilocytic astrocytomas were considered grade I whereas protoplasmic, fibrillary and gemistocytic astrocytomas as grade II, anaplastic astrocytomas as grade III and glioblastoma multiforme as grade IV. Similarly, oligodendrogliomas and mixed gliomas were grade II while ependymomas correspond to grade I or rarely II. Anaplastic oligodendrogliomas were assigned grade III and anaplastic ependymomas to grades III and IV.

The main contribution of this classification was the achievement of a reasonable measure of uniformity in the technologies and concepts related to the common brain tumors and their variants. Therefore, it gained initial acceptability by many neuropatholgists. Rubinstein[15] and Russell and Rubinstein[12] also adopted this 1979 WHO classification with relatively minor modifications.

However, over the years, it elicited much critique among recognized authorities in the field.[10,11,39,40]

Some of the important drawbacks of this classification related to the glioma group are mentioned below :

1. The histological typing was based on the predominant cell type in the tumor This therefore introduced an element of subjectivity and hence inter-observer variation in deciding which cell type was in majority in any particular tumor.[40]

2. Naming tumors after the predominant cell type was complicated by the presence of mixed gliomas.[40] These were put under the category of oligodendroglial tumors which seemed inappropriate. Further, there were no clear cut guidelines regarding the definitive proportions of the two cell types which should be present in a glioma before labeling it as mixed. This again introduced a lot of inter-observer variation in diagnosis. Some neuropathologists considered the presence of even 10% of oligodendrocytes in an astrocytic tumor as evidence of mixed glioma.

3. Tumor malignancy was not clearly defined. Most categories of gliomas had one malignant counterpart. A 4-step grading system was mentioned in the discussion of the individual tumors but not in the basic outline of classification. The grade had been assigned to each tumor regardless of the cell type but the specific criteria for grading were not clearly stated.[10,11,40]

4. Glioblastomas were put in the group of poorly differentiated and embryonal tumors while astroblastomas were put along with astrocytic tumors.[10,11,39]

5. The histological criteria to clearly distinguish between anaplastic astrocytoma and glioblastoma were not well defined. The WHO monograph stated under anaplastic (malignant) astrocytoma that "it may be difficult focally to distinguish from glioblastoma". Again this introduced an element of subjectivity and inter-observer variation.

6. Tumor site and age received no attention in this classification.[10,11,39,40]

GRADING SCHEME OF DAUMAS DUPORT ET AL (ST. ANNE / MAYO, 1988)

The St. Anne Mayo scheme detailed by Daumas-Duport et al[23] names four histologic features to be used in grading and has four grades of astrocytoma.[1-4] This system uses a simple tabulation of the four features as present or absent in a standardized and objective numeric scale: cellular pleomorphism, mitotic activity, microvascular proliferation and necrosis. Each feature present receives one scoring point, and the grade for an individual tumor is based solely on the cumulative numeric score. Thus, astrocytomas with none of the four features are grade 1, those with one feature are grade 2, those with two features are grade 3, and those with three or four features present are grade 4.

The concordance among individual pathologists who use the St Anne Mayo system is reported to be as high as 94%.[41] Their grade 1 tumor however, is exceedingly rare (<0.25% in one series), and the other three grades produce three distinct survival curves; thus, this scheme may also be considered practically as a three-tiered scale.

In the original study of St Anne Mayo system published by Daumas-Duport et al,[23] and in the validation study of Kim et al,[41] the mean survival time for grade 2 astrocytomas was about 45+ months, for grade 3 astrocytomas about 6+ months and for grade 4 astrocytomas about 8+ months.

COMPUTERIZED GRADING USING TESTAST 268

Schmitt[42] and Schmitt and Oberwittler[43] developed a malignancy classifier system "TESTAST 268" which provided a numerical four step grading akin to the Kernohan grading on the basis of 8 classification variables — 5 histologic and 3 clinical.

They claim that this more objectivized semi-quantitative classification system was a more reproducible ruling out inter and intra-observer variabilities. An added advantage was the inclusion of clinical variables in assessment of tumor grading The grading done with TESTAST 268 showed good correlation with mean survival time except that there was lack of a significant prognostic difference between grades III and IV.

REVISED WHO CLASSIFICATION (KLEIHUES ET AL, 1993)

Taking the criticisms into consideration as well as the advances that had occurred in neuro-oncology, a revised WHO classification was formulated in 1993.[43] This proposed WHO classification pertaining to gliomas is shown below :

1. Astrocytic tumors
    (a) Astrocytoma
        Variants — fibrillary, protoplasmic, gemistocytic
    (b) Anaplastic astrocytoma
    (c) Subependymal giant cell astrocytoma
    (d) Glioblastoma
        Variants —Giant cell glioblastoma and gliosarcoma
    (e) Pleomorphic xanthoastrocytoma

2. Oligodendroglial tumors
    (a) Oligodendroglioma
    (b) Anaplastic oligodendroglioma

3. Ependymal tumors
    (a) Ependymoma
        Variants — Cellular, papillary, epithelial, clear cell, mixed
    (b) Anaplastic ependymoma
    (c) Myxopapillary ependymoma
    (d) Subependymoma

4. Mixed gliomas
    (a) Mixed oligoastrocytoma

(b) Anaplastic oligoastrocytoma
(c) Others

Certain major improvements in this classification with regard to gliomas were :

1. Glioblastomas were included in the category of astrocytic tumors because of 2 main reasons :
   (i) Recognition of the principal neoplastic cell as astrocyte by electron microscopy and immunohistochemistry.
   (ii) Glioblastomas are now considered to result from malignant anaplastic transformation of astrocytes rather than as "embryonal tumors".
2. Astroblastomas were now grouped in the category of 'neuroepithelial tumors of uncertain origin' and not under astrocytomas.
3. Mixed gliomas were put in a separate category.
4. Anaplastic astrocytomas were considered as an intermediate group between astrocytomas and glioblastomas being characterized by increased cellularity, variation in size, shape and chromatin content of nuclei, some mitotic activity but no tumor necrosis or marked endothelial proliferation of blood vessels.

REVISED WHO CLASSIFICATION (KLEIHUES AND CAVENEE, 2000)

The new grading and classification system of gliomas by WHO adopted in 2000[8] is listed in **Table 3.** This takes care of most of the criticisms lavelled against previous WHO classification schemes.

The important histopathological features for the characterization of the various grades of astrocytic tumors in this classification is shown in **Table 4.**

The comparison of this new WHO grading system for astrocytic tumors is shown in **Table 4.**

Using this grading system, typical ranges of survival was more than 5 years for grade II, 2-5 years for grade III and <1 year for majority of GBMs.

(IV) CLASSIFICATION OF EMBRYONAL TUMORS OF THE CENTRAL NERVOUS SYSTEM

Special diagnostic and conceptual problems exist in the identification and classification of CN neoplasms because in the central neuraxis, the cells of renewal that are the presumed target for neoplastic transformation are at their most abundant during development and consequently, the distinction between adult type and embryonal tumors may be blurred. The problem is further complicated by the frequency and multifocality of neoplastic progression, presenting the difficulty of distinguishing anaplastic from embryonal cell forms.

Table 4: Comparison of the World Health Organization (WHO) and St. Anne Mayo grading system of astrocytomas

Grade	WHO Designation	WHO Histological criteria	St. Anne / Mayo Designation	St. Anne / Mayo Histological criteria
I	Pilocytic astrocytoma	Biphasic pattern, bipolar cells, Rosenthal fibres, microcysts, giant cells, vascular proliferation, eosinophilic granular bodies		
II	Diffuse astrocytoma	Moderate cellularity, occasional nuclear atypia, absence of mitosis, necrosis and vascular proliferation	Astrocytoma grade 2	One criterion, usually nuclear atypia
III	Anaplastic astrocytoma	Increased cellularity and nuclear atypia; marked mitotic activity, no necrosis	Astrocytoma grade 3	Two criteria, usually nuclear atypia and mitotic activity
IV	Glioblastoma multiforme	Marked nuclear atypia, giant cells, brisk mitotic activity, prominent vascular proliferation and / or necrosis	Astrocytoma grade 4	Three criteria: nuclear atypia, mitoses, endothelial proliferation and / or necrosis

In 1973, Hart and Earle[45] proposed that certain supratentorial tumors of infancy should be classified as primitive neuroectodermal tumors (PNETs); these tumors were characterized microscopically by a high nuclear-to-cytoplasmic ratio and had greater than 95% of the tumor cells lacking differentiation by light microscopy.

Rorke[46] and Becker and Hinton[47] proposed that the term 'primitive neuroectodermal tumor' should be used to characterize all CNS neoplasms of embryonal neuroectodermal origin, including those tumors also known as medulloepithelioma, neuroblastomas, polar spongioblastoma, pineoblastoma, ependymoblastoma, retinoblastoma and olfactory neuroblastomas.

Indeed, the objectivity, reproducibility and ease of application inherent to the system led to an acceptance of the system among some pathologists who were no longer dependent on the site of origin to place a label on an otherwise undifferentiated neuroectodermal neoplasm. Furthermore, supporters of the system noted that it also encompassed the plasticity of the transformed embryonal neuroectodermal cell to express divergent neuroectodermal phenotypes.[48]

However, acceptance of this system was not universal. The most obvious defect with this classification was clinical. Up to 75% of medulloblastomas can be cured with a combined surgical, radiotherapeutic and chemotherapeutic approach and over 50% 5-year survival has been documented in patients with this tumor. In contrast, only 30% 5-year survival is usual for the tumors known as cerebral neuroblastomas[49] despite similarly aggressive therapy.[9]

Histologically, a number of factors distinguish these tumors at light microscopy level. Flexner-Wintersteiner rosettes are common in retinoblastomas, infrequent in pineoblastomas, rare in medulloblastomas and absent in cerebral neuroblastomas. Calcification is common in retinoblastomas and cerebral neuroblastomas but absent in medulloblastomas. 'Pale islands' of tumor cells are present in up to two thirds of childhood medulloblastomas and some medulloepitheliomas but lack in cerebral neuroblastomas, retinoblastomas and pineoblastomas. Finally, true Ependymal rosettes define the ependymoblastomas,[9,50] but are absent in the other neoplasms.

Hence in the most recent WHO classification,[8] the different entities of embryonal tumors are maintained due to distinctive histological features and genetic pathways.

In this text, we have maintained the cytogenetic approach used in previous editions to characterize these malignant embryonal tumors of neuroectodermal origin.

OUR CONTRIBUTIONS FROM AIIMS

1. In 1996, we reported a grading study of gliomas using computer aided malignancy classification and histologic morphometry.[51]

   Forty three cases of astrocytic tumors and mixed gliomas were studied with the aim of evaluating the reproducibility of the Kernohan grading system vis-à-vis (a) grading using computer aided malignancy classifier TESTAST 268 and (b) grading by quantitative morphometric evaluation of the various histological parameters of TESTAST 268. These patients' were then followed up for variable period with a maximum of forty months.

   Higher inter and intra-observer variability were observed in the Kernohan grading system. TESTAST 268 was found to be simpler, rapid and more reproducible. However, one drawback observed of this system was that it did not completely eliminate inter-observer variability because there was still some subjectivity in assignment of the categorical values against the histological features. Morphometric evaluation of the semi-quantitative assignment values of the 4 histological variables in the TESTAST 268 classifier using Zeiss Morphomat-30 revealed a statistically significant difference between the clusters of the measured quantitative values. A repeat grading using TESTAST 268 and categorical assignment values of histological features derived from the absolute values obtained from morphometry resulted in complete elimination of inter-observer variability. Thus, this study highlights the importance of objectivisation using TESTAST 268 and histologic morphometry in the grading of gliomas. However, since this is a preliminary study on a small number of cases, no cut off values of these measurements have been proposed.

2. We also published a study of gliomas using in vivo BrdU labeling index, WHO classification and computer aided malignancy grading.[52]

3. In 2000, we reported the results of a single center study on comparative survival evaluation along with assessment of inter-classification concordance in 102 cases of supratentorial astrocytic tumors in adults ($\geq$ 16 years of age).[53]

Hematoxylin and eosin (H&E) stained slides of these 102 cases were reviewed independently by two pathologists and each case classified or graded according to four different classification systems viz. Kernohan, Daumas-Duport (SAM-A), TESTAST-268 and WHO. The histological grading was then correlated with the survival curves as estimated by the Kaplan-Meier method.

The most important observation was that similar survival curves were obtained for any one grade of tumor by all the four classification systems. Fifty three of the 102 cases (51.9%) showed absolute grading concordance using all 4 classifications with maximum concordant cases belonging to grades 2 and 4. Intra-classification grade-wise survival analysis revealed a statistically significant difference between grade 2 and grades 3 or 4, but no difference between grades 3 and 4 in any of the classification systems.

It is apparent from the results of this study that if specified criteria related to any of the classification systems is rigorously adhered to, it will produce comparable results. Hence, preferential adoption of any one classification system in practice will be guided by the relative ease of histologic feature value evaluation with maximum possible objectivity and reproducibility. We recommend the Daumas-Duport (SAM-A) system since it appears to be the simplest, most objectivized for practical application and highly reproducible with relative ease.

REFERENCES :

1. Bigner DD, McLendon RE, Bruner JM: Russell and Rubinstein's Pathology of Tumors of the Central Nervous System. 6th Ed. Arnold, London, 1998.
2. CBTRUS: First Annual Report, 1995. Central Brain Tumor Registry of the United States. 1996.
3. CBTRUS: Annual Report Central Brain Tumor Registry of the United States. 1997.
4. Davis FG, Preston-Martin S: Epidemiology: Incidence and Survival. In: Russell and Rubinstein's

Pathology of tumors of the nervous system. 6th Ed. Edt Bigner DD, McLendon RE and Bruner JM. Arnold, London, 1998;5-45.

5. Parkin DM, Muir CS, Whelan SL, Gao YT, Ferlay J, Powel J: Cancer incidence in five continents Vol. VI. IARC Scientific Publication 120. International Agency for Research on Cancer, Lyon 1992.

6. Bailey P, Cushing H: A classification of the tumors of the glioma group on ahistogenetic basis with a correlated study of prognosis. J.B. Lippincott, Philadelphia, 1926.

7. Zulch KJ: Histologic typing of tumors of central nervous system. International Histologic Classification of Tumors, No. 21, Geneva : *WHO*; 1979;14-50.

8. Kleihues P, Cavenee WK: Pathology and genetics of tumors of the nervous system. WHO classification of tumors. International Agency for Research on Cancer Press Lyon, 2000.

9. McLendon RE, Enterline DS, Tien RD, Thorstad WL, Bruner JM: Tumors of central neuroepithelial origin. In : Russell and Rubinstein's Pathology of tumors of the nervous system. 6th Ed. Edt Bigner DD, McLendon RE and Bruner JM. Arnold, London, 1998;307-317.

10. Rorke LB: Classification and grading of childhood brain tumors. Overview and statement of problem. *Cancer* 1985;56:1848-49.

11. Rorke LB, Gills FH, Davis RL, Becker LE: Revision of World Health Organization classification of brain tumors for childhood brain tumors. *Cancer* Suppl. 1985;56:1869-86.

12. Russell DS, Rubinstein LJ: Tumors of central neuroepithelial origin. In : Pathology of Tumors of the Nervous System. London; Edward Arnold, 1989;83-247.

13. Gilles FH: Classification of childhood brain tumors. *Cancer* 1985;56:1850-57.

14. Rubinstein LJ: Tumors of central nervous system. In: Atlas of Tumor Pathology, Series 2, Fascicle 6, Washington : *Armed Forces Institute of Pathology* 1972;19-50:55-85.

15. Rubinstein LJ: Supplement to tumors of the central nervous system. Atlas of Tumor Pathology, second series, fascicle 6. Washington DC, *Armed Forces Institute of Pathology* 1982.

16. Kernohan JW, Mabon RF, Svien HJ, Adson AW: A simplified classification of gliomas. *Proc Staff Meet Mayo Clin* 1949;24:71-75.

17. Ringertz J: Grading of gliomas. *Acta Pathol Microbiol Scand* 1950;27:51-64.

18. Kernohan JW, Sayre GP: Tumors of central nervous system. Fascicle 35, Atlas of Tumor Pathology. Wastington: *Armed Forces Institute of Pathology* 1952;17-129.

19. Zulch KJ: Atlas of the histology of brain tumors. Berlin: Springer. 1971.

20. Wier B, Grace M: The relative significance of factors affecting operative survival in astrocytomas, grades I & II. *Can J Neurol Sci* 1976;3:47-50.

21. Fazekas JT: Treatment of grade I and II brain astrocytomas: The role of radiotherapy. *Int J Radiat Oncol Biol Phys* 1977;2:661-66.

22. Scanlon PW, Taylor WF: Radiotherapy of intracranial astrocytomas. Analysis of 417 cases treated from 1960 through 1969. *Cancer* 1979;5:301-07.

23. Daumas-Duport C, Sheithauer B, O'Fallon J, Kelly P: Grading of astrocytomas: A simple and reproducible method. *Cancer* 1988;62:2152-65.

24. Wier B: The relative significance of factors affecting postoperative survival in astrocytoma grade III & IV. *J Neurosurg* 1973;38:448-52.

25. Nelson JS, Tsukada Y, Schoenfeld D, Fulling K, Lamarche J, Peress N: Necrosis as a prognostic criteria in malignant supratentorial astrocytic gliomas. *Cancer* 1983;52:550-54.

26. Burger PC, Vogel FS, Green SB, Strike TA: Glioblastoma multiforme and anaplastic astrocytoma: Pathological criteria and prognostic implications. *Cancer* 1985;56:1106-11.

27. Garcia DM, Fulling KH, Marks JE: The value of radiation therapy in addition to surgery for astrocytomas of adult cerebrum. *Cancer* 1985;55 919-27.

28. Garcia DM, Fulling KH: Juvenile pilocytic astrocytoma of cerebrum in adults. *J Neurosurg* 1985;63: 382-86.

29. Krouwer HGJ, Davis RL, Silver P, Prados M: Gemistocytic astrocytomas : a reappraisal. *J Neurosurg* 1991;74:399-406.

30. Kepes JJ: Xanthomatous lesions of central nervous system. Definition, classification and some recent observations. In: Progress n Neuropathology (Ed. Zimmerman, H.M.) Raven Press, New York, 1979;179-213.
31. Kepes JJ, Rubinstein LJ, Eng LF: Pleomorphic xanthoastrocytoma, a distinctive group of young subjects with relatively favourable prognosis. A study of 12 cases. *Cancer* 1979;44:1839-52.
32. Palma L, Maleci A, DiLorenzo N, Lauro SM: Pleomorphic xanthoastrocytoma with 18 year survival, case report. *J Neurosurg* 1985;63:808-10.
33. Laws ER, Taylar WF, Clifton MB, Okazaki H: Neurosurgical management of low grade astrocytomas of cerebral hemispheres. *J Neurosurg* 1984;61:665-73.
34. Mork SJ, Halvorsen TB, Lindegaard KF, Eide GE: Oligodendroglioma : Histologic evaluation and prognosis. *J Neuropathol Exp Neurol* 1986;45:65-78.
35. Hoshino T, Rodriguez LA, Cho KC, Lee KS, Wilson CB, Edwards MSB, Levin VA, Davis RL: Prognostic implications of the proliferative potential of low grade astrocytomas. *J Neurosurg* 1988;69:839-42.
36. Mork SJ, Boken AC: Ependymoma. A follow up study of 101 cases. *Cancer* 1977;40:907-15.
37. Ross GW, Rubinstein LJ: Lack of histopathologic correlation of malignant ependymomas with post-operative survival. *J Neurosurg* 1989;70:31-36.
38. Nelson DF, Nelson JS, Davis DR, Chang CH, Griffin TW, Pajak TF: Survival and prognosis of patients with astrocytoma with atypical or anaplastic features. *J Neuro Oncol* 1985;3:99-103.
39. Rorke LB: Classification of central nervous system tumors in children. *Progress in Experimental Tumor Research* 1987;30:57-60.
40. Becker LE: An appraisal of the World Health Organization classification of tumors of the central nervous system. *Cancer* 1985;56:1858-64.
41. Kim TS, Halliday AL, Hedley-Whyte ET, Convery K: Correlates of survival and the Daumas Duport grading system for astrocytomas. *J Neurosurg* 1991;74:27-37.
42. Schmitt HP: Numerical classification of malignancy in astrocytomas : an attempt to improve and to validate the classifier TESTAST. Studien Zur Klassifikation Frankfurt; Indek Verlag: 1989;361-68.
43. Schmitt HP, Oberwittler Ch: Computer aided assessment of malignancy in gliomas: Contemporary approach to the problem of tumor grading. 9th International Congress of Neurological Society, New Delhi, 1989 (Abst).
44. Kleihues P, Burger PC, Scheithauer BW: Histological typing of tumors of the central nervous system. WHO International Histological Classification of Tumors. 2nd Ed. Springer Verlag, Berlin, 1993.
45. Hart MN, Earle KM: Primitive neuroectodermal tumors of the brain in children. *Cancer* 1973;32: 890-97.
46. Rorke LB: The cerebellar medulloblastoma and its relationship to primitive neuroectodermal tumors. Presidential Address. *J Neuropathol Exp Neurol* 1983;42:1-15.
47. Becker LE, Hinton D: Primitive neuroectodermal tumors of the central nervous system. *Human Pathol* 1983;14:538-50.
48. Sarkar C, Roy S, Tandon PN: Primitive neuroectodermal tumors of the central nervous system: An electron microscopic and immunohistochemical study. *Ind J Med Res* 1989;90: 91-102.
49. Bennett JP, Rubinstein LJ: The biological behaviour of primary cerebral neuroblastomas. A reappraisal of the clinical course in a series of 70 cases. *Ann Neurol* 1984;16:21-27.
50. Mork SJ, Rubinstein LJ: Ependymoblastoma: A reappraisal of a rare embryonal tumor. *Cancer* 1985;55:1536-42.
51. Sharma S, Karak AK, Sarkar C, Gomathi G, Banerji AK, Schmitt HP: A grading study of gliomas using computer aided malignancy classification and histologic morphometry. *J Neuro Oncol* 1996;27: 75-85.
52. Sharma S, Karak AK, Singh R, Mehta VS, Sarkar, C, Schmitt, HP: A correlative study of gliomas using in-vivo bromodeoxyuridine labeling index and computer aided malignancy grading. *Pathol Oncol Res* 1999;5;134-41.
53. Karak AK, Singh R, Tandon PN, Sarkar C: A comparative survival evaluation and assessment of interclassification concordance in adult supratentorial astrocytic tumors. *Pathol Oncol Res* 2000;6: 46-52.

3
Tumor Markers: An Overview

C Sarkar, MC Sharma, S Mukhopadhyay

INTRODUCTION

Tumor markers are biochemical indicators of the presence of a tumor. They include cell surface antigens, cytoplasmic proteins, enzymes and hormones. In clinical practice, the term usually refers to a molecule that can be detected in the blood or other body fluids.[1] The main utility of tumor markers in clinical medicine has been as a laboratory test to support a diagnosis or to follow the response of tumors to therapy.

However, tumor markers can also be defined more broadly as substances that make possible either a qualitative diagnosis of neoplasia or a quantitative estimate of tumor burden.[2] Used in this sense, tumor markers may be demonstrable not only in the blood, CSF, etc, but also in tissue sections. Here the issue is not the presence of a tumor but the characterization of the tumor in terms of lineage/differentiation, proliferative potential and prognosis. We will use the term "tumor marker" in this broader sense. We will address the topic of tumor markers in brain tumors under two main categories: markers used in diagnosis and markers used in assessment of prognosis.

TUMOR MARKERS USEFUL IN THE DIAGNOSIS OF BRAIN TUMORS

1. Markers detected in Serum or Cerebrospinal Fluid

Unfortunately, brain tumors are, with one exception, bereft of tumor markers that can be detected in the serum or cerebrospinal fluid (CSF). Although many potential candidate markers have been evaluated, none have met the criteria of useful tumor markers, i.e. none of them are produced only by neoplastic tissue and released into an easily accessible fluid compartment in measurable quantities during an early stage of tumor development. This is particularly so in glial neoplasms, which constitute the bulk of central nervous system (CNS) neoplasia.

The only exception to this bleak scenario is seen in primary germ cell tumors of the CNS. These rare tumors arise in midline structures including the pineal and sellar regions, third ventricle, hypothalamus and spinal cord.[3] Because most pineal neoplasms arise from misplaced germ cells, germ cell tumors are common in this region. Pineal germ cell tumors are histologically identical to their gonadal counterparts and secrete the same repertoire of tumor markers that can be assayed in the serum and the CSF.

The primary CNS germ cell tumors include germinomas, embryonal carcinomas, yolk sac tumors, choriocarcinomas, teratomas and mixed germ cell tumors. The markers that are most important in their diagnosis are alpha-fetoprotein, human chorionic gonadotropin and placental alkaline phosphatase.

Alpha-fetoprotein (AFP)

Alpha-fetoprotein is an oncofetal glycoprotein (MW 70,000) that belongs to the alpha-globulin group of serum proteins. In the fetus it is a major plasma protein, peaking in the third and fourth months of gestation. AFP levels fall progressively after this period and settle down at normal adult levels by the end of infancy. The major sources of this protein are the liver and the visceral endoderm of the yolk sac. It is invariably present in yolk sac tumors and also in a high proportion of other germ cell tumors that contain yolk-sac elements. However, it lacks specificity, its levels being elevated in a host of other neoplastic and non-neoplastic conditions, chief of which are diseases of the liver and testis.

Since AFP is a marker for yolk-sac elements, an elevated CSF AFP level should prompt consideration of testicular germ cell tumors, 16-25% of which metastasize to the CNS. Only when this possibility has been excluded should primary intracranial germ cell tumors be considered.

AFP measurement has been helpful in the diagnosis and management of primary intracranial germ cell tumors.[4,5] In particular, the presence of an elevated CSF AFP level or an abnormal CSF-to-serum gradient has been uniformly associated with primary intracranial *non-germinomatous* germ cell tumors.[6-8] These tumors typically arise in the pineal region. The presence of elevated CSF AFP values seems to be a more reliable and sensitive marker for these tumors than CSF cytology. If AFP is elevated and hCG is normal, the most likely diagnosis is a yolk-sac tumor, although an undifferentiated malignant germ cell tumor or a malignant teratoma may also be considered. Germinomas, choriocarcinomas and the other remaining germ cell tumors do not produce AFP.[2]

The preoperative evaluation of pineal tumors should include CSF cytology and measurement of a panel of tumor markers both in the serum as well as in the CSF. The standard panel would consist of PLAP, hCG and AFP. Lumbar CSF levels of these markers generally exceed ventricular levels.[4] Because pineal surgery has become safer over the past decade, and because the histology of pineal tumors is often mixed, the pendulum has swung against test doses of radiotherapy in favor of open biopsy.[6,8] A possible exception may be a patient who presents with disseminated CNS disease and has a very high AFP level. This patient would almost certainly have a yolk-sac tumor.[6,7] Following surgery, tumor-marker levels should be checked on a regular basis to monitor the response to therapy and to help identify tumor recurrence early.[4,9,10] Again lumbar CSF appears to be more sensitive than ventricular CSF. Monitoring serum markers also seems to be helpful in detecting extra-neural metastases and monitoring their response to therapy.[11]

Human chorionic gonadotropin (hCG)

HCG (MW 45,000) is a placental glycoprotein that is normally elaborated by the placenta and found in the serum of pregnant and postpartum women, as well as in fetal blood. It has two subunits named alpha and beta. The alpha subunit is related to luteinizing hormone (LH), and is secreted by the placental trophoblast. The beta subunit, when elevated in the absence of pregnancy, is a sensitive marker for disease. Classically, it is diagnostic of uterine choriocarcinomas, specifically the syncytiotrophoblastic elements in that tumor. However, elevated hCG levels have also been found in patients with testicular tumors, ovarian tumors, and cancers of the stomach, prostate, liver and breast.

CSF hCG is usually 0.5-2% of the serum hCG level in non-CNS tumors. The presence of a significantly higher CSF hCG is usually a reliable diagnostic sign of a metastatic choriocarcinoma or of a primary choriocarcinoma in the pineal or suprasellar region.[12-14] CSF hCG elevation also occurs with embryonal carcinoma, malignant teratoma, undifferentiated malignant germ cell tumor, and to a lesser extent, germinoma.[6-8,15] Elevation of hCG values has also been reported with benign teratoma.[16]

Placental Alkaline Phosphatase (PLAP)

PLAP is another placental protein (see hCG) found normally in the serum of pregnant women. It is also found in the serum of women with trophoblastic tumors and in a small percentage of patients with non-trophoblastic tumors. CSF PLAP is often positive in patients whose tumors stain positively for PLAP.[10] However, CSF PLAP can also be elevated in patients with choriocarcinoma, yolk sac tumor, embryonal carcinoma and malignant teratoma.[5] Parallel to AFP in yolk sac tumor, and hCG in choriocarcinomas, very high PLAP levels correlate strongly with germinoma.

All these markers (PLAP, hCG and alpha-fetoprotein) can be demonstrated immunohistochemically in tissue sections. The interpretation of positivity in tissue is similar to that in serum or CSF. In brief PLAP-positivity indicates germ-cell differentiation, alpha-fetoprotein indicates the presence of yolk-sac elements, and hCG positivity indicates the presence of syncytiotrophoblastic elements in primary intracranial germ cell tumors.

Finally, it must be appreciated that tumor markers *alone* are inadequate to establish a diagnosis in CNS neoplasia. Conversely, their absence does not rule out germ cell tumors, since some of the malignant germ cell tumors do not secrete any markers.[5]

Other markers

Attempts to use CSF GFAP levels as a marker for glial tumors have not been consistently successful. Although the highest CSF GFAP levels to date have been from patients with malignant gliomas, a wide spectrum of nonglial tumors and nonneoplastic neurological diseases have also shown consistently elevated CSF GFAP levels. The diagnostic usefulness of CSF GFAP still needs to be defined, although it seems that its value as a tissue marker of glial lineage far outweighs its use as a CSF tumor marker.[2]

CEA (carcinoembryonic antigen) has been tried as a marker for carcinomas metastatic to the brain but has not proved to be consistently reliable.[17] NSE (Neuron-specific enolase) was originally thought to be a marker for neuroendocrine tumors, but instead a raised CSF NSE level appears to be a nonspecific marker for neurological damage in a variety of other disease processes.[18] Other substances that have been measured in the CSF are LDH,[19] beta-glucuronidase, polyamines,[20,21] desmosterol and beta-2-microglobulin.[22,23] However, none of these substances have been found reliable enough to use diagnostically as tumor markers.[2]

2. Markers detected in Brain Tissue

A. Intermediate filament proteins

The intermediate filament proteins (so named because their diameter is intermediate between actin filaments and microtubules) are intracellular filaments measuring approximately 10 nm in diameter. They comprise a family of proteins that are grouped into six classes based on biochemical and structural parameters[24] **(Table 1)**. These markers are partially lineage-specific and are also markers of particular developmental stages in the nervous system. The study of the expression of these markers in the nervous system, therefore, gives an estimate of how primitive/mature a cell is and what lineage it is attempting to differentiate towards.

GFAP (Glial Fibrillary Acidic Protein)

GFAP (MW 48,000 to 52,000) is one of the five major types of cytoplasmic intermediate filaments[25,26] **(Table 1)**. It is present in normal, reactive and neoplastic astrocytes as well as developing and neoplastic oligodendrocytes[27] and represents the principal cytoskeletal constituent elaborated by human astrocytes.[28] However, expression of this marker is not totally specific for cells of glial origin, since it has also been documented in tumors of salivary glands and sweat glands.

Antibodies to GFAP stain the cytoplasm and cytoplasmic processes of astrocytes. Neoplasms derived from astrocytes (astrocytomas) are characterized by a felt work of fine, GFAP-positive astrocytic cell processes that give the background a fibrillary appearance. Similarly, pilocytic astrocytomas are composed of bipolar cells with long, thin, "hair-like" processes that are GFAP-positive. GFAP expression tends to decrease during glioma progression. Therefore, staining is scant in protoplasmic astrocytomas, inconsistent in anaplastic astrocytoma and highly variable in glioblastomas. In glioblastomas, foci of GFAP-negative anaplastic tumor cells suggest the formation of a new clone with additional genetic alterations.[29] However, there is no indication that the extent of GFAP expression is a prognostic factor in gliomas.

Most glial neoplasms can be confidently diagnosed on morphologic grounds on H&E-stained sections alone, without the need for supplemental immunohistochemistry. However, variants of anaplastic astrocytoma and glioblastoma multiforme may occasionally take on bewildering appearances that necessitate the use of immunohistochemical markers. A case in point is the "giant cell glioblastoma", originally thought to be a sarcoma due to its cytologic features and reticulin-rich matrix.[30] GFAP-positivity has shown this tumor to be glial in origin. The same is true for spindle cell and "xanthosarcomatous", lipid-rich variants of glioblastoma.[31,32] Many high-grade astrocytic neoplasms may therefore be misclassified as sarcomas if GFAP is not included in the diagnostic armamentarium.

Another problem is posed by glial tumors that appear to be cohesive (in nests and sheets), contain glands,[33] undergo squamous metaplasia[34] assume clear cell change (due to cytoplasmic lipidization), or label for cytokeratin or EMA.[35] Any of these features could mislead a pathologist into thinking of an epithelial malignancy. Glioblastomas may also mimic medulloepitheliomas by virtue of primitive-appearing papillary structures.[36] In each of these cases, immunohistochemical staining for GFAP can be the decisive investigation in unmasking the glial nature of the neoplasm.

A few words of caution are, however, appropriate in the interpretation of GFAP staining in tissue sections. Firstly, the difficulties in the technical aspects of GFAP staining and the variability in the assorted anti-GFAP sera available mean that the disparity in GFAP interpretation among labs can be enormous.[37] Secondly, the distinction between GFAP-positive *neoplastic* astrocytes and entrapped/peritumoral GFAP-positive *reactive* astrocytes can be very difficult.[38,39] These limitations must be appreciated to put GFAP expression in perspective.

GFAP expression may also be seen in oligodendrogliomas, not only in the reactive astrocytes but also in the neoplastic oligodendroglial cells.[40] However, GFAP is not a consistent marker for cells of oligodendroglial origin. Indeed, there is no immunohistochemical marker to date that would allow the sensitive and specific detection of human oligodendroglial tumor cells. In oligoastrocytomas, GFAP positivity is found more consistently in the astroglial component, compared with a more variable expression in the oligodendroglial component.

About 50% of ependymomas can be shown immunohistochemically to contain GFAP focally, especially in the pseudorosettes. Staining for GFAP is usually positive in myxopapillary ependymomas, and this may help to distinguish these tumors from paragangliomas and schwannomas. GFAP is absent from the normal choroid plexus, but occurs focally in 25-55% of choroid plexus papillomas and about 20% of choroid plexus carcinomas. Other tumors that can be GFAP-positive include astroblastoma, ganglioglioma (glial component), desmoplastic cerebral astrocytoma of infancy, desmoplastic infantile ganglioglioma and medulloblastomas (where GFAP expression signifies a poor prognosis).

Neurofilament proteins (NFP)

Neurofilaments represent the intermediate filaments of neurons and their processes **(Table 2)**. The expression of NFP in CNS tumors is a specific indicator of commitment to the neuronal lineage. NFPs are composed of three subunits, which are catalogued into three classes named NF-L (low MW), NF-M (medium MW), and NF-H (high MW) respectively. The relevance of this classification is that the higher MW isoforms tend to be expressed to a higher degree in more advanced stages of neuronal development.[41] Neurofilaments-positive tumors include ganglioneuroblastomas, ganglioneuromas, neurocytomas, medulloblastomas and pineoblastomas.[39]

Vimentin

Vimentin is one of the six major types of cytoplasmic intermediate filaments **(Table 1)**. It is characteristic of cells of mesenchymal derivation, such as endothelial cells, fibroblasts, and vascular smooth muscle cells.[42] However, it is not constrained to cells of mesodermal origin but is occasionally also expressed in tumors of epithelial or neural nature, not infrequently in conjunction with keratin or GFAP.[43]

Vimentin can be demonstrated in astrocytic neoplasms, following a pattern similar to that of GFAP **(Table 2)**. However, vimentin-positive cells may lack GFAP expression. In terms of developmental stage, the expression of vimentin is similar to nestin, both being expressed relatively early in astrogliogenesis.[44] Its presence in astrocytomas, therefore, could indicate a lower degree of differentiation. Indeed, there is a tendency for vimentin to be expressed more consistently in high-grade astrocytomas.[45] Vimentin is, for example, consistently expressed in anaplastic astrocytomas.

Vimentin, however, is notoriously non-specific. A wide assortment of CNS tumors can be vimentin-positive, limiting its diagnostic value. CNS tumors that show vimentin positivity include oligodendrogliomas,[46] astroblastomas, medulloblastomas, atypical rhabdoid tumors (with smooth muscle actin coexpression), meningiomas, mesenchymal non-meningothelial tumors (such as tumors derived from adipose tissue, muscle and blood vessels), hemangiopericytoma (usually coexpresses CD34), Langerhans cell histiocytosis (with S-100 and CD1a coexpression) and stromal cells in hemangioblastomas (with NSE coexpression). The coexpression of vimentin with the more lineage-distinctive NFP and GFAP in CNS tumors, and the observation that it often becomes the predominant IFP in in-vitro cultures, has lead to its being called the "default" IFP in CNS tumors.[39, 44]

Alpha-internexin and peripherin

These recently characterized IFPs **(Table 1)** are neuron-specific in that they accompany the down-regulation of nestin and the commitment of cells towards neuronal differentiation. They have been demonstrated in neuroblastomas, ganglioneuroblastomas and ganglioneuromas.[47]

Table 1: *The intermediate filament proteins (IFP)*

Intermediate filament type	Molecular mass (kDa)	Class	Tissue distribution
Keratin (acidic)	40-68	I	Epithelium
Keratin (basic)	40-68	II	Epithelium
GFAP and peripherin	50-52 (GFAP)	III	Mesenchyme, primitive neuroepithelial cells, astrocytes, developing neurons (GFAP)
	57 (peripherin)		Neurons (peripherin)
NFP and alpha-internexin	68-200 (NFP)	IV	Neurons
	66 (alpha-internexin)		
Lamin	60-70	V	Nuclear membrane
Nestin	210-240	VI	Primitive neuroepithelium, developing astrocytes, neurons, schwann cells

Adapted from Wikstrand et al[39]

Nestin

Nestin is the largest and the most recently described member of the family of intermediate filaments **(Table 1)**. It is a marker for primitive neuroepithelial stem cells and disappears towards the end of gestation from all cells except endothelial cells of the CNS and Schwann cells of the peripheral nervous system.[48] Antibodies to nestin help in the differential diagnosis of choroid plexus carcinoma and medulloepithelioma. The latter tumor is characteristically positive for this marker, while the former is nestin-negative.[44] However, the use of nestin as a marker in CNS tumors is limited by its lack of specificity. Nestin expression or coexpression has been documented not only in medulloblastomas but also in gliomas of all grades, meningiomas, and even metastatic melanomas.

CK (Cytokeratin) and EMA (Epithelial Membrane Antigen)

The cytokeratins are the most complex of the intermediate filament proteins **(Table 1)**. The utility of these markers has traditionally been to differentiate between primary brain tumors (CK-negative) and metastatic carcinomas (CK-positive). This interpretation is now known to be overly simplistic.[49] To overcome the pitfalls associated with CK-positivity in some primary brain tumors, it has been recommended that CK immunohistochemistry to distinguish between primary and metastatic carcinomas must be performed in conjunction with markers for NFP and GFAP.[39]

Although EMA is not an IFP, it is included here because its expression closely parallels that of CK. EMA is a glycoprotein present in human milk fat globule membranes;[50] it is probably analogous to the antigen demonstrated by the antisera raised against the casein fraction of human milk. EMA is an excellent marker for most normal and neoplastic epithelia but is not restricted to them.[51] It can also be expressed by mesotheliomas, meningiomas, a variety of mesenchymal neoplasms, and even some malignant lymphomas.[52, 53] It has also been found to be a marker of normal and neoplastic perineurial cells.

EMA and CK are both positive in the majority of metastatic carcinomas. The differential diagnosis with choroid plexus neoplasms (which share these markers) is discussed elsewhere in this chapter. Another use of CK/EMA positivity is in the differential diagnosis between metastatic renal cell carcinomas (CK/EMA positive) and hemangioblastomas (CK/EMA negative).

Both EMA and CK are expressed by virtually every choroid plexus papilloma. CK is also expressed in choroid plexus carcinomas. This helps in the differential diagnosis with medulloepitheliomas, which are characteristically CK-negative.[54]

EMA may be expressed in atypical rhabdoid tumors. CK is expressed by craniopharyngioma, reflecting its epithelial derivation. Meningiomas typically exhibit EMA-positivity, a feature that helps in the differential diagnosis with hemangiopericytoma.

B. Neuronal and neuroendocrine markers

The concept of neuroendocrine cells arose from the observation that hypothalamic neurons synthesized oxytocin and vasopressin and secreted them into the bloodstream in an *endocrine* fashion, as opposed to the usual *neuronal* transmission via synaptic contacts. This initial observation was buttressed by similar observations on a diverse range of neurons, including the adrenal medulla, APUD cells of the gastrointestinal tract and several others. Apart from their common argentaffin properties, argyrophilia, dense-core secretory granules, and similar phenotypes, neuroendocrine cells and tumors arising from them share the expression of synaptophysin, chromogranin and neuron-specific enolase. These neuroendocrine markers will be considered in the following paragraphs.

Synaptophysin, Chromogranin, Neuron Specific Enolase (NSE)

The chromogranin family is composed of acidic glycoproteins (MW 20,000 to 100,000) located in the soluble fraction of neurosecretory granules.[55] The most abundant is chromogranin A (MW 75,000).[56] Two others have been named chromogranin B (secretogranin I) and chromogranin C (secretogranin II).[57] Nearly all types of neuroendocrine tumors are reactive, so that chromogranin has become the most widely used "pan-endocrine" marker.[58] Synaptophysin is a transmembrane glycoprotein (MW 38,000) that has been isolated from neuronal presynaptic vesicles. It is expressed in normal, reactive and neoplastic cells of neuroectodermal and neuroendocrine types. Neuron-specific enolases are the gamma-gamma and alpha-gamma isoenzymes of the enzyme enolase. The name reflects the finding that these isoenzymes are found preferentially in neurons and neuroendocrine cells. NSE is found in the majority of neuroectodermal and neuroendocrine neoplasms. Unfortunately, this high sensitivity is offset by its low specificity, stemming from its detection in several other types of normal and neoplastic cells.[59]

Brain tumors that are positive for neuronal markers **(Table 2)** include gangliocytomas

Table 2 : *Expression of the major immunohistochemical markers in the most common primary CNS tumors*

Tumor	Nestin	Vimentin	Synaptophysin	NFP	GFAP
Astrocytomas	+/-	+	-	-	+
Ependymomas	+	+/-	-	-	+
Medulloblastomas	+/-	+	+	+/-	+/-

Adapted from Wikstrand et al[39]

and gangliogliomas,[60] desmoplastic infantile ganglioglioma (neuronal component), dysembryoplastic neuroepithelial tumors, central neurocytomas (synaptophysin is the most reliable marker),[61] paragangliomas (NSE is the most sensitive marker),[62] pineoblastomas and pineocytomas (variable), central neuroblastomas and ganglioneuroblastomas, medulloblastomas (synaptophysin immunoreactivity is characteristic),[63] supratentorial PNETs (synaptophysin is consistently identified) and endothelial cells in hemangioblastomas (with CD31 and CD34 co-expression). On the other hand, astrocytic tumors, meningiomas, ependymomas and choroid plexus papillomas are negative for neuroendocrine markers.[44]

S-100

Many benign peripheral nerve-sheath tumors express Schwann cell characteristics, including the presence of S-100 antigen and the potential for melanocytic differentiation.

Some, but not all, malignant peripheral nerve-sheath tumors are immunoreactive for S-100 protein. This S-100-positivity, along with CK-negativity helps to distinguish epithelioid malignant peripheral nerve-sheath tumors from tumors of epithelial origin, to which they are histologically similar.

Tumor cells in low-grade diffuse astrocytomas usually show immunoreactivity to S-100 protein in the nucleus and cell processes, but this feature has no diagnostic relevance.[64] S-100 is also consistently expressed in anaplastic astrocytomas.

Other tumors that express S-100 include oligodendrogliomas (a property it shares with many other neuroectodermal tumors), ependymomas,[64] choroid plexus tumors (90% of cases), astroblastomas, meningiomas (variable and usually not prominent), melanocytic lesions (with HMB-45 coexpression) and Langerhans cell histiocytosis (with CD1a coexpression).

C. Leukocyte common antigen (LCA/CD 45 RB), T-cell markers (CD3) and B-cell markers (CD19/CD20)

The majority (approximately 98%)of primary brain lymphomas are of B-cell origin. A benign mixed T- and B-cell infiltrate, which often contains a plasmacytic component, can also be found adjacent to lesions. Reed-Sternberg cells in Hodgkin's lymphomas usually mark with CD15 and CD30 and are non-reactive for LCA.

D. Progesterone

Meningiomas often harbor progesterone receptors, and rapid growth during pregnancy has been reported. Progesterone receptors can be demonstrated immunohistochemically, and positivity status correlates with recurrence. The absence of progesterone receptors in meningiomas correlates with shorter disease-free intervals.[65]

E. Other antibodies

The antibodies HEA 125 and BerEP4 may be helpful in the distinction between choroid plexus carcinomas and metastatic carcinomas. These antibodies label more than 95% of the latter, but only 10% of the former.[66] In addition, expression of CEA (carcinoembryonic antigen) suggests metastatic carcinoma, although occasional choroid plexus carcinomas are also positive.[67] Ubiquitin and crystallin are two other immunohistochemical markers that have been studied in brain tumors.[68] However, they have not found widespread acceptance or applicability. Transthyretin is a useful marker for normal and neoplastic choroid plexus epithelia, although up to 20% of choroid plexus papillomas may be negative, and in rare instances, other brain tumors and metastatic carcinomas are also positive.[69,70] Transthyretin is more often absent in choroid plexus carcinomas than in choroid plexus papillomas. The photoreceptor cell markers retinal S-antigen (or arrestin), rod-opsin, and inter-photoreceptor retinoid-binding protein (IRBP) are expressed in subsets of pineal parenchymal tumors and medulloblastomas, suggesting that these subsets exhibit phenotypic features of the photoreceptor lineage.[71]

F. Tumors for which no specific marker is available

These include oligodendroglioma and polar spongioblastoma.

TUMOR MARKERS USEFUL IN THE ASSESSMENT OF THE PROGNOSIS OF BRAIN TUMORS

Ki-67 (MIB-1)

Ki-67[72] and MIB-1[73] are commercially available monoclonal antibodies that recognize a cell cycle-associated, nonhistone nucleoprotein. They therefore are markers of cells that are actively cycling. Whereas Ki-67 antibodies require frozen sections for their application, MIB-1 is applicable to formalin-fixed, paraffin-embedded material.

Although conventional histologic examination remains the only method routinely applied to the prognostic assessment of astrocytomas, proliferation markers may prove to be more accurate than histologic grading in the prediction of a given lesion's clinical evolution. There appears to be a clear correlation of proliferation with clinical outcome.[74] Ki-67 has been used to discriminate between grades of anaplasia in astrocytomas as well as in other types of brain tumors.[75]

Tumors which usually show a low Ki-67 labeling index (LI) (LI = the fraction of Ki-67-positive tumor nuclei) include pilocytic astrocytoma, oligodendroglioma, gemistocytic astrocytoma (LI usually < 4%), oligoastrocytoma, ependymomas, choroid plexus papillomas (LI=0.2-6%), gangliogliomas (1.1–2.7%), desmoplastic infantile astrocytoma and desmoplastic infantile ganglioglioma (LI=0.5–5%), central neurocytoma (LI=< 2%), classical meningioma (mean LI=0.7%) and atypical meningioma (mean LI=2.1%).

In contrast, a Ki-67 LI > 3% has been associated with the potential for aggressive growth in supratentorial fibrillary astrocytomas.[72] The LI in anaplastic astrocytomas is usually 5-10%[76] and the mean LI in glioblastomas is 15-20%.[77-79] A Ki-67 LI >7.5% is associated with higher histological grade and poorer survival in astrocytic tumors. Significantly, this index has been found by some workers to be more significant statistically than histologic grading.[72] In another study, a Ki-67 LI >5% was found to constitute a threshold value for the prediction of shorter survival.[79] There is some evidence that in low-grade astrocytomas, the proliferative potential correlates with survival and time to recurrence.[72, 80-82] However, in glioblastomas, there is general consensus that that the MIB-1 LI and related proliferation markers do not allow prognostication in individual patients.[83]

Other tumors with a high Ki-67 LI include choroid plexus carcinomas (LI=7.3-60%), medulloblastomas (LI often>20%), atypical rhabdoid tumors, anaplastic meningiomas (mean LI=11%), hemangiopericytomas (mean LI=10%) and lymphomas (LI=19-47%).

Prognostic connotations of Ki-67 LI in non-astrocytic tumors

In oligodendrogliomas, it has been shown that the Ki-67 LI may provide additional prognostic information, independent of patient age, tumor site and histologic grade.[84] Increased frequency of MIB-1 and p53 labeling may be associated with aggressive behavior and tumor recurrence in gangliogliomas,[85] Neurocytomas with a MIB-1 LI >2% have a significantly shorter recurrence-free interval than lesions with an index of <2%.[86]

p53

p53: the gene and the protein

The *protein* p53 is named for its biochemical nature (*p*hosphoprotein) and its molecular mass (**53** kilo Daltons). It was discovered in 1979 by three independent groups using different approaches[87-89] to the study of tumorigenesis.

The p53 protein is coded for by a tumor-suppressor gene (the p53 *gene*) on chromosome 17p13.1. It plays a role in several cellular processes, including cell cycle control, response of cells to DNA damage, cell death, cell differentiation, and neovascularization.[90]

Prognostic significance of p53 in low-grade diffuse astrocytomas

The predictive role of p53 mutations in glioma progression is still poorly understood.

1. Recent studies indicate that in low-grade astrocytomas that progress to

glioblastoma, the frequency of p53 mutations is very high, of the order of 58-83%.[91]

2. Neoplasms with p53 mutations also appear to progress more frequently than neoplasms without the mutation, but evidence for this correlation remains circumstantial.[92-94]
3. The time-interval to progression appears to be shorter in patients with low-grade astrocytomas that carry a p53 mutation.[95]
4. Low-grade astrocytomas that carry a p53 mutation appear to be associated with a shorter survival.[96]

Hence, it seems that low-grade astrocytomas with p53 mutations progress to glioblastomas faster and more often than tumors without these mutations. The effect on patient survival, however, is still murky, with some studies showing that p53 mutations have no effect on clinical outcome.[93, 97, 98]

p53 mutations in other brain tumors

In contrast to diffusely infiltrating astrocytomas, p53 mutations do not seem to play a significant role in the evolution of other tumors such as pilocytic astrocytomas,[99,100] ependymomas,[101] pineocytomas/ pineoblastomas[102] and medulloblastomas (in whom p53 mutations are found in only 5-10% of cases).[103]

EGFR (EPIDERMAL GROWTH FACTOR RECEPTOR)

EGFR (MW 170,000) is a receptor for epidermal growth factor (EGF) and transforming growth factor-beta (TGF-beta). In addition to its transmembrane (receptor) domain, it also has tyrosine kinase activity. The gene (on chromosome 7) for this receptor accounts for the majority of gene amplifications seen in high-grade astrocytomas.[104,105] In all, it is amplified in one-third to one-half of glioblastomas and in a few anaplastic astrocytomas.[106] This gene amplification constitutes a hallmark of primary glioblastomas, more than 60% of which show EGFR expression.[107] All glioblastomas with EGFR amplification also show simultaneous loss of chromosome 10.[108] EGFR, like p53, is amenable to demonstration by immunohistochemical methods on tissue sections.

OTHER ONCOGENES AND RELATED GENE PRODUCTS

Several other oncogenes and related gene products have been characterized in brain tumors. The prognostic significance of many of these moieties is as yet unclear. Although the overall incidence is low, myc is the most common oncogene amplified in medulloblastomas.[109] Whether myc expression will be of prognostic utility remains to be determined. The nerve growth factor receptor (NGF) was the first growth factor to be associated with neuroectodermal cells. It is expressed in subsets of astrocytomas, neuroblastomas and medulloblastomas, and in the last, appears to connote a favorable prognosis.[110] Platelet-derived growth factor (PDGF) was the first cellular growth factor shown to correspond to a known viral oncogene. The receptor for this growth factor (PDGFR) is expressed at higher levels in glial tumors than in normal human tissue. As discussed above, PDGFR amplification is a common event in primary glioblastomas. However, the ubiquitous expression of this receptor precludes its use in a diagnostic setting.[39]

OUR CONTRIBUTIONS FROM DEPARTMENT OF PATHOLOGY, AIIMS (REFERENCES 111 – 150)

1. We have studied tumor markers to assess their diagnostic and prognostic utility as well as to establish the histogenesis and differentiation potential in a wide variety of brain tumors such as primitive neuroectodermal tumors, xanthomatous glial tumors, mixed gliomas, medullomyoblastoma, neurocytoma, gliofibromas, gliosarcomas and nerve sheath tumors.

2. We have also studied the hormonal pattern of pituitary adenomas in India.

3. We have studied the proliferation potential of all types of brain tumors using in vivo and in vitro BrdU, PCNA and MIB-1.

4. Recently, the oncogenes like C-myc and tumor suppressor genes viz. p53 and retinoblastoma gene have been studied with special reference to gliomas and medulloblastomas.

It is obvious from our studies[111-150] that tumor markers are very helpful to supplement the diagnosis as well as help to assess the histogenesis and differentiating potential of many CNS tumors. The parameters of cell proliferation and its molecular controls are interlinked and their alterations are broadly reflected in histological changes and clinical outcome. Hence, it can be concluded that the study of a combination of clinical, histological, cell proliferation and molecular markers will help in better assessment of biological aggressiveness and also in better patient management and therapy.

REFERENCES:

1. Pamies RJ, Crawford DR: Tumor markers: an update [review]. *Med Clin North Am* 1996;80:185-99.
2. Bullard DE, Koeleveld RF, Snodgrass SM, Schold SC: Tumor markers. In: Wilkins RH, Rengachary SS, editors. Neurosurgery. 2nd ed. New York: McGraw-Hill; 1996;15-20.
3. Bjornsson J, Scheithauer BW, Okazaki H, Leech RW: Intracranial germ cell tumors: pathobiological and immunohistochemical aspects of 70 cases. *J Neuropathol Exp Neurol* 1985;44:32-46.
4. Allen JC, Nisselbaum J, Epstein F, Rosen G, Schwartz MK: Alphafetoprotein and human chorionic gonadotropin determination in cerebrospinal fluid: an aid to the diagnosis and management of intracranial germ-cell tumors. *J Neurosurg* 1979; 51:368-74.
5. Baumgartner JE, Edwards MS: Pineal tumors [review]. *Neurosurg Clin North Am* 1992;3:853-62.
6. Edwards MS, Hudgins RJ, Wilson CB, Levin VA, Wara WM: Pineal region tumors in children. *J Neurosurg* 1988;68:689-97.

7. Ho DM, Liu HC: Primary intracranial germ cell tumor. Pathologic study of 51 patients. *Cancer* 1992;70:1577-84.
8. Hoffman HJ, Otsubo H, Hendrick EB, Humphreys RP, Drake JM, Becker LE: Intracranial germ-cell tumors in children. *J Neurosurg* 1991;74:545-51.
9. Arita N, Ushio Y, Hayakawa T, Mori S, Bitoh S, Hasegawa H: Role of tumor markers in the management of primary intracranial germ cell tumors. *Prog Exp Tumor Res* 1987;30:289-95.
10. Shinoda J, Yamada H, Sakai N, Ando T, Hirata T, Miwa Y: Placental alkaline phosphatase as a tumor marker for primary intracranial germinoma. *J Neurosurg* 1988;68:710-20.
11. Watterson J, Priest JR: Control of extraneural metastasis of a primary intracranial non-germinomatous germ-cell tumor: case report. *J Neurosurg* 1989;71:601-4.
12. Schold SC, Bullard DE: Cerebrospinal fluid analysis in central nervous system cancer. In: Wood JH, editor. Neurobiology of cerebrospinal fluid Vol. 2. New York: Plenum; 1980;549-59.
13. Seidenfeld J, Marton LJ: Biochemical markers of central nervous system tumors measured in cerebrospinal fluid and their potential use in diagnosis and patient management: a review [review]. *J Natl Cancer Inst* 1979;63:919-31.
14. Wasserstrom WR, Schwartz MK, Fleischer M, Posner JB: Cerebrospinal fluid biochemical markers in central nervous system tumors: a review [review]. *Ann Clin Lab Sci* 1981;11:239-51.
15. Cohen AR, Wilson JA, Sadhegi-Nejad A: Gonadotropin-secreting pineal teratoma causing precocious puberty. *Neurosurgery* 1991;28:597-603.
16. Lippe BM, Edwards MS, Braunstein GD, Halks-Miller M: A nonmalignant teratoma secreting hCG: expanding the spectrum of ectopic hormone production. *J Pediatr* 1984;105:765-68.
17. Suzuki Y, Tanaka R: Carcinoembryonic antigen in patients with intracranial tumors. *J Neurosurg* 1980;53:355-60.
18. Jacobi C, Reiber H: Clinical relevance of increased neuron-specific enolase concentration in cerebrospinal fluid. *Clin Chem Acta* 1988;177:49-54.
19. Lampl Y, Paniri Y, Eshel Y, Sarova-Pinhas I: LDH isoenzymes in cerebrospinal fluid in various brain tumors. *J Neurol Neurosurg Psychiatry* 1990;53:697-99.
20. Marton LJ, Edwards MS, Levin VA, Lubich WP, Wilson CB: CSF polyamines: a new and important means of monitoring patients with medulloblastoma. *Cancer* 1981;47:575-60.
21. Phillips PC, Kremzner LT, De Vivo DC: Cerebrospinal fluid polyamines: biochemical markers of malignant childhood brain tumors. *Ann Neurol* 1986;19:360-64.
22. Mavlight GM, Stuckey SE, Cabanillas FF, Keating MJ, Tourtelotte WW, Schold SC: Diagnosis of leukemia or lymphoma in the central nervous system by beta 2-microglobulin determination. *N Engl J Med* 1980;303 718-22.
23. Nagelkerke AF, vanKamp GJ, Veerman AJP, de Waal FC: Unreliability of beta-2 –microglobulin in early detection of central nervous system relapse in acute lymphoblastic leukemia. *Eur J Cancer Clin Oncol* 1979;21: 659-63.
24. Weber K, Osborn M: Cytoskeleton: definition, structure and gene regulation. *Path Res Pract* 1982;175:128-45.
25. Clark HB: Immunohistochemistry of nervous system antigens. Diagnostic applications in surgical neuropathology. *Semin Diagn Path* 1984;1:309-16.
26. De Armond SI, Eng LF, Rubinstein U: The application of glial fibrillary acidic (GFA) protein immunohistochemistry in neurooncology. *Pathol Res Pract* 1980;168:374-94.
27. Perentes E, Rubinstein U: Recent applications of immunoperoxide histochemistry in human neuro-oncology. An update. *Arch Pathol Lab Med* 1987;111:796-812.
28. Cosgrove M, Fitzgibbons PL, Sherrod A, Chandrasoma PT, Martin SE: Intermediate filament expression in astrocytic neoplasms. *Am J Surg Pathol* 1989;13:141-45.
29. Schmitt HP: Rapid anaplastic transformation in gliomas of adulthood. "Selection" in neurooncogenesis. *Pathol Res Pract* 1983;176:313-23.
30. Margetts JC, Kalyan-Raman UP: Giant-celled glioblastoma of brain. A clinicopathological and radiological

study of ten cases (including immunohistochemistry and ultrastructure). *Cancer* 1989;63: 524-31.
31. Kepes JJ, Rubinstein LJ: Malignant gliomas with heavily lipidized (foamy) tumor cells. A report of three cases with immunoperoxidase study. *Cancer* 1981;47:2451-59.
32. Rosenblum MK, Erlandson RA, Budzilovich GN: The lipid-rich epithelioid glioblastoma. *Am J Surg Pathol* 1991;15:95-34.
33. Kepes JJ, Fulling KH, Garcia JH: The clinical significance of "adenoid" formations of neoplastic astrocytes imitating metastatic carcinoma in gliosarcomas. A review of five cases. *Clin Neuropathol* 1982;1:139-50.
34. Mork SJ, Rubinstein LJ, Kepes JJ, Perentes E, Uphoff DF: Patterns of epithelial metaplasia in malignant gliomas. II. Squamous differentiation of epithelial-like formations in gliosarcomas and glioblastomas. *J Neuropathol Exp Neurol* 1988;47:101-18.
35. Hitchcock E, Morris CS: Cross reactivity of an anti-epithelial membrane antigen monoclonal for reactive and neoplastic glial cells. *J Neurooncol* 1987;4:345-52.
36. Mork SJ, Rubinstein LJ, Kepes JJ, Perentes E, Uphoff DF: Patterns of epithelial metaplasia in malignant gliomas. I. Papillary formations mimicking medulloepithelioma. *J Neuropathol Exp Neurol* 1988;47:93-100.
37. Kleihues P, Kiessling M, Janzer RC: Morphological markers in neuro-oncology. *Curr Top Pathol* 1987;77:307-38.
38. Coffin CM, Mukai K, Dehner LP: Glial differentiation in medulloblastomas. Histogenetic insight, glial reaction, or invasion of brain? *Am J Surg Pathol* 1983;7:555-65.
39. Wikstrand CJ, Fung KM, Trojanowski JQ, McLendon RE, Bigner DD: Antibodies and molecular immunology. Immunohistochemistry and antigens of diagnostic significance. In: Bigner DD, McLendon RE, Bruner JM, editors. Russell and Rubinstein's Pathology of tumors of the nervous system. 6th ed. London: Edward Arnold; 1998;251-304.
40. Kros JM, Van Eden CG, Stefanko SZ, Waayer Van Batenburg M, van der Kwast TH: Prognostic implications of glial fibrillary acidic protein containing cell types in oligodendrogliomas. *Cancer* 1990;66: 1204-12.
41. Nixon RA: The regulation of neurofilaments protein dynamics by phosphorylation: clues to neurofibrillary pathobiology. *Brain Pathol* 1993;3:29-38.
42. Leader M, Collins M, Patel J, Henry K: Vimentin: An evaluation of its role as a tumor marker. *Histopathology* 1987;11:63-72.
43. Azumi N, Battifora H: The distribution of vimentin and keratin in epithelial and nonepithelial neoplasms. A comprehensive immunohistochemical study on formalin- and alcohol-fixed tumors. *Am J Clin Pathol* 1987;88:286-96.
44. Tohyama T, Lee VM, Rorke LB, Martin M, McKay RD, Trojanowski JQ: Nestin expression in embryonic human neuroepithelium and in human neuroepithelial tumor cells. *Lab Invest* 1992;66:303-13.
45. Kleihues P, Davis RL, Ohgaki H, Cavenee WK: Low-grade diffuse astrocytomas. In: Kleihues P, Cavenee WK, editors. Pathology and genetics. Tumors of the nervous system. Lyon: International Agency for Research on Cancer; 1997; p.12.
46. Cruz Sanchez FF, Rossi ML, Buller JR, Carboni P, Fineron PW, Coakham HB: Oligodendrogliomas: a clinical, histological, imunocytochemical and lectin-binding study. *Histopathology* 1991;19:361-67.
47. Foley J, Witte D, Chiu FC, Parysek LM: Expression of the neutral intermediate filament proteins peripherin and neurofilaments-66/alpha-internexin in neuroblastoma. *Lab Invest* 1994;71:193-99.
48. Yachnis AT, Rorke LB, Lee VM, Trojanowski JQ: Expression of neuronal and glial polypeptides during histogenesis of the human cerebellar cortex including observations on the dentate nucleus. *J Compar Neurol* 1993;334:356-69.
49. Franke FE, Schachenmayr W, Osborn M, Altmannsberger M: Unexpected immunoreactivities of intermediate filament antibodies in human brain and brain tumors. *Am J Pathol* 1991;139:67-79.

50. Pinkus GS, Kurtin PJ: Epithelial membrane antigen. A diagnostic discriminant in surgical pathology. Immunohistochemical profile in epithelial, mesenchymal, and hematopoietic neoplasms using paraffin sections and monoclonal antibodies. *Hum Pathol* 1985;16:929-40.
51. Thomas P, Battifora H: Keratins versus epithelial membrane antigen in tumor diagnosis. An immunohistochemical comparison of five monoclonal antibodies. *Hum Pathol* 1987;18:728-34.
52. Sloane JP, Ormerod MG: Distribution of epithelial membrane antigen in normal and neoplastic tissues and its value in diagnostic tumor pathology. *Cancer* 1981;1786-95.
53. Swanson PE, Manivel CJ, Scheithauer BW, Wick MR: Epithelial membrane antigen reactivity in mesenchymal neoplasms. An immunohistochemical study of 306 soft tissue sarcomas. *Surg Pathol* 1989;2:313-22.
54. Rubinstein LJ: Cytogenesis and differentiation of primitive central neuroepithelial tumors. *J Neuropathol Exp Neurol* 1972;31:7-26
55. Lloyd RV: Immunohistochemical localization of chromogranin in normal and neoplastic endocrine tissues. *Pathol Annu* 1987;22(2):69-90.
56. Hagn S, Schmidt KW, Fischer-Colbrie R, Winkler H: Chromogranin A, B and C in human adrenal medulla and human endocrine tissues. *Lab Invest* 1986;55:405-11.
57. Lloyd RV, Kano M, Rosa P, Hille A, Hunter WB: Distribution of chromogranin A and secretogranin I (chromogranin B) in neuroendocrine cells and tumors. *Am J Pathol* 1988;130:296-304.
58. Larsson L, Alumets J, Eriksson B, Hakanson R, Lundquist G, Oberg K, et al: Antiserum directed against chromogranin A and B (CAB) is a useful marker for peptide hormone-producing endocrine cells and tumors. *Endocr Pathol* 1992;3:14-22.
59. Pahlman S, Esscher T, Nilsson K: Expression of g-subunit of enolase, neuron-specific enolase, in human non-neuroendocrine tumors and derived cell lines. *Lab Invest* 1986;54:554-60.
60. Diepholder HM, Schwechheimer K, Mohadjer M, Knoth R, Bolk B: A clinicopathologic and immunomorphologic study of 13 cases of ganglioglioma. *Cancer* 1991;68:2192-201.
61. Hassoun J, Soylemezoglu F, Gambarelli D, Figarella Branger D, von Ammon D, Kleihues P: Central neurocytoma: a synopsis of clinical and histological features. *Brain Pathol* 1993;3:297-306.
62. Kliewer KE, Cochran AJ: A review of the histology, ultrastructure, immunohistology and molecular biology of extra-adrenal paragangliomas [published erratum appears in *Arch Pathol Lab Med* 1990;114(3):308]. *Arch Pathol Lab Med* 1990;113:1209-18.
63. Gould VE, Wiedenmann B, Lee I, Schwechheimer K, Dockhorn-Dworniczak B, Eadosevich JA: Synaptophysin expression in neuroendocrine neoplasms as determined by immunocytochemistry. *Am J Pathol* 1987;126:243-57.
64. Kimura T, Budka H, Federsppiel S: An immunocytochemical comparison of the glia-associated proteins glial fibrillary acidic protein (GFAP) and S-100 protein (S-100P) in human brain tumors. *Clin Neuropathol* 1986;5:21-27.
65. Brandis A, Mirzai S, Tatagiba M, Walter GF, Samii M, Ostertag H: Immunohistochemical detection of female sex hormone receptors in meningiomas: correlation with clinical and histological features. *Neurosurgery* 1993;33:212-17.
66. Gottschalk J, Jautzke G, Paulus W, Goebel S, Cervos Navarro J: The use of immunomorphology to differentiate choroid plexus tumors from metastatic carcinomas. *Cancer* 1993;72:1343-49.
67. Paulus W, Janisch W: Clinicopathological correlation in epithelial choroid plexus neoplasms: a study of 52 cases. *Acta Neuropathol Berl* 1990;80:635-41.
68. Kato S, Hirano A, Kato M, Herz F, Ohama E: Comparative study on the expression of stress-response protein (srp) 72, srp 27, alpha B-crystallin and ubiquitin in brain tumors. An immunohistochemical investigation. *Neuropathol Appl Neurobiol* 1993;19:436-42.
69. Albrecht S, Rouah E, Becker LE, Bruner J: Transthyretin immunoreactivity in choroid plexus neoplasms and brain metastases. *Mod Pathol* 1991;4:610-14.
70. Coffin CM, Wick MR, Braun JT, Dehner LP: Choroid plexus neoplasms. Clinicopathologic and immunohistochemical studies. *Am J Surg Pathol* 1986;10:394-404.

71. Korf HW, Korf B, Schachenmayr W, Chader GJ, Wiggert B: Immunocytochemical demonstration of interphotoreceptor retinoid-binding protein in cerebellar medulloblastomas. *Acta Neuropathologica* 1992;83:482-87.
72. Montine TJ, Vandersteenhoven JJ, Aguzzi A, Boyko OB, Dodge RK, Kerns BJ, et al: Prognostic significance of Ki-67 proliferation in supratentorial fibrillary astrocytic neoplasms. *Neurosurgery* 1994;34:674-79.
73. Onda K, Davis RL, Shibuya M, Wilson CB, Hoshino T: Correlation between the bromodeoxyuridine labeling index and the MIB-1 and Ki-67 proliferating cell indices in cerebral gliomas. *Cancer* 1994;74:1921-26.
74. Hoshino T, Ahn D, Prados MD, Lamborn K, Wilson CB: Prognostic significance of the proliferative potential of intracranial gliomas measured by bromodeoxyuridine labeling. *Int J Cancer* 1993;53:550-55.
75. Hsu DW, Louis DN, Efird JT, Hedley-White ET: Use of MIB-1 (Ki-67) immunoreactivity in differentiating grade II and grade III Gliomas. *J Neuropathol Exp Neurol* 1997;56:857-65.
76. Coons SW, Johnson PC: Regional heterogeneity in the proliferative activity of human gliomas as measured by the Ki-67 labeling index. *J Neuropathol Exp Neurol* 1993;52:609-18.
77. Burger PC, Shibata T, Kleihues P: The use of the monoclonal antibody Ki-67 in the identification of proliferating cells: application to surgical neuropathology. *Am J Surg Pathol* 1986;10:611-17.
78. Deckert M, Reifenberger G, Wechsler W: Determination of the proliferative potential of human brain tumors using the monoclonal antibody Ki-67. *J Cancer Res Clin Oncol* 1989;115:179-88.
79. Jaros E, Perry RH, Adam L, Kelly PJ, Crawford PJ, Kalbag RM, et al: Prognostic implications of p53 protein, epidermal growth factor receptor, and Ki-67 labelling in brain tumours. *Br J Cancer* 1992;66:373-85.
80. Hoshino T, Rodriguez LA, Cho KG, Lee KS, Wilson CB, Edwards MS, et al: Prognostic implications of the proliferative potential of low-grade astrocytomas. *J Neurosurg* 1988;69:839-42.
81. Ito S, Chandler KL, Prados MD, Lamborn K, Wynne J, Malec MK, et al: Proliferative potential and prognostic evaluation of low-grade astrocytomas. *J Neurooncol* 1994;19:1-9.
82. Schiffer D, Chio A, Giordana MT, Leone M, Soffietti R: Prognostic value of histologic factors in adult cerebral astrocytoma. *Cancer* 1988;61:1386-93.
83. Kleihues P, Burger PC, Plate KH, Ohgaki H, Cavenee WK: Glioblastoma. In: Kleihues P, Cavenee WK, editors: Pathology and genetics. Tumors of the nervous system. Lyon: *International Agency for Research on Cancer*, 1997;19-24.
84. Kros JM, Hop WC, Godschalk JJ, Krishnadath KK: Prognostic value of the proliferation-related antigen Ki-67 in oligodendrogliomas. *Cancer* 1996;78:1107-13.
85. Prayson, Khajavi, Comair: Cortical architectural abnormalities and MIB1 immunoreactivity in gangliogliomas. a study of 60 patients with intracranial tumors. *J Neuropathol Exp Neurol* 1995;54:513-20.
86. Soylemezoglu F, Kleihues P, Esteve J, Scheithauer BW: Atypical central neurocytoma. *J Neuropathol Exp Neurol* 1997;56:551-56.
87. Lane DP, Crawford LV: T antigen is bound to a host protein in SV-40 transformed cells. *Nature* 1979;278:261-64.
88. DeLeo AB, Jay G, Appella E, Dubois GC, Law LW, Old LJ: Detection of a transformation-related antigen in chemically induced sarcomas and other transformed cells of the mouse. *Proc Natl Acad Sci U S A* 1979;76:2420-24.
89. Linzer DIH, Levine AJ: Characterization of a 54K Dalton cellular SV40 tumor antigen present in SV-40 transformed cells and uninfected embryonal cells. *Cell* 1979;17:43-48.
90. Bogler O, Huang HJ, Kleihues P, Cavenee WK: The p53 gene and its role in human brain tumors. *Glia* 1995;15:308-27.
91. Reifenberger J, Ring GU, Gies U, Cobbers L, Oberstrass J, An HX, et al: Analysis of p53 mutation and epidermal growth factor receptor amplification in recurrent gliomas with malignant progression. *J Neuropathol Exp Neurol* 1996;55:822-31.

92. Iuzzolino P, Ghimenton C, Nicolato A, Giorgiutti F, Fina P, Doglioni C, et al: p53 protein in low-grade astrocytomas: a study with long-term follow-up. *Br J Cancer* 1994;69:586-91.
93. Kraus JA, Bolln C, Wolf HK, Neumann J, Kindermann D, Fimmers R, et al: TP53 alterations and clinical outcome in low grade astrocytomas. *Genes Chromosomes Cancer* 1994;10:143-49.
94. Rasheed BK, McLendon RE, Herndon JE, Friedman HS, Friedman AH, Bigner DD, et al: Alterations of the TP53 gene in human gliomas. *Cancer Res* 1994;54:1324-30.
95. Watanabe K, Sato K, Biernat W, Tachibana O, von Ammon K, Ogata N, et al: Incidence and timing of p53 mutations during astrocytoma progression in patients with multiple biopsies. *Clin Cancer Res* 1997;3:523-30.
96. Chozick BS, Pezullo JC, Epstein MH, Finch PW: Prognostic implications of p53 overexpression in supratentorial astrocytic tumors. *Neurosurgery* 1994;35:831-37.
97. al Sarraj S, Bridges LR: p53 immunoreactivity in astrocytomas and its relationship to survival. *Br J Neurosurg* 1995;9:143-49.
98. Schlengel U: p53 an important or most overvalued tumor gene? *Laryngorhinootologie* 1994;73:651-53.
99. Ohgaki H, Eibl RH, Schwabb M, Reichel MB, Mariani L, Gehring M, et al: Mutations of the p53 tumor suppressor gene in neoplasms of the human nervous system. *Mol Carcinog* 1993;8:74-80.
100. Patt S, Gries H, Giraldo M, Cervos J, Navarro H, Martin H: p53 gene mutations in human astrocytic brain tumors including pilocytic astrocytomas. *Hum Pathol* 1996;27:586-89.
101. Fink KL, Rushing EJ, Schold SC, Nisen PD: Infrequency of p53 gene mutations in ependymomas. *J Neurooncol* 1996;27:111-15.
102. Tsumanuma I, Sato M, Okazaki H, Tanaka R, Washiyama K, Kawasaki T, et al: The analysis of p53 tumor suppressor gene in pineal parenchymal tumors. *Noshuyo Byori* 1995;12:39-43.
103. Adesina AM, Nalbantoglu J, Cavenee WK: p53 gene mutation and mdm2 gene amplification are uncommon in medulloblastoma. *Cancer Res* 1994;54:5649-51.
104. Libermann TA, Nusbaum HR, Razon N, Kris R, Lax I, Soreq H: Amplification, enhanced expression, and possible rearrangement of EGF receptor gene in primary human brain tumors of glial origin. *Nature* 1985;313:144-47.
105. Wong AJ, Bigner SH, Bigner DD, Kinsler KW, Hamilton SR, Vogelstein B: Increased expression of the epidermal growth factor receptor gene in malignant gliomas is invariably associated with gene amplification. *Proc Natl Acad Sci U S A* 1987;84:6899-903.
106. Hurtt MR, Moossy J, Peluso M, Locker J: Amplification of epidermal growth factor receptor gene in gliomas: histopathology and prognosis. *J Neuropathol Exp Neurol* 1992;51:84-90.
107. Watanabe K, Tachibana O, Sata K, Yonekawa Y, Kleihues P, Ohgaki H: Overexpression of the EGF receptor and p53 mutations are mutually exclusive in the evolution of primary and secondary glioblastomas. *Brain Pathol* 1996;6:217-23.
108. von Deimling A, Louis DN, von Ammon K, Peterson I, Hoell T, Chung RY: Association of epidermal growth factor receptor gene amplification with loss of chromosome 10 in human glioblastoma multiforme. *J Neurosurg* 1992;77:295-301.
109. Wasson JC, Saylors RL, Zeltzer P, Friedman HS, Bigner SH, Burger PC, et al: Oncogenic amplification in pediatric brain tumors. *Cancer Res* 1990;50:2987-90.
110. Segal RA, Goumnerova LC, Kwon YK, Stiles CD, Pomeroy SL: Expression of the neurotrophin receptor TrkC is linked to a favorable outcome in medulloblastoma. *Proc Natl Acad Sci US A* 1994;91:12867-71.
111. Sarkar C, Roy S: Immunohistochemistry - a new tool in diagnosis. In: Bijlani V, Wadhwa S, Tandon PN, editors. Techniques in basic neurosciences. Practical Manual of the First Course in Neurobiology for Postgraduates in Clinical Neurosciences.1987;33-40.
112. Roy S, Chowdhury C, Tandon PN: Differentiating medulloblastomas – A light microscopic, electron microscopic and immunohistochemical study (Abstr). *J Neuro Onc* 1984;33.

113. Chowdhury C, Roy S, Mahapatra AK, Bhatia R: Medullomyoblastoma- a teratoma. *Cancer* 1985;55: 1495-1500.
114. Sarkar C, Roy S, Sharma S, Dinda AK, Tandon PN: Markers in the study of nerve sheath tumors and peripheral tumors of neuron series. *Neurol India* 1988;36:361-67.
115. Sarkar C, Roy S, Dinda AK, Tandon PN: Hormones as markers in pituitary adenomas. *Neurol India* 1988;36:357-60.
116. Roy S, Sarkar C, Tandon PN: Use of markers in the study of brain tumors. *Neurol. India* 1988;36: 351-56.
117. Sarkar C, Roy S, Tandon PN: Oligodendroglial tumors: An immunohistochemical and electron microscopic study. *Cancer* 1988;61:1862-6.
118. Sarkar C, Roy S, Tandon PN: Primitive neuroectodermal tumors of the central nervous system - An electron microscopic and immunohistochemical study. *Ind J Med Res* 1989;90:91-102.
119. Sharma S, Sarkar C, Mathur M, Dinda AK, Roy S: Benign nerve sheath tumors: A light microscopic, electron microscopic and immunohistochemical study of 102 cases. *Pathology* 1990;22:191-95.
120. Sarkar C, Roy S, Kochupillai N, Gupta N, Tandon PN: A clinicopathologic study of pituitary adenomas. *Ind J Med Res* 1990;92:315-23.
121. Sarkar C, Roy S and Bhatia, S: Xanthomatous change in tumors of glial origin. *Ind J Med Res* 1990;92:324-31.
122. Roy S, Sarkar C: Some recent advances in neuro-oncology with a particular reference to newer techniques for diagnosis and prognostication. *Ind J Pathol Microbiol* 1990;33:195-209.
123. Dinda AK, Sarkar C, Roy S: Rosenthal fibres: an immunohistochemical, ultrastructural and immunoelectron microscopic study. *Acta Neuropathol* 1990;79:456-60.
124. Sarkar C: Pituitary adenomas in India - A clinicopathological study (Abstr). *Endocr Pathol* 1992;3.
125. Kharbanda K, Karak AK, Sarkar C, Dinda AK, Mathur M, Roy S: Prediction of biologic aggressiveness in human meningiomas: A cell kinetic study using bromodeoxyuridine on cells of primary explant culture. *J Natl Cancer Inst* 1992;84:194-95.
126. Kharbanda K, Karak AK, Sarkar C, Dinda AK, Mathur M, Roy S: A sequential cell kinetic study of meningioma cells in primary explant culture using bromodeoxyuridine. *J. Neuro Onc* 1993;16:117-23.
127. Dinda AK, Kharbanda K, Sarkar C, Roy S, Mathur M, Banerji AK: In-vivo proliferative potential of primary human brain tumors: its correlation with histological classification and morphological features: II. Nonglial Tumors. *Pathology* 1993;25:10-14.
128. Dinda AK, Kharbanda K, Sarkar C, Roy S, Mathur M, Banerji AK: In-vivo proliferative potential of primary human brain tumors: Its correlation with histological classification and morphological features: I. Gliomas. *Pathology* 1993;25:4-9.
129. Kharbanda K, Sarkar C, Dinda AK, Karak AK, Mathur M, Roy S: Morphological appearance, growth kinetics and glial fibrillary acidic protein (GFAP) expression in primary in vitro explant culture of astrocytic neoplasms. *Acta Oncol* 1993;32:301-06.
130. Ghosh M, Dinda AK, Chattopadhyay P, Sarkar C, Bhatia S, Sinha S: Rearranged p53 gene with loss of normal allele in a low grade non-recurrent glioma. *Cancer Genet Cytogenet* 1994;787:68-71.
131. Kharbanda K, Dinda AK, Sarkar C, Karak AK, Mathur M, Roy S: Cell culture studies on human nerve sheath tumors. *Pathology* 1994;26:29-32.
132. Chattopadhyay P, Banerji M, Sarkar C, Mathur M, Mahapatra AK, Sinha S: Infrequent alteration of the c-myc gene in human glial tumors associated with increased number of c-myc positive cells. *Oncogene* 1995;11:2711-14.
133. Kharbanda K, Dinda AK, Sarkar C, Karak AK, Dhir R, Mathur M, et al: A correlative study of in vivo and in vitro labelling index using bromodeoxyuridine in human brain tumors. *J Neuro Onc* 1995;23: 185-90.
134. Banerji M, Dinda AK, Sinha S, Sarkar C, Mathur M: C-myc oncogene expression and cell proliferation in mixed oligo-astrocytomas. *Int J Cancer* 1996;65:730-33.

135. Sarkar C, Rathore A, Kamarajan P, Mathur M, Chattopadhyay P, Sinha S: Alterations in p53 expression in astrocytic tumors (Abstr). *Neurol India* 1997;45 (Suppl. 1):9.
136. Sharma MC, Sarkar C: Subependymal giant cell astrocytomas. Immunohistochemical and proliferative markers. *Neurol India* 1997;45 (Suppl. 1):35.
137. Sinha S, Dil-Afroze, Misra A, Sulaiman IM, Sarkar C, Mahapatra AK, et al: Changes in brain tumor genomes identified by RAPD analysis (Abstr). *FASEB Journal* 1997;11(9):2313.
138. Sarkar C, Sharma MC, Sudha K, Gaikwad S, Varma A: A clinico-pathological study of 29 cases of gliosarcoma with special reference to two unique variants. *Ind J Med Res* 1997;106:229-35.
139. Chattopadhyay P, Rathore A, Mathur M, Sarkar C, Mahapatra AK, Sinha S: Loss of heterozygosity of a locus on 17p13.3, independent of p53, is associated with higher grades of astrocytic tumors. *Oncogene* 1997;15:871-74.
140. Sinha S, Sarkar C, Mahapatra AK, Husnain S: Genetic alterations in brain tumors identified by RAPD analysis. *Gene* 1998;206:45-48.
141. Sudha K, Karak AK, Sharma MC, Mathur M, Sarkar C: Assessment of proliferative potential of human meningiomas using PCNA LI and AgNOR counts. *Ind J Pathol Microbiol* 1998;41:323-30.
142. Sharma MC, Rathore A, Karak AK, Sarkar C: A study of proliferative markers in central neurocytoma. *Pathology* 1998;30:355-59.
143. Sharma MC, Gaekwad S, Mehta VS, Dhar J, Sarkar C: Gliofibroma: Mixed glial and mesenchymal tumor. Report of 3 cases. *Clin Neurol Neurosurg* 1998;100:153-59.
144. Rathore A, Kamarajan P, Mathur M, Sinha S, Sarkar C: Simultaneous alterations of retinoblastoma (Rb) and p53 protein expression in astrocytic tumors. *Pathol Oncol Res* 1999;5:21-27.
145. Sarkar C, Sharma MC, Gaekwad S, Sharma C, Singh VP: Choroid plexus papillomas. A clinico-pathological study of 23 cases. *Surg Neurol.* 1999;52:37-39.
146. Sharma MC, Sarkar C, Karak AK, Gaekwad S, Mahapatra AK, Mehta VS: Intraventricular neurocytoma: A clinico-pathological study of 20 cases with review of the literature. *J Neurosurg Sci* 1999;6:319-23.
147. Sharma S, Karak AK, Singh R, Mehta VS, Sarkar C, Schmitt, HP: A correlative study of gliomas using in-vivo bromodeoxyuridine labeling index and computer aided malignancy grading. *Pathol Oncol Res* 1999;5:134-41.
148. Sarkar C, Rathore A, Chattopadhyay P, Mahapatra AK, Sinha S: Role of 17p13.3 chromosomal region in determining p53 protein immunopositivity in human astrocytic tumors. *Pathology* 2000;32: 84-88.
149. Sharma MC, Arora R, Lakhtakia R, Mahapatra AK, Sarkar C: Ependymoma with extensive lipidization mimicking adipose tissue: a report of five cases. *Pathol Oncol Res* 2000;6:136-40.
150. Misra A, Chattopadhyay P, Dinda AK, Sarkar C, Mahapatra AK, Hasnain SE, et al: Extensive intra-tumor heterogeneity in primary human glial tumors as a result of locus non-specific genomic alterations. *J Neurooncol* 2000;48:1-12.

4
Role of Radiotherapy in Malignant Gliomas

GK Rath, DN Sharma

INTRODUCTION

Gliomas are the most common primary brain tumors. Malignant gliomas (MG) account for about 50% of all primary brain tumors. Of them, 30% are glioblastoma multiforme (GBM) and 10% are anaplastic astrocytomas (AA). According to Delhi Cancer Registry[1] report (1994-95), brain tumors are the third commonest cause of cancer death in males after lung & esophagus. The incidence is likely to rise further due to improving life span, better availability of imaging modalities like CT/MRI Scan and better health awareness.

Radiation therapy is one of the most important treatment modalities for malignant gliomas, mainly because these tumors are usually localized. The patient's head can be immobilized, allowing for good targeting of the tumor. As a result of technological advances, the range of treatment options using radiation therapy has increased. Photon, electron, neutron and proton radiation beams are all in clinical use. The accuracy of the treatment delivery has also improved, with the development of three-dimensional techniques and image fusion. Our understanding of both the radiation sensitivity of gliomas, and the tolerance of normal brain tissue is better, allowing the design of novel fractionation schemes. Radiation is being combined with both local and systemic chemotherapy, offering hope for patients in terms of improved tolerance to treatment while maintaining a good quality of life. This chapter will discuss the basic principles of radiotherapy, and the indications, techniques and outcomes for the radiation therapy of malignant gliomas in adults.

BASIC PRINCIPLES OF RADIOTHERAPY

An understanding of the basic principles of radiotherapy is essential to the successful use of radiation therapy. These principles include:

1. The higher the dose of the radiation delivered to the tumor, higher the probability of the local control of the tumor. Hence, generally, radiation oncologists aim to deliver the maximum dose to the tumor without causing undue toxicity to the surrounding normal tissues.

2. The lower the dose to the surrounding normal tissues, the lower the associated morbidity, hence radiation oncologists use multiple beams, optimized treatment planning, shielding, brachytherapy and other techniques to limit the dose to the surrounding normal tissues, thereby minimizing the morbidity.
3. Larger tumors require higher doses of radiation for control. Conversely, small or microscopic tumors require lower doses for control.
4. Hypoxic tumor cells (usually in he center of the tumor) are relatively radioresistant and require higher doses of radiation to achieve cell kill.. Surgical removal of the hypoxic cells decreases the radiation dose required and increases the probability of the local control.
5. The risk of morbidity increases if larger volumes are irradiated. On the other hand, smaller irradiated volumes can tolerate higher radiation doses with less potential morbidity. Hence, the aim is to minimize the volume of tissue irradiated without missing areas harboring the tumor.
6. Tumor cells usually proliferate faster than the normal tissues. Shortening the time interval between surgery and radiation therapy reduces the potential repopulation of tumor cells. Hence prolonged delays between surgery and start of radiation therapy should be avoided.

Physical basis of radiation therapy: Radiation therapy is treatment of diseases by the ionising radiation. The types of radiation are often X-rays and gamma rays. Besides these, other types of radiation in clinical use are electron, neutron and charged particle beams. The X-rays are generated from the Linear Accelerator **(Fig. 1)** and gamma rays from the decay of certain radioactive materials like cobalt-60 and cesium-137. The beam characteristics are different from different therapy machines and are chosen as per the requirement of the tumor. Broadly,

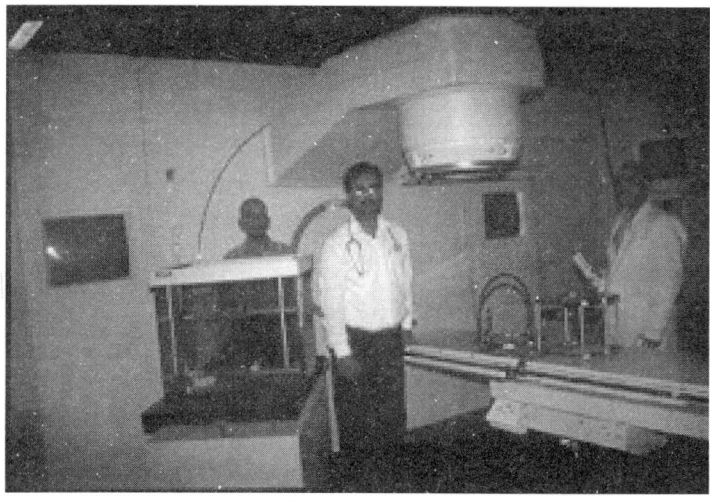

Fig. 1 : Linear Accelerator

the radiation therapy can be categorized into (1) External beam radiation therapy (EBRT) or teletherapy and (2) brachytherapy. EBRT delivers radiation from a distance while in brachytherapy the source of radiation is placed inside or close to the tumor tissue. Treatment with EBRT is usually the fractionated treatment. Standard fractionation involves daily fraction of 1.8–2.0 Gy, 5 times a week and usually for 5-7 weeks. Brachytherapy treatment is given over shorter period of time usually 2-7 days.

BIOLOGICAL BASIS OF RADIATION THERAPY

The ionizing radiations (X-rays and gamma rays) are capable of damaging the genetic material (DNA) resulting in reproductive death of tumor cell. **(Fig. 2)** They exert their effect indirectly through the hydrolysis of water and later produce free radicals for the ultimate DNA damage. Presence of the intracellular oxygen fixes the radiation damage, hence adequate hemoglobin is important for the lethal effect of radiation. The central portion of the tumors, which is relatively hypoxic, is generally resistant to the radiation therapy. Hypoxic cell sensitizers (primarily nitroimadazole derivatives like metronidazole and misonidazole) have been tried with limited success.

Radiotheray : Mechanism of Action

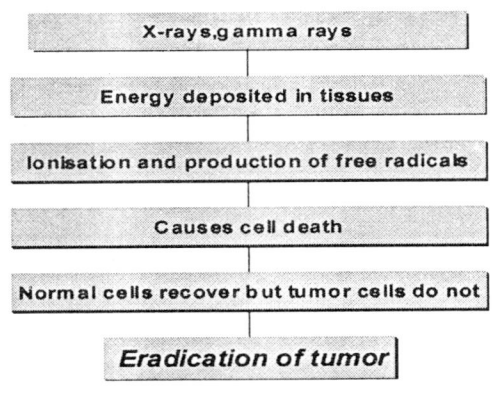

Fig. 2

The normal cell cycle consists of G0, G1, G2, M and S phase. The G2 and M phase of the cell cycle are very radiosensitive as compared to S phase. A tumor having maximum cells in the radiosensitive phase of cell cycle has much more probability of being controlled. The radiation is lethal to both tumor cells as well as normal tissue cells. The sublethal injuries to the normal cells are repaired within six hours but inefficiently in the tumor cells. Thus radiation selectively kills the tumor cells.

Commonly, radiation is delivered in multiple fractions in order to decrease toxicity to the normal tissue, while maintaining the same tumor response. Standard fractionation consists of the daily delivery of 180 to 200 cGy fractions of radiation. The time interval between fractions allows to "repair" of sublethal radiation damage, which is the first of what has been-termed the four "R's" of radiation biology.[2] These are repair of sublethal damage, reassortment, reoxygenation, and repopulation. The mechanism of repair in sublethal radiation damage is not well understood, but may possibly be related to the integrity of the cell cycle checkpoints in normal tissue, where normal checkpoints, e.g., p21, arrest the cell cycle in radiation damaged tissue, allowing time for repair to take place. Given the prevalence of checkpoint defects in tumors, radiation-damaged malignant cells may proceed through cell cycle and probably into an apoptosis (programmed cell death) pathway.[3]

EFFECT OF RADIATION ON THE NORMAL BRAIN TISSUE

The brain is unique due to the fact that the normal parenchymal cell populations are either static or slowly dividing. The deleterious clinical effects of radiation therapy on normal brain tissues are well known. In some early studies, when the whole brain, or large portions of the brain, were treated with a high radiation dose, significant morbidity occurred. Clearly, the effect is multi-factorial. Patient characteristics such as age, history of previous cerebral vascular events, presence of dementia and the region of the brain irradiated, affect the tolerance to radiotherapy. There are also treatment-related factors, which include dose per fraction, overall length of treatment, total dose; and volume of brain tissue being treated. Radiation side effects are divided into acute, subacute and chronic or late. With a daily dose of 1.8–2 Gy per fraction (Standard fractionation) and a total dose approaching 70-80 Gy, acute changes during the treatment course are very uncommon. Administration of a larger single dose of 35 Gy in monkeys is known to produce increased cerebral edema after 18-36 weeks.[4] Neuronal changes are followed by breakdown of the myelin, and proliferative and degenerative changes in astrocytes and other glial elements after 12-20 weeks. Symptoms include nausea, vomiting, headache, neurological deterioration and sleepiness. The effects occur during the first 6 months after irradiation, and include break down or decreased production of myelin due to damage to the oligodendroglial cells. This usually occurs approximately 2 months after irradiation and is generally reversible. The symptoms are similar to the acute effects and include somnolence, anorexia lassitude, irritability.[5] The resulting demyelination is usually visible on MRI and CT imaging. Enhancement is seen in the irradiated areas on T2-weighted MRI scans and slowing may be observed in the region, on EEG.[6] Other objective findings include pleocytosis or mild protein elevation in the cerebrospinal fluid.

Sheline et al[7] suggested that the normal brain tolerance was 52 Gy with 2 Gy per fraction (1980). The necrosis rate at this dose level was estimated to be 0.04-0.4%. The authors concluded that a dose of 60 Gy using 2 Gy per fraction was reasonably safe to use in clinical practice. This is still considered to be the case. Different fractionation schemes have been used. The effects of these various fractionation schemes can theoretically be determined by using a formula calculating the biological equivalent dose or BED. This formula is based on what is termed the a/b ratio, which is different for early (high) and late (low) reacting tissue. The brain is considered to be a late reacting tissue and therefore has a ratio of 3. The ratio is used in the formula

$$BED = D[1+d/[a/b]]_2$$

where D is the total dose in Gy and d is the dose per fraction (the amount given at each treatment). This formula can be used to determine the biological equivalence between different treatment programs.

PROGNOSTIC FACTORS

The prognosis of MG continues to be dismal. The strongest prognostic factors (in order) are age, tumor type, performance status, and extent of the surgical resection.[8] **(Table 1)** Higher age is associated with poorer prognosis because of the increasing grade malignancy. In an analysis of 300 patients by Brain Tumor Study Group (BTSG)[9], a significant improvement in survival was noted in the patients who had minimal residual tumor following surgery. Simpson et al[10] reported a median survival of 11.3 months in patienys with GBN after total resection, 10.4 months with subtotal resection and 6.6 months with biopsy alone. Performance status is of prognostic value by virtue of its dependence on quality and quantity of neurological defects.

TREATMENT BY RADIATION THERAPY

The treatment of AA and GBM is similar. Radiation therapy is almost always indicated when a diagnosis of MG is made. Today, the recommended treatment of MG is surgical resection followed by radiation therapy. When resection is not possible due to difficult anatomical location of the tumor, large size of the tumor, contraindications to surgery because of associated medical illnesses or patients choice, definitive or palliative radiation therapy should be considered.

Table 1 :

Age < 45 yrs.	> 65 yrs.
Histopath : AA	GBM
KPS : 90-100	10-40
Duration of symp > 6 m	< 6 months
PostSx residuum : Nil	Substantial

Various studies have been published using radiotherapy for the treatment of malignant gliomas. **(Tables 2-3)**. In 1966, one of the first studies examining the long term results of radiation therapy was published.[11] Two studies[12,13] were published in 1978 where combined treatment with surgery and radiation therapy were used. In a study by Walker et al[12] in 1978, patients were randomized into four groups: Surgery alone, surgery and BCNU, surgery and radiotherapy, and surgery combined with chemoradiotherapy. The median survival time was best in the groups that included radiotherapy (14-19 weeks vs 35-36 weeks). In the same year, EORTC[13] in collaboration with BTSG published a study evaluating the early and late administration of CCNU with surgery and irradiation. The median survival time in the irradiated group was 43-62 weeks, superior to above mentioned 14-19 weeks median survival seen without irradiation. Another landmark study was performed by Brain Tumor Cooperative Group (BTCG)[14] in 1980 regarding the use of post operative radiotherapy in high grade gliomas. In his study, patients with MG were randomly selected to receive postoperative irradiation or to be observed. Ninety percent of the 222 patients had GBM. The median survival for the supportive care group was 14 weeks and for the postoperative irradiation group was 36 weeks. The one year actuarial survival was 3% for the non irradiated group and 24% for the postoperative irradiated group ($p = 0.001$). Since these original studies, many new techniques have been developed in an

Table 2 : *RT for Anaplastic Astrocytoma*

Author	Therapy	1-yr. Surv	5-yr. Surv
Stage et al	Sx	12	0
	Sx+RT	54	10
Sheline et al	Sx	—	22
Kramer et al	Sx+RT	—	19
Boyage et al	S+RT	50	19

Table 3 : *RT for GBM*

Author	Therapy	1-yr. Surv	2-yr. Surv
Taveras	Sx	5	0
	Sx+RT	30	8
Uihelen	Sx	15	8
	Sx+RT	41	6
Rutten	Sx+RT	32	18
Marsa	Sx+RT	31	8
Boyages	Sx+RT	17	0

attempt to improve local control and to reduce normal tissue toxicity.

TREATMENT PLANNING

Treatment planning for radiotherapy requires close cooperation among radiologist, radiation oncologist, medical physicist and radiation technologist. Radiologist helps in localizing and delineating the tumor volume while physicist performs the dosimetric calculations. Radiation technologist is the one who ultimately delivers the treatment and hence plays an important role. Radiation oncologist is the key person at every step of treatment planning. Currently radiation therapy can be delivered essentially by three means:

1. Fractionated external beam radiation therapy
2. Stereotactic irradiation
3. Interstitial brachytherapy

External Beam Radiation Therapy (EBRT): This is the commonest modality of the radiotherapeutic treatment. It is usually started 3-4 weeks after surgery to allow for the normal wound healing. Imaging with CT and or MRI scan has become routine for treatment planning. Contrast enhanced postoperative MRI Scan must be done within three days of resection to be able to define what is unresected tumor versus postoperative changes. After defining the site and extent of tumor, the surface marking is done on the scalp. Usually two parallel opposed lateral portals are used for tumors with central and deep location. Though treatment can be delivered on Co-60 teletherapy unit but Linear Accelerator is preferable. Different arrangement of portals with different energies can be chosen according to the location of tumor. If the tumor is lateralised, combinations of dual energies (6 MV and 20 MV) provide the better dose distribution yielding higher dose to the tumor and lesser dose to normal brain. Frontal lesions can be treated with anterior and lateral isocentric perpendicular beams and the dose distributions can be optimized with help of wedges. Midcerebral tumors are best treated with parallel opposed anterior and posterior portals as well as with lateral portals.

Volume of Irradiation: The treatment volume in malignant gliomas has been controversial but, of late, there seems to be consensus. In 1957, Bull and Rovit[15] and later Kramer[16] based on autopsy studies of patients irradiated for GBM, recommended that because of the diffuse nature of the tumors the entire brain should be irradiated. With the advent of CT scan, Hoschberg and Pruitt[17] reported that 78% of the recurrences of GBM were within the margin of the tumor bed and 56% were within 1 cm. These findings were confirmed by Wallner et al.[18] Currently, the whole brain radiotherapy for the treatment of GBM is no longer practiced. Localized field covering 2.0-3.0 cm area around the peritumoral edema is the standard portal up to 45 Gy. This is followed by the boost dose to the reduced field with 1.5-2.0 cm margin around the tumor.

DOSE OF IRRADIATION

The dose of RT too has been controversial. Several conflicting reports using different doses have been published. Walker et al[19] reported a dose response analysis in the patients treated on BTCG protocols. Dose ranged from 45 to less than 60 Gy using daily fractions of 1.7 to 2.0 Gy. Median survival increased from 28 to 42 weeks in group treated with the doses of 50 to 60 Gy. On the other hand, Chang et al found no significant difference in the patients treated with 60 Gy alone and the those treated with 60 Gy to whole brain followed by 10 Gy boost to the localized field. Radiation Therapy Oncology Group (RTOG) also reported almost similar finding. Based on these reports, a total dose of 60-64 Gy with daily fraction of 1.8 Gy to 2.0 Gy is adequate to treat most of patient with MG.

DOSE ESCALATION

Dose escalation studies using conformal EBRT to doses of up to 80 Gy have been conducted but no dramatic survival advantage has been observed. Dose escalation through hyperfractionation has allowed a dose as high as 70 Gy but no clear survival benefit has been demonstrated in clinical studies. Accelerated fractionation also not yielded any significant superiority over the conventional fractionation schedule.[20]

STEREOTACTIC IRRADIATION

Stereotactic irradiation is the delivery of a single high dose radiation fraction to a limited volume of tissue, using multiple arcs or beams from different directions. There are three conventionally used stereotactic radiosurgery methods. They include the gamma knife; a linear accelerator based stereotactic radiosurgery and proton beam stereotaxy. The gamma knife **(Fig. 3)** consists of multiple collimators in 3 helmets **(Fig. 4)** with 201 cobalt-60 sources, while the linear accelerator utilizes multiple sweeps or arcs. Both techniques are designed to deliver radiation from multiple directions to a targeted focus, thus sparing the surrounding normal tissue. Three-dimensional reconstruction of target areas is employed in both techniques. A similar dose delivery can be obtained with the proton beam stereotactic system.

INTRAOPERATIVE RADIATION THERAPY

Intraoperative radiotherapy (IORT) is the delivery of a single large dose of radiation using electron beam or low energy X-rays during surgery. A special cone is placed directly over the tumor or tumor-bed under direct visualization. It is indicated for recurrent tumors as well as primary tumors for supplementing the dose of EBRT. It provides a median survival of 12 months in recurrent tumors. This modality is available in only few centers of the world. A preliminary report on ten, patients treated with IORT at the time of wide excision for primary resection or

Fig. 3 : Gamma Knife

salvage resection has been published.[21] In this study, the IORT doses ranged from 15-25 Gy, depending on the, tumor volume and previous radiation therapy. After IORT, the Karnofsky performance status improved in four patients and was unchanged in the remaining six patients.

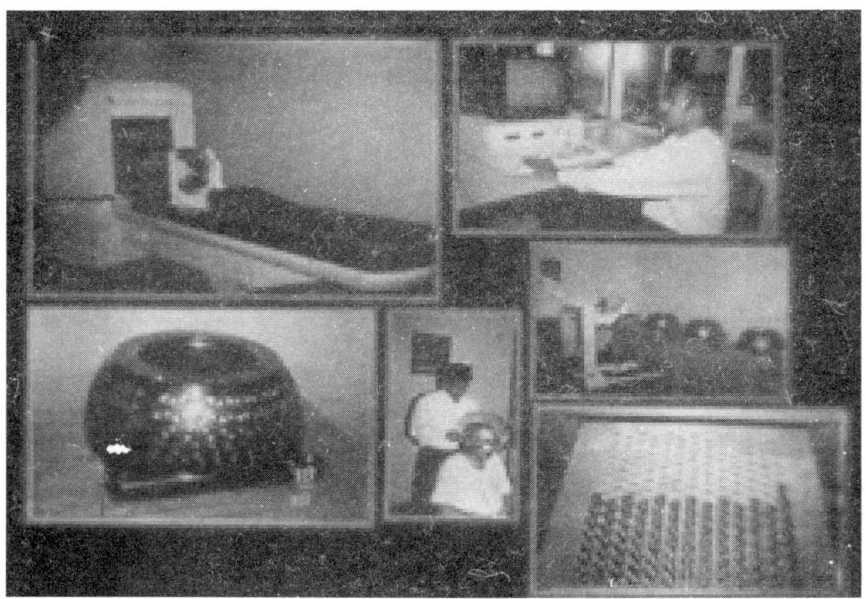

Fig. 4 : SRS Procedure

INTERSTITIAL BRACHYTHERAPY

It has been observed in many trials that there is no survival advantage with EBRT between doses of 60-80 Gy. But possibility of better results with doses above 80 Gy cannot be ruled out. This forms the basis of use of brachytherapy in MG. With brachytherapy (+EBRT) dose can be escalated to more than 100 Gy. This form of radiotherapy also known an interstitial implantation involves the direct placement of one or multiple radioactive sources in the area of the tumor, most frequently performed by a minimally invasive neurosurgical procedure. This allows the delivery of high doses of radiation to the target volume while minimizing radiation exposure to the surrounding structures. The radiation sources can be, implanted directly into the tumor cavity, either for; short time or indefinitely depending on the desire effect. This is limited by tumor size (the larger tumor the more difficult to implant the treatment volume completely and uniformly), and location, i.e. accessibility to treatment. Relatively few newly diagnosed patients (about one third) are candidates for this procedure and mainly indicated for post EBRT recurrences. The procedure is lot easier with the availability of stereotactic head ring. Iodine-125 is the most common employed isotope, but other isotopes such as iridium-192 have also been implanted. Scharfen et al[22] implanted 307 patients with high-activity removable iodine-125 interstitial implants at the University of California in San Francisco. Patients treated primarily also underwent external beam irradiation. The median survival time for patients with primary GBM was BR weeks. For high-grade (non-glioblastoma) glioma this increased to 142 weeks. For recurrent GBM the median survival time from the date of implant was 49 weeks. That corresponding time for recurrent high-grade non-glioblastoma tumors was 52 weeks. Forty percent of patients with high-grade glioma required re-operation with a median time of 33 weeks after implantation. The authors concluded that interstitial implantation is a well-tolerated procedure that prolongs survival in patients with primary and recurrent. GBM as evidenced by the 3-year survival rates of 22% and 15%, respectively.

In spite of the benefits reported by some investigators, the use of brachytherapy remains controversial. This is mainly because of some published reports, which do not show any survival advantage. Besides this, SRT is another alternative modality in such patients, which is simple and noninvasive. The results with SRT and brachytherapy are almost similar. **(Table 2)**

Sneed et al[23] reported a series of patients with GBM treated with temporary iodine-125 implants. Dose volume histograms could be obtained from 97 patients. In this group, a quadratic; relationship was found between a total biologically effective dose and survival, with a trend toward optimal survival probability at 47 Gy minimum brachytherapy dose (corresponding to about 65 Gy to 95% of the tumor volume); survival decreased with lower or higher radiation doses.

COMBINED RADIATION THERAPY AND CHEMOTHERAPY

Various studies have been published on the use of combination therapy involving radiation and chemotherapy. One of the earliest studies of randomized trials combining chemotherapy and radiation therapy was published in 1980 by WALKER et al.[14] There were four treatment arms: semustine radiotherapy carmustine and radiotherapy, and semustine and radiotherapy. No statistically significant benefit in survival was found for the combined treatment groups as compared, with the radiotherapy alone group. A large meta-analysis was published in 1993, reviewing the results from 16 randomized trials involving more than 3,000 patients.[24] The survival of patients treated with radiation therapy alone was compared with the survival of patients treated with radiation therapy and any single or combination chemotherapy regimen. The estimated increase in survival for patients treated with combination radiation and chemotherapy was 10.1% at year and 8.6% at 2 years.

FUTURE PROSPECTS

The high failure rate of conventional treatment has led to the development of many new approaches. Some are directed towards making the treatment more specific while others are aimed at making it more tumoricidal. They are still in the experimental phase and certainly have the potential to establish. Their description is beyond the scope of this chapter, but it is worth enumerating the list, which is as follows.

- Boron neutron capture therapy (BNCT)
- Proton beam therapy
- Radioimmunotheapy
- HDR brachytherapy

REFERENCES:

1. Delhi Cancer Registry (ICMR) report 1994-1995. New Delhi, India.
2. Hall EJ: Radiobiology for the radiologist. JB Lippincott Company, Philadelphia, 1995.
3. Waldman T, Zhang Y, Dillehay L, Kinzler K, et al: Cell cycle arrest versuscell death in cancer therapy. *Nat Med* 1997;3(9):1034-6.
4. Caveness WF: Pathology of radiation damage to the normal brain of the monkey. *Natl Cancer Inst Monogr* 1977;46:57-76.
5. Freeman JE, Johnston PGB, Voke JM: Somnolence after prophylactic cranial irradiation in children with acute lymphoblastic leukaemia. *Br Med J* 1973;4:523-25.
6. Groothuis DR, Vick NA: Radionecrosis of the central nervous system: the perspective of the clinical neurologist and neuropathologist. In: Gilbert HA, Kagan AR (Eds.). Radiation damage to the nervous system. A Delayed Therapeutic Hazard 93-106, Raven Press, New York. 1980.
7. Sheline Ge, Wara WM, Smith V: Therapeutic irradiation and brain injury. *Int J Radiat Oncol Biol Phys* 1980;6(9):1215-28.
8. Nelson DF, Nelson JS, Davis DR, et al: Survival and prognosis of patients with nalignnt astrocytoma. *J neurooncol* 1985;3:99-103.

9. Wood JR, Green SB, Shapiro WR: The prognostic importance of tumor size in malignant gliomas. A CT Scan study by Brain Tumor Cooperative Group. *J Clin Oncol* 1988;6:338-343.
10. Simpson WJ, Horton J, Scott C, et al: Influence of location and extent of surgical resection on survival of patients with GBM: Results of three consecutive radiation therapy oncology group (RTOG) clinical trials. *Int J Radiat Oncol Biol Phys* 1993;26:239-44.
11. Bouchard J: Radiation therapy of intracranial tumors- long term results. *Acta Radiol Ther Phys Biol* 1966;5:11-16.
12. Walker MD, Alexander E Jr., Hunt WE, et al: Evaluation of BCNU and/or radiotherapy in the treatment of anaplastic glioma. A cooperative clinical trial. *J Neurosurg* 1978;49:333-43.
13. EORTC Brain Tumour Group. Effect of CCNU on survival rate of objective remission and duration of free interval in patients with malignant brain glioma – final evaluation. *Eur J Cancer* 1978;14:851-56.
14. Walker MD, Green SB, Byar DP, et al: Randomized comparisons of radiotherapy and nitrosoureas for the treatment of malignant glioma after surgery. *N Engl J Med* 1980; 303(23):1323-29.
15. Bull JWD, Rovit RL: The radiographic localisation of intracerebral gliomas. *J Fac Radiol Lond* 1957;8:147-53.
16. Kramer S: Tumor extent as a determining factor in radiotherapy of glioblastomas. *Acta Radiol (Oncol)* 1969;8:11-17.
17. Horchberg FH. Pruit A: Assumptions in radiotherapy of glioblastoma. *Neurology* 1980;30:907-11.

5

Stereotactic Radiosurgery for Malignant Gliomas

K Ganapathy

INTRODUCTION

A malignant glioma, whether an Anaplastic astrocytoma or a glioblastoma multiforme, is probably one of the most aggressive tumours, owning to its infiltrative nature and rapid growth.[1] By sheer numbers, this is the single largest group of intracranial tumours. Factors associated with decreased survival include older age, low performance status, larger tumour volume and higher grade. A multidisciplinary effort using a combination of different treatment options is required in the management of gliomas. Most patients with malignant gliomas, even after extensive treatment, still die of local recurrence Conventional fractionated external beam radiotherapy after surgery does not appear to *significantly* improve local tumour control or ultimate survival. **Stereotactic radiosurgery (SRS), used as a boost,** is now being considered for lesions persistent after adjuvant external beam radiotherapy. SRS is a reasonable alternative to invasive brachytherapy. Necessity for repeat surgery due to secondary radio necrosis after SRS is less than that after brachytherapy.

STEREOTACTIC RADIOSURGERY (SRS)

SRS is the term given to a non invasive method of delivering a high dose, single fraction of radiation, to a limited well defined target volume[2] sparing adjacent normal tissue. The source of radiation can be X-rays from a linear accelerator or gamma rays from a cobalt unit.[3] Radiosurgery is becoming an increasingly popular method for treating a variety of intracranial tumours[4] Stereotactic conformal radiotherapy allows a higher dose to the tumour while sparing uninvolved brain from radiation.[5] SRS is being used as an adjunct to surgery and external beam radiotherapy in the treatment of patients with malignant gliomas.[6] The current resurgence of interest in stereotactic radiosurgery is primarily due to the increasing availability of hardware and software to modify existing linear accelerators. This is a major advantage particularly in a developing country. Long term follow up, earlier diagnosis, sophisticated imaging, development of more accurate treatment planning and delivery systems, and lowered cost have made stereotactic radiosurgery a primary treatment option in the management of several intracranial lesions. The availability of the micro multi leaf collimator has made ultra precise conformal stereotactic irradiation with the linear accelerator a reality.

The best results in stereotactic irradiation are obtained when the right patient is chosen.[7] This presupposes a thorough knowledge of the disease being treated, and familiarity with the advantages and limitations of the various treatment options available. Reproducible Stereotactic accuracy, superb imaging modalities, an excellent treatment planning system, a system to precisely deliver the exact dose of radiation with adequate quality control measures, and a thorough knowledge of radiobiology is crucial. The source of radiation, namely charged particles, X-rays or gamma rays, does not by itself appear to be the primary determining factor in the ultimate outcome.[8]

PRINCIPLES OF RADIOSURGERY

The goal of radiosurgery is to inactivate and arrest tumour growth, rather than its physical removal. To achieve an effective "tumour kill", using conventional methods of radiation, the dose required is several times more than the maximum dose which the adjacent normal brain can tolerate. In order to protect critical structures of the brain the "safe" dose actually delivered to the tumour, in conventional radiotherapy, is usually less than the required tumouricidal dose. In the technique of radiosurgery a high dose of radiation (8 to 15 times that of a single conventional dose), that is, about 12 to 25 Gray (1200 to 2500 rads) is delivered to a precisely demarcated abnormal area in the brain as a single one-time treatment. Using advances in image technology, electronics and high speed computer graphics, it is ensured that the maximum dose of radiation is delivered precisely to the abnormal area. The fall out in adjacent areas is within acceptable limits. Biological effects produced by radiosurgery range from blood vessel thrombosis to reproductive cell death and frank necrosis within the treatment volume.

The brain is unique in that the normal cell population is essentially static. Radiation effects within the brain, therefore, are characterized by delayed reactions with very little in the way of dose limiting acute toxicity (other than edema). Like other late-reacting tissues, the normal brain parenchyma is very sensitive to the individual doses (fractionation) of radiation. This reflects a large capacity for repair at lower doses per fraction. Current treatment regimens for primary brain tumours is 55 to 60 Gy in 25 to 30 fractions, delivered over 5 to 6 weeks.

In malignant gliomas, tumour cells have been demonstrated in the hypodense "edematous region" in the CT scan and several millimeters beyond the abnormal signal areas in the MRI. Hence it follows that precise 3D delineation of the target volume, which is a prerequisite for any stereotactic radiosurgery, will not be accurate in malignant gliomas. Conventional radiotherapy always includes a safety zone of 2 – 2.5 centimeters beyond the limits seen on imaging studies. The follow-up of patients with CT, MRI, and isotope diagnostic examinations proves that the relapses take place in the vicinity of the primary site in most cases.[9]

SRS therefore should primarily be used as a stereotactic booster dose for small well defined residual lesions. **This presupposes that a dose of 5400 to 6000 Cgy has already been given with a safety zone.** Although there has been some success with enhanced surgical guidance and localized radiotherapy, current techniques are still unable to eradicate the infiltrative glioma cells.[10] The long-term beneficial results of radiosurgery in gliomas is yet to be unequivocally proven. Associated confounding factors make the exact contribution of radiosurgery to patient survival exceedingly difficult to quantify. In adults it can be used as a salvage adjuvant therapy. SRS has also been employed in the management of glial neoplasms of childhood

A possible development in SRS in the future is the use of drugs to sensitize the target and protect the adjacent brain.[11] 21 aminosteroids are thought to provide protection against radiation injury through inhibition of lipid peroxidation and selectivity for vascular endothelium. In the rat, the 21 aminosteroid U 74389G (a drug chemically related to tirlizad) provided protection against radiation induced vasculopathy and edema in the adjacent brain when a pre-radiation dose of 15mg/kg was used. Interestingly, necrosis occurred (using a 4 mm gamma knife collimator) at the targeted area with doses varying from 50 to 150 Gy.

RADIOBIOLOGY OF RADIOSURGERY FOR MALIGNANT GLIOMAS

The radiobiological basis for treatment of any tissue is the composition of the target tissue to be irradiated. In high-grade malignant gliomas, there is a heterogeneous cellular composition of solid tumour tissue, mixed tumour and brain tissue, and, at the margins, normal brain infiltrated by isolated tumour cells. At any given point in time, tumour cells exist in various stages of the cell cycle and have varying degrees of hypoxia; both of these factors influence radiosensitivity. While a single high-dose treatment has the greatest desirable biological effect on rapidly proliferating tissues, normal tissues too are at risk. For increasingly higher radiation doses, the distinction between normal and tumour tissue is reduced. Complications from adverse effects on normal tissue increase and the "therapeutic ratio" decreases.

Fractionation can increase the therapeutic ratio by decreasing the normal tissue effects and increasing tumour tissue effects for several reasons (the "4Rs"):

1. Repair of DNA damage between fractions is more efficient in normal tissues than in tumour cells.
2. Reoxygenation can occur between fractions, making cells more sensitive to the effects of radiation. Hypoxic cells are classically thought to be relatively resistant to these same effects.

3. Redistribution of cells into more sensitive phases of the cell cycle may result in increased damage to target cells.
4. Repopulating tumour cells will be irradiated with successive doses in a fractionated treatment plan.

The effects of radiation on normal and target tissue can also be better understood with a knowledge of the factors involved in an approximation of the biologically effective dose (BED).

The BED (n) = nd (1+d / (a/b), where n is the number of fractions, d is the dose per fraction, nd is the total dose, and a and b are linear quadratic model cell survival parameters like mathematical equation. For glioblastomas, a/b is approximately 10, and for normal tissue, a/b is 3. Defining the therapeutic ratio as BED (target): BED (normal) and assuming constant tumour control for a radiosurgical dose of 20 Gy, it has been calculated that radiation damage to normal tissue in the target volume or in the surrounding rim only was reduced by fractionation. However, the benefits of fractionation, in terms of reduced BED (normal) in the surrounding rim, were less than would have been expected. For small targets, the benefits of fractionation must be weighed against the potential disadvantages of tumour proliferation between treatments, and the time taken in a fractionated treatment schedule to achieve a higher dose in the target volume. One must also remember that the current practice is to use Radiosurgery to "boost" the enhancing portion of a newly diagnosed glioblastoma *before, or after, fractionated external irradiation and that Radiosurgery is in no way intended to replace normal fractionated radiation.*

The complication rate after Radiosurgery for malignant glioma is lower than might be expected, given the high doses delivered within the prescription isodose lines. Within the target volume, small increments in dose may have a disproportionate radiobiological effect and the steep dose gradient provided by Radiosurgery systems accentuates this. Beyond the prescription line, the radiobiological dose decreases faster than isodose curves suggest, possibly *limiting both the desired effect on infiltrating tumour cells and the adverse effects on normal tissue.* The observed acute toxicity was 0%–7%, and only 6% required re operation for symptomatic radiation necrosis.

So does Radiosurgery make sense from a radiobiological point of view as a boost for glioblastomas and recurrent malignant gliomas? *Yes;* in that a dose with a very high radiobiological effect can be safely delivered to a known target of solid tumour tissue, the site of the highest rate of recurrence. *No;* in that the infiltrated brain adjacent to solid tumour tissue requires some other mode of treatment and that the targeting issues for recurrent tumours, as presented by static imaging studies, are less than ideal.

PATIENT SELECTION

Patients eligible for Radiosurgery are highly selected and are those who might be expected to do well with any number of different conventional treatments. Most of these patients are physically and neurologically well, with small tumours not crossing the midline and without ependymal or leptomeningeal extension. Similar to the arguments that took place over brachytherapy, these "selection" factors favourably influence the expected outcome of Radiosurgery patients. Tumours selected for Radiosurgery are generally smaller; coupled with the steep dose gradient outside the target volume, this helps limit undesirable side effects in normal tissue.

CLINICAL SELECTION

1. Karnofsky Performance Status (KPS) score of >60.
2. Well-circumscribed tumour of < 4.0 cm.
3. No subependymal spread.
4. A location not adjacent to the brainstem or optic chiasm.

RADIOLOGICAL ANATOMICAL TARGET SELECTION

The target selected for Radiosurgery is the contrast-enhancing volume and/or the resection cavity of newly diagnosed malignant gliomas. For recurrent gliomas, the contrast-enhancing volume is, almost always selected. For recurrent gliomas with a low-grade histology initially and an enhancing component, in patients who are not candidates for reoperation, only the contrast-enhancing volume is targeted and chemotherapy is used for the tumour component demonstrated on T2-weighted magnetic resonance imaging (MRI). Fast FLAIR imaging enables the use of a single imaging sequence enabling one to load a reduced image amount into the radiotherapy planning software. This is time saving and reduces potential errors.[12]

Targets are usually outlined at the edge of the contrast enhancement on sequential axial MRIs. A highly conformal treatment plan is then designed limiting the high dose to the region of the solid tumour tissue thus limiting complications. The two variables associated with a **higher risk of acute toxicity** are: (1) a target volume of > 8.2 ml; and (2) a maximum dose: prescription dose ratio of ≥2, which is a measure of dose inhomogeneity.

PHYSIOLOGICAL TARGET SELECTION

Algorithms allow the surgeon to superimpose data from magnetic resonance spectroscopy (MRS), positron emission tomography (PET), and single photon emission computed tomography on MRI studies. Brain adjacent to the solid tumour tissue contains viable, infiltrating, tumour cells that are presumed responsible for marginal failures after focal treatments. MRS attempts to discriminate between solid tumour, necrosis, and normal brain tissue based on the relative levels of choline (Cho), creatine, N-acetylaspartate (NAA), lactate, and lipids. This technology has advanced rapidly and is now available on many modern MRI machines, allowing for multiple-voxel (rather than single-voxel) sampling.[1] Positron emission tomography data were successfully combined with magnetic resonance (MR) images to define the target volume for the radiosurgical treatment of patients with recurrent glioma or metastasis. This approach may contribute to optimizing target selection for infiltrating or ill-defined brain lesions. This could give a better understanding of the metabolic changes following radiosurgery. The ability to use PET data in GKS represents a crucial step toward further developments in radiosurgery,[13] Serial G PET could be a potential tool for predicting the outcome of radiosurgery for brain tumours by detecting hyper acute changes in tumour glucose metabolism.[14]

Patients with abnormal spectra outside the contrast-enhancing volume clearly require therapy designed to attack the infiltrating tumour cells in surrounding brain. For those with abnormal spectra limited to the enhancing volume, focal radiosurgical treatment makes the most sense. For patients with abnormal spectra outside the normally targeted contrast-enhancing volume, additional therapies are necessary to prolong time to recurrence, and to increase survival.

Increases in Cho correlated with poor radiologic response and suggested tumor recurrence, confirmed histologically. *Response within the gamma knife target* was observed as a *reduction of Cho* levels and an increase in lactate/lipid levels, typically within 6 months of treatment. The development of a spectral abnormality preceded a coincident increase in contrast enhancement by 1 to 2 months. Proton MR spectroscopic imaging provided diagnostic and monitoring information before and after radiosurgery. Evaluation of metabolic changes with proton MR spectroscopy and structural changes with MR imaging improved tissue discrimination and provided correlation with histologic findings. Positron emission tomography scanning, co-registered with magnetic resonance imaging, allows the 'boost' concept in radiosurgery to become a sophisticated and accurate reality.[15]

GENE THERAPY COMBINED WITH SRS

Experiments were carried out in a nude mouse model of human glioblastoma to determine whether gamma-knife radiosurgery combined with herpes simplex virus thymidine kinase (tk) suicide gene therapy and tumor necrosis factor alpha (TNF alpha) gene transfer provided an improved multimodality treatment of this disease. These findings suggest that gene therapy in combination with more conventional therapeutic methods may provide an improved strategy for extending the life expectancy of patients afflicted with this ultimately fatal disease.[16]

CHEMOTHERAPY COMBINED WITH SRS

A significant prolongation of the median survival time in rats was demonstrated when radiosurgery (25 Gy) was combined with BCNU / carboplatin administration. This may be an effective therapeutic regimen in the treatment of radioresistant human malignant gliomas.[17] Studies have been carried out using fractionated stereotactic radiosurgery and concurrent paclitaxel[18] and carmustine.[19]

TECHNICAL

Stereotactically-guided conformal radiotherapy is a practical technique for irradiating irregular lesions in the brain. Multi-leaf collimators with conformal blocks are now being regularly used.[20] The shaping of the conformal fields may be achieved using lead alloy blocks, a conventional multi-leaf collimator (MLC) or a mini/micro-MLC. Although the former gives more precise shaping, it is labour intensive. The latter methods are more practical as both, mould room and treatment room times are reduced, but the shaping is limited by the finite leaf-width.[20]

Intensity-modulated radiotherapy (IMRT) offers the potential to more closely conform dose distributions to the target, and spare organs at risk (OAR). Its clinical value is still being defined[21] Hyperfractionated radiation therapy has been used for brain stem gliomas in children and adults[22] and for recurrent glioblastoma multiforme[19] Feasibility, and toxicities of an accelerated treatment program by using a concomitant stereotactic radiotherapy boost given weekly during a course of standard external-beam irradiation (EBXRT) in patients with malignant gliomas has also ben studied.[23] Multiple fractions per day using external beam radiotherapy for adults with supratentorial malignant gliomas[24] has also been tried out.

PROCEDURAL STEPS

Prior to treatment, informed consent is obtained and should include a discussion of possible acute, early, and late-delayed radiation side effects[25] that may follow Radiosurgery. After planning is complete, IV dexamethasone (10mg) is given in some centers. At 2-3 months a follow up imaging study is obtained. Analysis of dose-effect relationships in radiotherapy of malignant anaplastic gliomas[26, 27] has been published. The dose has to be individually tailored depending on the volume, the location, previous RT dose, grade of tumour, and age of patient.

RESULTS OF SRS IN MALIGNANT GLIOMAS

Multivariate analyses have shown that increased survival was associated with five variables: lower pathologic grade, younger age, higher Karnofsky performance status (KPS), smaller tumour volume, and unifocal tumour. Survival was not found to be significantly related to radiosurgical technical parameters (dose, number of isocenters, prescriptions isodose percent, inhomogeneity) or extent of preradiosurgery surgery.[28] For patients with recurrent or persistent, small, malignant intracranial tumours, Radiosurgery has obviated the need for prolonged hospitalisation and has eliminated the risks associated with general anesthesia and open craniotomy.[29]

Shrieve et al[30] recently updated the Boston experience, reporting on 78 patients with glioblastoma multiforme. A single isocentre was used in 95% of cases, with a minimum peripheral dose of 6-24 Gy (median 12 Gy) for a median tumour volume of 9.4 ml. The median survival time was 19.9 months, with a 2-year survival rate of 35.9%. Fifty percent of patients underwent reoperation, at a median of 7.9 months following Radiosurgery; in 48.7% recurrent tumour was found. The median length of survival following reoperation was 12.4 months.

Sakaria et al[31] evaluated 115 patients from three institutions who received a Radiosurgery boost and found that, when compared to the RTOG results, *patients treated with Radiosurgery had significantly longer median and 2-year survival rates*, depending on prognostic class (P=0.01). *This benefit appeared to be greatest for the patients in the worst prognostic classes,*(3-6) with minimal benefit for patients in classes 1 and 2. The 2-year survival rates for the Radiosurgery versus the previously reported RTOG classes were 81% versus 76% for classes 1 and 2, 75% versus 35% for class 3, 34% versus 15% for class 4, and 21% versus 6% for classes 5 and 6. In the 64-glioblastoma multiforme patients, 45 were treated as part of a boost protocol that included post diagnosis conventional irradiation plus Radiosurgery or three cycles of chemotherapy plus irradiation and Radiosurgery. The mean interval from diagnosis to Radiosurgery was 6.2 months and the median survival time in this group was 20 months, with a 2-year survival rate of 41%. There was no acute toxicity. Twelve patients (19%) required reoperation at a mean time of 5 months after Radiosurgery. In 21 patients the median survival time was 56 months and the 2-year survival rate was 88%. Twenty-three percent underwent reoperation, at a mean of 8 months post-Radiosurgery. With RTOG standard treatment group, there appeared to be a survival benefit for classes 3-5. The 2-year survival rates for the Radiosurgery group versus the RTOG groups were: class 3, 73% versus 35%; class 4, 24% versus 15% and class 5, 26% versus 6%.

Several other publications discuss the role of SRS in the management of malignant gliomas.[31, 32, 33] Adjuvant stereotactic radiosurgery for anaplastic ependymoma[34] has also been reported.[35] *The proliferative potential of malignant astrocytic tumors in the radiosurgically treated area is reduced after SRS. Radiological enlargement of enhanced lesions on MR images is due to propagation of the residual tumor cells that were not covered by radiosurgical target volume or to radiation necrosis.* In glial tumors radiosurgery helped either to "sterilize" the tumor bed after removal or to treat remnants of the lesions in critical areas.[36]

COMPLICATIONS

Complications that occur after Radiosurgery for primary malignant gliomas can be grouped by time intervals into acute (hours to days), early-delayed (weeks to months), and late-delayed (months to years). Acute and early-delayed reactions are nearly always transient and can be treated with medical therapies.[25] Acute side effects are rarely serious, given the small target volumes in these selected cases, and are rarely reported as such Despite increased utilization of fractionated stereotactic radiation therapy (SRT) or stereotactic radiosurgery (SRS), the incidence and nature of immediate side effects (ISE) associated with these treatment techniques are not well defined.[25] In one series 35% experienced one or more ISE. Most of the ISE (87%) were mild, and consisted of nausea, dizziness/vertigo, seizures, and new persistent headaches[25] Intracranial Papillary endothelial hyperplasia (PEH) that occurred after surgery and SRS for glioma has been reported.[37]

FRACTIONATED STEREOTACTIC RADIOTHERAPY

SRT is the term used when **stereotactic irradiation is delivered over several sittings:**[38] This modification is particularly suitable for larger lesions especially when they are abutting critical areas.[39] A relocatable stereotactic frame ensures the same stereotactic accuracy as in single dose SRS. Reducing the dose per sitting ensures safety to adjacent critical areas[40] Data from experimental studies provide a strong radiobiological rationale for the use of fractionated RS in the treatment of tumors located near critical normal structures, including visual pathways. The sparing effect of fractionated RS is greater for late-responding tissues, relative to the rapidly proliferating tumor tissues[41] SRT is particularly used in sellar lesions where the optic nerve is within 5 mm.[15] Split-course fractionated gamma knife radiosurgery (FSRS) has recently been tried in combination with conventional external-beam radiation therapy (CEBRT).[42]

DECISION MAKING IN RADIOSURGERY

It has to be understood that treatment is not to reverse existing neurological deficit but to contain further growth of the tumour. The necessity for follow up imaging studies has to be stressed. A comprehensive up to date knowledge of the long-term natural history of the untreated disease is imperative. Radiosurgery is a rapidly evolving subspeciality where indications keep changing. The practice of radiosurgery demands a good knowledge of neuroimaging. One must be familiar with anatomy, physiology and pathology of the brain. Understanding the radiobiologic effects of high dose irradiation is essential. Knowledge of conventional radiotherapy dose planning methods is desirable. Proficiency with stereotactic techniques and computer literacy are mandatory. Equally important is a thorough knowledge of alternative treatment modalities.

PRESENT DAY RADIOSURGERY

The question is often raised whether there is any role for such high tech therapy in a developing country like India where cost containment is the order of the day. 22 years ago, when the first CT scan was introduced the same question was asked. Today, there are at least 900 CT and 120 MRI scanners. Within six years of the first SRS unit being established in India by the author, ten more units have been established. Radiosurgery, like many new ideas, was ignored and misunderstood for several years. The pendulum has swung and today the euphoria occasionally leads to exaggerations, over statements, and unrealistic expectations. Results can be known only months and years later - thus there is a potential for beneficial and harmful results alike. Proof is critical. Long-term safety and efficacy of radiosurgery is slowly but surely being established.

The trend in medicine worldwide is towards minimally invasive or non-invasive procedures. Radiosurgery, augurs well for the future. However as Lars Leksell the Father of Radiosurgery pointed out 25 years ago, "Fools with tools are still fools". The single most important factor in stereotactic radio surgery is not whether the source of radiation is x rays or gamma rays but whether the clinical features, natural history and the pathology of the lesion warrants stereotactic irradiation. Selection of stereotactic irradiation as a treatment option presupposes a mature head on young shoulders - that is familiarity with the old and the new! There is a very real danger of misusing this tool and adopting the hammer nail approach (When one has a hammer to be used, everything around looks like a nail !). The more sophisticated the technology, and the less the morbidity, the more critical should be the selection of cases. Realisation of the limitations of this excellent tool is important. Dispassionate long-term follow-up and meticulous recording of side effects is crucial.

Acknowledgments: I wish to thank Mrs. Vijayalakshmi Ganapathy and Mr. G. Murali Krishnan for secretarial assistance.

REFERENCES:

1. Rand SD, Krouwer HG: MRS imaging of gamma knife treatment for gliomas: from metabolite ratios to therapeutic rationale. *Am J Neuroradiol* 2001;22(4):598-99.
2. Varady P, Dheerendra P, Nyary I, Vajda J, Ladislau S: Neurosurgery using the Gamma Knife. *Orv Hetil* 1999;140(7):331-45.
3. Ganapathy K: Radiosurgery in Textbook of Neurosurgery. 1996, 2nd edition Vol. 2 p 1220-26 Editors Ramamurthi B and Tandon PN.I.Churchill Livingstone.
4. Friedman WA, Foote KD: Linear accelerator radiosurgery in the management of brain tumours, *Ann Med* 2000;32(1):64-80.
5. Gildenberg PL: Multimodality program involving stereotactic surgery in brain tumor management Stereotact *Funct Neurosurg* 2000;74(3-4):179-84.
6. Gannett D, Stea B, Lulu B, et al: Stereotactic radiosurgery as an adjunct to surgery and external

beam radiotherapy in the treatment of patients with malignant gliomas. *Int Radiat Oncol Biol Phys* 1995;33:461-68.
7. Ganapathy K: Decision making in Radiosurgery in "Neurosciences Today" 1997;1:72-77C.
8. Ganapathy K: Radiosurger: Advances in Clinical Neurosciences. 1995;5:369-85.
9. Mangel L, Julow J, Major T, Valalik I, Horvath A, Kiss T, Somogyi A, Nemeth G: CT- and MRI-guided conformal interstitial and external beam radiotherapy of primary brain tumors: prospects in Hungary, *Orv Hetil* 2000;30:1703-09
10. Alexander E 3rd, Loeffler JS: The role of radiosurgery for glial neoplasms. *Neurosurg Clin N Am* 1999;10(2):351-58.
11. Ganapathy K: Newer concepts in stereotactic radiosurgery. *Progress in Clin neurosci* 1997;12: 125-36.
12. Essig M, Debus J, Schlemmer HP, Hawighorst H, Wannenmacher M, van Kaick G: Improved tumor contrast and delineation in the stereotactic radiotherapy planning of cerebral gliomas and metastases with contrast media-supported FLAIR imaging *Strahlenther Onkol* 2000;176:84-94.
13. Levivier M, Wikier D, Goldman S, David P, Metens T, Massager N, Gerosa M, Devriendt D, Desmedt F, Simon S, Van Houtte P, Brotchi J: Integration of the metabolc data of position emission tomography in the dosimetry planning of Radiosurgery with the gamma knife: early experience with brain tumours. Technical note. *J Neurosurg* 2000;93 Suppl 3: 233-38.
14. Maruyama I, Sadato N, Waki A, Tsuchida T, Yoshida M, Fujibayashi Y, Ishii Y, Kubota T, Yonekura Y: Hyperacute changes in glucose metabolism of brain tumours after stereotactic Radiosurgery: a PET study. *J Nucl Med* 1999;40(7):1085-90
15. Sims E, Doughty D, Macaulay E, Royle N, Wraith C, Darlison R, Plowman PN: Stereotactically delivered cranial radiation therapy: a ten-year experience of linac-based radiosurgery in the UK, *Clin Oncol (R Coll Radiol)* 1999;11(5):303-20.
16. Niranjan A, Moriuchi S, Lunsford LD, Kond ziolka D, Flickinger JC, Fellows W, Rajendiran S, Tamura M, Cohen JB, Glorioso JC: Effective treatment of experimental glioblastoma by HSV vector-mediated TNF alpha and HSV-tk gene transfer in combination with radiosurgery and ganciclovir administration. *Mol Ther* 2000;2(2):114-20.
17. Khil MS, Kolozsvary A, Apple M, Kim JH: Increased tumor cures using combined radiosurgery and BCNU in the treatment of 9l glioma in the rat brain. *Int J Radiat Oncol Biol Phys* 2000;47(2):511-16.
18. Lederman G, Wronski M, Arbit E, Odaimi M, Wertheim S, Lombardi E, Wrzolek M: Treatment of recurrent glioblastoma multiforme using fractionated stereotactic radiosurgery and concurrent paclitaxel. *Am J Clin Oncol* 2000;23(2):155-59.
19. Curran WJ Jr, Scott CB, Weinstein AS, et al: Survival comparison of radiosurgery-eligible and ineligible malignant glioma patients treated with hyperfractionated radiation therapy and carmustine: a report of Radiation Therapy Oncology Group 83-02. *J Clin Oncol* 1993;11:857-62.
20. Adams EJ, Cosgrove VP, Shepherd SF, Warrington AP, Bedford JL, Mubata CD, Bidmead AM, Brada M: Comparison of a multi-leaf collimator with conformal blocks for the delivery of stereotactically guided conformal radiotherapy. *Radiother Oncol* 1999;51(3):205-9.
21. Khoo VS, Oldham M, Adams EJ, Bedford JL, Webb S, Brada M: Comparison of intensity-modulated tomotherapy with stereotactically guided conformal radiotherapy for brain tumors. *Int J Radiat Oncol Biol Phys* 1999;45(2):415-25.
22. Shrieve DC, Wara WM, Edwards MSB, et al: Hyperfractionated radiation therapy for brain stem gliomas in children and adults. *Int J Radiat Oncol Biol Phys* 1992;24:599-610.
23. Cardinale RM, Schmidt-Ullrich RK, Benedict SH, Zwicker RD, Han DC, Broaddus WC: Accelerated radiotherapy regimen for malignant gliomas using stereotactic concomitant boosts for dose escalation. *Radiat Oncol Investig* 2998;6(4):175-81
24. Halperin EC: Multiple fraction per day external beam radiotherapy for adults with supratentorial malignant gliomas. *Journal Neurooncology* 1992;14:225-62.
25. Werner-Wasik M, Rudoler S, Preston PE, Hauck WW, Downes BM, Leeper D, Andrews D, Corn BW,

Curran WJ Jr.: Immediate side effects of stereotactic radiotherapy and radiosurgery. *Int J Radiat Oncol Biol Phys* 1999;43(2):299-304.

26. Walker MD, Strike TA, Sheline GE: An analysis of dose-effect relationship in radiotherapy of malignant anaplastic glioma. *J of Neurosurgery* 1978;49:333-43.

27. Wallner KE, Galicich JH, Kroi G: An analysis of dose-effect relationship in radiotherapy of malignant gliomas. *Int J Radiat Oncol Biol Phys* 1979;5:1725-31.

28. Larson DA, Gutin PH, McDermott M, Lamborn K, Sneed PK, Wara WM, Flickinger JC, Kondziolka D, Lunsford LD, Hudgins WR, Friehs GM, Haselsberger K, Leber K, Pendl G, Chung SS, Coffey RJ, Einapoli R, Shaw EG, Vermeulen S, Young RF, Hirato M, Inoue HK, Ohye C, Shibazaki T: Gamma Knife for glioma: selection factors and survival. *Int J Radiat Oncol Biol Phys* 1996;36(5):1045-53.

29. Coffey RJ, Lunsford LD, Flickinger JC: The role of Radiosurgery in the treatment of malignant brain tumours. *Neurosurg Clin N Am* 1992;3(1):231-44.

30. Shrieve DC, Alexander E 3rd, Black PM, Wen PY, Fine HA, Kooy HM, Loeffler JS: Treatment of patients with primary glioblastoma multiforme with standard postoperative radiotherapy and radiosurgical boost: prognostic factors and long-term outcome. *J Neurosurg* 1999;90(1):72-7.

31. Sarkaria JN, Mehta MP, Loeffler JS, et al: Radiosurgery in the initial management of malignant gliomas: survival comparison with the RTOG recursive partitioning analysis. Radiation Therapy Oncology Group. *Int J Radiat Oncol Biol Phys* 1995:32:931-41.

32. Mehta MP, Masciopinto J, Rozental J, et al: Stereotactic radiosurgery for glioblastoma multiforme: report of a prospective study evaluating prognostic factors and analysing long-term survival advantage. *Int J Radiat Oncol Biol Phys* 1994;30:541-49,

33. Kondziolka D, Flickinger JC, Bissonette DJ, et al: Survival benefit of stereotactic radiosurgery for patients with malignant glial neoplasms. *Neurosurgery* 1997;41:776-85.

34. Jawahar A, Kondziolka D, Flickinger JC, Lunsford LD: Adjuvant stereotactic radiosurgery for anaplastic ependymoma. *Stereotact Funct Neurosurg* 1999;73:23-30.

35. Loeffler JS, Alexander E 3rd, Shea WM, Wen PY, Fine HA, Kooy HM, Black PM: Radiosurgery as part of the initial management of patients with malignant gliomas. *J 28.Clin Oncol* 1992;10(9):1379-85.

36. Raco A, Raimondi AJ, D'Alonzo A, Esposito V, Valentino V: Radiosurgery in the management of pediatric brain tumors. *Childs Nerv Syst* 2000;16(5):287-95.

37. Hagiwara A, Inoue Y, Shakudo M, Wakasa K, Sato K, Ohata K, Daikokuya H, Yamada R: Intracranial papillary endothelial hyperplasia: occurrence of a case after surgery and radiosurgery. *J Comput Assist Tomogr* 1999;23:781-5. Review.

38. Regine WF, Patchell RA, Strottmann JM, Meigooni A, Sanders M, Young AB: Preliminary report of a phase I study of combined fractionated stereotactic radiosurgery and conventional external beam radiation therapy for unfavorable gliomas. *Int J Radiat Oncol Biol Phys* 2000;48:421-6.

39. Tokuuye K, Akine Y, Sumi M, Kagami Y, Murayama S, Nakayama H, Ikeda H, Tanaka, M, Shibui S, Nomura K: Fractionated stereotactic radiotherapy of small intracranial malignancies. *Int J Radiat Oncol Biol Phys* 1998;42(5):989-94.

40. Brada M, Baumert B: Focal fractionated conformal stereotactic boost following conventional radiotherapy of high-grade gliomas: a randomized phase III study. A joint study of the EORTC (22972) and the MRC (BR10). *Front Radiat Ther Oncol* 1999;33:241-3.

41. Kim JH, Khil MS, Kolozsvary A, Gutierrez JA, Brown SL: Fractionated radiosurgery for 9L gliosarcoma in the rat brain. *Int J Radiat Oncol Biol Phys* 1999;5:1035-40.

42. Regine WF, Patchell RA, Strottmann JM, Meigooni A, Sanders M, Young B: Combined stereotactic split-course fractionated gamma knife radiosurgery and conventional radiation therapy for unfavorable gliomas: a phase I study. *J Neurosurg* 2000;93 Suppl 3:37-41.

6

Stereotactic Biopsy: Techniques, Indications, Limitations and Complications

V Kulkarni, V Rajshekhar

INTRODUCTION

The development of stereotactic biopsy has made the practice of empirical therapy for deep-seated, 'inaccessible' intracranial masses obsolete.[1] Even with sophisticated imaging techniques such as CT and MRI, the pathological diagnosis and thus management of these lesions is relatively imprecise.[1-3] A CT guided stereotactic biopsy makes a safe and rational treatment of inaccessible intracranial mass lesions possible.

TECHNIQUE

Simply put stereotaxy involves the ability to precisely locate and access any point in a defined three-dimensional space–i.e. stereotactic volume or cranium. To reach any point precisely, it needs to be registered in Cartesian XYZ coordinates (or arch-radius coordinates) in a particular three-dimensional space. The frame thus forms the very basis of stereotaxy. The prototype of CT guided **frame based stereotactic system** are BRW (Brown-Roberts-Wells system), CRW system (Cosman-Roberts-Wells system) and Leksell System.

A) Instruments

A stereotactic procedure involves a few but indispensable gadgets. In the CRW system these include–

1. Nickel-plated aluminium **head-ring**, that is suspended with 4 carbon fiber posts and van be fixed to the skull with plastic and steel-set pins.
2. **Localizer ring** with 6 vertical and three diagonal carbon fiber rods.
3. **Arc-guidance system with probe carrier.**
4. **Phantom base**-base ring equivalent of head ring with a movable phantom tip- (phantom target).

5. A dedicated computer with software package.
6. **Biopsy probe**—A 1.2 mm diameter cannula and a blunt stillete with guide and flexible **cup biopsy forceps**

B) Procedure

The stereotactic biopsy procedure primarily involves two stages. Initially the stereotactic coordinates of the lesion are obtained and the second stage involves the actual biopsy procedure guided by those coordinates.

(I) Obtaining the stereotactic coordinates involves six basic steps as described below:

a) Local analgesia

Stereotactic procedure in adults is most often undertaken in an awake cooperative patient, to whom the procedure has been explained earlier. Local analgesia is usually adequate for adult patients. The frame fixation and twist drill craniostomy is performed after infiltrating local analgesic like 2% lignocaine. A mild sedation may be needed in case of an agitated patient. General anaesthesia is required for children and some uncooperative adults.

b) Head Ring fixation

A rigid frame fixation is the most important step in this procedure, since all the coordinates are taken in reference to this frame. Movement of the frame (with respect to head) even by a few millimeters anytime during the stereotaxy procedure can cause gross errors in target localization, necessitating refixing of the frame and repeating the CT scan for obtaining fresh coordinates.

The head ring is fixed to patient's head, suspended by carbon fiber posts which are rigidly fixed to the skull by 4 self-penetrating screws. The screws should ideally penetrate only the outer table. (In children, penetration of inner table can give rise to CSF leak). The lateral parts of foreheads and mastoid regions are ideally suited for these since it avoids the temporalis muscle—(which prevents bony engagement of the screws) and important neurovascular pedicles of the scalp. This also allows adequate low fixation of the frame. The metal frame and the screws cause metal artifacts in the corresponding CT slice and can hamper proper visualization of the target. A knowledge of approximate vertical location of the target is thus essential for the frame fixation, so that the target falls either above the level of the screws or between the screws and the base ring, as far as possible. A target which is below the base ring cannot be approached and hence for brain-stem targets the base ring has to be fixed as low as possible.

c) **Stereotactic imaging – CT/MR**

The localizer ring is then mounted on the base-ring and a high-resolution contrast-enhanced CT scan is taken. An MRI based stereotactic imaging needs MR compatible frame. A stereotactic image shows an axial cranial slice surrounded by circular cross-sections of 9 rods of the localiser ring surrounding the head.

d) **Calculation of coordinates from CT/MR or a dedicated computer**

An appropriate CT slice is selected demonstrating the target point within the lesion. **(Figure)** The pixel coordinates of the target and the nine localizer rods in that slice are noted and entered into the computer. The computer program is then run, which provides target and arc coordinates.

e) **Target coordinates set on the Arc System**

The arc system is set according to the stereotactic coordinates obtained from the computer program. In CRW frame the arc coordinates are same as target coordinates—AP, lat and vertical. The frame is then transferred onto the phantom base.

f) **Target simulation on the Phantom Base**

The phantom target on the phantom base is then set according to the

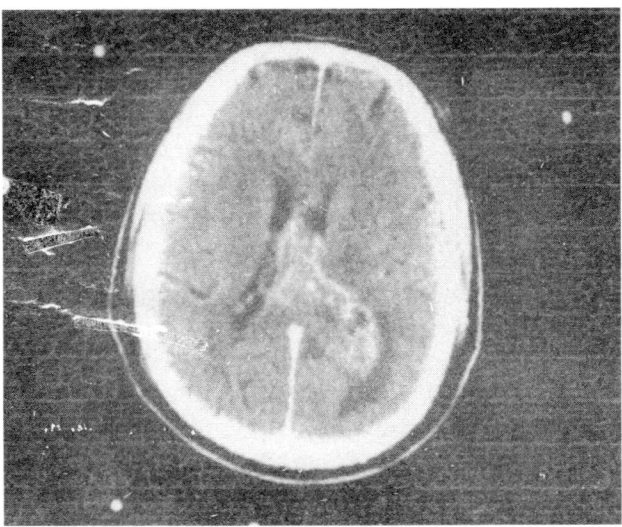

Fig. 1 : A contrast enhanced computerised tomographic axial image of the head on which stereotactic head-ring is fixed and the localiser ring is mounted. A contrast enhancing diffuse lesion is seen inside the lateral ventricle. The cross-sections of nine localiser rods are seen surrounding the head, which serve as localizing coordinates.

actual target coordinates obtained from the computer program. A probe is then passed through the probe carrier on the arc system to see if it accurately reaches the phantom target. This way the proposed stereotactic surgery is simulated outside the body (the target on the phantom base simulates the target inside the head). This eliminates errors in data acquisitions or data-transfer, faulty software, wrong settings of coordinates on the frame etc. Phantom base is thus an effective quality assurance method in stereotactic surgery and its use is always advised before actually targeting the lesion. However not all stereotactic systems provide a phantom base.

(II) Stereotactic biopsy procedure

a) Arc System fixed on Head Ring

The arc-system is then transferred to the head-ring fixed to the head and locked in position.

b) Cranial entry – twist drill craniostomy

The proposed area of probe entry is shaved and thoroughly cleaned with betadine and draped. A twist drill craniostomy is made at the entry point with the twist drill provided in the stereotactic instrument set. The probe is provided with the guard which prevent accidental plunge of the instrument. Once the inner table is perforated the dura is entered by advancing the twist drill tip by a few millimeters. The twist drill is then removed and the cannula with probe is passed through the craniostomy. The length of the probe is previously adjusted with a guard to the target depth coordinate. At this point the tip of the probe and the cannula lies within the chosen target.

c) Biopsy with cup forceps (1 x 2 mm)

The stylete or the probe is then removed gently. After assuring that there is no blood coming out of the cannula, a flexible cup biopsy forceps is passed through the cannula. The size of the biopsy forceps is limited to 1x2 mm, to minimize tissue damage. One or two bites of the target tissue are obtained. the biopsy forceps is then removed. The tissue is then transferred on to a pad of moistened gelatin sponge.

d) Intraoperative smear feedback

The bits of tissue are immediately handed over to and processed by neuropathologist for rapid smear preparation. The success of the entire procedure is based on positive biopsy. Therefore the smear report is awaited while the cannula is kept in situ-within the target. An inconclusive report demands, obtaining more tissue for biopsy. If the smear suggests periphery of the lesion a judicious advancement of the cannula may be required to obtain deeper tissue. In certain cases, choosing an alternate target in same or different CT slice and removing and resetting the arc system coordinates according to new target may need to be undertaken. Re-fixation of head-ring is not required.

e) Post operative CT scan

After the positive biopsy is obtained the arc system and the head ring are removed. A post-procedure plain CT scan is then done to rule out an intracranial hematoma. A delayed CT scan after 4-6 hours is preferred. However in case of intra-procedure bleed or clinical deterioration an immediate CT scan is indicated.

f) Discharge

After the procedure the patient is kept under observation for 24 hours and discharged thereafter if there are no complications.

INDICATIONS

A stereotactic biopsy is always undertaken for intra-axial (and intraventricular) lesions only. (A stereotactic biopsy for extra-axial lesions involves considerable risk of intracranial bleed as described subsequently.)

Stereotactic biopsy is indicated in cases of –

1. **Multiple intra-axial masses**[4]

   Multiple intracranial lesions poses a diagnostic dilemma where differential diagnosis include metastasis, inflammatory lesions, lymphoma, multicentric glioma etc. The morbidity of a diagnostic craniotomy can be avoided by resorting to stereotactic biopsy. Since the diagnosis is obtained on the spot the treatment can be started soon after the procedure.

2. **Diffuse ill-defined intra-axial masses**[4]

   Diffuse ill-defined intra-axial lesion with mass effect seen on CT scan are often unsuitable cases for surgical resection. In such cases the distinction between normal and abnormal tissue may not be obvious even per-operatively. In cases of significant mass effect the morbidity of craniotomy may be quite high. A stereotactic biopsy allows precise targeting of the abnormal area with minimal morbidity.

3. **Deep seated intra-axial lesions**

   Surgical treatment of deep-seated, 'inaccessible' lesions, i.e. thalamic, ganglionic, brainstem[5]—is technically difficult and often results in significant postoperative complications. To avoid this many such lesions were treated empirically in the past, based on CT morphology. Stereotactic biopsy of these lesions offers the rational and safe management of these lesions.

4. **Lesions located in eloquent regions**

   Lesions located in eloquent area–like motor or speech area can be safely biopsied by stereotactic procedure to avoid the commonly encountered post-operative deficits associated with craniotomy and open biopsy.

LESIONS TO AVOID: CONTRA-INDICATIONS

The commonest procedural morbidity of stereotactic biopsy is intracranial bleed. This can be minimized by judiciously avoiding lesions prone to this complication.[4]

1. **Extra axial lesions**

   Intracranial haemorrhage commonly occurs from pial vessels when the stereotactic probe enters the pial surface to reach an intra-axial lesion. Reaching a deep extra-axial mass would involve breach of two pial surfaces thus increasing the chances of intracranial haemorrhage.

2. **Vascular lesions – giant aneurysm, cavernous angioma**

   Stereotactic biopsy is contraindicated for any lesion suggestive of an obvious vascular etiology, on CT imaging. A high index of suspicion for cavernous angioma or a giant aneurysm can save the patient from catastrophic and fatal intracranial bleed during the biopsy procedure.

3. **Superficial cortical lesions**

   Biopsy of superficial cortical lesions is more likely to cause haemorrhage from vessels in the pia.

4. **Lesions close to the sylvian fissure, suprasellar region**

   Lesions close to sylvian fissure, suprasellar cistern or posterior third ventricular region often distort the anatomy and can have large cerebral arteries or veins or major branches of circle of Willis and deep veins draped around them. Inadvertent pulling or shift of these structures can cause catastrophic intracranial haemorrhage. Therefore choice if the target and trajectory in these lesions should be carefully done.

RESULTS

Stereotactic biopsy procedure has a high **positive yield of more than 90%.** The diagnostic accuracy of stereotactic biopsy (concordance with the final diagnosis on larger paraffin histological sections) in experienced hands is reported to be in the range of 90 to 95% in most series.[6-7] An **intra-operative smear feedback** is very important for the success of the procedure. The choice of the target is usually based on CT morphology. Sometimes a gliotic area or area in the periphery of the lesion or necrotic is targeted. This can result in inconclusive or even incorrect biopsy results. This can be avoided with intra-operative communication with the pathologist about the suspected diagnosis based on clinical and radiological features. If the smear feedback is unsatisfactory or inconclusive or does not correlate with the clinical and radiological features, an alternative or deeper target is chosen and biopsied. This way an intra-operative smear feedback improves yield of stereotactic biopsy.

RESULTS: CMC SERIES

When nondiagnostic biopsies were analyzed in a consecutive series of 407 patients undergoing computerized tomography (CT)–guided stereotactic biopsies at CMC hospital nineteen biopsies (4.7%) were negative (normal tissue or nonspecific pathology) and 10 (2.4%) were inconclusive (definitive diagnosis could not be made although representative tissue was obtained), giving an overall nondiagnostic biopsy rate of 7.1% (29 of the 407 cases).[7] The **positive yield was found to be as high as 92.9%.** The neoplastic masses suspected from CT morphology (390 cases) had a high yield of positive biopsy – 94.4.% i.e. out of 390, 368 tumors were correctly diagnosed. On the other hand in only 58.5% of the cases suggestive of inflammatory masses, a definitive and etiologic diagnosis could be obtained. However in all cases of inflammatory masses, a neoplasm was conclusively ruled out with stereotactic biopsy.

When the lesions were classified in four groups depending on their CT density and contrast enhancement, the yield of positive biopsy was found to be **independent of their CT morphology.** The yield was also found to be independent of the operator experience. It was found that the adherence to certain basic principles in patient and target selection would ensure a reasonable percentage of positive yield with stereotactic biopsy procedures even if the surgeon is relatively inexperienced, i.e. the **learning curve is sharp** in the performance of CT-guided stereotactic biopsies.

LIMITATIONS

Common areas of diagnostic confusion and inconclusive biopsy that decrease the yield of a stereotactic biopsy are discussed below.

1. **Firm masses**

    Sometimes small firm lesions with rubbery consistency (e.g. chronic inflammatory lesion) may resist efforts of cannula penetration and tissue retrieval, migrating through the softer brain stroma, or getting pushed away as the stereotactic probe approaches. This often results in biopsy of the peripheral gliotic brain leading to inconclusive biopsy. Such problems can also occur with ventricular masses. An open biopsy may be the only option in those cases. Firm lesions are also difficult to smear.

2. **Inflammatory masses**[7]

    Though a tumor can be usually differentiated from granulomatous lesion on a stereotactic biopsy, etiologic differentiation of chronic granulomatous lesions can often be difficult e.g., tuberculoma, cysticercous granuloma, aspergilloma, sarcoidosis etc. The difficulty arises often due small size of sample, firmness of lesion making it difficult to smear and inadequate sample for special staining and cultures. An open biopsy may be needed in some of these cases.

3. **Heterogenous masses (pineal masses)**

   Tumors such as mixed germ cell tumors and teratoma may have different components within them and biopsy of one area might not be representative of the whole tumor.

4. **Undergrading of gliomas**

   Astrocytomas are known to show areas of varying degrees of de-differentiation. Areas of very low grade malignancy and very high grade malignancy have been demonstrated in the same tumor. Therefore the biopsied area may not be representative of the entire lesion.[8-9] A glioma diagnosed as astrocytoma grade 2 on stereotactic biopsy may show aggressive behavior, by virtue of focal high grade region which was not biopsied. Therefore it is important to biopsy the enhancing portion of the tumor. Presence of necrosis on CT scan would suggest a high grade glioma, even if the biopsied specimen shows features of lower grade tumor.

On the other hand, the differentiation between a low grade astrocytoma and reactive gliosis can also be very difficult. A non-neoplastic proliferation of reactive astrocytes may occur around non-neoplastic lesions like multiple sclerosis and even neoplastic lesions like epidermoid, craniopharyngioma, malignant lymphoma. A biopsy taken from the periphery of lesion may be result in erroneous diagnosis.

COMPLICATIONS

A) Evident hemorrhage

Significant bleed from the stereotactic cannula during biopsy often gives rise to panic reaction, prompt removal of the probe and abandonment of the procedure. It is important to avoid this mistake. A small bleed can occur during biopsies, and the best course of action is to keep the cannula in place without the stillette so that intracranial blood is vented out of cannula. A patient wait will often lead to gradual decrease in the bleeding, as the bleeding point often gets sealed off by a clot. The cannula is then removed without the stillete.

A gentle irrigation with fluid might help in certain cases. In case the bleeding continues for considerable amout of time, a bipolar or monopolar coagulation can be tried by passing the probe through the stereotactic cannula. An injection of gelfoam or coagulant is the last resort in case of persistent bleeding.

B) Hidden hemorrhage

Sometimes the bleeding may not be evident at the time of procedure. However a delayed post-procedure CT scan might show intracranial hematoma at the site of biopsy. If there is a significant mass effect or patient shows clinical deterioration open surgical evacuation may be needed.

COMPLICATION AVOIDANCE

Stereotactic biopsy is usually undertaken for deep, inaccessible and eloquent masses. Understandably the biopsy related haemorrhage in these locations can be devastating. The best way to deal with this complication is by avoiding it by following simple guidelines.

a) **Choice of probe track:** Care should be taken while approaching targets in the locations like insular/sylvian region, pineal and suprasellar region. A trajectory that will avoid major vessels in these regions should be chosen.

b) **Violate only one pial plane**

c) **Avoid suspected vascular lesions**

d) **Limit number of biopsy specimens**

e) **Use narrow probe and biopsy instruments**

f) **Avoid cortical targets**

COMPLICATION RATES

A stereotactic biopsy is a relatively safe procedure and most series report mortality rate of 0-2.6 % and morbidity of 0-13%. Out of 732 patients who underwent stereotactic biopsy at CMC hospital between 1987 and 1994, procedure related major morbidity was seen in 18 patients (2.5%), while 14 (1.9%) suffered minor morbidity. The procedure related mortality was only 0.4% (3 patients).[4]

CONCLUSION

A CT guided stereotactic biopsy can offer a safe and rational management of intracranial masses. Judicious selection of cases, proper target selection, intraoperative smear feedback and following certain basic principles will result in a high diagnostic yield with minimal morbidity.

REFERENCES:

1. Rajshekhar V, Abraham J, Chandy MJ: Avoiding empiric therapy for brain masses in Indian patients using CT guided stereotaxy. *Br J Neurosurg* 1990;4:365-71.
2. Selvapandian S, Rajshekhar V, Chandy MJ, Idicula J: Predictive value of computerized tomography diagnosis of intracranial tuberculomas. *Neurosurgery* 1994;35:845-50.

3. Rajshekhar V, Chandy MJ. CT–guided stereotactic management of intracranial tuberculomas. *Br J Neurosurg* 1993;7:619-24.
4. Ranjan A, Rajshekhar V, Chandy MJ: CT–guided stereotactic surgery : Overview of 600 stereotactic procedures. *Neurol India* 1993;41:193-97.
5. Rajshekhar V, Chandy MJ: Computerized tomography-guided stereotactic surgery for brainstem masses: a risk-benefit analysis in 71 pateints. *J Neurosurg* 1995;82:976-81.
6. Chacko G, Chandi SM, Chandy MJ: Smear Diagnosis of central nervous system lesions: a critical appraisal. *Neurol India* 1998;46:115-18.
7. Ranjan A, Rajshekhar V, Joseph T, Chandy MJ, Chandi SM: Nondiagnostic CT –guided stereotactic biopsies in a series of 407 cases: influence of CT morphology and operator experience. *J Neurosurg* 1993;79:839-44.
8. Kepes JJ: Pitfalls and problems in the histopathologic evaluation of stereotactic needle biopsy specimens. *Neurosurgical Clinics of North America* Vol. 5, 1994;(1):19-33.
9. Burger PC, Kliehues P: Cytologic composition of untreated glioblastoma for evaluation of needle biopsies. *Cancer* 1989;63:2014-23.

7
Brain Tumour Experimental Trial Designs

NJ Laperriere

There exists a large number of phase II studies in the brain tumour literature for which there is no control group and these are studies that often report encouraging results for various experimental or novel treatment approaches. However, in most instances, it is usually the efffect of selecting patients with favourable prognostic factors that is most often responsible for the encouraging results described.

As an example, there were several phase II studies reporting an improvement in survival with brachytherapy as a boost to external radiation therapy for patients with malignant gliomas as compared to historical controls.[1-3] Florell et al reviewed 100 consecutive patients with malignant gliomas that had been managed at their institution with post-operative external beam radiation therapy and systemic chemotherapy on whom they had complete follow-up.[4] They visited the brachytherapy teams in San Francisco and Toronto. Approximately 30% of patients were eligible for an implant, with good concurrence between the two brachytherapy centres, and found that patients who were brachytherapy eligible had a median survival of 13.9 months as compared to 5.8 months for the patients not eligible for brachytherapy, possibly inferring that the prior results suggesting an improvement in survival with brachytherapy may have been related to favourable prognostic factors rather than the experimental therapy. Subsequently Laperriere and colleagues published a randomized study of brachytherapy in 140 with malignant gliomas.[5] There was no improvement in survival for the brachytherapy arm of the study, with a median survival of 13.8 months for the brachytherapy arm as opposed to 13.2 months ($p = 0.49$) for the non-brachytherapy arm, thereby confirming the suggestion put forth by Florell and colleagues.

In an attempt to try to account for the impact of favourable prognostic factors in a phase II study, the Radiation Therapy Oncology Group (RTOG) performed a recursive partitioning analysis on a large database of patients with malignant gliomas who had been part of prior studies by the RTOG.[6] The analysis looked at duration of symptoms, presenting neurologic abnormalities, age, tumour grade, and post-operative performance status, and this yielded 6 subgroups of patients with median survivals ranging from 4.7 to 58.6 months. This exercise clearly demonstrated how a particular collection of patients with variable prognostic factors could impact on median survival, and in most instances in the published studies of therapy in malignant gliomas, probably most often overwhelmed any possible treatment effect of an investigative therapy. However, this tool now allows investigators to compare the outcome of their cohort of patients in a phase II study to the appropriate RTOG RPA class, and allows one to potentially tease out treatment effect from prognostic factor distribution effect. This approach also reduces the need to pertorm randomized phase III studies until there is some phase II study that shows a marked treatment effect as compared to a RTOG RPA class comparison.

There are also some difficulties in the performance and interpretation of randomized phase III studies. The issues that I will explore will include the size of studies, prognostic factors and the need for stratification, methods of assessing response, treatment at progression, and the composition of control arms.

One of the major difficutlies with many randomized studies in brain tumours is that most often the trials do not have an adequate number of patients to find a small possible treatment effect.[7] Whenever an experimental approach is considered, it is then applied to a heterogeneous population of patients, which would include many patients for which the cause of treatment failure that the experimental approach was designed to address is not necessarily the cause of failure.[8] As an example, if a randomized study to address hypoxia in malignant gliomas was designed, but the reality is that hypoxia is an operational issue in only 20% of patients with the selection criteria for the randomized study, then approximately 3000 to 4000 patients would need to be randomized on the study to be associated with a 80-90% likelihood of finding a small improvement in survival of the order of 10-15%. This makes the point that investigating a heterogeneous population of patients with an experimental therapy that is focussed on only one of several possible operative mechanisms associated with treatment failure, is higly unlikely to be associated with a positive outcome in a randomized study. We need better methods of selecting patients at diagnosis for a possible investigative therapy in which the selected therapy may make a significant impact.

There are several well established prognostic factors in malignant gliomas. Age is one of the most powerful factors, and **figure 1** demonstrates the impact of age on survival on a cohort of 3298 patients with malignant gliomas in the province of Ontario in Canada during the years 1982-1994.[9] There is a stepwise decrease in survival with increasing decade of age. There are several published studies that document the powerful prognostic effect of histologic variation in malignant gliomas.[10-12] It is well established that anaplastic gliomas do significantly better than patients with glioblastoma multiforme, and that patients with anaplastic gliomas that have an oligodendroglioma component do significantly better than

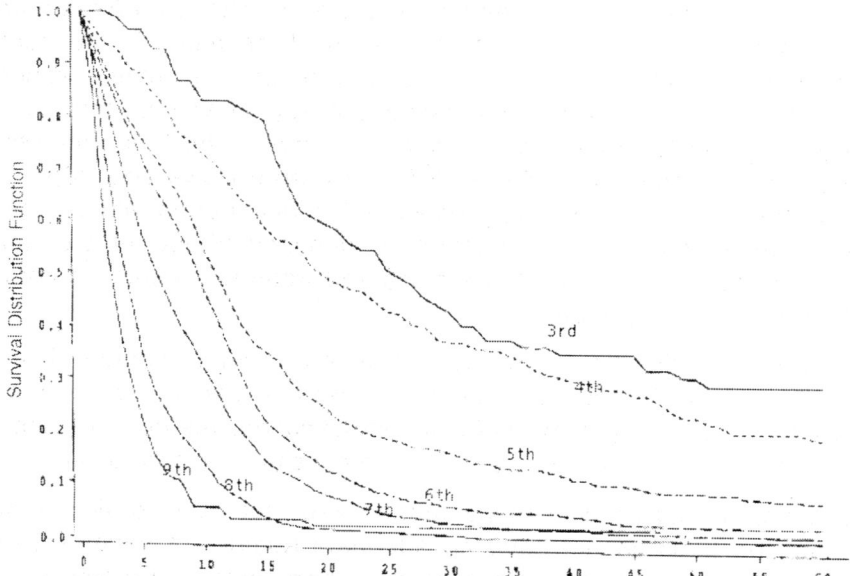

Fig. 1 : Survival in months on a cohort of 3298 patients with malignant gliomas in the province of Ontario in Canada during the years 1982-1994 stratified by decade of age.

patients without an oligodendroglioma component.[13] Nelson et al reported on the adverse effect on survival that was associated with malignant gliomas with histologic evidence of necrosis as compared to malignant gliomas that did not have necrosis.[14] Patients with anaplastic astrocytomas with an oligodendroglioma component had a median survival 7.3 years of as compared to 3 years for those without an oligodendroglioma component.[15] As well Karnofsky performance status is well known to influence survival, and it is vitally important that any randomized study in malignant gliomas must either stratify their patients for these important prognostic features, or demonstrate that these factors were in fact balanced between the treatment arms.[16] A large (474 patients) Medical Research Council randomized study of radiation dose (45 Gy/20 vs 60 Gy/30) in malignant gliomas did not stratify for age, and there was a significant imbalance in age distribution between the randomized arms.[17] They attempted to control for this imbalance by statistically accounting for the effect of age on the whole group, and then applying a corrective factor to the survival to account for this imbalance, but this problem could have been prevented by stratifying the two treatment groups by age.

Assessing response in malignant glioma studies can be associated with significant difficulties. Progression-free survival, survival, quality of life, functional status, and imaging response are some of the commonest criteria utilized in brain tumour studies, and I will discuss some of the difficulties associated with these response criteria.

What does the enhancing region on CT or MRI scans actually represent in malignant gliomas? We know that surgery, radiation, and steroids can all affect the region of enhancement.[18,19] Reduction in enhancing mass may actually represent spontaneous resolution of post-operative changes, or response to radiation or chemotherapy, or may represent an increase in steroid dosage, many of which simultaneously occur in a typical patient. Ascribing a reduction in enhancing mass to the effects of therapy may in fact be erroneous in many cases. As well, it has been demonstrated that assessment of volumes is subjective and is associated with a significant intra-observer and inter-observer variation, and so may not be reliable in many instances.[20] Several newer imaging modalities being investigated in the study of brain tumours include Position Emission Tomography (PET), Single Photon Emission Computerized Tomography (SPECT), and MRI spectroscopy.[21-28] These modalities remain in the investigative category of imaging and following patients with brain tumours, and are likely to also be associated with some qualitative and quantitative difficulties in the forseeable future. There does not appear to be any gold standard imaging modality in the immediate future that is going to resolve these difficulties in following and assessing response in patients with malignant intra-axial brain tumours.

There are several quality of life modules available in the literature with reasonably well demonstrated reliability and fidelity.[29-31] There are obvious difficulties in measuring quality of life in patients with brain tumours, particularly in patients where their higher cognitive functions have been affected either by tumour or treatment such that they are no longer able to reliably fill out the quality of life assessments. In addition, there does not exist any standard way of analysing, reporting or interpreting quality of life data, so that the reader of studies where quality of life has been measured is to a certain extent a hostage of the interpretations and choices made by the investigators. In addition, there is a well known loss of patients over time in brain tumour studies, where either as a result of death, or commonly as a result of marked deterioration of the condition of patients with progressive disease, there is a rapid fall-off of the number of assessments of quality of life in the latter half of the studies.[31]

Several standardized scales of neurologic function and performance status (Karnofsky Performance Status, ECOG) exist, but they are associated with significant inter-observer variation, and often do not discriminate sufficiently between groups of patients for small differences in function.[32]

Treatment at progression can have a significant impact on ultimate survival in studies of patients with malignant gliomas. Chemotherapy and repeat surgical excision at recurrence are therapies that can extend life, and randomized phase III studies should report the proportion of patients who underwent any additional therapy at recurrence, as an imbalance in therapy at recurrence could create the erroneous conclusion that the improvement in survival in the experimental arm of a study was related to the experimental therapy.

A current review of the literature in malignant gliomas only supports the use of post-operative radiation therapy as initial treatment for this patient population. Although adjuvant chemotherapy does not confer any significant additional benefit in most randomized studies, it likely does confer some modest additional benefit.[33] Unfortunately, many control arms in randomized studies in malignant gliomas include chemotherapy as part of initial treatment, and while the benefit of chemotherapy is at best quite small, it may mask a small benefit of the experimental therapy.

In summary, the RTOG RPA class data is a way of assessing the results of phase II studies to account for the possible impact of prognostic factors on outcome, and reduces the need for the performance of larger and more expensive randomized phase III studies. One would only likely move on to a randomized study if the phase II study showed a significant improvement in survival as compared to the appropriate RTOG RPA class. We definitely need better ways to assess tumour response, quality of life, and functional status in our brain tumour studies, and need to pay particular attention to balancing the major prognostic factors in phase III studies

REFERENCES:

1. Loeffler JS, Alexander E, III, Wen PY, Shea WM, Col.: Results of stereotactic brachytherapy used in the initial management of patients with glioblastoma. *J Natl Cancer Inst* 1990;82:1918-1921.
2. Gutin PH, Prados MD, Phillips TL, Wara WM, Larson DA, Leibel SA, et al: External irradiation followed by an interstitial high activity iodine-125 implant "boost" in the initial treatment of malignant gliomas: NCOG study 6G-82-2. *Int J Radiat Oncol Biol Phys* 1991;21(3):601-6.
3. Lucas GL, Luxton G, Cohen D, Petrovich Z, Langholz B, Apuzzo ML, et al: Treatment results of stereotactic interstitial brachytherapy for primary and metastatic brain tumors. *Int Radiat Oncol Biol Phys* 1991;21:715-721.
4. Florell RC, Macdonald DR, Irish WD, Bernstein M, Leibel SA, Gutin PH, et al: Selection bias, survival, and brachytherapy for glioma. *J Neurosurg* 1992;76(2):179-83.
5. Laperriere NJ, Leung PM, McKenzie S, Milosevic M, Wong S, Glen J, et al: Randomized study of brachytherapy in the initial management of patients with malignant astrocytoma. *Int J Radiat Oncol Biol Phys* 1998;41(5):1005-11.
6. Curran WJ, Jr., Scott CB, Horton J, Nelson JS, Weinstein AS, Fischbach AJ, et al: Recursive partitioning analysis of prognostic factors in three Radiation Therapy Oncology Group malignant glioma trials [see comments]. *J Natl Cancer Inst* 1993;85(9):704-10.
7. Moher D, Dulberg CS, Wells GA: Statistical power, sample size, and their reporting in randomized controlled trials. *Jama* 1994;272(2):122-4.
8. Dorie MJ, Brown JM: Potentiation of the anticancer effect of cisplatin by the hypoxic cytotoxin tirapazamine. In: Vaupel PW, Kelleher DK, Gunderroth M, editors. Tumor Oxygenation. Stuttgart, Germany: Fischer-Verlag; 1995; p. 125-135.
9. Paszat L, Laperriere N, Groome P, Schulze K, Mackillop W, Holowaty E: A population-based study of glioblastoma multiforme. *Int J Radiat Oncol Biol Phys* 2001;51(1):100-7.
10. Sandberg-Wollheim M, Malmstrom P, Stromblad LG, Anderson H, Borgstrom S, Brun A, et al: A randomized study of chemotherapy with procarbazine, vincristine, and lomustine with and without radiation therapy for astrocytoma grades 3 and/or 4. *Cancer* 1991;68(1):22-9.

11. Prognostic factors for high-grade malignant glioma: development of a prognostic index. A Report of the Medical Research Council Brain Tumour Working Party. *J Neurooncol* 1990;9(1):47-55.
12. Trojanowski T, Peszynski J, Turowski K, Kaminski S, Goscinski I, Reinfus M, et al: Postoperative radiotherapy and radiotherapy combined with CCNU chemotherapy for treatment of brain gliomas [see comments]. *J Neurooncol* 1988;6(3):285-91.
13. Bauman GS, Cairncross JG: Multidisciplinary management of adult anaplastic oligodendrogliomas and anaplastic mixed oligo-astrocytomas. *Semin Radiat Oncol* 2001;11(2):170-80.
14. Nelson DF, Nelson JS, Davis DR, Chang CH, Griffin TW, Pajak TF: Survival and prognosis of patients with astrocytoma with atypical or anaplastic features. *J Neurooncol* 1985;3(2):99-103.
15. Donahue B, Scott CB, Nelson JS, Rotman M, Murray KJ, Nelson DF, et al: Influence of an oligodendroglial component on the survival of patients with anaplastic astrocytomas: a report of Radiation Therapy Oncology Group 83-02. *Int J Radiat Oncol Biol Phys* 1997;38(5):911-4.
16. Bleehen NM. Studies on high grade cerebral gliomas. *Int J Radiat Oncol Biol Phys* 1990;18:811-813.
17. Bleehen NM, Stenning SP: A Medical Research Council trial of two radiotherapy doses in the treatment of grades 3 and 4 astrocytoma. The Medical Research Council Brain Tumour Working Party. *Br J Cancer* 1991;64(4):769-74.
18. Cairncross JG, Macdonald DR, Pexman JH, Ives FJ: Steroid-induced CT changes in patients with recurrent malignant glioma. *Neurology* 1988;38(5):724-6.
19. Watling CJ, Lee DH, Macdonald DR, Cairncross JG: Corticosteroid-induced magnetic resonance imaging changes in patients with recurrent malignant glioma. *J Clin Oncol* 1994;12(9):1886-9.
20. Rovaris M, Rocca MA, Sormani MP, Comi G, Filippi M: Reproducibility of brain MRI lesion volume measurements in multiple sclerosis using a local thresholding technique: effects of formal operator training. *Eur Neurol* 1999;41(4):226-30.
21. van der Hiel B, Pauwels EK, Stokkel MP: Positron emission tomography with 2-[18F]-fluoro-2-deoxy-D-glucose in oncology. Part IIIa: Therapy response monitoring in breast cancer, lymphoma and gliomas. *J Cancer Res Clin Oncol* 2001;127(5):269-77.
22. De Witte O, Goldberg I, Wikler D, Rorive S, Damhaut P, Monclus M, et al: Positron emission tomography with injection of methionine as a prognostic factor in glioma. *J Neurosurg* 2001;95(5):746-50.
23. Ohtani T, Kurihara H, Ishiuchi S, Saito N, Oriuchi N, Inoue T, et al: Brain tumour imaging with carbon-11 choline: comparison with FDG PET and gadolinium-enhanced MR imaging. *Eur J Nucl Med* 2001;28(11):1664-70.
24. Mirzaei S, Knoll P, Kohn H: Diagnosis of recurrent astrocytoma with fludeoxyglucose F18 PET scanning. *N Engl J Med* 2001;344(26):2030-1.
25. Dowling C, Bollen AW, Noworolski SM, McDermott MW, Barbaro NM, Day MR, et al: Preoperative proton MR spectroscopic imaging of brain tumors: correlation with histopathologic analysis of resection specimens. *AJNR Am J Neuroradiol* 2001;22(4):604-12.
26. Croteau D, Scarpace L, Hearshen D, Gutierrez J, Fisher JL, Rock JP, et al: Correlation between Magnetic Resonance Spectroscopy Imaging and Image-guided Biopsies: Semiquantitative and Qualitative Histopathological Analyses of Patients with Untreated Glioma. *Neurosurgery* 2001;49(4):823-9.
27. Pirzkall A, McKnight TR, Graves EE, Carol MP, Sneed PK, Wara WW, et al: MR-spectroscopy guided target delineation for high-grade gliomas. *Int J Radiat Oncol Biol Phys* 2001;50(4):915-28.
28. Matheja P, Schober O: 123I-IMT SPET: introducing another research tool into clinical neuro-oncology? *Eur J Nucl Med* 2001;28(1):1-4.
29. Aaronson NK, Ahmedzai S, Bergman B, Bullinger M, Cull A, Duez NJ, et al: The European Organization for Research and Treatment of Cancer QLQ-C30: a quality-of-life instrument for use in international clinical trials in oncology. *J Natl Cancer Inst* 1993;85(5):365-76.
30. Weitzner MA, Meyers CA, Gelke CK, Byrne KS, Cella DF, Levin VA: The Functional Assessment of Cancer Therapy (FACT) scale. Development of a brain subscale and revalidation of the general version (FACT-G) in patients with primary brain tumors. *Cancer* 1995;75(5):1151-61.

31. Bampoe J, Laperriere N, Pintilie M, Glen J, Micallef J, Bernstein M: Quality of life in patients with glioblastoma multiforme participating in a randomized study of brachytherapy as a boost treatment. *J Neurosurg* 2000;93(6):917-26.
32. Schag CC, Heinrich RL, Ganz PA: Karnofsky performance status revisited: reliability, validity, and guidelines. *J Clin Oncol* 1984;2(3):187-93.
33. Fine HA, Dear KB, Loeffler JS, Black PM, Canellos GP: Meta-analysis of radiation therapy with and without adjuvant chemotherapy for malignant gliomas in adults. *Cancer* 1993;71(8):2585-97.

8

Planning and Executing a Classical Approach to Supratentorial Gliomas

AK Singh, V Gupta, V Dua

Supratentorial gliomas constitute a significant part of the clinical burden of an average neurosurgeon. Malignant brain tumours account for nearly half of all brain tumours. In major training centres around the country, unfortunately, gliomas are often considered as a burden and are often left to the junior persons to operate upon. This attitude is the result of a combination of excessive workload in most of these centres, the bias of senior surgeons in favour of benign and vascular pathologies, and the relatively bleak eventual outcome in malignant brain tumours. An unfortunate consequence of this is that some times craniotomies are placed wrongly or are of inappropriate size. This article is primarily meant to restate basic principles of classical operative neurosurgery, with special reference to supratentorial gliomas, based on personal experiences of the senior author.

The most common form of gliomas seen in clinical practice is the glioblasoma multiforme – accounting for more than half of all supratentorial gliomas – which carries a uniformly poor prognosis despite all the recent advances in surgical and adjuvant therapies. Bucy, in 1977, observed that we are no doing no better with the surgical treatment of this type of tumours than "did our pioneer neurosurgeon ancestors one hundred years ago".[1] 24 years later the situation is essentially unchanged.

Glioblastomas were recognised as a separate entity in 1924 and the surgical treatment then consisted of turning a flap and taking a small piece for histopathology.[2] Small amount of bone was removed beneath the temporal muscle for decompressive purposes. If brain swelling prevented replacement of bone flap the latter was removed sometimes.

This often led to temporary relief in symptoms but the protracted terminal illness of the patient with a repulsive local bulge was hardly a boon to the nascent field of neurosurgery or to the patient.

McKenzie, in 1936,[3] recommended internal decompression, which consisted of removal of the tumour and the surrounding edematous brain as much as possible. The dura was closed tightly and bone flap replaced. While this could lead to an earlier recurrence of symptoms, as it often did indeed, it also led to a shortening of the terminal stages of this hopeless illness, besides having a better aesthetic effect for all concerned. This technique has remained, for the most part, the staple of neurosurgical treatment of gliomas, including glioblastomas.

Yasargil introduced the next major changes in the management of gliomas, amongst other intracranial pathologies, by introducing and popularising the use of microscope and microsurgical techniques. He was able to demonstrate the complete removal of large intrinsic masses by utilizing potential subarachnoid spaces that cover the brain, without damaging the surrounding normal brain. An increasingly larger number of neurosurgeons are emulating his techniques and rightly so as these are less traumatic with less stormy post op courses. However, the effect on the eventual outcome has not changed significantly as has been brought out in the chapter on meta-analysis in this volume.

Management of gliomas have seen many technological advances in the past decade or so, amongst them being radiosurgical techniques and the possibility of various immunological interventions. However, till date the data from around the world indicates that grossly complete surgical extirpation remains the single most effective modality for treating these lesions - both for obtaining a long term recurrence free life and for palliation.

PLANNING SURGICAL TREATMENT OF GLIOMAS:

Surgical operations are like military operations (that is why both are called operations!)- the more meticulously planned they are the less the chances of unwanted or unanticipated problems. It is necessary to remind one self that each individual patient has a set of unique problems, no matter how "routine" the radiology looks like. The following aspects are important in planning a surgical strategy, for gliomas as much as for any other lesion.

1. Defining aim (s) of surgery.
2. Precise anatomical location of the lesion.
3. Relationship of the lesion to major cortical areas, vascular channels, and to major pathways.
4. Identifying safe corridors to the lesion.

AIM(S) OF SURGERY:

a) *Tissue confirmation:* Modern radiological investigations, like MRI, MRS and SPECT are often able to predict the pathological nature of an intracranial lesion with great degree of assurance. However, there are instances, rarely to be sure, of benign lesions like tuberculomas mimicking, radiologically, a mitotic pathology. In today's world of evidence based medicine and consumer protection laws, it is necessary to obtain a definitive confirmation of the pathology before advising a costly intervention like stereotactic radiosurgery which may not have been required at all. Besides, only histopathology can help a clinician prognosticate with any degree of accuracy and thus help a patient or his family make their future plans. The tissue can be obtained for histopathological examination during an open craniotomy or by using stereotacic techniques. The present chapter concerns those lesions which are supratentorial and are often causing mass effect; in these open surgical removal is considered the best therapeutic option.

b) **Reducing mass effect:** Most of the supratentorial malignant tumours in our country still present when they have attained significant size and where the size of the lesion is often of life threatening proportions. Debulking a tumour is needed not only for improving the neurological status of the patient but also for enhancing the efficacy of any adjuvant therapy that may need to be given to the patient.

c) *Increasing recurrence free survival:* Maximum surgical removal is still the single most important determinant of the length of recurrence free survival. While it is not possible to remove all the tumor, it is often possible to remove all of the grossly visible tumor.

d) **To not add to the existing neurological deficit:** The age old Hippocratic advice to aspiring physicians to not add to the misery of the patients is never more true than in neurosurgical practice because the adverse effects, if any, of an operative intervention have the potential of robbing a patient of some of the most salient features of our humanity – a loss which often seems heavier than death itself.

DEFINING ANATOMICAL LOCATION AND RELATIONSHIPS OF THE LESION:

Diagnosis of gliomas, like all intracranial tumours, has become simpler following the advent of non invasive imaging modalities. MRI, not only delineates the exact anatomical relationships of a lesion it also indicates the pathology of the lesion with reasonable degree of certainty, specially if MRS has been added to the armamentarium. However, this same ease of diagnosis has given rise to its own problems because of some teleological differences between radiologists and neurosurgeons, as also by the fact that an increasing number of neurosurgeons are paying less attention to the radiographs and more to the report of the radiologist.

A source of major confusion is the tendency of radiologists to report that a lesion is fronto-parietal or fronto-temporal or parieto-temporal (**Figs. 1 & 2**).

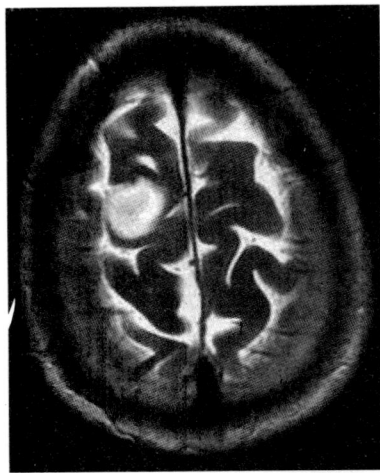

Fig. 1: reported as lesion in "high parietal" region

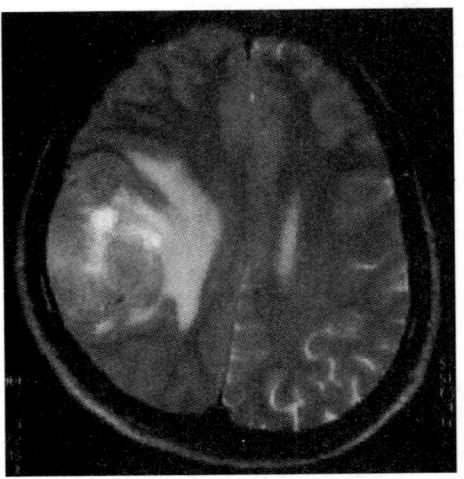

Fig. 2: Described as a "fronto-parietal" lesion

What is lost sight of by radiologists, while writing these reports, is the fact that intrinsic brain tumours by and large respect the sulcal boundaries and do not cross pial barriers. Central to safe extirpation of lesions, be they tumors or AVMs, has been the location of motor strip.

What appears as a tumour in fronto-parietal region on a CT or MRI will, therefore, mostly be a tumour that either grew in the frontal lobe and went posteriorly pushing the central sulcus, or else started in the parietal lobe and grew anteriorly pushing the central sulcus forwards. The exact site where the tumour started from would be indicated by the neurological signs; thus a tumour starting in the frontal lobe and going backwards is likely to produce significant motor deficit with minimal or no sensory deficit; on the other hand, a tumour starting in parietal lobe and growing forwards will show a preponderance of parietal lobe dysfunction signs. A tumour which truly crosses the central sulcus, i.e. the barrier between frontal and parietal lobes, will invariably have complete mono or hemiplegia on the contra-lateral side. The same rule applies to lesions near the Sylvian sulcus, except in this case the clinical signs may not be as helpful.

The first, and the most consistently reliable, system of surface delineation of major sulci were the Taylor Haughton lines, and they remain still the most widely used and important method of surface marking (**Figs. 3 & 4**). The central sulcus, as defined by the Taylor Haughton lines, is almost always angled $45°$ to the orbitomeatal line, with its origin at the pterion and termination at the vertex, 2 inches behind coronal suture (**Figs. 5a & b**).

A CT scan shows the relationship of the coronal suture, and some times the pterion, to the tumor but that is about all that can be made out on a standard CT.

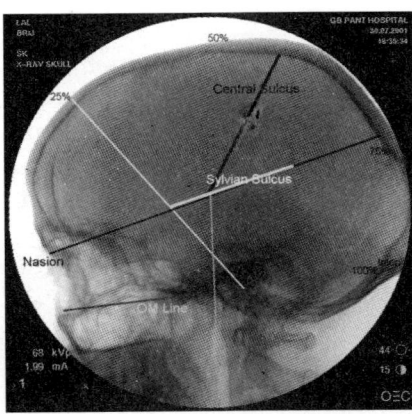 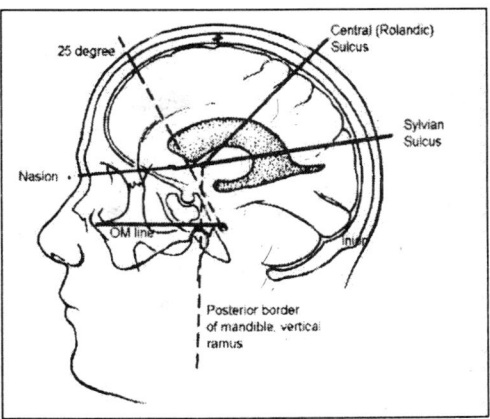

Figs. 3 & 4: Taylor Haughton Lines

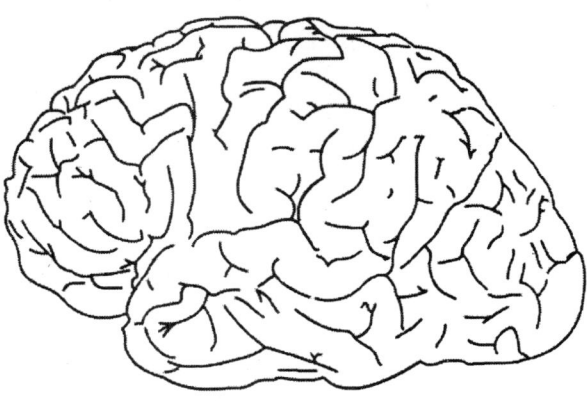 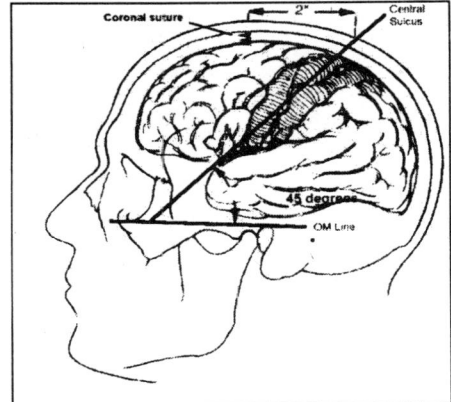

Figs. 5(a -b) Taylor Haughton Lines

For further localizing information it will be necessary to put some markers, like a metal clip, on the scalp and repeat the CT study.

An MRI, however, shows the central sulcus with a reasonable degree of accuracy **(Fig. 6)**. The central fissure is reliably located just posterior to the junction of the superior frontal sulcus and the first sulcus it abuts in the frontoparietal region. For masses >3 cm size, their relationship to the external auditory meatus, the pterion, the central sulcus, and the coronal suture is mostly sufficient. For lesions <2 cm size and with indistinct margins stereotactic localization utilizing intraop navigation may be very useful.

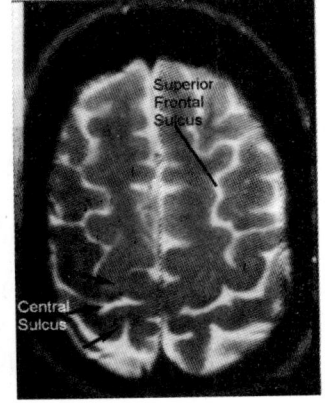

Fig. 6 : Central Sulcus

IDENTIFYING SAFE CORRIDORS:

Further more MRI, on account of its capability of providing direct multiplanar images and of providing differentiations between various structures of the brain, shows the relationships of a lesion to all the important white matter and central gray matter structures. The exact relation of the mass to the various deep basal ganglia, the internal capsule etc. is very clearly brought out by MRI (**Figs. 7 & 8**).

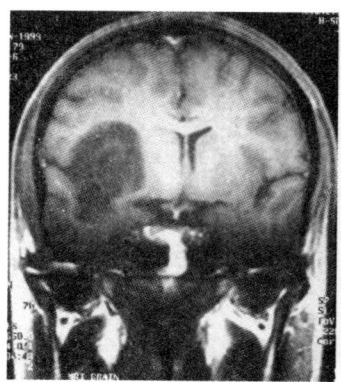
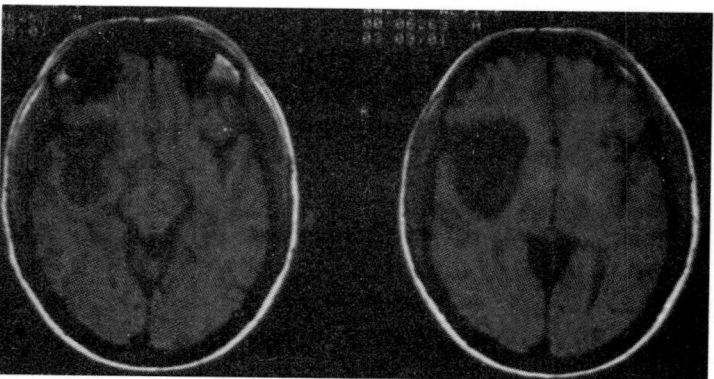

Fig. 7 Fig. 8

The information provided by the MRI coupled with knowledge of 3-D anatomy of various structures of the brain helps a neurosurgeon decide on the safe corridors that can be taken. For surface lesions, obviously, knowledge of the eloquent areas suffices but for deep seated lesions it is necessary to be aware of the various fibre tracts, specially the internal capsule, the optic radiations and other site specific pathways.

EXECUTING A SAFE OPERATION

Attention to minor details is the only way of ensuring safety to each patient that we undertake to operate upon. Dr. Dave, the senior author's venerated teacher, would never tire of repeating that there is no such thing in neurosurgery as "unimportant". After planning the type of surgery that is contemplated there are many steps one must take before actually beginning surgery.

PRE OP PREPARATION:

1. **Medication:** Other than the drugs required for pre-anaesthetic preparation, there are some important drugs that need to be given from the point of view of neurosurgeons. These include corticosteroids, whose value has often been debated and which have been abused by overusage. However, a judicious use is often helpful in preventing or limiting damage to cells during surgery. It is

adequate to start the patient on 4 mg TDS 24 hrs prior to surgery. Antiepileptics must be continued and the past dose given on the morning of surgery. If the lesion is very close to known epileptogenic areas, it is wise to give an additional dose just prior to surgery. We prefer diphenyl hydantoin 100 mg IV I 8 hrly dose in the peri operative phase. The use of prophylactic antibiotics in clean elective surgery has been a hotly contested and investigated matter. We prefer giving a shot of a broad spectrum antibiotic at the start of induction of anaesthesia. In the post op period the preferred drug combination is chloramphenicol 500mg QID and erythromycin 500 mg TDS for 7 days – we prefer playing safe given our a state of less than perfect sterility in the theatre. Chloromycetin, till date, is the drug which crosses the blood brain barrier the best. Erythromycin is preferred because it is the drug of choice for URI which is common after intubation and is a probable source of infection. If the patient is diabetic, he is converted to injectable medication well before surgery as that permits a surer control of DM. Anti-hypertensives, for hypertensive patients, are given until the morning of surgery.

2. *Hair shaving:* A number of studies, class II evidence type, have demonstrated that shaving hair does not reduce the incidence of infection. In fact if performed more than 24 hrs prior to surgery it may lead to an increased incidence of infections. However, most of the neurosurgeons, including the present authors, generally are more at ease with a shaven head as it permits them to mark out the flaps with greater accuracy. Considering the importance of correct positioning of the scalp flap, the temporary loss of hair is a bearable one. However, if one is sure of one's landmarks or the patient is keen on retaining the hair, then a thorough shampoo one day prior provides adequate safety.

3. *Counselling:* Though not strictly a matter of neurosurgical planning, it is a vital step of any therapeutic intervention that a clinician may be undertaking and unfortunately one that is often neglected. It is well to remember that a] the patient is as human as we ourselves with similar relationships, b] the concerns of a patient are genuine and central to the process of healing, and c] even the most meticulously planned and executed operations may be attended by complications. Gaining the confidence of a patient and his / her relatives by adequate and honest interaction is also the best way of preventing mal- practice suits.

Operative steps:

1. *Positioning the patient:* An improper positioning has often been the bane of many a surgeons. Generally the tumour bearing site for a surface lesion or the proposed site of entry for a deep seated lesion should be the highest point after the patient has been fixed. The exceptions being when an inter-hemispheric approach is planned or when using the semi sitting position for lesions located

in the parieto occipital regions. The head end should be raised about 30 degrees relative to the chest to aid venous drainage, ensuring that there is no kinking of neck vessels. The limbs should be comfortably positioned and supported, taking precautions that neurovascular structures are not compressed.

2. **Marking a flap:** Perhaps this is the single most important step in ensuring a safe and successful outcome. The flap, as a rule, for gliomas should be adequate bordering on being generous rather than the minimalist – unlike the situation while operating upon, say, the aneurysms. The tumour or the proposed site of entry – for deep seated lesions – should be in the center of the flap. While it is desirable that the flap should be cosmetically acceptable, i.e. as far as possible it should be inside the hairline, it is even more important that it should be adequate for the primary purpose at hand and that it should be broad enough at the base not to suffer from lack of blood supply. A flap should be based, as far as possible, on one of the larger scalp vessels and the base should be at least as broad as the dome. Different types of classical flaps for malignant gliomas are shown in **(Figs. 9-14)**.

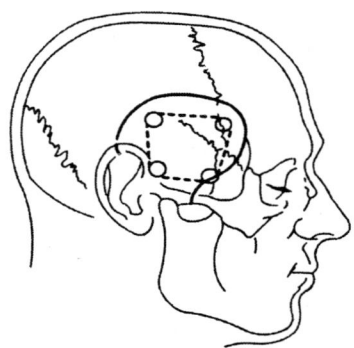

Fig. 9: Question mark incision
For mid & post temporal

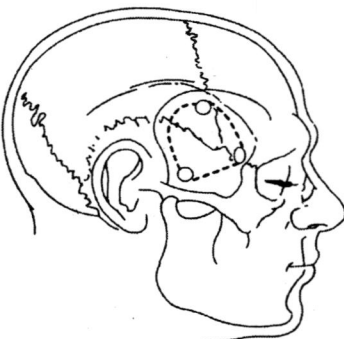

Fig. 10: Reverse question mark
For anterior temporal & frontal

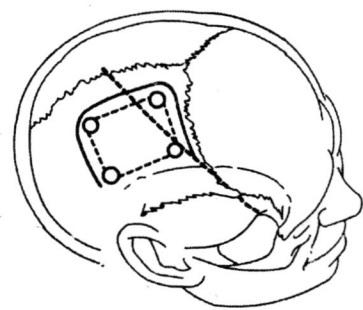

Fig. 11: For anterior parietal

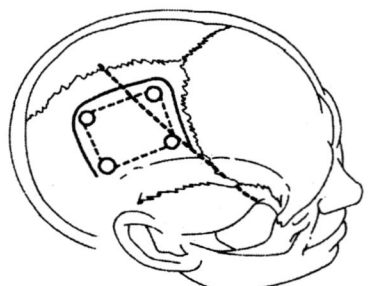

Fig. 12: For posterior frontal

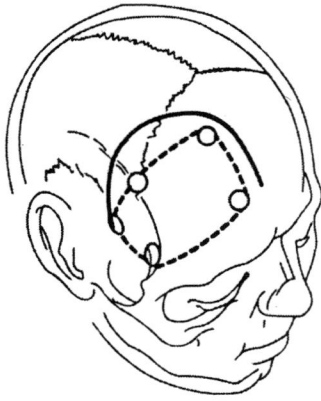 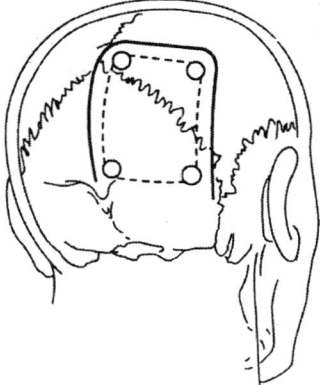

Fig. 13: For anterior and mid frontal   Fig. 14: For parieto occipital

3. **Craniotomy:** We generally infiltrate the scalp with 100 ml of a mixture of 25 ml 2% xylocaine, 75 ml N. saline, and 0.5 ml adrenaline (giving an adrenaline strength of 1:200,000). The infiltration should be in the layer above galea as that is where the blood vessels are and this can be tough as the galea is rather well attached to the connective tissue overlying it. Pressure with fingers, a minimal use of hemostats or scalp clips, and of diathermy help in reducing bleeding from the scalp as well as preventing unnecessary damage to the blood supply.

The burr-holes, or the single burr-hole if powered craniotomy is being done, should be placed as far away as possible so as to get the maximum exposure of brain. The connecting cuts in the bone should begin at the outer ends of the burr-holes again to get the maximum opening. It is a good practice to place dural hitch sutures before opening dura as there is a real danger of troublesome epidural bleeding because the brain is likely to be rather lax after the offending lesion has been adequately removed. Surgicel or, preferably spongestan, can be put between dura and skull at sites where there is troublesome bleeding. The senior author does not use preop mannitol in surface gliomas as a matter of rule for it can give a false impression of the adequacy of decompression besides being un-necessary as the lesion is superficial and surrounding edematous brain is also to be sacrificed anyway. However, if the dura is tense or if the lesion is deeply located per op mannitol is given in the does of 1 – 1.5 Gm / kg body weight. The senior author prefers to open the dura using a # 11 knife and a dura forceps. Some surgeons prefer to use a hitch suture to elevate the dura and then make a small slit in the latter through which a narrow cottonoid is pushed and dural opening enlarged using a fine Matzenbaum scissor. If the brain is tense, one can also pass a Penfield dissector # 3 under the dura and then slit the dura on the dissector using the knife. The dural flap should, unlike the bone flap, be just adequate for reaching the tumour. Unopened dura provides the best safety to the underlying brain. The flap should be based towards a sinus if one is close by or on a large vessel. It is necessary to keep all the exposed tissue well covered with moistened cotton and to frequently irrigate the opened brain and surrounding tissue. Dessicated and coagulated tissue is dead tissue which can act as nidus for infection. Movement of personnel in the theatre leads to dust particles being carried on the eddies so created and these then can settle down on the exposed tissue leading to infection later on. Frequent irrigation washes these away from the wound.

A surface lesion will now be visible and corticectomy all around the visible edges using diathermy and sharp dissection should be done before the tumor is debulked. A lesion just under the surface will be apparent by flattening of the overlying gyrus, and by a change in the vascularity - usually a paucity of vessels leading to paleness of the overlying cortex. If not so apparent, then a gentle palpation using a moistened finger may provide clues. A glioblastoma is usually softer – leading to a "dimpling" effect, whereas a lower grade glioma or an oligodendrogloma will give a firmer feeling under the finger. If still in doubt it is a good practice to pass in a ventricular needle after making a small corticectomy. The change in resistance will indicate the presence of the tumour. The availability of intra operative ultrasound, CT or even MRI in some centers have made this part of the job more accurate and easier though costlier. A navigational system makes it even more easy, though again it adds to the cost significantly. At any rate for lesions larger than 2 cms, if the craniotomy has been placed correctly there should be no problem in locating a mass lesion. After the lesion has been locate vis-à-vis the dural opening, further approach is determined by whether the lesion is on surface, under a gyrus, in the depth of a sulcus, or deeper still, and by its proximity to eloquent areas of the brain. For lesions in the non eloquent areas and on surface or under a gyrus, it is appropriate and adequate to remove the overlying cortex, and debulk the tumour from inside out as fast as possible. Mostly, the worrisome bleeding from inside the tumour stops almost mysteriously after the tumor is adequately removed. For lesions in the sulci, or for smaller lesions, it is preferable to work around them so as to preserve the transit vessels – described by Yasargil - which might be supplying normal brain. If available, ultrasonic cavitatining aspirator is useful, specially when the tumour is not soft in consistency. Haemostasis should be achieved with the help of bipolar cautery as far as possible and use of films of cellulose(Surgicel) should be restricted to minimum. Senior author has employed the valsalva manuvore after haemostasis for many years now with satisfactory outcome. Our practise is to obtain postoperative CT scans both plain and contrast next post op day to document the extent of resection.

Patient is usually kept on oral or intravenous dehydrants for next forty eight hours. Sutures are removed on seventh post op day.

REFERENCES:

1. Kahn AE, Krosby EC: Gliomas of cerebral hemispheres. In Correlative Neurosurgery, Third edition Schneider RC, Kahn EA, Crosby EC, Taren JA (Ed) Charles C Thomas Publishers Springfield USA 1982.
2. Betty MJ: Quality of survival in treated patients with supratentorial gliomata, *J Neurol Neurosurg Psychiatry* 1964:27:556-61.
3. Bucy PC: Intracranial tumor: Where we have been and where are we going. Clin Neuro Surg 1977:305-09.
4. Mckenzei KG: Glioblastoma: A point of view concerning treatment. *Arch Neurol Psychait* 1936;36:542-46.
5. Salcman M: Intrinsic cerebral Neoplasms In Brain Surgery complication avoidance and management Ed Appuzzu MJ Churchill Livingstone New York 1993.

9

What is Adequate Surgery for Glioma?– The Rationale for Cytoreduction

SS Praharaj, S Dubey, KVR Sastry

The history of glioma surgery is replete with periods of great enthusiasm for its surgical removal that have alternated with periods of conservatism. Most controversies were in the past related to technical limitations of operating deep-seated gliomas and those in the eloquent cortex. Many of these have now become a secondary issue, with the advent of microscope, MR imaging, neuronavigation, various intraoperative monitoring devices, and etc. However controversy exists regarding the extent of surgery and the length and quality of survival. Most of the literature is retrospective in nature, and contaminated by the effects of other treatments such as radiotherapy and chemotherapy. Also, the extent of tumour removal is based almost exclusively upon the surgeon's intraoperative impression. The results of multicentre trials have been obscured by the variability of surgery offered in different institutions ranging from biopsy to total resections. Against the backdrop of these limitations we have tried to analyse the role of cytoreductive surgery in the management of low grade and malignant gliomas.

LOW GRADE GLIOMAS

The management of patients with low grade gliomas continues to be controversial, especially pertaining to the timing of surgical intervention, the extent of tumor resection and the need for adjuvant therapy. The timing of surgery varies in literature from immediate surgery to observation and delayed surgery to conservative management.[1,2,3] The extent of surgical resection is also controversial. No prospective randomized clinical studies are available to analyze the impact of extent of tumor resection on recurrence patterns and survival rates. The assessment of extent of resection is based on the surgeon's impression, which is not quantitative and often not correlated with postoperative imaging studies. In light of the above the pros and cons of surgical versus conservative management of low grade gliomas is discussed.

EARLY SURGERY FOR LOW-GRADE GLIOMAS

Low-grade gliomas include a number of tumors of different histological types. CT scan and MRI may not give accurate pathological diagnosis. McDermott et al reviewed contrast enhanced CT scan of gliomas and compared them with their labeling index (LI) as measured by BudR labeling.[15] In anaplastic group, 28.6% had no enhancement on CT scan. In the low grade glioma group 36.1% showed contrast enhancement. This showed that CT scan error was high for both low-grade gliomas and anaplastic astrocytomas. Histopathology also did not always correlate with the biological behaviour of the tumors. In patients with low-grade gliomas on histopathology, BudR was high (>1%) in 39%. These had earlier progression and tended to die sooner, and suggested that additional aggressive therapy may be required in these cases. Ito et al[16] in 87 patients with low-grade glioma on histopathology did LI on them and found that LI ranged from <1% to 9.3% (mean 1.3%). The extent of surgery and LI were both predictive of survival in all patients. Knowing the LI, the neurosurgeon is better equipped to talk to the patient regarding the need for radiotherapy and chemotherapy. This is only possible to determine after surgery with tissue sampling. Kondziolka et al in 20 patients with suspected low-grade glioma on clinical features and MRI, found that only 10 had low-grade glioma on histopathology, 9 had anaplastic astrocytoma and 1 had encephalitis.[17] Thus nearly 50% were wrongly diagnosed. These make a strong case for surgery with tissue sampling.

How much to remove has been a controversial subject. Several retrospective studies have suggested that survival in patients with low-grade gliomas may be improved when extensive surgical resection is carried out.[2, 7, 8, 9, 10] However the extent of resection was determined by the surgeon's operative impression or based on postoperative CT scans. To eliminate bias, Berger et al used MRI volumetric image analysis on pre- and postoperative scans.[18] They found that 20% of their low-grade gliomas had recurrence. Most recurrent tumors showed a higher histological grade. The time it took to develop a recurrence and the histology at that time was critically dependant upon the amount of tumors present before and after surgery. Thus the extent of resection was an important factor in predicting recurrence and histology. Patients who had a preoperative tumor volume of less than 10 cm^3 never developed a recurrence when followed to the end of the analysis, about 50 months, while those with initial tumor volume of 30 cm^3 and greater had a recurrence within 30 months of surgery and all were of malignant histological type.[13] 13 to 85% of low-grade gliomas recur at a higher histological grade.[14]

CASE FOR CONSERVATIVE MANAGEMENT

There exists no dispute regarding early surgery in low-grade gliomas, which present with raised intracranial pressure, progressive neurological deficits or intractable seizures, or those situated in the polar regions. Diffuse gliomas of the cerebral hemispheres and in the eloquent areas, presenting not with raised intracranial pressures but with a single seizure, which is controlled on anticonvulsant medication, constitutes a controversial area.[4,5] These are usually fibrillary astrocytomas, oligodendrogliomas or mixed gliomas. On contrast CT and MRI, they are non-enhancing. Surgery in these instances are not curative, and may be associated with unacceptable morbidity, in spite of sophisticated imaging and neurophysiological monitoring. Most studies on the treatment of low-grade gliomas, all of which are retrospective, have concluded that time to tumors progression and survival are prolonged by gross total excision and in case of incomplete resection, by postoperative irradiation.[2,7,8,9,10] However, retrospective analyses do not have the scientific validity as those obtained by randomized studies or even case controlled trials.[11] Moreover, the indolent nature of these tumors makes it difficult to evaluate the benefits of any treatment and also leaves ample time for the toxic effects of treatment (such as radiotherapy and chemotherapy) to become manifest. Stereotactic biopsy has less morbidity (less than 5%) than extensive resection and gives the diagnosis, but has a 5 to 20% misdiagnosis rate.[6] While it can be argued that Stereotactic biopsy may permit identification of non-enhancing anaplastic tumors that were thought to be low grade on CT scan or MRI, careful clinical and radiological follow up of such patients can identify those with an accelerated course (which are anaplastic) and require aggressive treatment. One important factor quoted in favor of early total resection is the potential of these low-grade tumors to recur at a higher histological grade.[12,13] However the percentage in which it occurs varies from 15 to 80% in literature. Recht and colleagues[19] studied two groups of patients with low-grade gliomas: one (biopsy proven) with early surgery and radiotherapy and the other with suspected low-grade glioma followed clinically and radio logically, delaying surgery to a later date when and if deterioration occurs. The quality and length of life, rate of malignant conversion were identical in the two groups. But the latter group requires good follow up and surgery should be undertaken before irreversible deficits start.

Thus as a neurosurgeon it is necessary to present to the patient family the data as they are presently known. One must honestly evaluate one's own chances of carrying out a successful gross total resection and the risks involved in his hands before offering a gross total resection or conservative management.

HIGH GRADE GLIOMAS

The prognosis of patients with glioblastoma has improved very little, even with technological advances that have improved the odds for patients with other tumors types.[20,21] Adjuvant radiation therapy, either alone or in combination with chemotherapy, can improve survival[22,23] but the role of surgery as a determinant of outcome has not been established.[20,21] Several series have shown an association between extensive resection and longer survival in malignant gliomas,[22,23,24] whereas other studies show no benefit whatever from radical resection.[25,26] The pros and cons for radical resection and conservative approaches are discussed below:

CASE FOR RADICAL RESECTION OF HIGH GRADE GLIOMAS.

Several studies have shown that even patients with good prognostic features do not achieve long-term survival in large numbers if aggressive resection is not under taken.[22, 23, 24] This is very true of tumors outside the CNS where survival is directly related to the feasibility of wide excision. This has also been shown to be the case in a variety of benign and malignant tumors of the brain such as medulloblastomas, low grade gliomas, meningiomas, etc.,[27, 28] and also of metastatic tumors.[29] It is logical to expect therefore, that extensive resection would lengthen the survival of patients with high grade gliomas too. This has been borne out by various prospective and retrospective trials also.[30, 31] The extent of resection has been found to correlate with the quality of survival, amelioration of neurological deficits,[34] fewer postoperative complications and shorter stay in the intensive care.[34, 35] The amount of residual tumor in postoperative images has also been shown to correlate with length of survival in various prospective and retrospective studies.[32, 33] Recently, in works based on volumetric analysis using computerized image analysis technique,[36] it has been shown that patients who underwent extensive resection had a significantly longer time to tumors progression and survival.

Radical resection has several benefits other than alleviation of mass effect, such as removal of radioresistant cells, breakdown of blood brain barrier enabling chemotherapeutic agents to reach the site, a decrease in the tumors burden of the immunocompromised host, and a redistribution of noncycling cells into a more metabolically active state by reduction in population, pressure, etc.[37, 38] It also potentiates the effects of adjunctive therapies such as radiation, chemotherapy and immunotherapy.[39]

CASE FOR CONSERVATIVE APPROACHES FOR GLIOBLASTOMAS

For glioblastomas which are asymptomatic, do not have significant mass effect clinically and radiologically, are deep seated or in the eloquent areas in the dominant hemispheres, surgery may not be recommended especially if high risk of significant morbidity is likely. Intraoperative brain mapping, computer based stereotactic surgery with volumetric methods have been advocated for such tumors. However, there is little if any data comparing morbidity after conventional surgery to tumors resection with such procedures.[40] Since the latter also require an open route to the lesions it stands to reason that there would be attendant morbidity in tumors in eloquent areas. Most studies in glioblastomas emphasise on survival statistics without examining the short- or long-term patient function and the quality of life which is now assuming more importance.

Gross total excision in glioblastoma has a very different meaning since the tumors cells infiltrate far beyond the visible tumors margins. Kelly[41] suggested that less than 50% of actual tumors volume may be delineated on the contrast CT scan and MRI. Gross infiltration of the brain tissue by high grade gliomas is also suggested by the fact that a lobectomy has not been clearly shown to increase a patient's survival in comparison to more limited techniques.[40] There is no agreement in literature as to whether cytoreductive surgery helps to increase survival.[26, 40]

In patients who are symptomatic because of the tumors, surgery may be indicated to relieve the raised intracranial pressure. But the temporary benefits of such procedure must be weighed against the patient's age, Karnofsky's scale and the functional status, the location of the lesion, and also the preoperative response of the patient to steroids and antioedema measures. Adjuvant radiotherapy and chemotherapy would not be indicated in patients who continue to remain in poor Karnofsky's scale (of less than 70).

CELL KINETICS OF GLIOMAS

There are two main pools of tumor cells in Gliomas, like in every tumor. The proliferating pool, where the tumor cells are dividing with a certain cell cycle time and the non-proliferating pool when the time cells are resting. The ratio of the cells in the proliferating pool to that of the total tumor cell population is called the growth fraction. There is also cell loss in tumor tissue. The rate of cell loss is represented as a fraction of the rate at which cells are added to the total tumor cell population by mitosis. This is called the cell loss factor.[42] There is a constant movement of cells from the non-proliferating pool to proliferating pool in malignant tumors.

In Glioblastoma the cells cycle time is about 2-3 days and the growth fraction in the viable part is 30-40%. The cell loss factor is about 80-85%.

In Astrocytomas, the cell cycle time is longer and growth fraction markedly low.[43]

Presume that the symptoms due to a tumor appear when it is 10 gram (app. 2.72 cms diameter on CT). The life becomes, threatened, when it reaches – 100 gram (approximate 5.85 cms diameter on CT).[44]

With surgery alone, in most cases, only 50% of the tumor is removed (0.5 log/cell). If 90% of the tumor is removed it achieves 1 log kill and 2 log kill, if the tumor resection is 99%. The tumor will regain its original size in 8-11 weeks even after surgery and radiation therapy. Hence, adjuvant therapy should be administered repeatedly within this period.

In our own experience, it was found that it is not possible always to diagnose the grade of the tumor on CT morphology or duration of the disease. In a retrospective study of 172 patients with gliomas, 18 out of 26 Astrocytoma grade I and II, and 7 out of 50 Astrocytoma IV and Glioblastoma were hypodense on CT pre operatively **(Table 1)**. There was evidence of calcification in 9 out of 26 grade I and II astrocytoma and 5 out of 50 grade IV astrocytoma and GBM. Variegated appearance was seen on CT scan in 2/26 grade I and II astrocytomas and 34/50 grade IV astrocytomas and GBM.

Table 1 : CT Morphology

	Astro I&II G.gliomas n = 26	Astro III n = 38	Astro IV & GBM n = 50	Oligo & Mixed n = 38	M. Oligo & Mixed n = 20
Ill defined	10	10	4	7	2
Hypodense	18	15	7	14	3
Isodense	5	12	8	12	11
Calcification	9	11	5	14	3
Enhancement	5	15	25	21	10
Edema	7	22	36	10	14
Variegated	2	18	34	10	14

Similarly, 12 out of 26 grade I and II, and 12 out of 50 Grade IV and GBM presented after more than 2 years of symptoms. Our policy is to achieve as much as possible, cyto reduction at surgery. When we analyzed the time of recurrence of tumor in 57 patients, with grade IV astrocytoma and GBM, who had post operative CT documentation of extent of tumor resection at surgery and post op radiotherapy, the tumor recurred, after a significantly longer period in patients in whom the tumor was "totally" resected **(Fig. 1)**. This

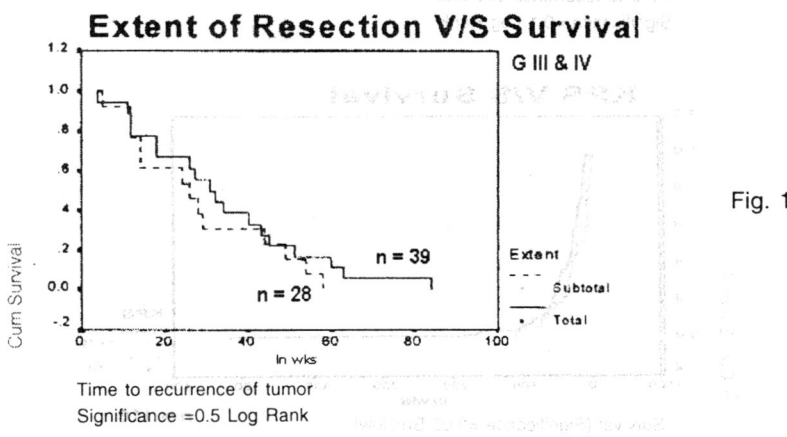

Fig. 1

Time to recurrence of tumor
Significance =0.5 Log Rank

is similar to Lacroix's study.[45] Similarly, those patients who had chemotherapy at the second surgery, for recurrence of the tumor, had longer survival, compared to these who did not have chemotherapy post operatively **(Fig. 2)**. The other factor which had influence of survival where patients age and KPS **(Figs. 3 & 4)**.

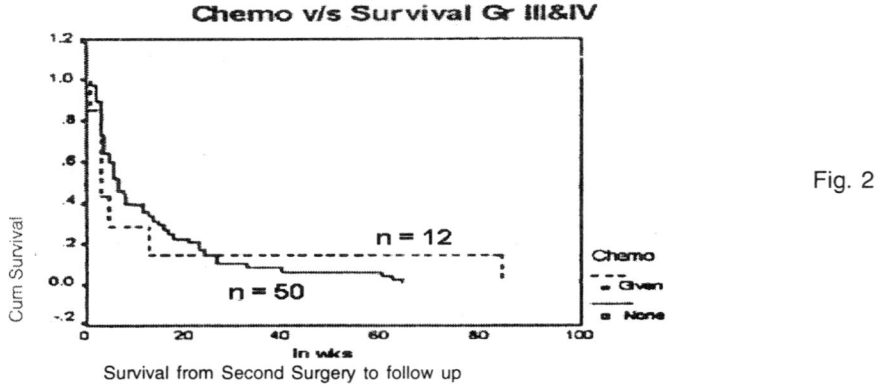

Fig. 2

Results

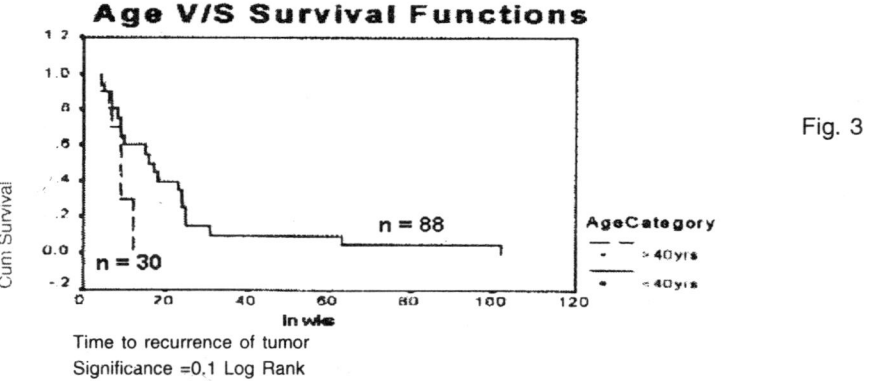

Fig. 3

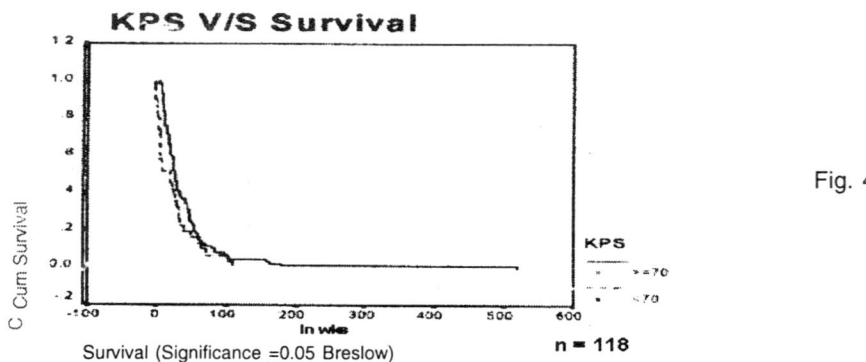

Fig. 4

References:

1. Guthrie BL, Laws ER: Supratentorial low-grade gliomas. *Neurosurg Clin N Am* 1990;1:37-48.
2. Laws ER, Taylor WF, Clifton MB, et al: Neurosurgical management of low grade astrocytomas of the cerebral hemispheres. *J Neurosurg* 1989;61:665-73.
3. Hirsch JR, Rose CS, Pierre Khan A, et al: Benign astrocytic and oligodendrocytic tumors of cerebral hemispheres in children *J Neurosurg* 1989;70:568-72.
4. McDermott MW, Krouwer HG, Asai A, Ito S, Hoshino T, Prados MD: A comparison of CT contrast enhancement and BUDR labeling indices in moderately and highly anaplastic astrocytomas of the cerebral hemispheres. *Can J Neurol Sci* 1992;19(1):34-39.
5. Ito J, Tanaka R, Kato T: Glia maturation factor (GMF) and the malignant proliferation of glioma cells. *Tanpakushitsu Kakusan Koso.* 1984;29(14):1955-63.
6. Kondziolka D, Lunsford LD: The role of stereotactic biopsy in the management of gliomas. *J Neurooncol* 1999;42(3):205-13.
7. Fazekas JT: Treatment of grade I and II brain astrocytomas: The role of radiotherapy. *Int J Radiat Oncol Biol Phys* 1977;2:661-66.
8. Gol A: The relatively benign astrocytomas of the cerebrum. *J Neurosurg* 1961;18:501-06.
9. Leibel SA, Sheline GE, Wara WM, et al: The role of radiation therapy in the treatment of astrocytomas. *Cancer* 1975;31:1551-57.
10. McCormack BM, Miller DC, Budzilovich GN, et al: Treatment and survival of low-grade astrocytomas in adults – 1977-1988. *Neurosurgery* 1992;31:636-42.
11. Berger MS, Rostomily RC: Low grade gliomas: functional mapping resection strategies, extent of resection and outcome. *J Neurooncol* 1997;34(1):85-101.
12. Berger MS, Ojemann GA: Techniques of functional localization during removal of tumors involving the cerebral hemispheres. In: Traynelis VC, Loftus CM (eds): Intraoperative monitoring techniques in Neurosurgery. New York. McGraw Hill 1994;113-17.
13. Afra D, Muller W, Benoist G, et al: Supratentorial recurrences of gliomas, results of reoperation of astrocytomas and oligodendrogliomas. *Acta Neurochir (Wein)* 1978;43:217-27.
14. Piepmeir JM: Observations cn the current treatment of low–grade astrocytic tumors of the cerebral hemispheres. *J Neurosurg* 1987;67:177-81.
15. Recht LD, Lew R, Smith TW: Suspected low-grade glioma: Is deferring treatment safe? *Ann Neurol* 1992;31:431-36.
16. Cairncross JG, Laperriere NJ: Low-grade gliomas: To treat or not to treat? (A reply). *Arch Neurol* 1990;47:1139-40.
17. Chandrasoma PT, Smith MM, Apuzzo MLJ: Stereotactic biopsy in the diagnosis of brain masses: Comparison of results of biopsy and resected surgical specimens. *Neurosurgery* 1989;24:160-65.
18. Laws ER, Taylor WF, Bergstralh: The neurosurgical management of low-grade astrocytoma. *Clin Neurosurg* 1985;33:575-88.
19. Recht LD, Lew R, Smith TW: Suspected low-grade glioma: is deferring treatment safe? *Ann Neurol* 1992;31(4):431-36.
20. Nazzaro JM, Neuwelt EA: The role of surgery in the management of suptratentorial intermediate and high grade astrocytomas in adults. *J Neurosurg* 1990;73:331-44.
21. Salcman M: Supratentorial gliomas: Clinical features and surgical therapy. In Wilkins RH (ed): Neurosurgery, Vol. I. New York, McGraw Hill, 1985;579-90.
22. Walker MD, Alexander E Jr, Hunt WE, et al: BCNU and/or radiation in the treatment of anaplastic gliomas. A cooperative clinical trial. *J Neurosurg* 1978;49:333-43.
23. Chang CH, Horton J, Schonfeld D, et al: Comparison of postoperative radiotherapy and combined postoperative radiotherapy and chemotherapy in the multidisciplinary management of malignant gliomas: A joint Radiation Therapy Oncology Group and Eastern Cooperative Oncology Group Study. *Cancer* 1983;52:997-1007.

24. Davis L, Martin J, Goldstein SL: A study of 211 patients with verified glioblastoma multiforme. *J Neurosurg* 1949;6:33-44.
25. Huber A, Beran H, Becherer A, et al: Supratentorial gliomas: Analysis of clinical and temporal parameters in 163 cases. *Neurochirurgia* 1993;36:189-93.
26. Kreth FW, Warnke PC, Scheremet R, et al: Surgical resection and radiotherapy versus biopsy and radiation therapy in the treatment of glioblastoma multiforme. *J Neurosurg* 1993;78:762-66.
27. Park TS, Hoffman HJ, Hendrick EB, et al: Medulloblastoma: Clinical presentation and management: Experience at the hospital for sick children, Toronto, 1950-1980. *J Neurosurg* 1983;58:543-52.
28. Adegbite AB, Khan MI, Paine KWE, et al: The recurrence of intracranial meningioma after surgical resection. *J Neurosurgery* 1983;58:51-56.
29. Patchell RA, Tibbs PA, Walsh JW, et al: A randomized trial of surgery in the treatment of single metastasis to the brain. *New Eng J Med* 1990;322:494-99.
30. Cohadon F, Aouad N, Rougier A, et al: Histologic and nonhistologic factors correlated with survival time in supratentorial astrocytic tumours. *J Neuro-Oncol* 1985; 3:105-11.
31. Nelson DF, Nelson JS, Davis DR, et al: Survival and prognosis of patients with astrocytoma with atypical and anaplastic features. *J Neuro-Oncol* 1985;3:99-103.
32. Ammirati M, Vick N, Liao Y, et al: Effect of the extent of surgical resection on survival and quality of life in patients with supratentorial glioblastomas and anaplastic astrocytomas. *Neurosurgery* 1987; 21:201-06.
33. Fadul C, Wood J, Thaler H, et al: Morbidity and mortality of craniotomy for excision of supratentorial gliomas. *Neurology* 1988;38:1374-79.
34. Ciric I, Ammirati M, Vick N, et al: Supratentorial gliomas: Surgical considerations and immediate postoperative results. Gross total resection versus partial resection. *Neurosurgery* 1987;21:21-26.
35. Wood JR, Green SB, Shapiro WR: The prognostic importance of tumour size in malignant gliomas: A computed tomographic scan study by the Brain Tumour Cooperative Group. *J Clin Oncol* 1988; 6:338-43.
36. Rostomily RC, Spence AM, Duong D, et al: Multimodality management of recurrent adult malignant gliomas: Results of a phase II multiagent chemotherapy study and analysis of cytoreductive surgery. *Neurosurgery* 1994;35:378-88.
37. Hoshino T: A commentary on the biology and growth kinetics of low-grade and high-grade gliomas. *J Neurosurg* 1984;61:895-900.
38. Salcman M: Surgical decision making for malignant brain tumours. *Clin Neurosurg* 1989;35:285-313.
39. Tel E, Hoshino T, Barker M, et al: Effect of surgery on BCNU chemotherapy in a rat brain tumor model. *J Neurosurg* 1980:52:529-32.
40. Nazzaro JM, Neuwet EA: The role of surgery in the management of supratentorial intermediate and high-grade astrocytomas in adults. *J Neurosurg* 1990;73:331-44.
41. Kelly PJ: Sterotactic technology in tumor surgery. *Chir Neurosurgery* 1987;35:315-253.
42. Sano K: Integrative treatment of gliomas. *Clinical Neurosurgery* 1982;30:93-124.
43. Hoshino T, Barker M, Wilson CB, Boldrey EB, Fewer D: Cell kinetics of human gliomas. *J Neurosurg* 1972;37:15-26.
44. Hoshino T, Wilson CB, Rosenblum ML, Barker M: Chemotherapeutic implications of growth fraction and cell cycle time in glioblastomas. *J Neurosurg* 1975;43:127-35.
45. Lacroix M, Abi-Said D, Fourney DR, et al: A multivariate analysis of 416 patients with glioblastoma multiforme: prognosis, extent of resection and survival. *J Neurosurg* 2001;95:190-98.

10
Surface vs Deep seated Lesion different?/ Same Surgical Strategies

AK Reddy, P Tripathy, M Panigrahi

INTRODUCTION

Neuro oncology has come a long way since the first brain tumor operation done by Rickman Godlee in 1884. Both diagnosis and treatment have improved substantially. Many tumors, like deep seated lesions that were once considered un resectable are now excised with the help of advanced neurosurgical techniques with less morbidity and mortality.[33]

The primary goal in the management of a patient with a glioma is not only to prolong survival but also to protect neurological function to make longer survival worthwhile. The role of aggressive surgery in achieving these goals is highly dependent on a number of factors including the histology of tumor, the location and margin of the tumor, presence of mass effect, the functional capacity of the patient and ultimately, the anticipated impact that treatment will have on the quality of life. Because this constellation of variables are different for each patient, treatment strategies must be modified to match the patients need within the boundaries of medical care.[23]

HISTORY

In early part of last century brain tumors were diagnosed either by pneumo-encephalogram introduced by Walter Dandy in 1918 or by Arteriography introduced by Antonio Egas Moniz in 1931. Sir G.N. Hounsfield in 1967 introduced the first functional CT scan which gave major break through in diagnosing brain tumor. A decade later Moorey & Hinshaw in 1979 introduced the first MRI which could delineate the tumor tissue characteristic in addition to localizing it. In subsequent years MR angiogram, MR Spectroscopy and functional MRI were introduced, which further helped the clinician in identifying the histological grading of the tumor, tumor surrounding, tumor vasculature etc. Similarly positron emission tomography predict the metabolism of tumor.[33]

Using microscope by Yasargil in 1957, Cavitron ultrasonic aspirator by Epstein in

1983, laser by Tew in 1983, and introduction of intra operative endoscopy to brain surgery by Oppel in 1987 made the surgery more comfortable with less tissue injury.[33]

Recently the concept of cortical brain mapping, neuro navigational technique and intra operative ultra sonogram have aided the operating surgeon in resecting the tumor maximally in a safer corridor with less morbidity.[34]

Inspite of all these advancement in neuro imageology, neurosurgical techniques and intra operative monitoring devices the goal of achieving complete cure in glioma is not achieved.

AIMS OF TREATMENT

Like any other brain tumor, achieving complete excision without morbidity remain the first and foremost aim of treating glioma. Prolonging the survival with preserved functional abilities makes the survival worthwhile. Treatment strategies should be such that it should prevent or delay recurrence of glioma.

CURRENT MANAGEMENT STRATEGIES OF GLIOMA

The therapeutic modalities available currently to treat gliomas are surgery, radiotherapy, chemotherapy and other adjuvant therapy.[33] The primary indication for performing surgery on a suspected glioma are diagnosis, decompression or cytoreduction and whenever possible total excision. An accurate diagnosis is of course the initial step. Along with confirming the actuality of an existing glial tumor, the surgeon also must determine on the basis of tissue samples the tumor grade and consistency of it. When the tumor because of size, swelling or hydrocephalus, produces a significant neurological deficit and the degree of oedema around the tumor is too severe to treat with steroids alone, decompression of the tumor is performed. The goals are to lower intracranial pressure, reverse the patients symptoms and prevent new deficits. Cytoreduction, a rather controversial concept, appears to reduce the tumor burden and increase the effectiveness of adjuvant therapies by decreasing the number of tumor cells that must be treated, altering the cell kinetics, removing radio resistant hypoxic cells, and removing areas of the tumor inaccessible to chemotherapy. When possible a long disease free interval or cure after complete excision is the goal of surgery.[34]

FACTORS THAT ALTER THE GOAL

Several factors alter the goal of management of glioma.

They are:

1. Age
2. Neurological status
3. Tumor characteristics:

a) Location – Surface vs deep seated
b) Size – Small vs large
c) Margin – circumscribed vs diffuse
d) Histological grade – low vs high
e) Consistency – cystic vs solid.

Jules M Nazzaro et al 1990 opined that age of the patient, neurological status and tumor characteristics are the major factors that alter the surgical outcome. According to the same authors tumor location may be defined in multiple ways. Here we have concentrated on surface vs deep seated glioma.

SURFACE GLIOMA

Gliomas that are located in cerebral hemispheres, lobar in nature, close to surface, can be in eloquent or in non eloquent area.

DEEP SEATED GLIOMA

Gliomas that are located deep in cerebral hemisphere, sub cortical, and central in position. Included in this group are gliomas involving thalamus, basal ganglia etc. Thalamic tumors constitute 1 to 5% of all intracranial tumor. Michael E. Cohen et al 1997.

DISTRIBUTION

Overall the site of origin of the tumors was found to be 92% in surface and 5% in deep seated, A. Peraud et al 1999.

SURGICAL STRATEGIES

Surface glioma

The surgical strategies for surface gliomas are planned depending in whether it is small or big in size, circumscribed or diffuse, in eloquent or non eloquent area and low or high grade. Technical refinements in surgery with subpial dissection or transsulcal approaches with brain mapping and awake craniotomy for eloquent areas has helped in reducing morbidity. Preoperative functional MRI has helped in surgical planning of eloquent area glioma. The use of operating microscope, ultra sonic aspirator, and per operative ultrasonogram have greatly helped in better surgical outcome.

Low grade glioma in non eloquent area, small in size, well circumscribed the option are:

a) Observation
b) Biopsy and observation
c) Excision

Cairncross JG et al 1999 opined that small low grade gliomas who are asymptomatic may be conservatively treated followed with serial CT or MRI scans. Similarly a policy of observation is supported by Recht et al (1992) who in a retrospective review of 46 patients with radiographic evidence of low grade gliomas, found no difference in the rate of malignant transformation, survival and quality of life between 26 patients who had pathological diagnosis and treatment deferred until evidence of radiographic or clinical progression and 20 patients who had early diagnosis and treatment.

Morantz 1987, Shapiro 1992, Kordziolka et al 1993 suggest early biopsy to establish pathologic diagnosis of all radiographic lesion resembling low grade glioma. Berger et al 1994, Laws et al 1986, Morantz 1987, favours early excision for all patients with low grade gliomas, including those with small tumors.

Low grade gliomas in non eloquent area large in size circumscribed surgical option is to excise the tumor. If large but diffuse in nature, patient is neurologically stable cytoreduction and if unstable decompression is the surgical option. Ciric et al (1987) have examined the relationship between the extent of surgical resection and the immediate post operative results for 42 patients undergoing resection of cerebral gliomas. A gross total or near total tumor resection was achieved in 85% of patients, without mortality and with acceptable medical and surgical morbidity. No patients with gross total or near total resection suffered permanent neurologic deterioration. At discharge 55% of all patients were neurologically stable and 41% were improved. Large tumors are more likely to have anaplastic foci (Russel and Rubinstein, 1989), or develop malignant phenotypes, and for these reason, early and extensive surgery may influence favourably recurrence rates, possibly modify tumor behaviour, and ultimately prolong over all survival (Berger et al 1993).

Pilocytic tumors of the carebrum differ from fibrillary or protoplasmic counterparts by being relatively non infiltrating and well demarcated from surrounding normal brain (Clark et al 1985). Total resection of these tumor is often possible and can result in cure, Forsyth et al 1993.

Non eloquent high grade gliomas well circumscribed, if patient is neurologically stable macro or microscopic excision with or without lobectomy is the surgical option. If unstable cytoreduction is the option. H.G. Hollerhage et al 1991 concluded that microsurgical complete tumor resection combined with postoperative radiotherapy is the best treatment for high grade glioma with regard to both early results and long term survival. Additional lobectomy seems not necessary in completely resected tumors. However in incompletely resected tumors it may even worsen early outcome and shorten survival time. Mechanical cytoreduction is the most rapid and effective means known of removing large numbers of cells potentially resistant to radiation or chemotherapy, Hoshino, 1984. Non eloquent high grade glioma diffuse in nature, options depend on the neurological status of the patient. If stable cytoreduction and if unstable decompression should be the choice. JR Simpson et al 1993.

In eloquent area, low grade gliomas small or large well circumscribed, options depends on patient. If cooperative microsurgical excision with awake craniotomy and brain mapping, is the surgical option, Michael D Taylor et al 1999. In un cooperative patient excision with the help of neuronavigation is the treatment of choice Zakhary R et al 1999. Similarly the options for high grade glioma in eloquent areas, well circumscribed, is microsurgical excision or decompression using awake craniotomy and brain mapping of the lesion. But in diffuse variety the surgical strategies would be to do cytoreduction. If necessary doing an awake craniotomy with brain mapping or using neuro navigation. Watanabe et al 1991.

History of management of deep seated glioma

Earlier authors stressed the inoperability of thalamic tumors due to their poor accessibility. Only observation with or without radiotherapy was the mode of treatment, as per Smyth and Stein 1938, Hyndman and Van Epps 1939. Tor Kildsen (1948) reported 2 cases of thalamic lesions treated with a shunting procedure and was in favour of post operative radiotherapy. Arsenin (1958) reported on results of direct operative approach on ten patients with radical removal of the tumor. A point was made for direct attack on tumors situated in the dorso medial area, particularly when they are well circumscribed. Later reports have concluded that direct operative attack is not the treatment of choice in view of high morbidity and mortality, Tovi et al 1961. Thus empirical radiotherapy without tissue diagnosis was recommended by Cheek W R et al 1966. Later a plea for early diagnosis by needle biopsy and radiotherapy was recommended by Hirose G et al 1975. George M. Gree et al in 1988 recommended stereotactic biopsy and radiotherapy. One year later Patrick J Kelly advocated stereotactic resection of thalamic tumor. In 1991, James M Drake et al first did a computer & robot assisted resection of thalamic astrocytomas in children. H J Steiger et al 2000 reported short result of microsurgical resection of thalamic astrocytomas.

Surgical strategies to deep seated glioma

Open craniotomy for biopsies and resection of thalamic tumors has in the past been associated with high morbidity and mortality. Recent approach to the resection of thalamic tumors is based on computer assisted stereotactic open microsurgical technique (Kelly, 1989). According to Kelly, a stereotactic approach permits the surgeon to maintain an appropriate spatial orientation so that deep seated lesions may be correctly located.

Low grade glioma small circumscribed lesion in the thalamus may do very well for years without treatment, so observation only is better. Mc.Cormack et al 1992, Shaw et al 1993. Low grade glioma small or large circumscribed, cystic or solid, volumetric stereotactic resection is the surgical option, Kelly et al 1989. Deep seated low grade glioma diffuse in nature, patient's are neurologically intact stereotactic resection is appropriate, Salcman M 1990a.

Deep seated low grade glioma diffuse in nature stereotactic resection/ decompression should be considered in patient's with progressive neurological symptoms. Show et al 1993.

Deep seated high grade glioma small or large well circumscribed, patient is neurologically stable the controversies whether stereotactic biopsy or resection gives better result not settled yet. According to Coffey et al 1988 stereotactic biopsy followed by RT or CT can produce results comparable to those achieved by extensive procedures. According to Hitchock ER et al 1989, preferred approach in patients who have high grade glioma circumscribed has been stereotactic resection and radiation therapy.

For high grade glioma diffuse in nature patient is neurologically stable biopsy only and if unstable stereotactic resection followed by radiotherapy is best option, Kelly 1989. However for Ramana R et al 1999, stereotactic biopsy followed by radio and or chemotherapy is the best option for thalamic or brain stem anaplastic and glioblastoma.

MODERN APPROACH TO DEEP SEATED GLIOMA

a) Stereotactic volumetric resection with neuronavigation

Inspite of stereotactic volumetric resection of glioma in basal ganglia and thalamus, it has failed to prolong the survival of patients with glioblastoma beyond an average of 37 weeks (Kelly et al 1986). It is expected that the safety and precision with which craniotomies are carried out will be increased by the routine application of frameless stereotactic methods and intractive image technology (Watanabe et al 1991).

b) Robotic surgery

A robot is defined as a "reprogrammable multifunctional manipulator designed to move materials, parts, tools or specialized devices through variable programmed motions to perform a variety of tasks".[34]

The evolution of robotic and its application to neurosurgery have been very rapid since first computer and robot assisted resection of thalamic astorcytoma, done in children by James M Drake et al 1991.

The main consideration regarding a robot suitable for stereotactic neurosurgery are substantial maneuvering capability, an appropriate reach to allow an un obstructing support of the probe over the entire area, sturdiness, and sufficient fail safe features to avoid causing damage in the event of some malfunction, Drake JM et al 1991.

c) **Microsurgical approaches**

   1. **Trans-sylvian trans insular approach**

      Dan S Heffez, 1997 used this approach to resect deep seated lesions. He suggested that this approach can be practiced without inducing permanent neurological deficits in deep seated gliomas relying primarily on precise intra operative lesion localization. This approach permits exploration of the lesion through a limited exposure and by using a lateral, rather than a pterional craniotomy which appears to limit anatomic shifts, an absolute requirement of image guided surgery.

   2. **Trans callosal approach**

      Aida T et al 2000, presented their experience with trans callosal approach for deep seated lesions. They have preferred anterior trans callosal approach for basal ganglia and posterior transcallosal approach for thalamic tumors. According to them this approach is a safe and feasible approach to deep seated lesions but complications related to this like venous infarction, subdural collection should always be kept in mind.

   3. **H J Steiger et al 2000,** described two main surgical approaches for thalamic, and basal ganglia tumor. For small tumors arising from pulvinar or habenula, contralateral infratentorial supracerebellar approach and for extensive tumors arising from pulvinar extending laterally, a parieto occipital trans cortical transventricular approach has been executed. The authors have concluded that thalamic gliomas can be excised through this microsurgical approaches with an acceptable risk.

SUMMARY

In summary, irrespective of histological grade, surgical resection results in better prognosis in surface glioma. Awake craniotomy gives a good result in surface glioma of eloquent areas irrespective of histological grade. In the present day with the availability of microsurgical operative techniques, advanced imageology with neuro navigation aid, deep seated gliomas can be excised with less morbidity and mortality.

CONCLUSION

The surgical strategies differ in the management of glioma, depending on its location, though the aim of management does not change.

REFERENCES:

1. Peraud A, Ansari H, Bise K, Reulen HJ: Clinical outcome of Supratentorial Astrocytoma WHO-grade II. *Acta Neurochir (Wien)* 1998;140:1213-22.

2. Aida T, Abe E, Iwasaki, Y, et al: Transcallosal approach to paraventricular tumor, Article in Japanese. PMID- 6483101, 2000.
3. Arseni C: Tumors of the basal ganglia, their surgical treatment. Arch Neurol Psychait (Clinic) 1958;80:18-24.
4. Berger MA, Deliganis J, Dobbins, Keles G: The effect of extent of resection on recurrence in patients with low grade cerebral hemisphere gliomas. Cancer 1994;74:1784-91.
5. Cairncross J, Laperrier N: Low graded gliomas, to treat or not to treat? Arch Neurol 1990;47: 1139-40.
6. Cheek WR, Taveras JM: Thalamic tumors. J Neurosurg 1961;18:730-40.
7. Ciric LM, Ammirati N Vick, Michael M: Supratentorial gliomas: Surgical considerations and immediate post operative results, Gross total resection versus partial resection. Neurosurg 1987;21:21-26.
8. Clark GJ, Henry, Mckeever P: Cerebral pilocytic Astrocytoma. Cancer 1985;56:1128-33.
9. Coffey RJ, Lunsford LD, Taylor FH: Survival after stereotactic biopsy of malignant gliomas. Neurosurgery 1988;22:465-73.
10. Douglas B: Kirkapatrick "The first primary brain tumor operation". J Neurosurgery 1981;61:809-13.
11. Dan S Heffez: Stereotactic trans-sylvian transinsular approach for deep seated lesion. Surg Neurol 1997;48:110-12.
12. Fosyth PE, Shaw B, Scheithauer JO, Fallo DJ Layton, Kztzmann J: Supratentorial pilocytic astrocytomas. A clinicopathological, prognostic and flow cytometric study of 51 patients. Cancer 1993;72: 1335-42.
13. George M, Greenc Patrick W, Hitchon Robert L, Schelpet, et al: Diagnostic yield in C.T guided stereotactic biopsy of gliomas J Neurosurgery 1989;71:494-97.
14. Hyndman OR, Van Epps C: Tumors of the thalamus, a ventriculo graphic entity. Arch Surg 1939;39: 792-97.
15. Hirose G, Lombroso CT, Eisenberg H: Thalamic tumors in childhood. Arch Neurol 1975;32:740-44.
16. Hollerhage HG, Zumkeller M, Becker M, Dietz H: Influence of type and extent of surgery on early results and survival time in glioblastoma multiforme. Acta Neurochirugica 1991;113:31-37.
17. Hoshino T: A commentary on the biology and growth kinetics of low grade and high grade gliomas. J Neurosurgery 1984;61895-900.
18. Steiger HJ, Gotz C, Schmid R, Elsaesser, Stummer W: Thalamic astrocytomas, surgical anatomy and results of a pilot series using maximum microsurgical removal. Acta Neurochirurgia 2000; 142:1327-37.
19. Hitchcock, ET ISSQ IMA, Sotelo MG: Stereotactic excision of deeply seated intracranial mass lesions. Br J Neurosurgery 1989;3:313.
20. James M Drake, Michael Joy, Andrew Goldenberg David Kreindler: "Computer and robot assisted resection of thalamic astrocytomas in children". Neurosurgery Vol. 29 No.1 1991; 27-33.
21. Jules M Nazzaro, Neuwett EA: "The role of surgery in the management of Supratentorial intermediate and high grade astrocytomas in adults. J Neurosurgery 1990;73:331-44.
22. Simpson JR, Horton J, Scott C, Curran WJ, et al: "Influence of location and extent of surgical resection on survival of patients with glioblastoma multiforme. Results of three consecutive radiation therapy oncology group (RTOG) clinical trials. J Radiat Oncol Biol physi Vol. 26, 1993;239-44.
23. Joseph M Piepmeier: Primary brain tumor-The case for resective surgery. Clinical Neurosurgery Vol.45 1992;211-25.
24. Kondoziolka D, Lunsford L, Martinez A: Unreliability of contemporary neurodiagnostic imaging in evaluating suspected adult Supratentorial (low grade) Astrocytoma. J Neurosurgery 1993;79: 533-36.
25. Kelly PJ: "Computer assisted stereotaxis, new approaches for the management of intra-axial tumor. Neurology 1986;36:535-41.
26. Laws EW, Taylor E, Bergstralh H Okazai, Ciffon M: The neurosurgical management of low grade Astrocytoma of the cerebral hemispheres. J Neurosurgery 1986;61:575-88.

27. Michael E, Cohen, Las zlo Mechtlen, Patricia K Duffner: "Thalamic, hypothalamic and otpic pathway tumors" Hand book of clinical neurology Vol. 24 (68), Chap. 3 Neuro-oncology Part II Page 63-76.
28. Michael Salcman: "High grade gliomas" chapter 4 hand book of clinical neurology Vol. 24 (68). Neuro-oncology Part II, 1997;87-122.
29. Michael D, Taylor K, Mark Bernstein: Awake craniotomy with brain mapping as the routine surgical approach to treating patients with Supratentorial intraaxial tumors: a perspective trial of 200 cases". *J Neurosurgery* 1999;90:35-41.
30. Mc. Cormack BM, Miller DC, Budzilovich GN, et al: "Treatment and survival of low grade Astrocytoma in adults. Neurosurgery. 1992;31:636-42.
31. Morantz R: "Radiation therapy in the treatment of cerebral Astrocytoma. *Neurosurgery* 1987;20: 975-82.
32. Patric J Kelly: "Stereotactic Biopsy and Resection of thalamic Astrocytoma. *Neurosurgery* Vol. 25 No. 2 1989;185-95.
33. Philip H Goutin, Jerome B Posner: "Neuro oncology, diagnosis and management of cerebral gliomas-past present, and future". *Neurosurgery* Vol. 47 No.1, 2000.
34. Raymond Sawaya, Wiliam M, Ramboo Jr, Maarouf A Hammoud, B Lee ligon: Advances in surgery for brain tumors: *Neurologic Clinic* Vol. 13, No. 4, 1995;757-70.
35. Ramina R, Neto MC, Meneses M, et al: Management of deep seated glioma" *Critical Reviews in Neurosurgery* 1999;269(11):34-40.
36. Russel D and L: Rubinestein "Pathology of tumors of the nervous system, Baltimore. Williams and Wilkins (1989).
37. Salcman M: "Malignant Glioma management *Neurosurgery clinics of North America* 1, 1990a;49-63.
38. Shaw E, Scheithauerand B, Fallon JO: "Management of Supratentorial low grade gliomas. *Oncology* 1993;76:97-102.
39. Smyth GE, Stern K: "Tumors of the thalamus a clinicopathological study. *Brain* 1938;61:339-74.
40. Shapiro W: "Low grade gliomas when to treat? *Ann Neurol* 1992;31:437-38.
41. Torkildsen A: Should extirpation be attempted in cases of neoplasm in or around the third ventricle of brain? *J Neurosurg* 1948;5:249-75.
42. Recht L, Lew R, Smith T: Suspected low grade glioma is deferring treatment safe *AN Neurol* 1992;31:431-36.
43. Tovi D, Schisano G, Lijequist B: "Primary tumors of the region of the thalamus. *J Neurosurgery* 1961;18:730-40.
44. Watanabe E, Mayanagi Y, Kosugi Y, et al: Open surgery assisted by neuronavigator, a stereotactic articulated, sensitive arm. *Neurosurgery* 1991;28:752.
45. Zakhary R, Keles GE, Berger MS: Intra operative imaging techniques in the treatment of brain tumors: *Curr Op in Oncol II* 1999;152-56.

11
Surgery for Eloquent Area Glioma

S Sinha, D Singh, A Goyal

Gliomas may be benign indolent low grade or malignant-anaplastic or glioblastoma multiforme. Low grade glioma may be transformed into malignant nature in later stage. Even low grade located in eloquent area of brain can cause compression or destruction leading to neurological deficit. Moreover, Gliomas do not respect the boundaries between the brain's eloquent areas. Resection of functional cortex during glioma resection has the same effect as a stroke in the functional region. It cause an irreversible loss of functional cortex which can be noticed in immediate post operative period. The primary objective of in surgical management of eloquent glioma is to accurately localize and then completely remove without causing neurologic impairments. An important factor of the risk of surgery is the relationship of the glioma to eloquent regions because injury to these eloquent areas can cause irreversible neurologic deficit. The location of various eloquent brain areas can be determined using anatomic landmarks but individual variations may occur and presence of local lesion can distort landmarks making localization of functional cortex imperfect. A variety of methods have been employed to maximize surgical resection of glioma while minimizing risk of injury to adjoining functionally important brain. Cause of injury to functional cortex is related to unawareness on proximity of dissection around eloquent area and manipulation of eloquent cortex.

The goal of surgery is to obtain precise tissue diagnosis as well as to safe and complete excision of tumour without damaging to surrounding functional areas. Partial excision is sometimes required to prevent injury to eloquent area. However, surgery is not curative but tumour resection may improve pressure related symptoms and provide time for additional therapy (radio and chemotheraphy) to work. The surgeon needs to be especially meticulous and gentle when manipulating tumour tissue in proximity to tissues in anatomic locations usually functionally eloquent.

Eloquent brain consist of speech area (motor and sensory speech area), primary motor cortex, adjoining supplementary motor cortex, primary somatosensory cortex, visual and auditory cortex. Localization of these eloquent area of brain is important to maximize resection of glioma and to minimize morbidity.

SURGICAL PLANNING

- Computerized tomography (CT)

- Magnetic resonance imaging (MR) including gadolinium-diethylene-penta-acetic acid (GDPA) enhancement, MR spectroscopy, functional MR.

MR is most sensitive imaging test **(Fig. 1)**. GDPA enhancement correlate with volume of solid tumour tissue that can be safely resected.[1,2] MR spectroscopy is helpful in low grade glioma which can be localized on the basis of choline elevation when other imaging method provide inadequate information. Tumour demarcation is also possible as elevation of choline in glioma tend to occur in marginal rather than central part.

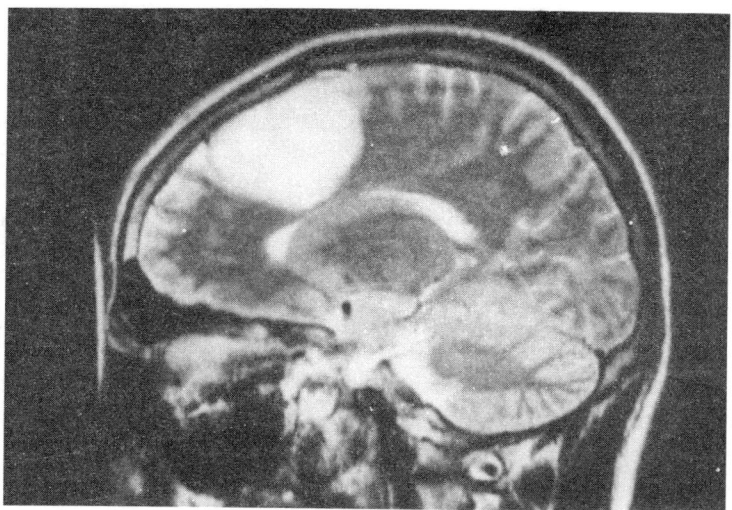

Fig. 1 : T-2 weighted sagittal MR of glioma showing hyperintense mass in the area of sensory motor cortex.

A functional MR may be taken to define proximity of glioma to eloquent areas, to demonstrate infilteration and/or displacement of these areas. This can be used non-invasively to localize cortical function with respect to lesion. The 3-D demonstration of cerebral anatomy and cortical function can lead to more accurate preoperative surgical planning. Demonstration of eloquent area can be helpful in selecting safe cortical incision. Functional MR is also helpful in determining cerebral hemisphere dominance by localizing the language function.

Fluorodeoxy glucose Positron emission tomography (FDP PET) may be helpful to determine metabolic activity of tumour to differenciate slow vs aggressive nature of glioma. Hypometabolism is suggestive of low grade glioma and hypermetabolism as malignant evolution. However, it may be unreliable in determining change to new focus of anaplastic degeneration within a low grade glioma, indicating new focus as hypometabolic when histologic sample obtained via a biopsy has become anaplastic.[3]

SURGICAL PROCEDURES

- Conventional classical craniotomy (placed over the lesion as defined topographically by preoperative radiographic studies)
- Frame based or frameless stereotactic volumetric resection allow tumour removal using minimally invasing route which is accurate, safe and cause less pain with least injury to neural tissue.

Localization of the eloquent area of the brain is important by intraoperative mapping methods to minimize injury to functional cortex.[4,5] Excision of the glioma should continued only after intraoperative functional brain mapping techniques have shown that the infilterated cortex is silent neurophysiologically. The primary motor and somatosensory areas collectively form the rolandic cortex and is bordered in front by supplementary motor area. Posterior to the coronal suture, the premotor sulcus is nearly always found within 2.5 to 3.0 cm.[3] In anaesthetized patients, electrophysiological moniterings like Somatosensory evoked potential (SSEP) and Motor evoked potential (MEP) can be determined for functional assessment. Recording of SSEP generated from stimulation of contralateral median nerve localizes primary sensory cortex. Phase reversal will determine central sulcus. The motor strip may be identified in awake or asleep patient using current between 2 mA and 16 mA, depending on anaesthetized condition of the patients. The current is delivered via a bipolar stimulating electrode attached to a constant current generator, which yields biphasic square wave pulses.[6] In the dominant hemisphere, the areas responsible for and subserving various language function may also be be identified with stimulation mapping. This include Broca's area, which is located anterior to the inferior aspect of the face motor cortex, and reading and naming cortical sites, which may be located within regions of the temporal lobe including the anterior 3 to 4 cm, inferior parietal and posterior frontal lobes.[7,8]

Sometimes errors in interpretation of monitoring data can lead to miscalculation and resection of functional cortex. Awake craniotomy, patient is sedated (Inj. Diprivan (propofol) and Inj. Medazolam but awake,under local anaesthesia can be performed to overcome these fallacies.[9] Intraoperative monitoring can be undertaken by using bipolar focal stimulation of 2-4 mA over the cerebral cortex and observing perception of stimulus. Direct electrical stimulation can demarcate sensorymotor strip. For language mapping, patient is instructed to say some words and interruption can be observed on cortical stimulation.

There may be various locations of lesion in eloquent area.

- Tumour may be entirely upon surface

- Only small portion of tumour present on surface
- Tumour is below surface but present within a sulcus
- Tumour is completely subcortical

The route of entry to debulke tumour should be such which causes least cortical damage. Tumours which are entirely upon the surface or only small portion of tumour is present on the surface, can be attacked directly, debulked and removed from underlying cortex.[10] However, tumours which are below the surface or subcortical, can be removed by sulcotomy[10] to open sulcus and limits of tumour are identified. The tumour's vascular supply is isolated and the tumour is subsequently removed **(Fig. 2)**. All types of sulci have complex anatomical shapes within their depths. Small finger like projections can occur at the bottom of given sulcus. These can often be found to have extensions of several centimeters and can indent proximal and even distal gyri at the bases. The arachnoid extends across the crest of gyrus and then split into two layers in passing across a sulcus. One layer extend across the top of a sulcus and other arachnoidal layer follows the pia down the walls and floor of the sulcus. The vasculature within the arachnoid cistern of one sulcus may contribute to the perfusion and drainage of several gyri. Thus, when opening the arachnoid of a given sulcus one must pay special attention to avoid injury to the vessel of that sulcus. The cortical vessels normally traverse the free surface of a gyrus before entering the sulcus to supply that sulcal region. However, an artery within the depth of a given sulcus or finger extension of a given sulcus may enter another communicating sulcus without

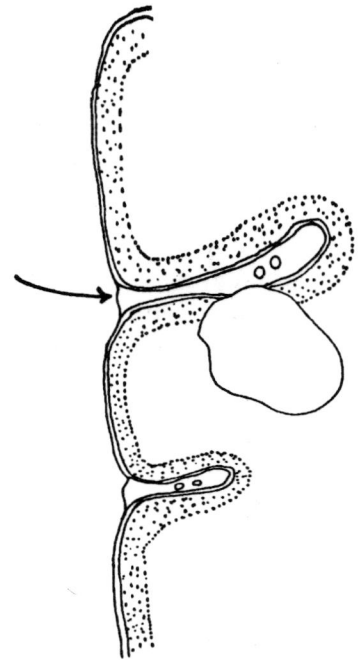

Fig. 2 : Transsulcal approach to remove glioma situated below the cortical surface

reappearing on the free surface. Thus adequate exposure of sulcal vessels is important to avoid injury to blood supply to adjacent gyri.

Tumours involving the white matter of frontal, temporal, parietal and occipital lobes are resected using sulcal-cisternal approach.[10] The appropriate sulcus-cistern is opened over a length which will allow adequate visualization of the lesion. The sulcal vasculature is then exposed and protected and feeding vessels to the tumour are subsequently eliminated. The lesion is then removed using microsurgical techniques in a piecemeal fashion.

Injury to eloquent cortex can result in a permanent or transient neurological deficit. Loss of functional cortex can result in irreversible neurological deficit. However, focal edema and ischaemia will cause reversible deficit which can be managed conservatively. Resolution of edema and restoration of normal oxygen and electrolytes may result in improved neurological function. Haematoma into tumour bed around eloquent area may also result in increased intracranial pressure or focal neurological deficit, which requires an immediate surgical intervention. Moreover, the incidence of clinically significant tomoural bed haematoma is minimal because removal of tumour reduce soft tissue mass effect which can accommodate a large volume of haematoma without causing symptomatic compression of brain.

Various surgical tools, useful to minimize neurological deficit in removal of glioma in eloquent area, are ultrasonic aspirator, LASER, intraoperative ultrasound, intraoperative CT and intraoperative MR including functional MR. Ultrasonic aspirator cause minimal traction and damage to surrounding tissue. Carbon dioxide laser is extremely useful in vapouring deep seated tumour with limited access. Intraoperative ultrasound and CT have advantages to verify complete resection prior to closure. Intraoperative MR confirms location of lesion, reconfirms optimal surgical approach and verify complete removal of glioma. However, intraoperative functional MR defines exact location and extent of glioma in relation eloquent cortex.

In conclusion, removal of glioma around eloquent brain require special precautions with meticulous microsurgical technique, MR and functional MR are helpful to localize eloquent brain and exact location of tumour, radical excision may be enhanced with aid of intraoperative mapping to identify sensory, motor and language cortex and finally awake craniotomy minimizes risk of damage to eloquent brain.

REFERENCES:

1. Kelly PJ, Daumas-Duport C, Kispert DB, et al: Imaging based stereotactic serial biopsies in untreated intracranial glial neoplasms. *J Neurosurg* 1987;66:865-74.

2. Kelly PJ, Daumas-Duport C, Scheithauer BW, et al: Stereotactic histologic correlation of computed tomography and magnetic resonance imaging defined abnormalities in patients with glial neoplasms. *Mayo Clin Proc* 1987;62:450-59.
3. Berger MS: Surgery of low grade glioma-Technical aspect. *Clin Neurosurge* 1996;44:161-80.
4. Sheng-Hong Tseng, Hsiu-Jung Wu, Ham-Min Tseng, Swei-Ming Lin: Functional brain mapping-assisted resection of gliomas in dominant hemisphere. *Formosan J Surg* 2000;33:157-65.
5. Berger MS: Functional mapping guided resection of low grade glioma. *Clin Neurosurg* 1994;42:437-52.
6. Berger MS, Ojemann GA, Lettich E: Neurophysiological monitoring during astrocytoma surgery. *Neurosurge Clin N Am* 1990;1:65-80.
7. Ojemann GA, Ojeman J, Lettich E, et al: Cortical language localization in left dominant hemisphere. *J Neurosurge* 1989;71:316-26.
8. Haglund MM, Berger MS, Shamseldin M, et al: Cortical localization of temporal lobe language site in patients with glioma. *Neurosurgery* 1996;34:567-76.
9. Meyer FB, Bates LM, Goerss SJ, Friedman JA, Windschitl WL, Duffy JR, Perkins WJ, O'Neill BP: Awake craniotomy for aggressive resection of primary gliomas located in eloquent brain. *Mayo Clin Proc* 2001;76:677-87.
10. Yasargil MG, Cravens GF, Roth P: Surgical approaches to inaccessible brain tumours. *Clin Neurosurg* 1988;34:42-110.

12

Stereotactic Biopsy and Radiotherapy for Supratentorial Eloquent Region Gliomas

A Shankar, V Rajshekhar

INTRODUCTION

From the time Cushing[1] declared that, "The larger and more formidable the lesion appears to be, the more radical is the effort made to remove its major portion..." the aim of every neurosurgeon has been to radically excise supratentorial gliomas. However, the eloquence of some regions of the brain, precluding their destruction by surgical removal, has continually thwarted them.

Even in the relatively non-eloquent regions, the surgeon has to decide whether he would be able to totally excise the enhancing component of high-grade gliomas, as it has been shown that performing a simple stereotactic biopsy carries less morbidity and mortality than attempting a partial excision of these lesions.[2]

Early in the course of the disease, most astrocytomas have normally functioning cortex within (macroscopically) abnormal-appearing brain tissue. Two series (of 28 and 14 patients) by Ojemann et al[3-4] have shown that intrinsic brain tumors can invade cortical and subcortical structures without disrupting function. In their studies, direct stimulation mapping carried out during surgery demonstrated functioning cortices within grossly abnormal-appearing brain.

Gliomas can be infiltrative without a well-defined plane between the tumor and the normal brain surrounding it. This precludes radical excision of such tumors located in eloquent regions of the brain. The strategy for the management of such lesions therefore would involve minimally invasive techniques and adjuvant therapy. Stereotactic biopsy and radiation therapy would be one such strategy for gliomas in eloquent regions of the brain.

INDICATIONS

Stereotactic procedures for suspected gliomas in the supra-tentorial compartment are generally reserved for the following:

1. Eloquent area lesions such as those involving

- The basal ganglia and thalamus.
- The insula.
2. Ill defined or deep hemispheric lesions.[5-6]
3. Bilateral lesions.
4. Diffuse lesions such as gliomatosis cerebri.[7]

CONTRAINDICATIONS

Gliomas that are located in eloquent cortical or subcortical regions, are well defined, focal in nature or produce severe mass effect are generally not treated by stereotactic biopsy, as these should be radically excised.[2]

TECHNICAL DETAILS

Most stereotactic biopsies in adults can be performed under local analgesia. A frame-based system such as the Cosman-Roberts-Wells (CRW) or the Leksell system are easy to use and are versatile. A frontal, pre-coronal approach is suited for the biopsy of over 90% of eloquent region supratentorial masses.[8-9] A twist drill craniostomy is used in our institution as this avoids the use of the operating room. This not only reduces costs, but also eliminates the need for sutures in most instances. Therefore, radiation therapy can begin even the day after the biopsy.

An intra-operative feedback from the pathologist with regard to the histological diagnosis improves the yield of a stereotactic biopsy procedure.[10] A smear preparation, as used by our pathologist, has proved to be highly reliable and accurate.

A plain CT scan done within four hours of the biopsy would reveal the presence or absence of any significant hemorrhage. The site of the biopsy is also frequently identifiable by the presence of a small quantity of air or a dot of blood.

RESULTS OF BIOPSY

Stereotactic biopsy of a supra-tentorial eloquent region glioma will yield a positive diagnosis in over 90% of patients.[7-8] In patients with non enhancing low grade gliomas the yield may be lower as it is difficult, on occasion, to differentiate between a low grade astrocytoma and gliosis on a small biopsy sample. There is a also a problem of under-grading gliomas on stereotactic biopsies and hence the complete clinical and radiological data should be considered together with the histological diagnosis before deciding on management.

COMPLICATIONS

The complications of stereotactic biopsies in patients with eloquent region tumours may be slightly higher than for masses in other locations, as even a slight increase in mass effect in an eloquent region will produce an appreciable neurological deficit or worsen a pre-existing deficit.[2] However, overall, the complication rate is around 3%. Mortality is generally restricted to patients with malignant gliomas, especially in deep-seated locations such as the thalamus.

RADIOTHERAPY

Both for malignant and benign gliomas of the eloquent regions, radiation therapy is recommended.[11-12] While focal radiotherapy with margins of 1-2 cm will suffice for benign gliomas, malignant gliomas require wider margins, which in larger masses will amount to whole brain irradiation.

Stereotactic radiation therapy (SRT) may be suitable for smaller (<3 cm) benign gliomas. For lesions up to 4 cm, both benign and malignant, SRT and stereotactic radiosurgery (SRS)[13] techniques can be used to deliver boost doses of around 10-15 Gy.

The benefits of radiotherapy either as an adjunct to surgery, or on its own, have been documented by various studies. It has also been demonstrated that radiotherapy alone is more effective than subtotal resection of a tumor.[14-17] A review of 19 articles by Nazzaro et al[18] showed a consistent association between external beam radiation and prolonged survival in patients with supra-tentorial astrocytomas. Shaw et al[11] found significant improvement following high dose radiation (greater than or equal to 5300 cGy) in patients with non-pilocytic astrocytomas. However, surgery followed by radiotherapy was found to have an even better prognosis. Coffey et al[6] reported improved survival in patients with high-grade gliomas who had undergone radiotherapy following surgical decompression.

RESULTS OF STEREOTACTIC BIOPSY AND RADIOTHERAPY

In a retrospective analysis of 39 patients who had undergone stereotactic biopsy for malignant gliomas, Ranjan et al[19] reported a mortality rate of 2.6%. They had one procedure-related death, but no morbidity. The median survival was 13 months, and the only prognostic factor was age at presentation, with those older than 40 years having a poorer outcome. A better prognosis for younger patients had also been reported in other studies.[17]

North et al[20] studied the effect of stereotactic biopsy followed by radiotherapy in patients with low-grade cerebral astrocytomas. They reported five and 10-year survival rates of 55% and 43% respectively. However, some of their patients had also undergone sub-total resection of the tumors before radiation. These patients generally had tumors that were smaller, more accessible and less invasive on imaging.

The outcome of stereotactic biopsy and radiotherapy in patients with non-anaplastic, non-pilocytic astrocytomas was studied by Lunsford et al.[21] Their series of 35 patients had a median survival of 9.2 years, and 16 patients showed regression of the tumor in follow up scans. The rest showed stabilization of tumor volume. Only three patients required cytoreductive surgery later on. Complete control of the seizures was obtained, using anti-convulsants, in 20 out of 22 epileptics.

Evaluation of 30 patients with low grade insular astrocytomas who had undergone stereotactic biopsy followed by radiation therapy at the author's institution showed an improvement in tumor volume and seizure control, with no change in memory or language functions, as well as in the Karnofsky Performance Scale. There were no procedure related mortalities or morbidity, and the patients were followed up for a median of 23 months.

CONCLUSION

Stereotactic biopsy followed by radiation therapy is a safe and effective alternative to radical surgery in selected cases of supra-tentorial gliomas involving eloquent regions of the brain.

REFERENCES:

1. Cushing H: Intracranial tumors: notes upon a series of 2000 verified cases with surgical-mortality percentages pertaining thereto. Philadelphia, CC Thomas & Sons, 1932.
2. Black PM: Surgery of cerebral gliomas: Past, present and future. *Clin Neurosurg* 1999;47:21-45.
3. Ojemann JG, Miller JW, Silbergeld DL: Preserved function in brain invaded by tumour. *Neurosurg* 1996;39:253-59.
4. Skirboll SS, Ojemann GA, Berger MS, Lettech E, Wivin R: Functional cortex and subcortical white matter located within gliomas. *Neurosurg* 1996;38:678-85.
5. Ostertag CB, Mennel HD, Kiessling M: Stereotactic biopsy of brain tumors. *Surg Neurol* 1980;14:275-83.
6. Coffey RJ, Lunsford LD, Taylor FH: Survival after stereotactic biopsy of malignant gliomas. *Neurosurg* 1988;22:465-73.
7. Salcman M: Supratentorial gliomas: Clinical features and surgical therapy: In Neurosurgery (Vol. I), Wilkins RH, Rengachary SS, (Ed): McGRaw Hill, 783, 1996.
8. Couldwell WT, Apuzzo MLJ: Initial experience related to the use of the Cosman-Roberts-Wells stereotactic instrument. *J Neurosurg* 1990;72:145-48.
9. Apuzzo MLJ, Sabshin JK: Computed tomographic guidance stereotaxis in the management of intracranial mass lesions. *Neurosurg* 1983;12:277-85.
10. Apuzzo MLJ, Chandrasoma PT, Cohen D, Zee C-S, et al: Computed imaging stereotaxy: Experience and perspective related to 500 procedures applied to brain masses. *Neurosurg* 1987;20:930-37.
11. Shaw EG, Daumas-Duport C, et al: Radiation therapy in the management of low-grade supratentorial astrocytomas. *J Neurosurg* 1989;70:853-61.
12. Gutin PH, Leibel SA, Wara WM, Choucair A, et al: Recurrent malignant gliomas: Survival following interstitial brachytherapy with high activity iodine-125. *J Neurosurg* 1987;67:864-73.
13. Black PM, Alexander E III, Tarbell N, Shrieve D, Goumnerova LC, Leffler JS: Linear accelerator based radiosurgery for malignant tumors of the central nervous system. *Crit. Rev. Neurosurg* 1996;6:225-31.
14. Vertosick FT Jr, Selker RG, Arena VC: Survival of patients with well differentiated astrocytomas diagnosed in the era of computed tomography. *Neurosurg* 1991;28:496-501.

15. Sheline GE: The role of radiation therapy in the treatment of low-grade gliomas. *Clin Neurosurg* 1986;33:563-74.
16. Morantz RA: Radiation therapy in the treatment of cerebral astrocytoma. *Neurosurg* 1987;20: 975-82.
17. Morantz RA: Low grade astrocytomas. In Neurosurgery (Vol I), Wilkins RH, Rengachary SS, (Ed): McGRaw Hill, 789, 1996.
18. Nazzaro JM, Neuwelt EA: The role of surgery in the management of supra-tentorial intermediate and high grade astrocytomas in adults. *J Neurosurg* 1990;73:331-44.
19. Ranjan A, Rajshekhar V, Chandy MJ: Outcome in patients with supra-tentorial malignant gliomas: Results of stereotactic biopsy and radiotherapy. *Neurol Ind* 1994;42:191-95.
20. North CA, North RB, Epstein JA, et al: Low grade cerebral astrocytomas – Survival and quality of life after radiation therapy. *Cancer* 1990;66:6-14.
21. Lunsford LD, Somaza S, et al : Survival after stereotactic biopsy and irradiation of cerebral non anaplastic, non-pilocytic astrocytoma. *J Neurosurg* 1995;82:523-29.

13
Surgical Strategies for Intraventricular Gliomas

CE Deopujari

Tumors arising from the lateral and third ventricle are frequently benign. Some of them are slow growing, low-grade malignancy lesions, especially in younger age group. The tumors grow to a large size before becoming symptomatic usually due to resultant hydrocephalus. Rarely direct pressure on surrounding structure may be the first presentation. The surgical strategy therefore, relates to reduction of raised intracranial pressure by treating the hydrocephalus as well as complete or radical excision of the tumor.

Pre-operative evaluation is directed towards answering the following questions; (i) What is the nature of the lesion? (ii) Is it growing from outside or from inside the ventricular chamber? (iii) Where is it attached? (iv) What is its relation to the optic nerve and the chiasm, the ventricular wall, and the basal vessels? (v) What is the safest corridor to excise it and unblock the cerebrospinal fluid (CSF) pathways hydrocephalus present? (vi) How is the hypothalamo-pituitary axis functioning?

Advances in imaging have made it much easier in recent times to anatomically localize these lesions. Malignant astrocytomas and meningiomas can be readily diagnosed with good quality CT and MR images. However, many of the tumors under the broad category of gliomas may not have the same pathology.

A break-up of tumors in lateral and anterior third ventricle taken up for surgery in the last six years in our experience is given below:

Intraventricular glial tumor	= 30
Lateral ventricle	= 12
Anterior and middle III ventricle	= 20
Histology	**No.**
Malignant atrocytoma	7
Low grade astrocytoma	10
Hamartoma	2
Subependymoma	1
Ependymoma	2
Germinoma	2
Epithelial tumor	2
Lymphoma	4
PNET	1
Neurocytoma	1

The various approaches for tackling these tumors are outlined as following:

(I) LATERAL VENTRICLE

A) MIDVENTRICULAR REGION (Fig. 1)
- Transcallosal
- Stereotactic

B) TRIGONAL REGION. (Figs. 2 & 3)
- Lateral tempo-parietal lobe incision.
- Middle temporal gyrus incision.
- Occipital incision/lobectomy.
- Superior parieto-occipital incision.
- Transcallosal.
- Trans temporal horn-occipito-temporal gyrus incision.

C) FRONTAL HORN.
- Middle frontal gyrus.
- Transcallosal.

D) TEMPORAL HORN.
- Middle temporal gyrus.
- Temporal lobectomy.

(II) THIRD VENTRICLE

A) ANTERIOR.
- Transcallosal.
- Superior
- frontal gyrus.
- Trans lamina terminalis.
- Corridors- Subchoriodal.
   Suprachoriodal
   Interforniceal.

B) MIDDLE. (Fig. 4)
- Transcallosal.
- Stereotactic.

C) POSTERIOR
- Transcallosal-
Post interhemispheric
- Stereotactic
- Supracerebellar
- Suboccipital-transtentorial

The surgical procedures carried out by us are summarized as follows:

1. Stereotactic Biopsy +/- VP Shunt	12
2. Transcallosal Excision	13
3. Translamina Terminalis	2
4. Trans cortical Excision	
a) Frontal	2
b) Parietal-occipital	3
c) Temporal	1

The surgical strategies for intraventricular gliomas should be directed towards making a specific histological diagnosis, to relieve pressure from adjacent neural structures and to unblock CSF pathways.

Transcallosal excision for complete or radical removal of the tumor in the anterior

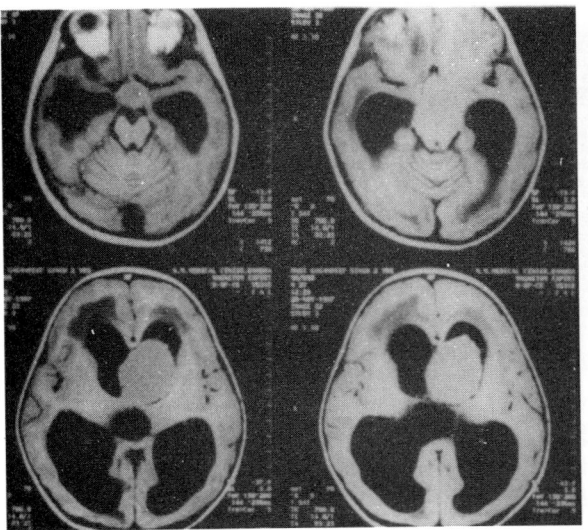 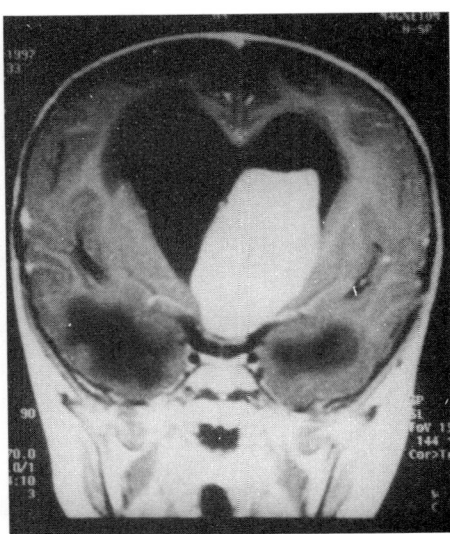

Fig. 1A : Pilocytic astrocytoma arising from the septum presenting in third as well as left lateral ventricle causing obstructive hydrocephalus
1) Axial T1 weighted images.
2) Coronal contrast image.

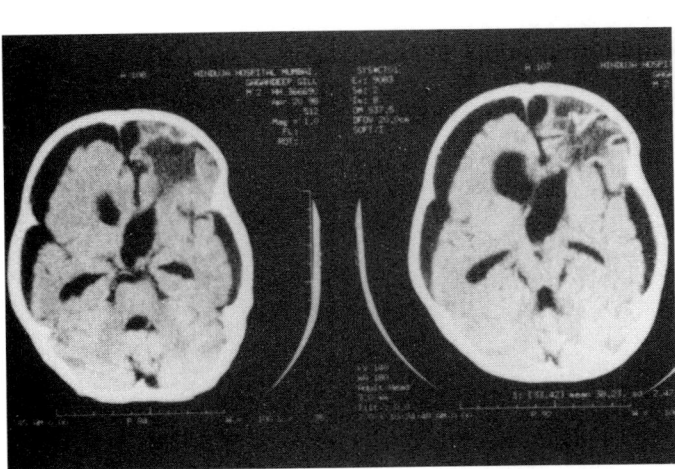 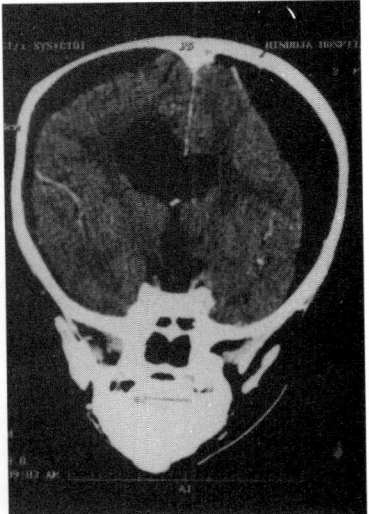

Fig. 1B : Post operative images of the same patient after transcallosal total excision. Subdural effusions gradually resolved without any intervention. Patient is doing well three years post operatively without recurrence

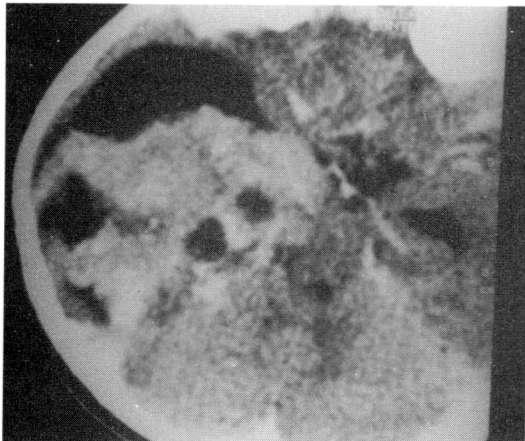 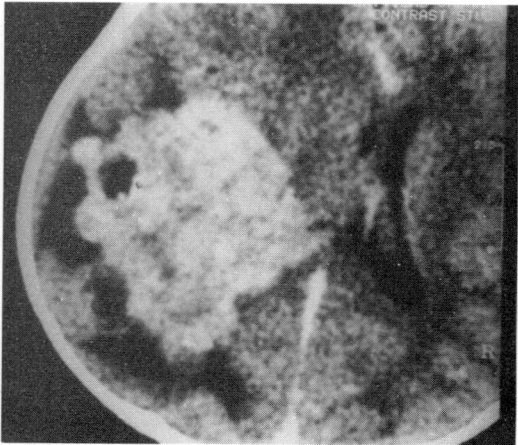

Fig. 2 : (A-B) Ependymoma presenting in the trigone and temporal horn in a one-year-old child. This was approached by temporal craniotomy and through the temporal horn for a subtotal excision. Child receiving chemotherapy post operatively

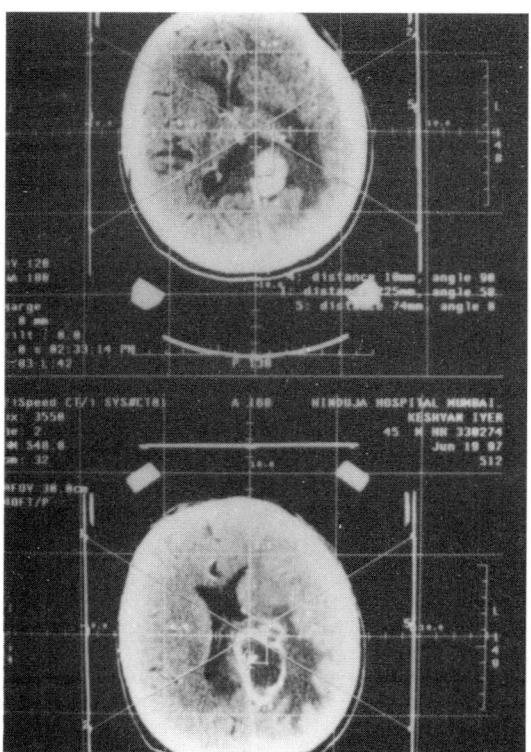

Fig. 3: Stereotactic biopsy of a heterogeneous lesion indenting on the left trigone. Tuberculous granuloma with abscess (positive for AFB) was found

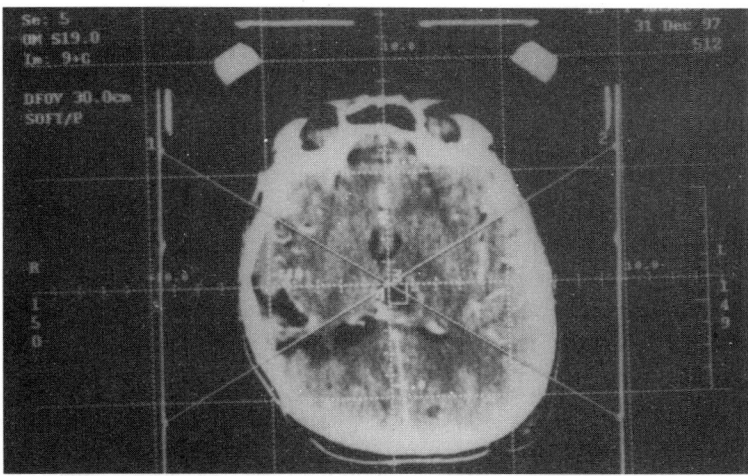

Fig. 4 : Stereotactic biopsy of a heterogeneous mass in the mid third ventricle/thalamic area. Histopathology: malignant astrocytoma

third ventricle and frontal horns was the most satisfactory. Stereotactic biopsy with or without a CSF diversion procedure was carried out in the mid-ventricular lesion, which are usually in the basal ganglia. Trans cortical excision was carried out in the trigone or lesions placed in the temporal horn. Transcallosal approach seems to be the most satifactory for removal of benign lesions such as low grade astrocytoma, neurocytoma and hamartomas.

Intra-operative complications can be avoided by placing the bone flaps away from eloquent cortex, preserving important drainage veins, obtaining proper relaxation of the hemisphere during the transcallosal approach, maintaining control of bleeding at all times and delivering the tumor in the field of view rather than using retraction. We have used pre-operative steroids, elevation of patient's head to 15-30 degrees to help in reduction of venous as well as intracranial pressure.

Anti-convulsant medication is used pre-operatively and continuously for atleast three months post-operatively Whenever in doubt about adequacy of excision or about clearance of CSF pathways, we have left a ventricular drain connected to a subcutaneous reservoir for drainage as and when necessary. Perforating the septum and irrigating the ventricles continuously intraoperatively and following the size of the ventricles post-operatively with periodic scans are the other measures taken to avoid complications related to hydrocephalus.

Complications to be expected in this surgery (as per literature) include visual field defect 20%-64%, hemiparesis 8%-30%, speech deficit 8%-36%, subdural haematoma 11%, death 12%-75%, seizures 29%-70%, infection-unknown, incomplete resection 33%-50% and persistent hydrocephalus 12%-33%.

Subdural collection is quite common though asymptomatic and was the only complication noted by us apart from persistent hydrocephalus requiring shunt in two cases each of lateral and third ventricle tumors.

REFERENCES:

1. Lobo J, Antunes J, Almeida Lima: Management of Third Ventricular Tumors; *Neurosurgery Quarterly*, Vol. 9, No.1, 1999;1-20.
2. Aviva Abosch, Michael W. McDermott, et al; Lateral Ventricular Tumors; Operative surgical techniques edited by Andrew Kaye and Peter Black. Chapter 64.
3. Dennis D Spencer, William Collins, et al: Surgical Management of Lateral Intraventricular Tumors; Chapter 50; Operative Neurosurgical Techniques ed. Sweet, Schmideck.
4. Michael LJ, Apuzzo; N. Scott Litofsky, Surgery in and around the Anterior Third Ventricle, Vol. I, chapter 18; Brain Surgery: Complication, Avoidance and Management edited by Michael LJ, Apuzzo, MD and published by Churchil Livingstone, 1993;541-79.
5. Joseph M. Piepmeier, Dennis D. Spencer, Kimberlee J. Sass, Timothy M. George; Lateral Ventricular Masses, Vol. I, Chapter 19; Brain Surgery: Complication, Avoidance and Management edited by Michael LJ, Apuzzo MD and published by Churchill Livingstone, 1993;581-99.

14

Is Microsurgery required in every case of Intraparenchymal Glioma?

K Sridhar

In the early 1990s surgeon like Cushing, Dandy and Krause demonstrated surgical approaches to deep seated brain tumours and redefined neurosurgical principles. The surgeons removed brain tumours and lesions, with morbidity and mortality rate that was acceptable at the period of time, taking into consideration the lack of imaging, monitoring, anesthetic and surgical technology. The advent of new imaging technology in the last 20 years, has made possible precise tumour localisation, and characterisation. This, along with improved monitoring and anaesthesia, has put the onus back on the surgeon to find the means to tackle these lesions which are now so clearly seen and defined.

Intracerebral tumours present to the clinician due to either raised intracranial pressure, or the local effect of the lesion on the brain as seen by seizures or neurological deficit. The main aim of surgery for these lesions is radical excision, while procuring a histological diagnosis. Radical excision in turn benefits the patient by effecting a reduction in the intracranial pressure, and reducing the effects of the lesion on the surrounding brain. It is the duty of the surgeon to ensure that in doing so, the patient improves, or at the least, maintains his or her neurological condition. The goals in front of the surgeon are therefore radical excision of the tumour with a good neurological result. The use of microsurgical techniques at surgery greatly help the surgeon in achieving these goals and it is mandatory that surgeons use the microscope at all surgeries.

Experience with microsurgical techniques has brought in the knowledge that the anatomy of the brain is such that it allows access to deep structure. "The brain is not an isolated compact structure like an island unto itself, rather it is like a continent with a multifaceted coastline with many water inlets, it floats in a sea of CSF, with rivers of cerebrospinal fluid which allows access to its interior". (M G Yasargil Clin Neurosurg 1986). Knowledge of this sulcal, fissural and cisternal anatomy is essential for the proper surgical management of brain tumours. While every neurosurgeon is well versed with the gyral anatomy and the partition of the brain into supra and infratentorial compartments, the sulci, fissures and cisterns have generally been excluded from the surgeon's perspective.

The main reasons for morbidity following surgery on intraparenchymal lesions include:

1. Surgical trauma to the normal cortex or white matter
2. Retraction injury to the brain
3. Vascular injury

SURGICAL INJURY TO THE NORMAL BRAIN TISSUE

Intraparenchymal lesions may present in the following ways with respect to the brain surface:

1. Entire tumour surfacing
2. Small portion surfacing
3. Below the surface but present within a sulcus
4. Completely subcortical–not presenting either at surface or sulcus
5. Intra or paraventricular

The route of approach to the lesion should be chosen such that the trauma caused to the normal brain is minimal. When the entire tumour is surfacing the lesion is removed by going through the surfacing area without disturbing the surrounding normal cortex. When only a small portion of the lesion is surfacing, the surgeon must stay within the surfacing small area of the tumour and debulk the lesion so that the lesion becomes small enough to be removed through the small cortical opening. However when the lesion is not surfacing the surgeon must study the available radiology to decide on the shortest path through normal white matter. The approach then depends on where the lesion is with respect to the gyral and sulcal pattern of the brain. If the tumour is not surfacing but just beneath the surface on the crest of the gyrus, the best approach would be to go through the top of the gyrus. If however the lesion is further deep and is close to one of the banks of the sulcus, the trans-sulcus approach has to be used when the lesion is deeper or intra-ventricular. Similarly approaches through fissures and cisterns can take the surgeon to deep portions of the brain hitherto thought to be inaccessible.

The mainstay of the trans-sulcal approach is to use the shortest route between the surface of the brain and the tumour so that surgical trauma to the normal brain is minimised. While approaching the tumour the microsurgeon is also saddled with the responsibility of looking after the microvasculature of the normal cortex and maintenance of the arachnoidal vessels and layer, so that the normal CSF milieu of the surrounding brain is maintained.

RETRACTION INJURY

The advantage of the operating microscope goes beyond the basic gains of illusmination and magnification, allowing precise dissection. The greatest advantage of the microscope is in the capability of sharop stereoscopic focus in a narrow deep opening with adequate illumination. The microscope creates a telescope effect which allows the surgeon to work at a depth of 10-12 cm through a small opening 5 mm wide and 16 mm long. The ability to work in this narrow and deep field with depth perception and illumination is indispensable. This allows the surgeon to minimise both the opening made in the normal brain tissue and also the use of retractors during surgery. Retractors cause brain injury in two distinct ways. With increasing retractor pressure the brain tissue gets pulled apart displacing and destroying fibre tracts. Increasing retractor pressures can cause venous hemorrhage in the adjacent brain tissue due to impedance to venous flow. Microsurgical techniques used in the removal of intraparenchymal lesions involve staying within the confines of the lesion, progressive debulking and removal of the tumour inside out. Once the main bulk of the tumour without further retraction. Thus with the microscope the surgeon does not have to make space (by retraction) to remove a tumour. The surgeon makes use of the space that the tumour has occupied and is therefore able to remove tumours with minimal or no retraction.

VASCULAR INJURY

The vessels that lie in relation to a tumour require careful consideration from the surgeon. Venous channels require more respect than the arteries. It is of prime importance that no normal venous channel be obliterated during surgery. While it is expected that with the reduction of intracranial pressure after tumour removal, collateral venous channels open up, and therefore obliteration of one or two veins do not affect venous outflow, this may not always be the case. Venous infarcts will develop where crucial channels have been sacrificed. The surgeon should therefore ensure that no venous channel is sacrificed unless there is no other option left.

There are three types of arteries that may be seen in relation to a tumour normal arteries, transit arteries and feeder arteries. Normal arteries may lie close to the tumour margin especially when the tumour abuts a sulcal surface. The surgeon must be able to identity the normal arteries and preserve them. Transit arteries pass through the tumour or on the surface of the tumour and go on to feed normal areas distal to the lesion. During their passage they may send off branches to feed the tumour. Generally vessels greater that 1 mm do not act as terminal branches for a tumours. It is therefore essential that the surgeon carefully isolate and identify the transit vessel, and coagulate only the feeding branches while preserving the former main vessels. Preserving the normal vessels and resecting only the feeders maintains normal blood supply to the adjacent normal brain while depriving the tumour of its vascular input.

At the time of tumour removal it may occasionally be difficult to decide which is a transit vessel and which is a terminal feeder. Debulking the tumour and creating space around the margins will bring into view the anatomy of the vessel and help dissect transit vessels and identify the feeders.

CASE ILLUSTRATION:

12 years old boy presented with symptoms and signs of raised intracranial pressure and right hemiparesis. The CT scan (Fig. 1) showed a mixed density irregularly

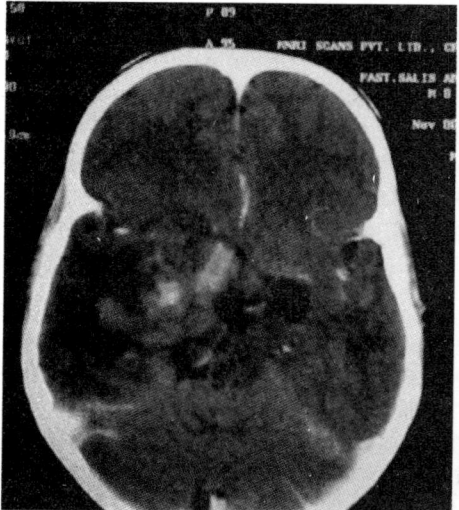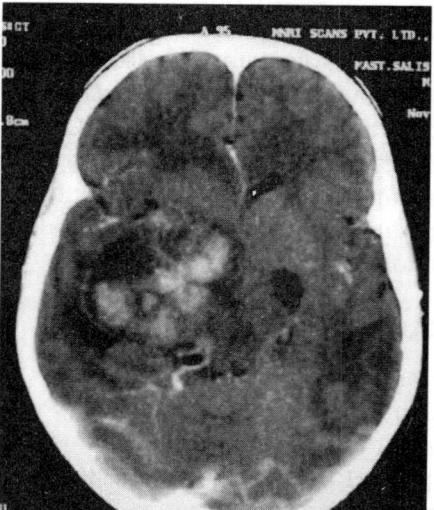

Fig. 1

enhancing lesion situated in the left medial temporal region bordering unto the ambient cistern medially and the region of the carotid bifurcation anteriorly. The lesion was approached through the inferior temporal sulcus (Fig. 2) and was removed radically expect for where the tumour was surrounding the branches of the Carotid, Middle Cerebral and Anterior cerebral arteries (Figs. 3, 4).

CONCLUSION:

Microsurgical techniques have to be employed to achieve the goal of a good neurological result with radical tumour excision. The use of the operating microscope cannot be restricted to the occasional difficult case. It must be made order of the day as very case must be given its due and every patient the best there is to offer. To give the advantage of precise dissection and preservation of function to

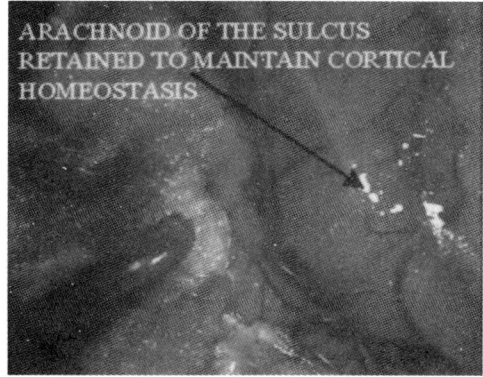

Fig. 2

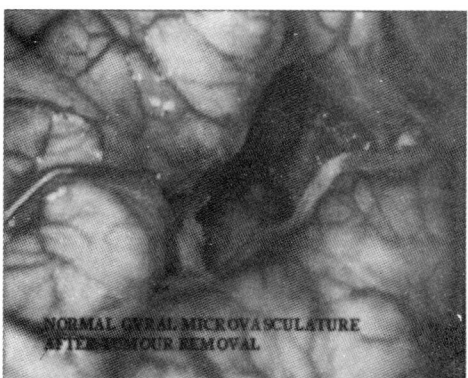

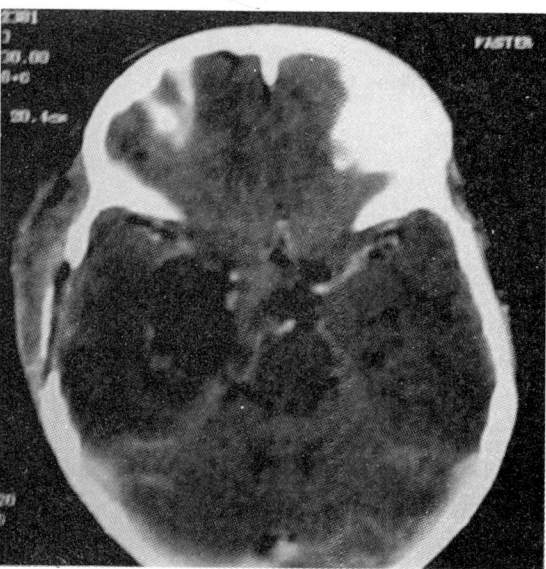

Figs. 3 & 4

benign tumours and to believe that it is not essential for malignant lesions of the brain is criminal. Only when the surgeon uses the microscope in all cases and at all times will the surgeon be able to confidently tackle the most difficult of situations. One cannot just climb onto the stage and perform. Practice makes perfect – Use the surfacing and easier tumours to practice microsurgical techniques – so that when the need arises in the case of deep seated lesions one is able to rise to the occasion.

The Operating Microscope does not only magnify the operative field but also the errors of the surgeon. The implication – the surgeon is forced to reduce the errors made.

> The result — A mediocre surgeon becomes
> a good surgeon !
> A good surgeon becomes
> an excellent surgeon !!

REFERENCES:

1. Andrew H Kaye, Edward R Laws: Brain Tumors—an encyclopedic approach. *Churchill Livingstone* 2001.
2. Laligam N Sekhar, Evandro De Oliveira: Cranial Microsurgery: Approaches and Techniques. Thieme Medical Publishers 1999.
3. Micheal Salcman: Intrinsic Cerebral Glioma in Brain Surgery: Complicaiton avoidance and mangement. (Ed.) MLJ Apuzzo. *Churchill Livingstone* 1993
4. Tandon PN: Supratentorial Astrocytoma in Textbook of Neurosurgery (Eds.) Ramamurthi B, Tandon PN, *BI Churchill Livingstone* 1996
5. MG. Yasargil: Surgical Approaches to Inaccessible Tumours. In Clinical Neurosurgery Vol. 34 William and Wilkins 1988

15
Imaging in Glioma

P Gulati, S Bankata

Tumors of neuroglial origin are known as gliomas. They are the largest group of primary CNS neoplasms. The three most common gliomas are astrocytoma, oligodendroglioma and ependymoma.

ASTROCYTOMAS

These are primary neoplasms that are composed of neoplastically transformed astrocytes. Astrocytomas are subdivided into fibrillary (diffuse) astrocytomas and circumscribed astrocytomas. Fibrillary astrocytomas are further divided into benign astrocytoma, anaplastic astrocytoma and Glioblastoma multiforme (Gliomatosis cerebri and Gliosarcoma). Circumscribed astrocytomas are divided into Pilocytic astrocytoma, Pleomorphic xanthoastrocytoma and Subependymal giant cell astrocytoma.

LOW GRADE ASTROCYTOMA

These affect a younger age group, have a more favorable prognosis and are less frequent in occurrence. These are unencapsulated tumors that may appear grossly circumscribed but are usually diffusely infiltrative. These are tumors of children and adults from 20 to 40 years of age.

On non contrast enhanced CT scanning these are seen as iso or hypodense lesions compared to the adjacent brain. Calcification may be seen in 15% to 20% of cases. They are usually non hemorrhagic with minimal or absent surrounding oedema. These show mild to moderate inhomogeneous contrast enhancement in about 40% of cases.

On MRI these are iso to hypointense as compared to adjacent brain on T1 weighted images and hyperintense on T2 weighted images. Edema, contrast enhancement and hemorrhage are uncommon. **(Figs. 1 & 2-2a)**

ANAPLASTIC/ MALIGNANT ASTROCYTOMA

These are among the most common primary malignant brain tumors. These may appear grossly circumscribed. Most of them are diffusely infiltrative with poorly

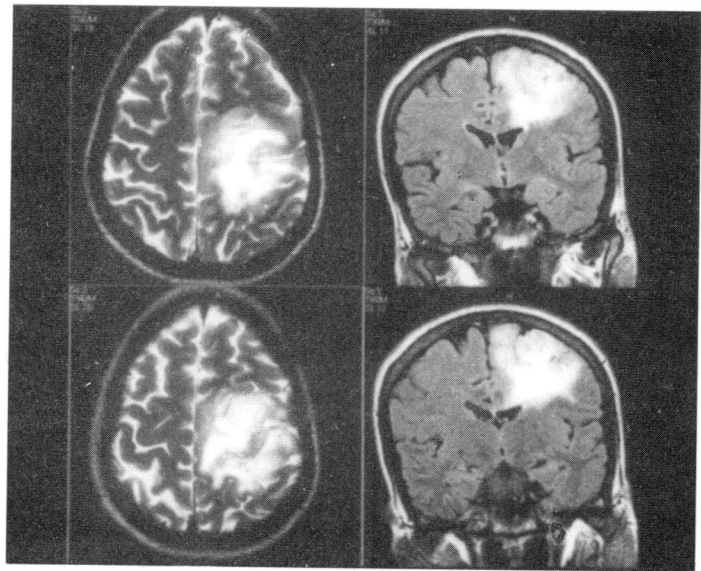

Fig. 1 : Axial T2 and coronal FLAIR scans showing left parietal lobe lesion with heterogeneous hyperintense signal on both the sequences. There is no significant midline shift. No area of necrosis. Biopsy–low grade astrocytoma

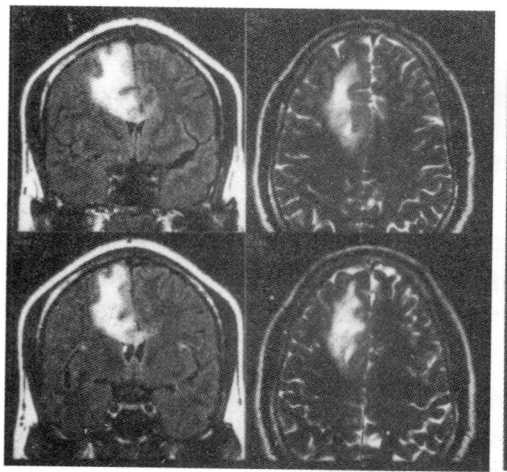 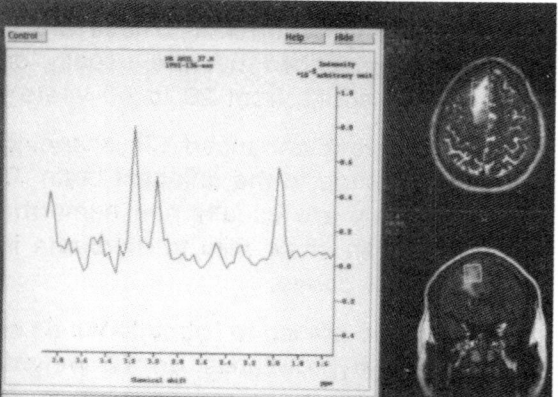

Fig. 2 : Axial T2 and coronal FLAIR scans showing right parietal lobe lesion with heterogeneous hyperintense signal on both the sequences. No area of necrosis. Biopsy–low grade astrocytoma

Fig. 2a : Single voxel spectroscopy of the same patient showing large choline peak with relatively low NAA. Biopsy low grade astrocytoma

delineated borders. Cystic areas are common. Hemorrhage can be seen but necrosis which is the hallmark of Glioblastoma multiforme is absent. Anaplastic astrocytomas compose approximately one quarter of all gliomas. These can occur in any age group but are typically found in older patients. The peak incidence is in the fifth and sixth decades of life. They are most commonly located in the frontal and temporal lobes.

Non contrast enhanced CT scans demonstrate them as inhomogeneous mixed density tumors. Calcification is uncommon. After administration of contrast they typically show strong and nonuniform enhancement. Irregular rim enhancement is also common. Varying amounts of peripheral edema might be seen. Some amount of intratumoral hemorrhage may be seen.

On MRI these are poorly delineated lesions that demonstrate heterogeneous signal intensities on both T1 and T2 weighted images. Mixed iso to hypointense areas are seen on T1 weighted images. Some hemorrhagic foci may be present. These have moderate mass effect. Marked and irregular peripheral ring like enhancement following contrast administration is seen. These may spread along the ependyma, leptomeninges and CSF. **(Figs. 3 & 4)**

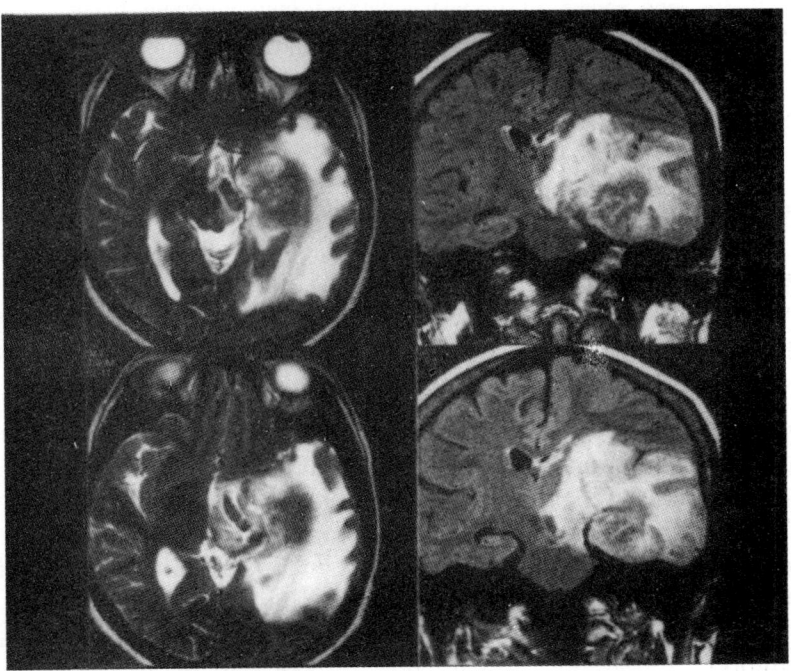

Fig. 3 : Axial T2 and coronal FLAIR images showing left deep temporal lobe lesion with heterogeneous hyper and isointense signal on both the sequences. There is associated perifocal edema showing hyperintense signal on both the sequences. There is associated mass effect and subfalcine herniation. Biopsy – anaplastic astrocytoma.

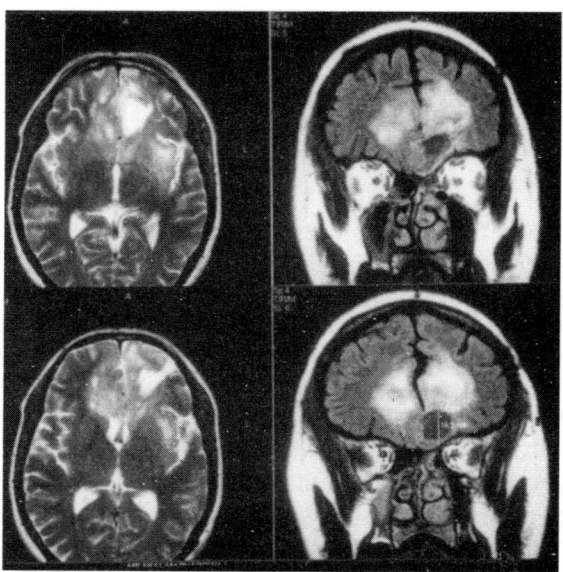

Fig. 4 : Axial T2 and FLAIR coronal scans in a follow up case of anaplastic astrocytoma in the left frontal lobe shows an irregular infiltrative heterogeneous mass in the left frontal lobe with transcallosal contralateral extension to right frontal lobe. There are associated post operative changes in the left frontal lobe.

GLIOBLASTOMA MULTIFORME

This is the most common of all primary intracranial CNS tumors comprising 15 to 20% of these tumors. 50% of astrocytomas are Glioblastoma multiformes. Glioblastoma multiforme usually occurs in patients over 50. These are usually located in the deep white matter, mostly in the frontal and temporal lobes. Multilobed and bihemispheric tumors that cross the corpus callosum are common. It is also the most malignant of all glial neoplasms. Necrosis is the hallmark of Glioblastoma multiforme.

They are usually large, heterogeneous masses with central necrosis, thick irregular walls and increased vascularity. Intratumoral hemorrhage is common. There is usually a lot of mass effect and edema. CNS metastases are common but distant metastases are rare. Glioblastoma multiformes spread widely and rapidly. The most frequent route is through extension along the white matter tracts. Bihemispheric spread across the corpus callosum, and anterior and posterior commisures is also seen. Extension along the internal and external capsules is common. Spread into the posterior fossa can occur down the cerebral peduncles and spinothalamic tracts. Satellite lesions often develop in patients with Glioblastoma multiforme that are grossly separated from the parent tumor but microscopically connected to it. Ependymal, subpial and CSF dissemination through the cerebral and spinal subarachnoid spaces occur at an advanced stage.

On non contrast enhanced CT scans these tumors are seen as markedly inhomogeneous masses. A central low density region is seen that represents necrosis or cyst formation in 95% of all Glioblastoma multiformes. Calcification is rare, hemorrhage of various ages is common. Peripheral edema surrounds the tumor and extends along central white matter tracts. There is moderate to strong but very inhomogeneous enhancement following intravenous contrast administration. Thick and irregular rim enhancement is a common finding.

On MRI scanning the heterogeneous nature of Glioblastoma multiforme is demonstrated. T1 weighted scans demonstrate a poorly delineated mixed signal mass with necrosis or cyst formation and a thick wall. Marked but inhomogeneous contrast enhancement is present in the majority of Glioblastoma multiformes. Prominent flow voids and hemorrhages of different ages are often present because of the highly vascular nature of these tumors. T2 weighted images reveal a very heterogeneous mass with mixed signal components. There is evidence of central necrosis which is seen as hyperintense signal on T2 weighted images. Hemorrhages of various ages are also seen. Peripheral edema is very striking. **(Fig. 5)**

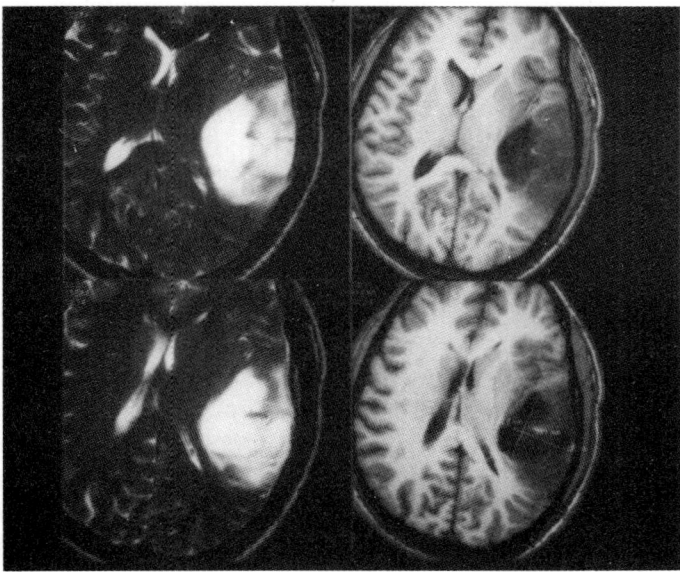

Fig. 5 : Axial T1 and T2W images showing heterogeneous mass lesion in the left temporal lobe with heterogeneous hyperintense signal on T2 and heterogeneous low intense signal on T1W images. There are areas of necrosis within the lesion seen as relatively very low signal on T1 and very bright signal on T2W images – Glioblastomamultiforme.

GLIOMATOSIS CEREBRI

This is a rare entity also known as diffuse cerebral gliomatosis. It is characterized by a diffusely infiltrative neoplastic overgrowth of large portions of the brain involving astrocytoma elements.

It is thought to represent an extreme in the spectrum of diffuse astrocytomas. Its markedly diffuse nature is thought to be a result of multicentric transformation and subsequent centrifugal spread. The underlying neuronal structures may be relatively spared. Gliomatosis cerebri diffusely enlarges the cerebral hemispheres, cerebellum or brain stem preserving normal anatomic landmarks.

The peak incidence is in the 2nd to 4th decades. The optic nerves and compact white matter pathways are the most common sites. The corpus callosum, fornices and cerebellar peduncles are also commonly involved. The clinical picture can be disproportionately mild compared to extent of involvement. The duration of the disease is variable and can last from months to years.

The most common imaging appearance is a diffusely infiltrating, non enhancing lesion that expands the cerebral white matter. A few contrast enhancing foci develop late during the course of the disease. On MRI there is presence of diffuse white matter signal hyperintensity on T2 weighted images. Usually there is no contrast enhancement.

Radiologically, the diffuse white matter changes may be difficult to differentiate from demyelinating (such as MS or PML) or dysmyelinating processes. Biopsy is sometimes necessary to make the distinction. In this case, mass effect argues for tumor. Lack of contrast enhancement argues against infection.

PILOCYTIC ASTROCYTOMA

These are 5% to 10% of all gliomas. They comprise one third of pediatric gliomas and are the second most common brain tumors in children. They are characteristically located around the third and the fourth ventricles. Nearly half are found in the optic chiasma and hypothalamus. Cerebellar vermian or hemispheric location is also common. Cerebral hemispheric and ventricular location is less common. These are well circumscribed but unencapsulated masses. Though cysts are common, necrosis is absent. Cerebellar astrocytomas typically have a large cyst with a small, reddish-tan mural nodule. Opticochiasmatic-hypothalamic Pilocytic astrocytomas are lobulated, grossly well circumscribed but microscopically infiltrating tumors in the floor or the walls of the third ventricle. Brainstem Pilocytic astrocytomas are generally uniform, non focal infiltrating neoplasms that diffusely expand the pons and medulla.

On non contrast enhanced scans Pilocytic astrocytomas are typically seen as round or oval sharply demarcated and smoothly marginated hypointense or isodense masses. Calcification occurs in approximately 10% of cases. After injection of intravenous contrast there is strong but variable enhancement. Some lesions enhance homogeneously in a solid manner while others have a small enhancing mural nodule in a large cyst. As the wall of most cystic astrocytomas is composed of nonneoplastic compressed brain it typically does not demonstrate enhancement. Occasionally some cases show mural enhancement. In a few cases, the cyst fluid enhances with dependent layering that creates a contrast-fluid level, particularly on delayed scans.

On MRI scans most cerebellar Pilocytic astrocytomas are cystic and appear hypointense or isointense on T1 weighted images and hyperintense on T2 weighted images. Mural nodules and solid tumors enhance strongly but inhomogeneously.

BRAINSTEM GLIOMAS

Brainstem gliomas are usually solid, infiltrating low grade nonpilocytic tumors. Pontine gliomas are usually diffusely infiltrating neoplasms that are homogeneously hypointense on T1 weighted images and hyperintense on T2 weighted images. Contrast enhancement is variable. The basilar artery may be encased by these tumors. Obstructive hydrocephalus is typically absent or mild even with advanced tumors. Medullary gliomas are low grade fibrillary or Pilocytic astrocytomas tht infiltrate and expand the medulla. Typical opticochiasmatic-hypothalamic Pilocytic astrocytomas are often solid. These are hypo to isointense with normal brain on T1 weighted images. They may occasionally attain a large size. Cyst formation and trapped pools of CSF may be prominent in some cases.

PLEOMORPHIC XANTHOASTROCYTOMA

This is a rare neoplasm. They are typically well demarcated, partially cystic tumors with a discrete mural nodule. Leptomeningeal adhesion or attachment is common. These are tumors of children and young adults with the average age being in the second or third decade. Typically a superficial cortical location is seen. The mural nodule usually abuts the leptomeninges. The temporal lobe is the most common site followed by parietal, occipital and frontal lobes.

On non contrast enhanced CT scans these are typically hypodense cystic appearing masses with distinct borders. A mural nodule that strongly enhances after contrast administration is a common finding.

On MRI most of these lesions are well delineated partially cystic masses that appear hypo or isointense to normal brain on T1 weighted images. A peripheral nodule is quite frequently seen. Increased signal intensity of both the nodule and cystic component are seen on proton density and T2 weighted images. Following contrast administration marked enhancement of the nodule can be seen.

SUBEPENDYMAL GIANT CELL ASTROCYTOMA

This is a circumscribed tumor that is associated with tuberous sclerosis. Subependymal hamartomatous proliferations along the caudothalamic groove are common in TS. These are typically sharply defined lobulated masses that are often calcified, cysts are common. These occur in 10% to 15% of patients with tuberous sclerosis. They are characteristically slow growing. They occur at the foramen of Monro and virtually never anywhere else in the brain.

On CT scan a focal mass is seen at the foramen of Monro with enlargement of lateral ventricles. Calcified Subependymal nodules along the striothalamic groove can also be identified in a majority of cases. These are heterogeneous neoplasms with mixed hypo and isodense regions. Calcification and cyst formation are common. Typically strong but inhomogeneous contrast enhancement is seen following intravenous contrast administration.

On MRI mixed signal intensity is observed on both T1 and T2 weighted images due to the heterogeneous nature of the mass lesion. Most of these lesions are hypo and isointense on T1 weighted images and iso hyperintense on T2 weighted images. There is strong but inhomogeneous enhancement.

OLIGODENDROGLIOMA

These are uncommon gliomas arising from the oligodendrocyte that make the CNS myelin. They constitute 5% of all gliomas. There is a high incidence (90%) of calcification (gyral brain calcification) in these tumors. They are predominantly lesions of the cerebral hemispheric white matter that grow outward from white matter into gray matter and are relatively avascular. Most frequently these tumors occur in young and middle-aged adults but are also found in children. The peak incidence is between 35 and 45 years of age. The most common site is the cerebral lobes of which 50% occur in the frontal lobe. The most common initial symptom is seizure. These tumors represent about 4% of all primary brain tumors. The pure form of oligodendrogliomas are rare. The mixed tumors contain both oligodendrocytes and astrocytes and are far more common. The grading system for oligodendrogliomas is on scale of A through D or I through IV, depending on the classification system used. The grade denotes the most malignant type of cell found in the tumor. Oligodendrogliomas are typically unencapsulated but well-circumscribed focal white matter tumors that may extend into the cortex and leptomeninges. Occasionally a poorly delineated, diffusely infiltrating mass lesion is seen. Foci of cystic degeneration are relatively common, but hemorrhage and necrosis are less frequent. Most oligodendrogliomas are slow growing neoplasms. These lesions are usually round / oval and hypodense and cause mass effect (75% of cases).

On CT scan oligodendroglioma is the most common intracranial tumor to calcify. Nodular or clumped calcification is seen in 70% to 90% of cases. On non contrast enhanced scans a partially calcified mixed density hemispheric mass that extends peripherally to the cortex is visualized. Cystic degeneration is commonly seen. Gross hemorrhage and edema are relatively rare. There is commonly no or minimal tumor enhancement (75%), however, there is pronounced enhancement in high-grade tumors. These tumors may be adherent to dura (mimicking meningiomas). There may or may not be erosion of inner table of skull.

MRI scans show mixed hypo and isointense areas on T1 weighted images and hyperintense foci on T2 weighted images. There is typically patchy and moderate enhancement. MRI is less sensitive than CT in detecting tumor calcification but delineates the tumor extent better.

EPENDYMOMA

Ependymomas comprise approximately 6% of intracranial gliomas. They comprise 8%–10% of childhood brain tumors and 63% of spinal intramedullary gliomas. Supratentorial ependymoma can occur at any age; posterior fossa ependymoma occurs at less than 10 years of age. The peak age range for ependymomas is 1 to 5 years. Approximately 60% of intracranial ependymomas are located in the 4th ventricle. 40% are located in the supratentorial compartment. Between two thirds and three quarters of supratentorial ependymomas are extraventricular.

Ependymomas are slow growing lobulated neoplasms that are often partly cystic. Calcification is common but gross hemorrhage is rare. When ependymomas occur in the fourth ventricle, they arise from the floor or roof and protrude through the outlet foramina into the adjacent CSF cisterns.

On CT scan the infratentorial ependymomas have a variable appearance. They may appear as sharply marginated isodense / slightly hyperdense mass lesions. Most are isodense on non contrast enhanced scans. There is presence of calcification in approximately 50% of ependymomas. There is presence of central necrosis in 50% of cases. In 70% of cases there is mild to moderate enhancement following intravenous contrast administration. There is presence of communicating hydrocephalus (100%) secondary to protein exudate clogging up resorption pathway. Characteristically there is expansion through foramen of Luschka into CPA or caudad into cisterna magna. There is presence of dissemination via CSF in 10%–33% of cases.

On MRI the classic appearance of posterior fossa ependymoma is a lobulated soft tissue mass that appears to form a cast or mold of the fourth ventricle and extrudes through its outlet foramina into the adjacent subarachnoid cisterns. The solid components are typically hypo or isointense compared to brain on T1 weighted images. They are hyperintense on T2 weighted images and proton density scans. The cystic portions are slightly hyperintense to CSF on T1 weighted images and hyperintense to brain on T2 weighted images. Intratumoral heterogeneity may represent necrosis, calcification, tumor vascularity or blood degradation products. Typically there is moderate inhomogeneous enhancement following contrast administration.

SUBEPENDYMOMA

These are rare benign CNS tumors occurring in middle aged and elderly adults. These arise most commonly from the lower medulla and project into the caudal fourth ventricle. Frontal horn of the lateral ventricle is another common location where they attach to the septum pellucidum. A few are found along the midbody of the lateral ventricle. Subependymomas are firm, well-delineated intraventricular masses that are attached to the septum pellucidum or inferior fourth ventricle.

On non contrast enhanced CT a well-delineated mass that is hypo or isodense to brain parenchyma is seen. Calcification is sometimes seen. On post contrast studies minimal or no enhancement is observed.

On MRI most subependymomas are homogeneously hypo to isointense masses on T1 weighted images. They may appear mildly hyperintense on T2 weighted images. There is evidence of some signal heterogeneity that reflects multiple small intratumoral cysts. Contrast enhancement is typically absent.

Spectroscopy is a noninvasive MR technique that gives the relative concentrations of certain chemical compounds within two to three cubic centimeters of tissues.

MRI has proved to be a very useful modality to the radiologist examining the brain for evidence of neoplasm. MR spectroscopy provides some special advantages in the imaging of brain neoplasm over that can be achieved by MR imaging alone. The advantages include –

- **It provides better differential diagnosis**
- **It provides a definition of tumor grade, aggressiveness and relevant biochemistry.**
- **Monitoring of a successful tumor response before its regression during non-surgical treatments and conversely the early definition of tumor recurrence.**

BASIC NEUROCHEMISTRY OF TUMORS

Five biochemical defects in varying degrees are common to the majority of brain tumors as measured by MRS.

- **Decreased NAA**
- **Increased Lactate**
- **Increased Lipid**
- **Decreased total creatine**
- **Increased choline**

Significance of Changes in Tumors Spectra

A reduction in or absence of NAA signifies the absence of neurons and axons from most tumors. Lactate is normally undetectable in brain Excess lactate is indicative of high grade active tumors because of the high ate of glycolysis in tumors. Lipid signals associated with tumor spectra have been ascribed to necrotic regions untreated tumors and treatment responsive necrosis in treated tumors. Decreased creatine is not a constant finding in tumors. When present it has been ascribed to the low energy status of glycolysing tumors. Excess choline is indicative of malignancy and is nearly always observed. This signifies increased cellular metabolism and rapid cell tumors. A reduction in all metabolites is indicative of radiation necrosis. Hence a distinction between recurrent tumors and radiation necrosis can be made.

REFERENCES:

1. Achten E, Jackson GD, Cameron JA, Abbott DF, Stella DL, Fabinyi GCA: Pre-surgical evaluation of the motor hand area with functional MR imaging in patients with tumors and dysplasic lesions. *Radiology* 1999;210:529-38.
2. Atlas SW: Adult supratentorial tumors. *Sem roentgenol* 1990;25:130-54.
3. Castillo M, Kwock L: Proton MR spectroscopy of common brain tumors Neuroimaging. *Clin North Am* 1998;8:733–52.
4. Dean BL, Drayer BP, Bird CR, et al: Gliomas: classification with MR imaging. *Radiology* 1990;174:411-15.
5. Earnest F IV, Kelly PJ, Scheitauer BW, et al: Cerebral Astrocytomas: Histopathologic Correlation of MR and CT contrast enhancement with stereotactic biopsy. *Radiology* 1988;166:823-27.
6. Forsting M, Wirtz CR, Tronnier VM, Staubert A, Kunze Synergy, Sartor K: Extirpation of glioblastoma: MR and CT follow-up residual tumor and re-growth patterns *Am J Neuroradiol* 1993;14:77-87.
7. Graif M, Bydder GM, Steiner RE, et al: Contrast- enhanced MR imaging of malignant brain tumors. *AJNR Am J Neuroradiol* 1985;6:855–62.
8. Lee Y-Y, Van Tassel P: Intracranial oligodendrogliomas: imaging findings in 39 untreated cases. *AJNR* 1989;10:119-27.
9. Lev MH, Rosen BR: Clinical applications of intracranial perfusion MR imaging. *Neuroimaging Chn North Am* 1999;9:309-31.
10. Lilja A, Bergstrom K, Spannare B, et al: Reliability of computed tomography in assessing histopathological features of malignant supratentorial gliomas. *J Comput Assist Tomogr* 1981;5:625-36.
11. Philippon JH, Clemenceau, SH, Fauchon FH, Foncin JF: Supratentorial low-grade astrocytomas in adults. *Neurosurg* 1993;32:554-59.
12. Sartor K: MR Imaging of the brain tumors. In Syllabus of MRI – From Basic Knowledge to Advanced Strategies, 2000;34-41.
13. Shin YM, Chang KH, Moon HH, et al: Gliomatosis cerebri Comparison of MR and CT features. *AJR Amj Roentgenol* 1993;161:859-62.
14. Spoto GP, Press GA, Hesselink JR, Solomon M: Intracranial ependymoma and subependymoma: MR manifestations, *AJNR* 1990;11:83-91.
15. Watanabe M, Tanaka R, Takeda N: Magnetic resonance imaging and histopathology of cerebral gliomas, *Neuroradiol* 1992;35:463-69.

16
Post-Operative Imaging– Which and When?

V Sawlani, S Jain, S Kumar

The general approach to the treatment of brain neoplasm is surgical resection of solitary lesions or limited disease followed by radiation therapy with or without chemotherapy. Following surgery for a brain tumor, complications such as intra or extra cerebral hematoma, cerebral adema and infection may develop. Tumour persistence or recurrence may follow. Some time focal radiological changes, particularly iatrogenic may simulate tumor growth.[1]

In the evaluation of these patients special care should be exercised in the performance of neuroradiologic examinations in order to demonstrate the pathologic process more precisely. Although the long term prognosis for patients with malignant brain tumours is limited the identification of patients who are likely to benefit from more aggressive therapy and the ability to make an early determination of treatment failure are important issues for pts. Management.[2] Treatment failure resulting from tumour recurrence is common and need to be differentiated from post radiotherapy changes for further appropriate treatment.

In the immediate post operative period the most common and important complications are hemorrhage, parenchymal swelling and intradural or extradural hemotoma. **(Fig. 1)** All of these can be life threatening because of mass effect.[3] Infarction or infection may develop in acute post operative period. The pt may be unconscious or under the influence of anesthesia and can not co-operate in this period. CT is the diagnostic study of choice for the initial evaluation in the immediate post operative period due to fast examination time, Lack of contraindication and high accuracy for detection of acute hemorrhage.[4] **(Fig. 2)** However numerous artifacts may be produced by metallic clips and post operative air. Every effort should be done to obtain high quality CT examination. In interpreting post-operative CT examination attention should be focused first on whether the purpose of surgery is accomplished and secondly whether there are associated complications. Comparison with pre-operative CT is essential.

MRI is not appropriate for evaluation of critically ill patients with a respirator or electronic monitors or pts. having intracranial implants. A longer examination time continues to be the major drawback to wide spread use of MRI in acutely ill

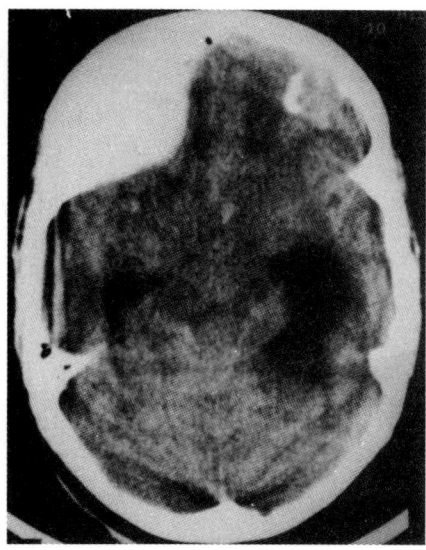 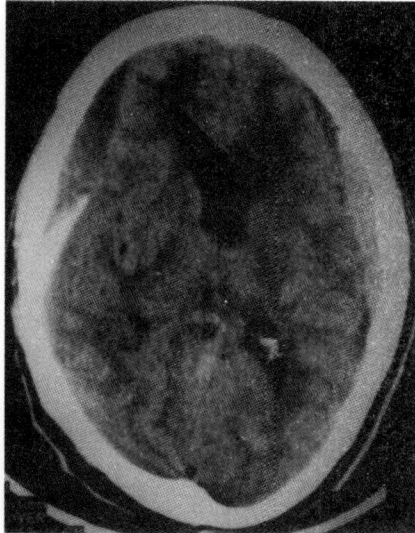

Fig. 1 : Immediate post-operative plain CT showing a large right frontal extradural hematoma with significant mass effect

Fig. 2 : Operated case of posterior fossa tumor. Immediate post-operative plain CT showing bilateral fronto-parietal subdural hematoma due to sudden decompression

patients, although this limitation has been largely over come recently with development of fast imaging techniques.

In the subacute and late post operative period the indications of neuroradiologic studies are to detect persence of infection, fluid collection, parenchymal loss, residual or recurrent tumours and radiation necrosis.

During the post-operative period focal iatrogenic radiological changes can simulate progressive tumour. The mechanical effect of excising the tumour causes a reversible break down of the blood brain barrier with appearance of enhancement at the limits of surgical cavity. This enhancement appears between 5th and 15th day and spontaneously regress with in several months. To avoid the diagnostic error of residual tumor the post operative baseline imaging should be done with in 96 hrs. of surgical intervention. Base line post-operative imaging is a standard reference for tumour progression and treatment monitoring. **(Figs. 3(a-b))**

The CT evaluation of recurrent or residual tumour may be inadequate because of lack of enhancement, subtle mass effect and inability to distinguish post operative encephalomalacia from tumour edema and post operative anatomic distortion. Hence CT sometimes under-estimate the volume or the spread of the tumour. MR imaging particularly contrast enhanced is much more sensitive than CT for detecting subacute and late post-operative changes due to inherent high

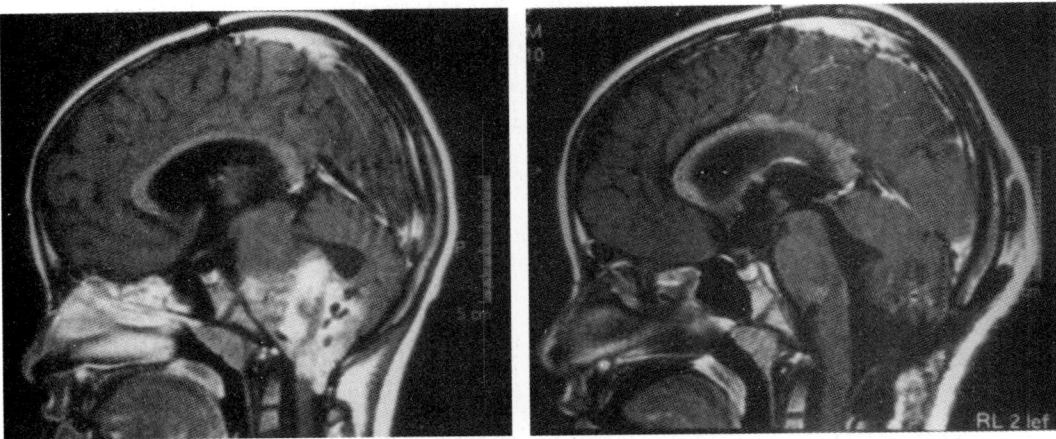

Figs. 3 : (a) Contrast enhanced T1W sagittal image showing enhancing 4th ventricular ependymoma (b) 3rd day post-operative baseline contrast enhanced MR study shows no enhancing residual lesion

contrast and spatial resolution, multiplannar capability and ability to evaluate temporal evolution of hematoma and hemorrhagic lesion.[5] **(Figs. 4(a-b))** T1 weighted MR images better demonstrate the soft tissue changes, mass effect and distorted Anatomy. T2 weighted MR images demonstrate the extent of tumour edema complex. MR imaging with Gd-DTPA (contrast enhanced MRI) is impor-

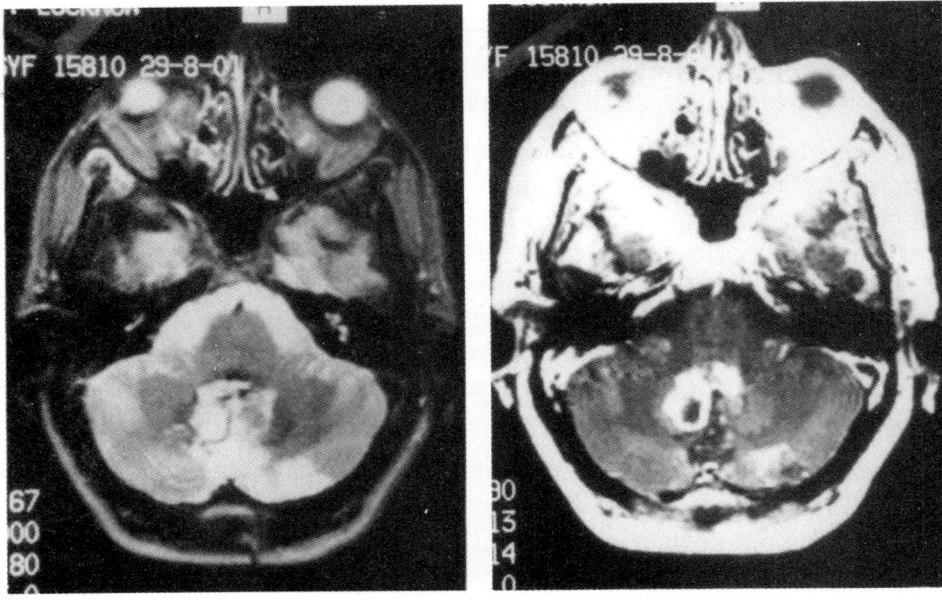

Figs. 4 : (a) Post-operative anaplastic cerebellar astrocytoma. Follow up MRI T2W axial (b) images shows residual enhancing tumor with evidence of haemorrhage

tant for evaluation of residual and recurrent tumours and Leptomeningeal and subependymal spreads.[6] Recurrent tumours may be diagnosed on the basis of either focal mass with anatomic distortion on serial scans or new enhancement in or around the site of the original tumour in comparison with the baseline post operative study. **(Figs. 5(a-b))**

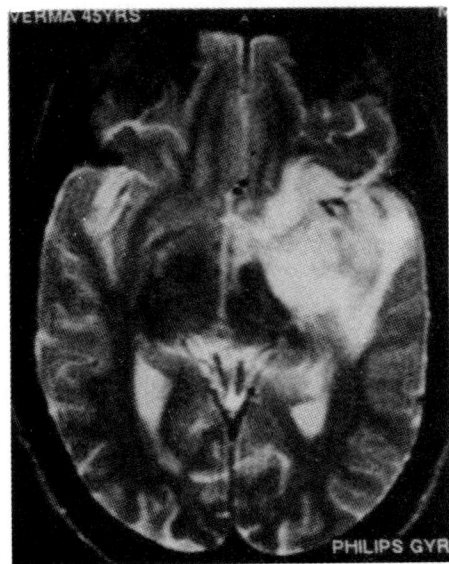
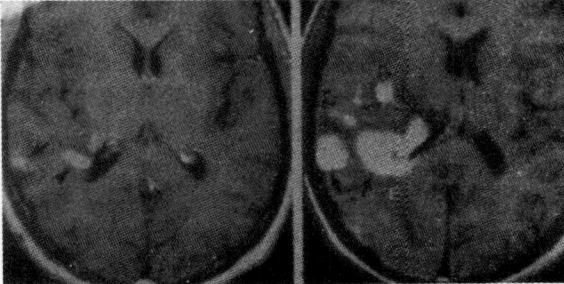

Figs. 5 : (a)T2W axial image shows left temporal lobe astrocytoma (b) 3 years later post-operative contrast enhanced T1W image shows recurrent enhancing tumor

Radiation necrosis closely resemble recurrent tumour because of following shared characteristics. They both originate at or close to the original tumour site, show contrast enhancement, grow over time and produce edema and mass effect. Kumar et al[7] has proposed some characteristic radiological features of radiation necrosis e.g. A new enhancing lesion with soap bubble or Swiss cheese pattern. Non enhancing tumour before surgery subsequently develop enhancing lesion. Enhancing lesion develops at a distance from primary glioma particularly in periventricular white matter.[7] Though CT and MR imaging provide good morphological information but still are ambiguous in distinguishing radiation necrosis from recurrence tumour.

MR spectroscopy may provide useful metabolic information which along with morphological images impart tissue discrimination and may help in differentiating recurrent tumour from radiation necrosis. The quantitative analysis of MRS data will assist in detecting early changes in cellular metabolism which corresponds to tumour recurrence and may also have a role in localized radiation therapies. The active tumour has high choline and reduced or absent NAA whereas radiation necrosis has low choline, creatinine and NAA levels.[8-9] **(Fig. 6)**

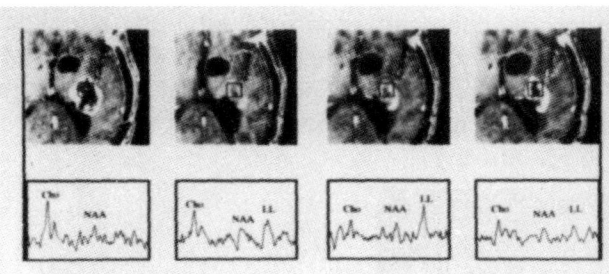

Fig. 6 : MRI and MR spectroscopy - post-operative, post-radiotherapy serial follow up of enhancing lesion shows decreasing choline level suggestive of radiation necrosis

However there are certain limitation of MR spectroscopy as there may be sampling error in single Voxel MRS. Hemorrhage calcification and surgical clips at surgical site warrant MR spectroscopy. Recently multi Voxel chemical shift imaging has shown promising results.[9] However it is time consuming and operator dependent.

MR perfusion imaging can reveal changes in tumour vascularity and may help in evaluation of tumour response and adequacy of therapy. Angiogenesis is one of the most malignant feature of recurrent high grade glioma. Dynamic contrast enhanced T2* W EPI imaging assess the relative central flow volume (rCBV) which is a direct measure of angiogenisis. The enhancing tumour recurrence show rCBV > 2.6 when as nonneoplastic contrast enhancing lesion show rCBV < 0.6.[10]

Positron emission tomography (18 FEG PET) can differentiate radiation injury from malignancy on the basis of difference in glucose uptake. Malignant tissue have high rate of aerobic glucose metabolism. The sensitivity of PET is 81-86% and specificity 50-94%.[11] However there may be false -ve in low histologic grade glioma and small tumour volume and false +ve in non malignant inflammatory process. Due to the high cost and non availability of PET particularly in developing countries, 201 Tl SPECT can be used instead which is widely available and economical **(Fig. 7)**. Both agents are markers of viable tumour and both techniques are equally sensitive for tumour recurrence with the lesion more than 1.5 cm in size.[11]

Computer – assisted stereotactic biopsy is a useful, safe and accurate means to obtain a pathological diagnosis in patients which irradiate gliomas who have evidence of disease progression. It is a invasive procedure but can be used when other imaging studies are equivocal.[12]

Post-operative imaging which and when? There is no single ideal post-operative imaging modality to answer all the quarries in the post-operative period. However depending on the post-operative period and pts. Condition the optimal imaging modalities are as follows. In immediate post-operative period plain CT is

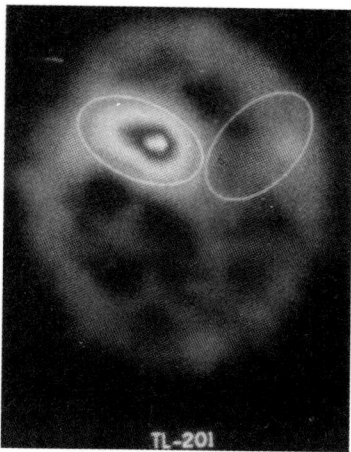

Fig. 7 : Follow up case of right frontal lobe glioma. 201 T1 SPECT scan shows increase uptake in right frontal region suggestive of recurrent lesion

the modality of choice as it shows the relevant important post-operative events in shortest possible time in critically ill patient. Then in acute post-operative period, contrast MRI should be done with in 96 hrs, of surgery as a baseline reference study for monitoring of disease process. Depending on pts clinical condition follow up sequential imaging may be done at every 8-12 weeks in the first year. (C EMR >CECT). Subsequent follow up imaging at 3-6 months interval depending on the initial tumour grade and patients deficit.

Contrast enhanced MRI or CT may not be adequate to differentiate recurrent tumour from therapy induced necrosis. In this situation advanced imaging techniques like MR spectroscopy, MR perfusion imaging, PET and SPECT scan or computer assisted steriotactic biopsy can be used for disease monitoring and appropriate treatment.

REFERENCES:

1. Horowitz NH, Rizzoli HV: Post-operative complications of intracranial neurological surgery, Baltimore, Williams and Wilkins, 1982;20-365.
2. Liebel SA, Scott CB, Loeffler JS: Contemporary approaches to the treatment of malignant gliomas with radiation therapy: *Seminar Oncol* 1994;21:198-219.
3. De La Paz RL, Davis KR: Post operative imaging of the posterior fossa. In taveras JM, Ferrucci JT (eds): Radiology: Diagnosis, Imaging, intervention, Philadelphia, JB Lipincott, 1988;1-11.
4. Jeffries BF, Kishore PRS, Singh KS, et al: Post-operative computed tomographic changes in the brain. *Radiology* 1980;135:751.
5. Lanzieri CF, Larkins M, Mancall A, et al: Cranial post-operative site: MR imaging appearance. *AJNR* 1998;9:27.
6. Elster AD, Di Periso DA: Cranial Post-operative site: Assessment with contrast enhanced MR imaging. *Radiology* 1990;174:93.

7. Kumar AJ, Leeds NE, Fuller NG, et al: Malignant gliomas: MR imaging spectrum of radiation therapy and chemotherapy induced necrosis of the brain after treatment. *Radiology* 2000;217:377-84.
8. Nelson SJ, Huhn S, Vigneron DB, et al: volume MRI and MRSI techniques for the quantification of treatment response in brain tumour. *JMRI* 1997;7:146-52.
9. Grevesee, Nelson SJ, Vigneron DB, et al: Serial proton MR spectroscopic imaging of recurrent malignant gliomas after gamma knife Radiosurgery. *AJNR* 2001; 22:613-24.
10. Cha S, Knopp AE, Johnson G, et al: Dynamic Contrast enhanced T2* W MR imaging of recurrent malignant gliomas. *AJNR* 2000;21:881-90.
11. Kahn D, Follett KA, Bushnell DL, et al: Diagnosis of recurrent brain tumour: value of 201 Tl SPECT Vs 18 F- Fluorodeoxyglucose PET. *AJR* 1994;163:1459-65.
12. Marc Levivier, Antonio Bacerra, et al: Radiation necrosis or recurrence. *J Neurosurgery* 1996; 84:148-49.

17
Closure Techniques and Controversies

S Kumar, V Gupta, D Singh

Principles of closure techniques are: Meticulous hemostasis, avoidance of dead spaces, sealing of opened paranasal sinuses and maintenance of thorough asepsis.

The control of hemostasis begins preoperatively with proper and physical examination especially suggestive of inherited coagulation and platelet deficiencies, liver disease, Vit.-K deficiency and history of anti-coagulation therapy. The correction of preoperative identifiable haemostatic defects can make closure a success.

Friable surgical bed, after adequate haemostasis should be covered with chemical hemostats like oxycel or surgicel or absorbable gelatin sponge. However, chemical haemostatic agents must not be treated as substitutes for meticulous surgical technique. Surgicel is acidic and reacts with blood to form an artificial clot and has also antibactering action due to low pH.[1,29] An excess quantity of surgicel can mimic tumour or hemorrhage on postoperative computed tomography.

DURAL CLOSURE

Planning of dural closure begins at the start of the case. Consideration should be given to preserving pericranium and temporal fascia. While making craniotomy tearing of dura should be avoided. Further bone should be removed until intact dura is seen. Dura should not be coagulated excessively; otherwise shrinkage and retraction will prevent subsequent closure.

The dura should be closed in as water tight as possible. The theoretical advantages of watertight closure of dura include a reduction in the incidence of aseptic meningitis, a decreased incidence of CSF fistulas and subgaleal CSF collections and facilitation of subsequent reoperation. For dural closure use of absorbable suture material such as polyglactin decreases risk of foreign body reaction and adhesion, which hinder reoperation.[30] As leak occurs within physiological range of pressure, in patients who are of significant risk of developing a CSF leak the complimentary use of fibrin glue over the suture line is recommended.[23,28] An opened frontal sinus is closed with attached pericranial graft after exantration of mucosa.

Hemostasis from epidural space is achieved by placing tacking sutures between dura and bony margins of craniotomy. Dissection of unexposed dura beyond the craniotomy margins should be avoided as it may result in epidural hematoma formation in spite of tacking sutures. The sutures should pass through only the outer layer of the dura otherwise injury to cortex or cortical vein may result in subdural hemotoma. If there is any doubt of such complication, dura should be opened and hemostasis assured. While making a drill hole in the bone, dura should be protected with a retractor or hole be made with bone punch. Tacking the surface of dura to the convexity of bone flap can obliterate the dead space between the dura and convex bone flap. Excessive dural traction during taking suture is avoided, as this may cause further bleeding.

When the dura is resected due to tumour invasion and whenever the dura opening can't be closed primarily, reconstruction is recommended especially when the defect is large, when reoperation is likely, when a paranasal or mastoid sinus is breached or when a csf fistula is anticipated. If watertight closure is not crucial, small defects can be covered with a gelatin sponge otherwise, duraplasty is recommended. Duroplasty with autologus tissue, such as pericranium or fascia lata is preferred over prosthetic or lyophilized cadaveric graft materials.[27] These substitutes incite considerable foreign body reaction and occasional hemorrhagic complications.[3,T10,19] Lyophilized dura substitutes are known to cause Creutzfeldt-Jacob disease.[15,33] Alloderm (acellular human dermis) is a reasonable alternative for available dural graft materials. Waren et al (2000) in a series of 200 cases of such procedure found three csf leaks and 4-wound infection. Sellar reconstruction with acellur dermal graft eliminates the need for sphenoid obliteration after transsphenoidal hypophysectomy, Citard et al 2000).

Anson and Marchand (1997) reported good or excellent outcome in 32/35 by using borine pericardium as dural substitute. It is relatively inexpensive, and requires no additional incision, it has low antigencity and toxicity, good strength and minimal elasticity.

Pericranial flap repair of anterior skull base defects has a 90% complication free and 95% overall success rate. It is simple, effective and in author's experience bone graft was not necessary.[21]

Poloxamer 407 has been used intradurally, experimentally to prevent spinal arachnoidal adhesion by 50%.[22]

CLOSING THE BONE FLAP

The objective is closing the craniotomy are:
i) To restore the shape of the cranial vault
ii) To protect underlying brain

iii) To prevent infection and facilitate healing of the skin flap

iv) To prevent hemispheric collapse, ventricular deformation and midline displacements (sinking skull – flap syndrome)

Properly planned craniotomy and adherence to basic principles of surgery usually easily meets the above objective. The craniotomy flap should be appropriately placed to prevent the need for additional bone removal. Edges of the flap should be bevelled to avoid future sinking of the flap. Bone wax be applied to the cut surface to control bleeding. Paranasal sinuses and mastoid air cells be avoided unless dictated by the necessity. Bone flap may be fixed with wire or synthetic suture, which should be passed through bone prior to tacking up the dura otherwise later is likely to be injured. If the wire is used, the ends of each stitch should be buried in drill hole to prevent skin irritation. Alternatively xy shaped mini plates provide adequate fixation and more rapid bone healing instead of a fibrous union.[25]

An epidural hemovac drain through a separate stab incision is left at craniotomy site just prior to replacement of bone flap. However, we do not practice this technique in our institution.

If there are more than one fragments of bone at craniotomy the pieces should be reconstructed meticulously prior to replacing the flap. Burr holes can be packed with bone dust or filled with synthetic bone plugs, which is not being practised by us and patients, have never complained about it. In frontotemporal craniotomies, temporalis muscle be sutured so as to cover the key burr hole.

Cranioplasty is indicated when bone flap cannot be replaced because of invasion by tumor, multiple fragments cannot be reconstituted or when infection is present. When bone is removed because of infection, minimum three months should pass from the time antibiotics have been discontinued before undertaking cranioplasty. Normalization of EEG, seizures, motor and speech function have been reported following cranioplasty, especially in calvarial defects exceeding 100 cm.[6,26] Brain protection and cosmesis are often consideration for cranioplasty.

Cranioplasty may be done with autogenous, homogenous or synthetic material.

Autogenous grafts have the advantage of being biologically compatibility and donates matrix allowing in growth of osteoblasts. Tibia, fibula, rib, iliac crest, scapula, sternum or outer table of skull has been used.[13] Babcock[4] used hetrogenous grafts using sheep or ox bone. Repair of craniotomy defects using bone marrow stromal cells was advocated by Krebsbach (1998) Osteogenesis repaired $99 \pm 2.20\%$ of original surgical defects within two weeks in mice.

Metals used in cranioplasty provide strength, malleiability and inertness. Gold, silver, aluminum, platinum, titanium, tantalum and steel have been used.[5,24,32] Ideal material for cranioplasty be inert, nontoxic, malleable, strong enough to protect underlying brain, a poor thermal conductor, radiolucent allowing subsequent X-ray and CT examination to be made without disturbing images.

Methyl methacrylate introduced in 1940 by Kleinschmidt[11] fulfils most of the criteria listed above and is widely used for repair of skull defects. It is supplied in a kit that contains a liquid monomer and powdered polymer.

Defects in the frontal region are bald area require reconstruction. Skulls defects larger than 2 to 3 cm in diameter should be considered for cranioplasty but such sized defects underlying thick muscle mass i.e. temporal and occipital region are not normally operated upon. In children less than 3 yrs. of the age the outer layer of the dura serves as periosteum and is capable of osseous regeneration. New bone formation molds well in respect to brain growth and total reossification occurs within 6-12 months. The introduction of rigid prosthesis offers counter compression but restricts brain growth and hence not recommended for children below the age of 3 years. Split rib graft serves the purpose better at this age.[31]

Intact cranium gives cosmetic effect and also protective role for brain growth and integrity. Erculei & Walker (1963) in a study of 342 war time injuries observed that cranioplasty shortens convalescence and improvement in seizures is attributable to simultaneous resection of cicatrix.

Other indications for cranioplasty include sense of insecurity caused by the presence of skull defects. Mount (1948) suggested tantalum discs to be filled it disfiguring anterior burr holes. In contaminated wounds or obvious infections one should wait for 6-12 months to monitor healing process.

The graft is secured with wire or any other suture material. Pericranium and galea are approximated for snug adherence. Perforation of the plate to promote in growth of connective tissue for fixation and drainage of underlying collection is advocated. Drains may be placed over the graft material.

REPAIRING THE SKIN FLAP

Aims are to restore cosmesis and minimize the possibility of skin infection. This can be achieved by skin apposition with eversion of epidermal edges to maximize skin healing and prevention of formation of postoperative hematomas and seromas. Scalp vessels need not be coagulated as hemostasis is easily achieved by good approximation of galeal edges. Skin is closed with interrupted nonabsorbable sutures are with staples. Extensive scalp may require a subgaleal drain for a period ofod of 24 hours to eliminate seroma. To avoid scalp necrosis, skin flap should be planned with consideration of previous skin incisions and scalp vascularity. Strangulation of scalp tissue should be avoided otherwise stitch abscess may form. Intraoperative scalp expansion for primary wound closure by placing a tissue expander subgaleally and intermittent air expansion carried out for 20-30 minutes was reported successfully in 9 cases by Onishi et al 1997.[19a]

REFERENCES:

1. Anand AG, Sawaya R: Intra-operative chemical hemostasis in neurosurgery. *Neurosurgery* 1986;18:223.
2. Anson JA, Marchand EP: Bovine pericardium for dural grafts: clinical results in 35 patients. *Neurosurgery* 1997;41:1446.
3. Awwad EE, Smith KR Jr, Martin DS, et al: Unusual hemorrhage with use of synthoic dural substitute: MR findings. *J Comput Assist. Tomogr* 1991;15:618.
4. Babcock WW: "Soup bone" implant for the correction of defects of the skull and face. *JAMA* 1917;69:352.
5. Black SPW, Kain CCM, Sights WP Jr.: Aluminium cranioplasty. Technical note. *Neurosurgery* 1968;29:562.
6. Blomstedt GC, Kytta J: Results of a randomized trial of vascomycin trial in craniotomy. *J Neurosurg* 1988;69:216.
7. Boop FA, Chadduck WW: Silastic duroplasty in pediatric patients. *Neurosurgery* 1991;29:785.
8. Erculei F, Walker AE: Post traumatic epilepsy and early cranioplasy. *J Neurosurg* 1963;20:1085.
9. Fisher WS, Braun W: Closure of posterior fossa dural defects using a dural substitute: Technical note. *Neurosurgery* 1992;31:151.
10. Fontana R, Talamonti G, D'Angelo V, et al: Spontaneous hematoma an unusual complication of silastic dural substitute: Report, 2 cases. *Acta Neurochir* 1992;115:64.
11. Klein Schmidt Quoted by Olin MS: Repair of defects of the skull in operation Neurosurgical Techniques pp 11.
12. Krebs Bach PH, Mankani MH, Satomura K, Kuznetsov SA, Robey PG: Repair of craniotomy defects using bone marrow stromal cells. *Transplantation* 1998;27:1272.
13. Kyoshima K, Gilio H, Kobayashi S, et al: Cranioplasty with inner table of bone flap. Technical note. *J Neurosurg* 1985;62:607.
14. Lane KL, Brown P, Howell DN, et al: Creutzfeldt – Jacob disease in a pregnant woman with an implanted dura mater graft. *Neurosurgery* 1994;34:737.
15. Laquerriere A, Yun J, Tiollier J, et al: Experimental evaluation of bilayered human collagen as a dural substitute. *J Neurosurg* 1993;78:487.
16. Lee KW, Sherwin T, Won DJ: An alternate technique to close neuro-surgical incisions using octylcynoacrylate tissue adhesive. *Pediatr Neurosurg* 1999;31:110.
17. Mount LA: Tantalum discs for covering trephine defects and tantalum clips for legation of internal carotid artery intracranially. *J Neurosurg* 1948;5:208.
18. Misra BK, Shaw JF: Extracerebral hematoma in association with dural substitute. *Neurosurgery* 1987;21:399.
18a. Onishi K, Maruyama Y, Sawaizumi M, Iwahira Y, Seiki Y: Usage of intra- operative scalp espansion for primary wound closure in cranio –facial operation. *No Shinkei Geka* 1997;25:795.
19. Palm SJ, Kirsch WM, Zhu YH, Peckham N, Kihara S, Anton R, Anton T, Balzer K, Eickmann T: Dural closure ith non penetrating clips prevents meningoneural adhesions: an experimental study in dogs. *Neurosurgery* 1999;45:875.
20. Price JC, Loury M, Carson B, Johns ME: The pericranial flap for reconstruction of anterior skull defects. *Laryngoscope* 1988;98:1159.
21. Reigel DH, Bazmi B, Shih SR, Marquardt: A pilot investigation of poloxamer 407 for the prevention of leptomeningeal adhesions in the rabbit. *Pediatr Neurosurg* 1993;19:250.
22. Shaffrey CJ, Sportnitz WD, Shaffrey ME, et al: Neurosurgical applications of fibrin glue: Augmentation of dural closure in 134 patients. *Neurosurgery* 1990;26:207.
23. Simpson D: Titanium in cranioplasty. *J Neurosurg* 1965;22:292.
24. Smith SC, Pelofsky S: Adaptation of rigid fixation to cranial flap replacement. *Neurosurgery* 1991;29:417.
25. Tabiaddor K, La Morgese J: Complication of a large cranial defect: Case report. *J Neurosurg* 1976;44:506.

26. Thammavaram KV, Benzel EC, Kesterson L: Fascia lata graft as a dural substitute in neurosurgery. *Smith Med J* 1990;83:634.
27. Toma AG, Fisher EW, Cheesman AD: Autogus fibrin glue in the repair of dural defects in craniofacial resections. *J Laryngol Otol* 1992;106:356.
28. Ulin AW, Gollub SS: Surgical bleeding. A Handbook for Medicine Surgery and specialties New York Mc Gram Hill, 1966; 404.
29. Vallfors B, Hasnsson HA, Svensson J: Absorbable or non-absorbable suture materials for closure of the dura mater? *Neurosurgery* 1981;9:407.
30. Ventureyra EC, Da Silva VF: Reduction Cranioplasty for neglected hydrocephalus. *Surg Neurol* 1981;15:236.
31. Weiford EC, Gardner WJ: Tantalum cranioplasty. Review of 106 cases in civilian practice. *J Neurosurg* 1949;6:13.
32. Yamada S, Aiba T, and Endo Y, et al: Creutzfeldt. Jacob disease transmitted by a cadaveric dura mater graft. *Neurosurgery* 1994;34:740.

18

Awake Craniotomy (Conscious sedation) for Mapping of Eloquent area of the Brain in Neurosurgical Practice

D Singh, S Sinha, A Goyal

Eloquent area of the brain is traditional described as motor, sensory, language and visual area of the cortex. Functional area of the brain exist on cortex as well as subcortical region.[1] The present concept of eloquent area is poorly understood. The intra operative damage to this area invariably occurs irrespective of seniority of the operating surgeon. There are a number of patient who have no eloquent area deficiency even in the presence of a large mass in eloquent area. On the other hand a very small lesion in eloquent area produces a major neurological deficit. Such observation can only be extrapolated into poor understanding of the concept of eloquent area.

There is a wide variance in language and motor site localization among individuals.[2] It is still not clear whether the entire eloquent brain area is so sanctum that nothing in this areas can be sacrificed or else there are eloquent sites within the eloquent area and possibly there are safe corridors within the eloquent area from where a safe entry can be made into subcortical region. Moreover the displacement of eloquent area does occur with mass lesion in the brain.[3,4] Switch over of the function to adjacent area has been reported in temporal lobe language sites in patient with glioma.[5]

Identification and preservation of eloquent area or eloquent sites of the brain is a desirable end result in neurosurgical practice. There are a number of methods and techniques to identify and preserve the eloquent area of the brain such as anatomical distribution of these areas,[6-8] I MRI,[9,10] electrophysiological monitoring,[11-13] PET/SPECT scan and image guided surgery (virtual reality). None of these methods can be translated into actual identification of eloquent sites intra operatively. Cortical motor evoked potential, SSEP are cumbersome and there interpretation may be misleading.[6,14-16]

Brain mapping of the motor, sensory and language area is feasible during surgery under awake craniotomy.[5] This is the only method, which enables identification of eloquent area before a corticotomy, or cortisectomy is made to remove a tumour. In the past awake craniotomy was reserved for epilepsy surgery[17-19] and removal of lesions from area of eloquent cortex[20-22] however it is now used as a routine surgical approach to treating patient with supratentorial intraaxial tumours whether eloquent or non eloquent.[23]

Awake craniotomy not only enables a safe corridor into the cortex but also facilitate a wider resection of the tumour as subcortical stimulation is also possible.[24-29] There are no complications of general anaesthesia, and the brain is adequately relaxed. Awake craniotomy can either be performed with Electrocorticography or application of low voltage current and monitoring a function depending upon area of the brain stimulated.

Low complication rate and shorter hospital stay enables it an economically viable option. The need for intensive care in the postoperative period is minimized.[30-32]

CASE SELECTION

Case selection involves both the surgeon and anaesthetist. Surgeon selection of the case involves the proximity of the tumour to the eloquent zone. These include a variety of tumour conditions. Only those cases are suitable who are conscious cooperative and fall into ASA grade 1 and 11. Patient should be emotionally stable and only those patient who are willing to cooperate during surgery are suitable. Patient with severe language problem, mental retardation, emotionally liable and with altered sensorium are not suitable for awake craniotomy Children and the cases that would require placement of the position other than supine or lateral are preferably excluded.

Procedure should be adequately explained to the patient and a good rapport requires to be established with the patient. Patient is explained that he will remain conscious during the procedure and would be asked to perform a desirable motor activity, strength testing during surgery. For mapping language area patient is observed for hesitancy, dysnomia, and speech arrests.

TECHNIQUE OF AWAKE CRANIOTOMY

Awake craniotomy can be performed with monitoring of desired eloquent site with either Electro corticography (ECoG) or bipolar /unipolar stimulation of cortex and watching directly a desired motor or language function.

Intraoperative use of neurophysiological technique—electrical stimulation of the motor area and recording of cortical somatosensory evoked potential (SSEP) after stimulation of median nerve (phase reversal)—allows in majority of the patient safe identification of sensory motor region and its preservation during surgery.[33-35]

Patient is placed in such a manner that there is adequate exposure of the face to place oxygen mask and to be able to put a endotracheal tube in case of requirement. Drape are such as to allow access of face, upper and lower limb of contralateral side to observe a motor activity after stimulation of cortex.

Informed conscent should be obtained from the patient and close relatives. Premedication is given with Benzodiazipines (Diazepam 10mgs) a night before surgery and 5mgm, 1 hour before surgery. Patient is placed so comfortably so as to allow the same degree of comfort in one position for a long time throughout surgery. Surgeon should be quick in making flap and tumour dissection so as to allow least difficulty to the patient. Intrvenous sedation with Midazolam 0.05 mg/Kg or Propofol with rapid recovery following discontinuation are now widely used.[36,37]

SITES OF LOCAL INFILTRATION:

A linear incision is most suited for awake craniotomy. Field block, which would adequately anaesthetize the desired incision line, is planned. Supraorbital, Suptratrochlear, Preauricular and Post auricular block with 2 ml of 0.5% sensoricain at each point is given. **(Fig. 1)** Line of incision is infiltrated with 10 ml of 1% Xylocain, 1:200000 adrenalin and 0.25% sensoricain. Caraniotomy is made as desired. Dura is instilled with cotton packs soaked with 1% xylocain.

IDENTIFICATION OF ELOQUENT SITES :-

Help of intra operative ultrasound can be taken to identify the tumour. Cortical stimulation is performed with the help of a bipolar cautery electrode **(Fig. 2)** with 4-5mm tip separation for 4-5 sec. Stimulation parameter are set at 60 Hz., Biphasic

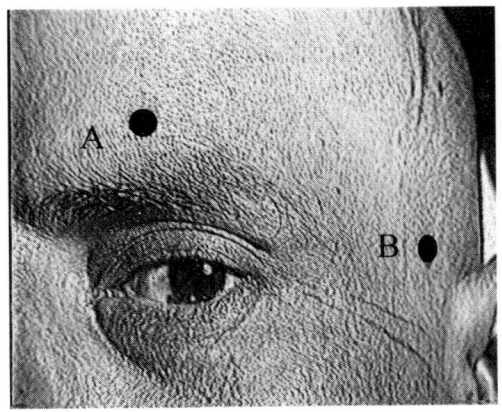

Fig. 1 : Sites of block with local infiltration at supra orbital, supratrochlear (A) and preauricular region. (B)

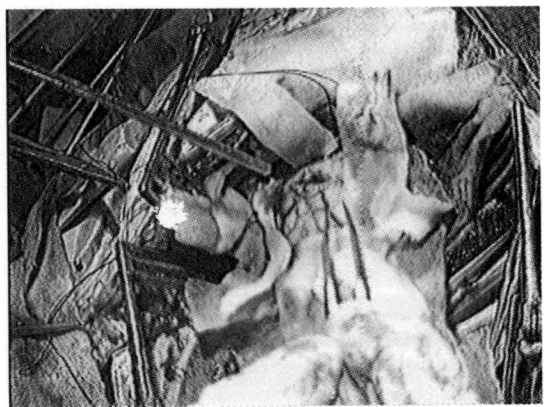

Fig. 2 : Identification of eloquent sites with bipolar current and corticotomy incision. (see text for details)

square wave pulse (1m sec/phase) with variable peak to peak current amplitude between 2 to 10 milliampere.[21] For the identification of motor cortex patient is observed for any movement of face, arm, or leg. A watch is kept on fasciculation of muscle of area stimulated. Strength testing of hand muscle is performed in majority of cases. Other specific finding include patients inability to perform a specific movement For localization of language sites, patient is asked to identify and name the list of objects (animals, flowers, etc.) and is observed for hesitation, dysnomia or speech arrest during stimulation of the cortex. The same procedure is repeated at various area of the cortex. Identified area can be marked with flag. **(Fig. 3)** to facilitate a safe corridor of entry.

After the eloquent sites have been identified cortcotomy or cortisectomy **(Fig. 4)** is made saving the eloquent sites thus entering into brain via safe corridors. Rest of the procedure is carried upon as in any other case. Additional help of various sedatives may be needed during closure of the wound.

RESULTS

Using bipolar stimulation method and observing the desired activities as described above we have performed 41 awake craniotomy from 1997-2001. There were 24 male and 17 females with age ranging from 20 to 58 years. There were 27 motor strip lesion and 24-language area involvement. Various pathological types included 30 glioma, 4 meningioma, 5 tuberculoma, and 2 dysembroblastic neuroectodrmal tumors.

Fig. 3 : Identified areas can be marked with flags for cortisectomy and dissection.

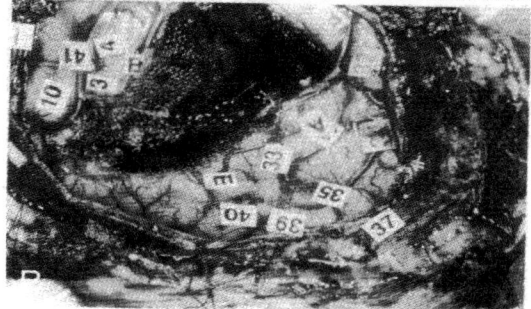

Fig. 4 : Cortisectomy and tumour removal.

Two patients did not cooperate and hence were converted into general anaesthesia. One case had excessive bleeding and tense brain, which required conversion into general anesthesia. There are 5 false negative results wherein some transient deficit occurred in post op period. There was no false positive result.

ADVANTAGES OF AWAKE CRANIOTOMY

Beside offering the only method to localize eloquent sites during surgery, there are a number of advantages of awake craniotomy over general anaesthesia. It's a safe method with adequate analgesia. Brain relaxation is satisfactory. Moreover no cumbersome and complicated monitoring is needed. Blood loss is minimal and there is no additional requirement of mannitol. Surgery time is lesser than conventional methods. OR time is also minimized since it does not require reversal from general anaesthesia. It minimizes the stay in ICU and even post op stay in the hospital is reduced. Eventually it turns out to be more economical than other methods of surgery.

Only disadvantage of awake craniotomy is difficult airway control, which can be minimized with proper draping of surgical site. Air embolism,[38] seizure and tense brain have been reported observations with awake surgery but use of medazolam or propofol has further reduced the rate of these complications It is not a suitable method in uncooperative patients and children. Surgeries of area of brain, which require prone position are relative contraindication as degree of discomfort overweighs the merit of awake craniotomy Patient may feel nausea vomiting during surgery which can be reduced with Inj. Perinorm. Instilling the dura with xylocain also reduces the intra operative nausea and vomiting.[21]

REFERENCES:

1. Skirboll SS, Ojemann GA, Berger MS, et al: Functional cortex and subcortical white matter located within glioma. *Neurosurgery* 1996;38:678-85.
2. Ojemann GA, Ojemann J, Lettich E, Berger MS: Cortical language localization in left dominant hemisphere. *J Neurosurg* 1989;71:316-26.
3. Berger MS, Kincaid J, Ojemann GA, Lettich BA: Brain mapping techniques to maximize resection, safety and seizure control in children with brain tumours. *Neurosurgery* 1989:25: 786-89.
4. Berger MS, Cohen WA, Ojemann GA: Correlation of motor cortex brain mapping data with magnetic resonance imaging. *J Neurosurg* 1990;72:383-87.
5. Haglund MM, berger MS, Shamseldin M, Lettich E, Ojelmann GA: Cortical localization of temporal lobe language site in patients with gliomas. *Neurosurgery* 1994;34:567-76.
6. Ahmadi J, Jong SH, Teal JS, Tsai FY, et al: Localisation of intracranial leision utilizing the coronal suture as the land mark on axial tomography. *Surgical Neurology* 1982;17:209-12.
7. Ebling U, Hueber P, Reulen HJ: Localisation of precentral gyrus in CT and its clinical appication. *J Neurol* 1986;233:73-76.
8. Ebling U, Rikli D, Hueber P, Reulen HJ: The coronal suture, a useful land mark in neurosurgery? Craniocerebral topography between bony landmarks on the skull and brain. *Acta Neurochir (Wien)* 1987;89:130-34.

9. Berger MMS, Cohen WA, Ojemann GA: Correlation of motor cortex brain mapping data with magnetic resonance imaging. *J Neurosurg* 1990;72:383-387.
10. Ebeling U, Steinmetz H, Huang Y, Kahn T: Topography and identification of precentral sulcus in MR-imaging. *AJNR* 1989;10:101-07.
11. Allison T, McCarthy G, Wood C, Darrey TM, Spencer DD: Human cortical potential evoked by stimulation of the median nerve. Cytoarchitectonomic ares generating short latency activities. *J Neurophysiolog* 1989;62:694-710.
12. Allison T, McCarthy G, Wood C, Darrey TM, Spencer DD: Human cortical potential evoked by stimulation of the median nerve. Cytoarchitectonomic ares generating short latency activities. *J Neurophysiolog* 1989;62:711-22.
13. Cushing H: A note upon the faradic stimulation of the precentral gyrus in conscious patients. *Brain* 1909;31:44-54.
14. Gergorie EM, Goldring S: Localization of function in the excision of leasion from the sensorimotor region. *J Neurosurg* 1984;61:1047-54.
15. King RB, Schell GR: Cortical localization and monitoring during cerebral operations. *J Neurosurg* 1987;67:210-19.
16. Le Roux PD, Berger MS, Haglund MM, et al: Resection of intrinsic tumors from nondominant face motor cortex using stimulation mapping: report of two cases. *Surg Neurol* 1991;36:44-48.
17. Archer DP, Jocelyne MA, McKenna, Lise Morin: Canadian Journal of Anaesthesia conscious-sedations analgesia during craniotomy for intractable epilepsy: a review of 354 consecutive cases.
18. Trop D: conscious - sedation analgesia during neurosurgical treatment of epilepsies - practice at Montreal Neurological Institute. In: Anaesthetic consideration for craniotomy in Awake Patients. Varkey G (Ed). International An anesthesiology clinics, No. 3 Little brown and company, Boston 1986;175-84.
19. Gignae E, manninen PH, Gelb AW: Comparison of fentanyl. Sufentanil and alfentanil during awake craniotomy for epilepsy, *Canadian J Anaesthesia* 1993;40:421-24.
20. Michel D Taylor, Mark Bernstein: Awake Carniotomy with brain mapping as the routine surgical approach to treating patients with Supratentorial intraaxial tumors: a prospective trial of 200 cases. *J Neurosurgery* 1999;90:35-41.
21. Ken B Johnson, Talmage D Egan: Remifentainl and Propofol combination for Awake craniotomy: case report with pharmacokinetic simulations. *J Neurosurg Anaesthesiology* Vol. 10, No. 1, 1999;25-29.
22. Berger MS: lesions in functional ("eloquent") cortex and subcortical white matter. *Clin Neurosurg* 1994;41:444-63.
23. Berger MS, Kincaid J, Ojelmann GA, Lettich E: Brain mapping techniques to maximum resection safety and seizure control in children with brain tumors. *Neurosurgery* 1989;25:768-92.
24. R. Andrew Danks, Malcolm Rogers, Linda S Angleo, et al: Patients tolerance of Craniotomy performed with the patients under local anesthesia and monitored conscious sedation. *Neurosurgery*, Vol. 42 No. 1, 1998.
25. Ebeling U, Schmid UD, Ying H, Reulen HJ: Safety Surgery of lesions near the motor cortex using intaoperative mapping techniques: A report on 50 patients. *Acta neurochir (Wien)* 1992;119:23-28.
26. Gregorie EM, Goldring S: Localization of function in the excision of lesions from Sensormotor region. *J Neurosurg* 1984;61:1047-54.
27. Haglund MN, BergerMs, Shamseldin M, Lettich E, Ojelmann GA: Cortical localization of temporal love language site in patients with gliomas. *Neurosurgery* 1994;34:567-76.
28. Block PMcl, Ronner SF: Cortical mapping for defining the limits of tumors resections, *Neurosurgery* 1987;20:914-19.
29. Walsh AR, Schmidt RH, Marsh HT: Cortical mapping and local anaesthetic resection as an aid to surgery of low intermediate grade glioma. *Br J Neurosurg* 1992;6:119-24.
30. Berger MS: Malignant astrocytomas: Surgical aspects *Semin Oncol* 1994;21:172-85.

31. Berger MS, Deliganis AV, Dobbins J, et al: The effects of extent of resection on recurrence in patients with low grade cerebral hemispheric gliomas. *Cancer* 1994;74:1784-91.
32. Berger MS, Ojemann GA: Introperative brain mapping techniques in neurooncology. *Steriotact Funct Neurosurg.* 1992;58:153-61.
33. Rostomily RC, Berger MS, Ojemann GA, Letticj E: Postoperative deficit and functional recovery following removal of tomour involving the dominant hemisphere supplementary motor ares. *J Neurosurg* 1991;75:62-68.
34. Schmid UD, Hess CW, Ludin HP: Somatosensory evoked potential following nerve and segmental stimulation do not confirm cervical radiculopathy with sensory deficit. *J Neurolg Neurosurg Psychiat* 1988;51:182-87.
35. Wood CC, Spencer DD, Allison T, McCarthy G, William PD, Gruff WR: localization of human sensorimotor cortex during surgery by cortical surface recordings of somatosensory evoked potentials. *J Neurosurgery* 1988;68:99-111.
36. Silbergeld DL, Mueller WM, Colley PS, et al: Use of Propofol for awake craniotomies - technical note *Surg-Neurolod* 1992;38:271-72.
37. Ian A Herrick, Rosemary A Craen, Adrian W, Gelt Richard S. Melachlan, John P Giroin, et al: Propofol sedation during Awake craniotomy for seizures. Electrocarticographic and epileptogenic effects, *Anaesth. Analgesia* 1997;84:1280-84.
38. Scuplak SM, Smith M, Harkness WF: Air embolism during awake craniotomy. *Anaesthesia* 1995;50(4):338-40.

19

Low Grade Gliomas—Case for Delayed Surgery

VK Jain, P Krishnan, S Behari

INTRODUCTION

Gliomas are primary tumors of glial cell origin. They span a spectrum from slow growing relatively indolent lesions (low grade gliomas) to aggressive rapidly growing lesions (gliobtastoma multiforme).

WHAT IS A LOW GRADE GLIOMA?

Low grade glioma is a pathological term signifying many histological entities Eg: Pilocytic astrocytomas, gangliogliomas, oligodendrogliomas, pleomorphic xanthoastrocytomas, sub ependymal gaint cell astrocytomas, low grade ependymomas, low grade astrocytomas etc. These exhibit considerable variation in their natural history–in the rate of progression and the tendency to undergo malignant transformation. Unfortunately not much can be said authoritatively regarding their natural behaviour due to tendency of histologically benign lesions to behave in a biologically malignant manner necessitating early intervention.

Most literature quotes low grade gliomas as comprising 25-35% of all primary intracranial gliomas and 10-15% of all brain tumours. Typically they occur in the young adult who may have two broad categories of presentation:

1. Features of raised intracranial tension with or without deficit due to SOL.
2. Progressive neurological deficit.
3. Seizures.

Based on preliminary data of our institute they comprise more than 1/3rd of all gliomas and approximately 5% of the neurosurgical case load.

TIMING OF SURGERY:

The vexing issue in the management of low grade gliomas is the timing of surgical intervention. This is due to–

1. Lack of well conducted prospective clinical trials.

2. Physician bias
3. Inadequate knowledge about biological growth pattern of the tumour.
4. Varied surgical modalities (biopsy–both open and stereotactic, subtotal removal, radical decompression) in Vouge
5. Armamentarium of adjuvant therapies available.

EARLY SURGICAL CANDIDATES

Most neurosurgeons agree that early intervention is a must for low grade gliomas causing significant mass effect with danger of herniation, patients with intractable seizure disorder, disabling neurological deficit, pilocytic variants and older age at diagnosis.

However there remain a significant subset of patients in whom no consensus is as yet reached.

WHY EARLY SURGERY?

Neurosurgeons who advocate early surgery often do so on grounds that a "wait and watch" policy has the risk of misdiagnosis and poorer outcome.

Misdiagnosis can be in two forms:

1. Underestimating the degree of malignancy
2. Mistaking a potentially curable benign lesion to be a glioma with attendant hazards of postponed treatment.

Poorer outcome can manifest as:

1. Worsening of neurological function
2. Shorter survival

UNDER ESTIMATION OF MALIGNANT POTENTIAL:

Accuracy and pitfalls of diagnosis by Imaging:

In the modern era diagnosis of low grade gliomas on imaging is by certain features they exhibit on CTIMRI. It is not uncommon to find low grade gliomas detected only on MRI with a normal CT. Hence MRI is the most sensitive test to diagnose these lesions. Classically it was thought that hypodensity on plain CT with poor absent contrast enhancement was a feature of a low grade glioma. However–

1. Silverman and Marks reported it has no prognostic value in patients with these tumours but
2. Peipmeier et al, claimed patients with enhancemt have a poorer prognosis even when standardized for age and

3. McCormack et al, showed enhancement on CT is associated with seven times increased risk of post op tumour recurrence relative to those showing no uptake.

Surgeons in favour of early intervention strongly believe that diagnosis based only on CT/MRI is often inadequate and incorrect. Three studies are summarized below:

Table 1 : *Error Rates of Image Diagnosis of LGG*

Series	Cases	LGG	Anaplastic	Others	Error Rate (%)
Kondziolka et al	20	10	9	1	50%
Bermstein and Guha	48	39	6	3	19%
Recht et al	20	19	0	1	5%

As is evident no acceptable figure has been reached to predict chance of false negative diagnosis with CECT.

Classically high grade gliomas enhance well on CECT and Gd DTPA scans on MRA. Perilesional edema is also more in high grade gliomas seen as hypodensity in plain CT and T2 signal prolongation an MRI. However, it is now accepted low grade gliomas span a continuum of poor to intense enhancement on imaging studies with variable amount of perilesional edema.

McDermott et al, University of San Francisco, California reviewed the results of 71 patients with features of low grade glioma on CECT and their histological correlates at operation.

Table 2 : *Enhancement and Grade of Tumour*

Enhancement	LGG (35 pts)	Anaplastic (36 pts)
-	63.9	28.6
+	36.1	71.4

These data again serve to suggest that imaging techniques are not foolproof in making a correct diagnosis.

For long the only determinant of increased enhancement wasthought to be neovascularisation–a sign of a malignant process. Current consensus suggest that the mass of the solid tumor tissue is also important–note the case of pilocytic astrocytomas.

In high grade gliomas hypodensity on CT usually represents edema which relates to the tumor cells that infiltrate the parenchyma but only occupy a small part of the total tumor volume. In low grade gliomas on the other hand hypodensity may be due to:

1. Edema
2. Tumour tissue
3. Cells infiltrated into adjacent parenchyma

The object of this long drawn out description is to emphasize that from growth patterns low grade gliomas can be classified into three types:

Type-I   Only tumour without surrounding infiltration. Eg: Pilocytic astrocytomas. **Complete surgical resection can theoretically cure this patients.**

Type-II   Tumour tissue plus surrounding parenchymal infiltration. Eg: Oligodendrogliomas. **Here the surgical benefit will only accrue to patient in relation to the proportion of lesional volume occupied by solid tumour.**

Type-III   No well made out tumour tissue, only diffuse parenchymal infiltration with tumour cells. Eg: Majority of low grade gliomas. **Resection of these lesions is in effect resection of viable through infiltrated brain and deficit might ensue post operatively**

From the Mayo Clinic Patrick Kelly et al, report a series of 178 low grade gliomas where multiple stereotactic biopsies were taken and correlate them with the CT appearances they had.

Table 3 : *Stereotactic Biopsy, Histological Nature and Image Findings–A Correlation*

Type	Hypodense	Isodense	Enhancing
No tumour	25	21	-
Tumour tissue	31	4	28
Infiltrating cells	377	99	-
Necrosis	1	-	-
Total	434	124	28

These figures show that the huge preponderance of low grade gliomas with a hypodense scan on plain CT not enhancing on contrast are diffused in nature. This would imply that the aim of achieving total surgical extirpation cannot possibly be achieved even with radical early surgery.

Over a time proponents of early surgery have begun to state that early surgery is not done with only curative intent in mind. There has been a shift in their view and they claim gross cyto reduction leads to –

1. Longer disease free interval (DFI)
2. Enhanced ability of immune cells to wipeout residual tumour.
3. Greater cell kill by post op radiotherapy if given
4. Chance of elimination of those cells in the tumour with increased malignant potential, i.e. cells capable of sustaining additional mutation causing progression to higher grade.

None of these contentions is valid.

1. The apparently longer DFI post surgery is explained by "lead time bias" not alteration of the natural history of disease.
2. There is still controversy on how immunologically privileged the brain is and the ability of the host defense mechanisms to fight let alone mop up tumour cells.
3. Post op adjuvant therapy tends to achieve only fractional cell kill and is not going to eliminate all malignant cells.

Further the concept of gross total or even radical resection of gliomas is tenuous at best. Astrocytomas often contain $3-6 \times 10^{10}$ tumour cells. A 99% extirpation still leaves 10^8 cells. So cytoreductive surgery alone never cures nonpilocytic glial neoplasms.

DELAYED SURGERY – THE RATIONALE:

One of the basic dictates of surgical oncology is that surgery should be done as early in the disease process as possible. But it has never been proved that earlier treatment of low grade gliomas increases life span. Further as more and more patients are being detected in a neurologically intact state the risk of producing deficit by surgery are beginning to weigh heavily on the minds of surgeons making them ask "should we delay surgery"?

The question of delayed surgery arises only if –

The biological behaviour of the tumour is such that waiting does not compromise survival. Alternative treatment strategies are available.

Alternate strategies are in form of–

1. Stereotatic biopsy and eradiation
2. Primary delay in surgery
3. Neo-adjuvant chemotherapy
4. Immunotherapy

5. Induced hyperthermia
6. Gene therapy

Of these the last four are yet in experimental stage and yet to be recognized as potential curative modalities.

As regards stereotactic biopsy it was once considered advisable

1. Only for patients whose medical condition was so poor as to preclude open surgical debulking.
2. Tumour in eloquent cortical areas.
3. Deep seated tumours in thalamus/brain stem/basal ganglia where extensive resection would lead to unacceptable neurological deficits.

While it is in more extensive use now opponents to the use of stereotactic biopsy voice two major concerns—

1. Possibility of sampling error.
2. Biopsy related complications.

The likelihood of sampling error has decreased with—

1. Development of new neuro imaging techniques that allow targeting of the most malignant part of the tumour.
2. Better software enabling selection of trajectories that void eloquent and vascular areas of the brain and permit multiple biopsies.
3. Better grading systems and newer diagnostic criteria using markers of cell proliferation to assess the biological growth potential with smaller specimen size.

Fears that stereotactic biopsy under estimates the malignant potential of a tumour falsely classifying it as a low grade glioma is based on the following data—

Table 4 : Histology–Stereotactic vs Open Biopsy(*)

Name	No. diagnosed LGG	No. biopsy proven anaplasia	Error in%
Scherer et al	18	13	72%
Russel and Rubenstein	55	44	80%

* Biopsy obtained at the time of definitive surgery

Miller et al, examined 72 patients whose initial diagnosis at open biopsy was low grade glioma. At recurrent (average 31 months) 14% were unchanged, 55%

were anaplastic, and 30% had GBM. The authors concluded that dedifferentiation occurs in 2/3rd of the cases. In 79 patients with low grade glioma Laws documented change to grade III/IV occurs approximately 50% of the time while Sofietti reports 79% have anaplastic change at reoperation/autopsy. These authors claim that presence of anaplastic areas found later in ttie course of disease is not due to initial sampling error rather it is due to tumour progression.

As regards complications related to procedure, these are primarily the risk of intracerebral bleed, subdural hematoma etc. Simple technical changes like avoidance of lateral trajectory, clear knowledge of vascular neuro anatomy, avoidance of repeated flushing in case of vascular puncture have all but eliminated these risks. Lunsford et al, Kelly et al, claimed complication rates of less than 1%.

On the other hand postponing the possibility of surgical misadventure that is morbidity and mortality from a craniotomy and decompression/radical removal is the single major benefit of delayed surgery. Three studies by Vertosick et al, McComack et al, and Bernstein and Guha show thirty day mortality as 1% and risk of significant neurological morbidity as 4%. While these would be considered low rates or GBM (where death is likely within one year) in low grade gliomas where average 5 years survival is 54.4% they are thought to represent a significant and (in the view of some) unacceptable morbidity.

Table 5 : *Surgical Mortality Rates for Astrocytomas (All procedures)*

Series	**Number**	**Mortality rate (%)**
Piepmeier et	60	3.3%
Phillipon et al	179	5%
McCormack et al	53	11.3%
Laws et al	499	7.6%
Fadul et al	109	3.3%
North et al	77	6.5%
Vertosick et al*	25	0
Lunsford et al*	35	0

* Only stereotactic biopsy

In 1988 Fadul reported on 109 patients who had undergone surgery for low grade gliomas with a mortality rate of 3.3%. He also had a significant post op neurological deficit rate of 2% with new deficits appearing in 13% patients who were normal preoperatively.

Other reasons for advocating delayed surgery in low grade gliomas are technical –

A surgeon's three dimensional orientation decreases as surgical exposure descends below the cortical surface. In a diffuse lesion (as low grade gliomas are prone to be) the surgeon may get lost in attempts to find the tumour.

Due to irregularity of shape maintaining balance within tumour extensions also becomes difficult.

Distinction between tumour and surrounding brain is often impossible at operation. So surgeons tend to be more conservative after "getting in" not wishing to inflict neurological damage with the result that early intervention in the form of radical surgery seldom serves its purpose.

Table 6 : *Predictors of Survival in LGG*

Study	No.	Grade	Age status	Performance resection	Extent of	RT
Laws et al	461	S	S	S	S	S
Garcia et al	86	S	NS	-	S	S
Piepmeier et al	60	-	S	-	NS	NS
Medbury et al	50	NS	S	-	NS	S
Shaw et al	126	S	S	-	S	S
Sofietti et al	85	ND	NS	S	S	NS
North et al	77	NS	S	S	S	NS

Vecht et al, in 1993 did a meta analysis of several studies to find out if extent of surgical resection is a significant independent predictor of survival in low grade gliomas.

These figures indicate that there is as yet no consensus or whether radical early surgery is a definitive independent predictor of improved outcome. While Sofietti et al, show longer survival in tumours with gross total removal Weir and Grace, Piepmier et al, and Medbury et al, claimed no correlation between length of survival and extent of resection.

How does stereotactic biopsy and radiotherapy, i.e. (the minimally invasive approach) fare in relation to open craniotomy and debulking?

Laws in 1975, produced a paper showing 5 year survival rates for patients with low grade gliomas as 32% for those who underwent biopsy, 44% for those who underwent subtotal resection and 61% for who under went radical resection. Reanalysis of this paper by Shaw eliminated the pilocytic variants and failed to show any significant benefit of extensive resection. It was contented that patients with subcortical/polar lesions generally undergo resection while stereotactic biopsy is reserved for patients with deep seated lesions who are thought to be poor candidates/elderly/poor performance scale preoperatively. Hence historically poorer outcomes with this modality lead to the thinking that it was inferior.

In 1982, L Dade Lunsford, started a prospective phase I – II study to determine the outcome of early stereotactic biopsy followed by external beam fractionated coned down radiation therapy. These were histologically verified non anaplastic, nonpilocytic, low grade astrocytomas. Long term follow-up evaluation was done for 35 patients (median 62 months, range 11-125 months). Both image defined response and clinical response were gauged. His results are compared with that of others below:

Table 7: *Survival Rates in Patients with Low Grade Gliomas*

Series	Number	Duration	Break up	5 yrs. Survival (%)	10 yrs. Survival (%)
Uihlein et al	83	55-59	33(S) 50(S+R)	65 54	– –
Stage and Stein	45	56-70	17(S) 28(S+R)	20 40	– –
Mersa et al	24	57-73	40(S+R)	41	22
Leibel et al	147	42-67	76(S) 71(S+R)	19 40	11 35
Laws et al	241	15-75	167(S) 74(S+R)	34 49	10 ?
Fazekas et al	68	58-74	23(S) 45(S+R)	32 54	26 32
Medbury et al	50	60-86	50(S+R)	45	32
McCormack et al	48	77-88	48(S+R)	64	–
North et al	66	75-84	66(S+R)	55	43
Phillippon et al	134	78-87	32(S) 102(S+R)	35 50	– –
Lunsford et al*	35	82-92	35(S+R)	81	51

* *Stereotactic biopsy followed by irradiation.*

Table 8 : *Median Survival in Different Series*

Series	Number	Duration	Median Survival (Yrs.)
Sofietti et al	85	50-82	3.2
Fazekas et al	68	58-74	< 5
Shaw et al	24	76-83	4.5
Piepmeier et al	50	75-85	7.5
Medbury et al	50	60-86	4.0
Vertosick et al	25	78-88	8.2
Phillipon et al	179	78-87	9.0
McCormack et al	53	77-88	7.25
Lunsford et al*	35	82-92	9.8

* Stereotatic biopsy followed by irradiation

As regards quality of life in 1992 Recht et al, University of Massachusetts published the results of a cohort style study in which two groups "wait" and "non-wait" were compared the quality of life, length of life and rate of malignant conversion were found to identical in both groups.

CONCLUSION

A "wait and watch" policy may be adopted for patients with gliomas who have a prolonged history with the characteristic imaging features that nequivocally suggest a LGG. A strategy of stereotaottc biopsy and radiotherapy may also be adopted. However, these patients must be kept on a regular follow up, and surgery should be immediately considered if there is an increase in symptoms or the appearance of fresh neurological deficits.

Early intervention should be restricted to those cases that cause significant mass effect with consequent risk of herniation and in those that cause intractable seizures or have disabling neurological deficit.

20
Low Grade Gliomas: Early Radiotherapy

NJ Laperriere

Low grade gliomas represent a heterogeneous population of patients with tumours with variable natural histories. Median survivals are highly variable and dependent on the distribution of prognostic factors within any individual cohort of patients. This has led to some degree of uncertainty in their management. Low grade fibrillary astrocytomas of the brain in adults remain incurable neoplasms, but these tumours are radioresponsive.[1]

Shaw et al reported on the survival experience of their patients with low grade gliomas managed over several decades at the Mayo Clinic.[2] Their survival curve for patients with fibrillary low grade gliomas demonstrated a median survival of about 5 years, and a constant downward trend of the survival curve over 15 years with no survival plateau evident.

The EORTC performed a randomized study of immediate versus delayed radiation therapy in 311 patients with low grade gliomas.[3] Five year survival was 63% versus 66% and 5 year progression-free survival was 44% versus 37% for immediate and delayed radiation therapy respectively. Approximately 1/3 of the delayed radiation therapy group never received radiation therapy at all, apparently as a result of poor performance status at the time of tumour progression.

So at the outset, the premise for selecting patients for early radiation therapy would not be on the basis of any survival benefit, but would be in an effort to forestall tumour progression and influence the quality of their subsequent survival in a positive way so that more time would be spent in a better overall functional status.

The major prognostic factors on which one woul base recommendations for early or delayed radiation therapy are age, histology, proliferative indices, and whether or not enhancement is present on their diagnostic imaging.

Age is a well recognized prognostic factor.[4,5] A randomized North American low grade glioma study of 203 patients demonstrated a marked worsening in survival in patients who were older than age 40.[6] An EORTC randomized radiation dose study in low grade gliomas demonstrated a stepwise decrease in survival when examining age < 30 versus age between 30 and 50, and age > 50.[3]

Several studies have demonstrated that the presence of an oligodendroglioma component in a low grade astrocytoma confers a significant favourable survival advantage over patients without an oligodendroglioma component in their tumour.[4,6]

In fact, when one combines age and the presence or absence of an oligodendroglioma component together, it was found that patients who were older than 40 and did not have an oligodendroglioma component had a median survival of approximately 2-3 years, a survival which is comparable to that seen in anaplastic astrocytoma, an illness where immediate radiation therapy is the standard recommended therapy.[6]

Several studies have documented the adverse effect of elevated measures of mitotic activity on survival in patients with low grade gliomas.[5,7-9] McKeever et al demonstrated that patients with low grade gliomas with a MIB-1 labelling index of > 2 and patients with a MIB-1 index of ≤ 2 had 5-year survivals of 36% and 90% respectively.[5]

It is now generally accepted that all supratentorial fibrillary astrocytomas in adults eventually progress to become malignant gliomas. Characteristically low grade gliomas are non-enhancing tumour on cross sectional imaging, and malignant gliomas vividly enhance, with or without a necrotic core. It has been observed that over time most low grade gliomas will acquire some degree of enhancement as they progress to a more malignant phenotype **(Fig. 1)**. Accordingly, it is not surprising that Bauman et al in a retrospective review of 401 patients with low

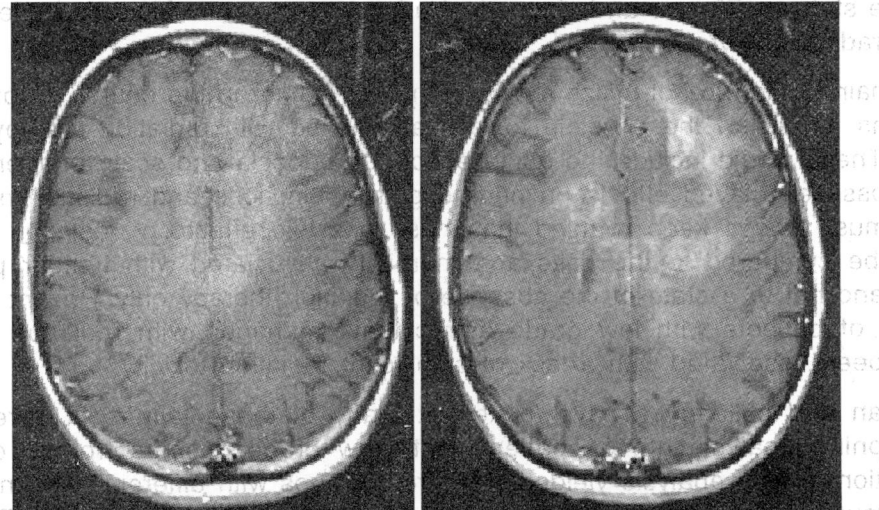

Fig. 1 : MRI images of a 24 year old female with a low grade astrocytoma. The left panel shows a T1 weighted image with gadolinium demonstrating no enhancement. The right panel shows a T1 weighted image with gadolinium demonstrating patchy enhancement 3 months later.

grade glioma found that the presence of enhancement was an adverse prognostic factor both in a univariate and multivariate analysis.[10]

Thus far, it has been demonstrated that the following prognostic factors are associated with a tendency to early tumour progression and a decreased survival: age > 40, absence of oligodendroglioma component histologically, elevated proliferation indices, and the presence of enhancement on diagnostic imaging.

Additionally, patients who present with tumours that are causing significant neurologic symptoms as a result of infiltration of eloquent areas of brain are patients in whom early radiotherapy would be advocated in an effort to prevent further neurologic deterioration. Many of these patients will experience a significant improvement in their neurologic deficit with tumour response over the subsequent months. Specific areas of involvement would include the brain stem, the motor cortex of the frontal lobes, and involvement of the speech areas of the brain.

As well, patients who present with significant mass effect are presenting later in the natural history of their neoplasm with greater involvement of brain than someone who presents earlier with a seizure. It has been well documented that patients presenting with seizures fare significantly better than patients presenting with either neurologic deficits or mass effect.[10]

Another way to analyse neurogic deficit and mass effect would be the analysis of performance status at presentation. Bauman et al did find that Karnofsky performance status < 70 was an independent adverse prognostic factor in patients with low grade astrocytoma.[10]

The main reason to not utilize radiation therapy earlier in the course of low grade gliomas relates to the possible toxicity associated with radiation therapy in the past. These would include the acute effects of alopecia and scalp erythema, and the possible delayed effects on neurocognitive functions and radiation necrosis. One must always keep in mind that these possible effects of radiation therapy must be compared to the risks and morbidity associated with tumour progression, and not to a state of the absence of radiation therapy only. Recent modern series of patients with low grade astrocytoma managed with radiation therapy have been associated with a lack of significant radiation toxicity.[11-13]

Bauman and colleagues have looked at many of these factors in a recursive partitioning analysis of 401 patients with low grade gliomas from 3 different institutions.[10] The analysis yielded 4 different groups with different median survivals: group I (KPS < 70, age > 40) median survival 12 months; group II (KPS ≥ 70, age > 40, enhancement present) median survival 46 months; group III (KPS < 70, age 18-40 or KPS ≥ 70 age > 40, no enhancement) median survival 87 months, group IV (KPS ≥ 70, age 18-40) median survival 128 months.

There have been several recent developments in the use of radiation therapy in patients with low grade gliomas that deserve mention. These relate to radiation dose and radiation volume, and I will discuss how reductions in both these factors will be associated with significant reductions on the possible late effects of radiation therapy in patients with low grade gliomas treated in the modern era.

The EORTC performed a randomized study of radiation dose in 379 patients with low grade gliomas.[3] They compared 45.0 Gy in 25 daily fractions to 59.4 Gy in 33 daily fractions, and found no difference in survival. The North American randomized study of dose (NCCTG-RTOG-ECOG) compared 50.4 Gy in 28 fractions to 64.8 in 36 daily fractions and also found no difference in survival.[6] The North American study found a significant difference in the incidence of radiation necrosis of 12% versus 2% in the high dose versus low dose cohorts respectively.

In addition, historically whole brain fields were used in these patients until the late 1980's when a move away from whole brain fields to more regional fields occurred. Today, patients are simulated with a CT simulator, and only the low density abnormality with a margin of about 2 cm will be treated to a lower total dose.

As a result of these treatment innovations of lower radiation dose and a significant reduction of the volume of brain treated to the prescribed dose, recently managed patients have experienced a significant reduction in the incidence of long term effects of radiation therapy.[13]

In summary, delaying the use of radiation therapy in a subgroup of patients with low grade glioma who have favourable prognostic factors is an accepted treatment approach. As described in this article, patients older than age 40, with no oligodendroglioma component histologically, with elevated proliferation indices, who have enhancement on their imaging, who present with neurologic deficits as a result of infiltration of eloquent areas of brain by their tumour, and who present with significant mass effect at presentation are patients that have a much higher likelihood of early tumour progression and reduced survival. It is patients with these latter characteristics in whom early radiotherapy is seriously considered. Early radiotherapy is not necessarily advocated for any single one factor, but when 2 or more of these factors are present in an individual case, early radiotherapy should be strongly considered.

REFERENCES:

1. Bauman G, Pahapill P, Macdonald D, Fisher B, Leighton C, Cairncross G: Low grade glioma: a measuring radiographic response to radiotherapy. *Can J Neurol Sci* 1999;26(1):18-22.
2. Shaw EG, Daumas-Duport C, Scheithauer BW, Gilbertson DT, O'Fallon JR, Earle JD, et al: Radiation therapy in the management of low-grade supratentorial astrocytomas. *J Neurosurg* 1989;70(6):853-61.

3. Karim AB, Maat B, Hatlevoll R, Menten J, Rutten EH, Thomas DG, et al: A randomized trial on dose-response in radiation therapy of low-grade cerebral glioma: European Organization for Research and Treatment of Cancer (EORTC) Study 22844. *Int J Radiat Oncol Biol Phys* 1996;36(3):549-56.
4. Sakata K, Hareyama M, Komae T, Shirato H, Watanabe O, Watarai J, et al: Supratentorial astrocytomas and oligodendrogliomas treated in the MRI era. *Jpn J Clin Oncol* 2001;31(6):240-45.
5. McKeever PE, Strawderman MS, Yamini B, Mikhail AA, Blaivas M: MIB-1 proliferation index predicts survival among patients with grade II astrocytoma. *J Neuropathol Exp Neurol* 1998;57(10):931-36.
6. Shaw E, Arusell R, Scheithauer B, Dinapoli R, et al: A prospective randomized trial of low- versus high-dose radiation therapy in adults with supratentorial low-grade glioma: initial report of a NCCTG-RTOG-ECOG study (Meeting abstract). *Proc Annu Meet Am Soc Clin Oncol* 1998;17.
7. Fujimaki T, Matsutani M, Nakamura O, Asai A, Funada N, Koike M, et al: Correlation between bromodeoxyuridine-labeling indices and patient prognosis in cerebral astrocytic tumors of adults. *Cancer* 1991;67:1629-34.
8. Lamborn KR, Prados MD, Kaplan SB, Davis RL: Final report on the University of California-San Francisco experience with bromodeoxyuridine labeling index as a prognostic factor for the survival of glioma patients. *Cancer* 1999;85(4):925-35.
9. Wakimoto H, Aoyagi M, Nakayama T, Nagashima G, Yamamoto S, Tamaki M, et al: Prognostic significance of Ki-67 labeling indices obtained using MIB-1 monoclonal antibody in patients with supratentorial astrocytomas. *Cancer* 1996;77(2):373-80.
10. Bauman G, Lote K, Larson D, Stalpers L, Leighton C, Fisher B, et al: Pretreatment factors predict overall survival for patients with low-grade glioma: a recursive partitioning analysis. *Int J Radiat Oncol Biol Phys* 1999;45(4):923-9.
11. Shaw EG: The low-grade glioma debate: evidence defending the position of early radiation therapy. *Clin Neurosurg* 1995;42:488-94.
12. Kiebert GM, Curran D, Aaronson NK, Bolla M, Menten J, Rutten EH, et al: Quality of life after radiation therapy of cerebral low-grade gliomas of the adult: results of a randomised phase III trial on dose response (EORTC trial 22844). EORTC Radiotherapy Co-operative Group. *Eur J Cancer* 1998;34(12):1902-09.
13. Hammack J, Shaw E, Ivnik R, Arusell R, Novotny P, O'Fallon J: Neurocognitive function in patients receiving radiation therapy (RT) for supratentorial low grade glioma (LGG): a North Central Cancer Treatment Group (NCCTG) prospective study (Meeting abstract). *Proc Annu Meet Am Soc Clin Oncol* 1995;14.

21
Delayed Radiation Therapy in Low Grade Gliomas

AK Bahadur

The management of adult supratentorial low grade glioma is still controversial. One of the issues concerning the treatment is that of timing of radiation therapy. After the tissue diagnosis and the maximum tumour resection whether the immediate postoperative radiation therapy should be given or should be deferred till the time of local recurrence?[1]

Where as the gross tumour resection is considered adequate treatment in pilocytic astrocytoma and the patients are subsequently kept on observation alone its not considered adequate treatment in case of astrocytoma, oligoastrocytoma and oligodendroglioma. Following a gross total or subtotal resection in astrocytoma, oligoastrocytoma and oligodendroglioma and subtotal resection or biopsy in pilocytic asrocytoma its not very clear whether to put the patient on radiation therapy or observation. As the radiation therapy to adjacent normal brain is not free from potential morbidity like cognitive dysfunction[2] and radionecrosis, and as these patients have long survivals,[3] an another option of delayed radiotherapy at the time of recurrence or progression is being studied. The preliminary hard data on this question of a place of delayed radiotherapy in treatment of low grade glioma are now available. One phase III trial has been completed by EORTC (The European Organisation for the Research Training of Cancer) addressing the issue of timing of radiotherapy. They have randomized 311 adults 16 to 65 years old between 1986 to 1997 with completely or incompletely resected WHO gradeII astrocytoma, oligoastrocytoma or oligodendroglioma (and incompletely resected pilocytic astrocytoma) to observation (i.e. delayed radiotherapy) or localized radiotherapy using 54 Gy in 30 fractions. The results showed no difference in 5-year overall survival rate, which was 63% with early radiotherapy and 66% with delayed radiotherapy.[4] Progression was documented by worsening of imaging studies with or without evidence of changes in neurological findings. The 5-year progression free survival was 44% with early postoperative RT versus 37% with delayed RT, the difference being statistically not significant. With nearly 5-year followup, 65% of patients in the delayed RT arm eventually required RT.

The concept of delayed radiation therapy is not new. The treatibility of a neoplastic disease is decided on the basis of natural history of the disease especially in terms of survival, symtomatology and effectiveness of particular modality of treatment. If the patient is asymptomatic, natural history of disease is long and there are no proven benefit from available treatment modalities then keeping the patient under observation may be a reasonable option. The patient may be subjected to further definitive treatment by radiation therapy at the time of first sign of recurrence or progression or on patient becoming symptomatic. The advocators of delayed radiation therapy further brings in acute and delayed effect of radiation to support the view as all the patients may not evan require radiation therapy.

It has been seen that the survival in low grade glioma is a function of age, histopathology and amount of surgical resection. In younger patients where resection is complete patients can be comfortably followed up by observation while in elderly patients with incomplete resection it would be appropriate to start immediate radiation therapy.

The 10-year survival of pilocytic astrocytoma is 80% hence observation alone after surgery in asymptomatic cases will be optimum requirement. However the 10-year survival in low grade astrocytoma, low grade oligoastrocytoma and low grade oligodendroglioma drops down to 17%, 33% and 19% respectively and therefore postoperative radiation therapy may offer survival benefit, relief in symptomatology and decrease in convulsions. The low grade gliomas are moderately radiosensitive. The measurement of volume overestimates response and therefore 50% decrease in volume is used to define response. The best way to asses response is still eluding us.

Based on the EORTC trial there does not appear to be survival advantage for early (versus delayed) RT for adults with supratentorial low grade glioma. The RTOG is coordinating an intergroup trial with WHO II astrocytoma, oligoastrocytoma and oligodendroglioma. Patients who are less than 40 years old and undergo a gross total resection are deemed to be at low risk for recurrence and are observed. Those with age of 40 years or greater or who undergo subtotal resection or biopsy are randomized to RT alone (54 Gy in 30 fractions to localized treatment fields) or RT followed by 6 cycles of procarbazine, CCNU and vincristine (PCV) chemotherapy. Central pathology review is required before randomization, and both tissue blocks and peripheral blood are being banked for future study.

Low grade gliomas comprises 30% to 40% of all paediatric central nervous tumours. At present the emphasis of treatment is on surgery Adjuvant therapies, such as radiation and/or chemotherapy are generally withheld until symptomatic or radiographic progression of disease is evident. The aim of surgery is gross total resection while preserving maximum neurological functions. The aim of radiation and chemotherapy is to relieve symptoms and tumour control with minimum radiation morbidity. Chemotherapy has the additional aim of deferring radiation to allow maximum development of childs brain. RT should be avoided below 2 years and dose should be decreased to 45-50 Gy below 5 years.

The futuristics of radiation therapy in low grade glioma will be seen in high tech radiation machines, imaging and its implication, radiobiology and advances in pathology, mixed beam RT, exact role of CT and gene therapy.

REFERENCES:

1. Leighton C, Fisher B, Bauman G, et al: Supratentorial LGG in adults: An analysis of prognostic factors and timing of radiation. *J Clin Oncol* 1997;15:1294-1301.
2. Hammack J, Shaw E, Ivnic R, et al: Neurocognitive function in patients receiving radiation therapy for supratentorial glioma: A North Central Cancer Treatment Group prospective study (abstract), *Proc ASCO* 1995;14.
3. Pipmeier J, Christopher S, Spencer D, et al: Variations in the natural history and survival of patients with supratentorial low grade astrocytoma. *Neurosurgery* 1996;38:872-79.
4. Karim AB, Cornu N, Bleechem D, et al: Immediate postoperative radiotherapy in LGG improves Progression free survival, but not overall survival: Preliminary results of an EORTC/MRC randomized phase III study *Proc ASCO* 1998 (absr);17:400a.

22

A Meta-Analysis of Low Grade Astrocytomas"

S Dua, A Mishra, A Kumar

In 1928, in his introduction to 'Sceptical Essays', Bertrand Russell wrote:

"The extent to which beliefs are based on evidence is very much less than believers suppose". Medical beliefs, and clinical practices that are based on them, are a case in point.

Meta Analysis is essentially synthesis of available literature about a topic to arrive at a single summary estimate. The processes of bringing together evidence from randomized controlled trials are called systemic reviews, meta-analysis or collaborative overviews depending upon the techniques used.

All relevant trials are identified and those of a satisfactory standard are reviewed.

In a meta-analysis, each trial is assessed separately and the summary statistics are then combined to give an overall result. Neither meta-analyses nor systematic reviews are a substitute for prospective clinical trials, rather, they are a complement to them.

The number of papers published on meta-analyses in medical research has increased sharply in the past 10 years **(Fig. 1)**. The merits and perils of the somewhat mysterious procedure of meta-analysis, however, continue to be debated in the medical community. What, then, is meta-analysis? A workable definition was given by Huque: "A statistical analysis that combines or integrates the

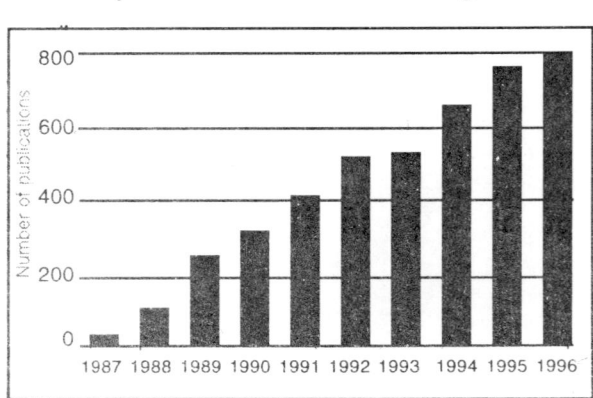

Fig. 1: Number of publications about meta-analysis, 1987-96

results of several independent clinical trials considered by the analyst to be 'combinable.

The terminology, however, is still debated, and expressions used concurrently include "overview," "pooling," and "quantitative synthesis.

"The term meta-analysis should be used to describe the statistical integration of separate studies.

The term makes sense. "Meta" implies something occurring later, more comprehensive, and is often used to name a new but related discipline designated to deal critically with the original one.

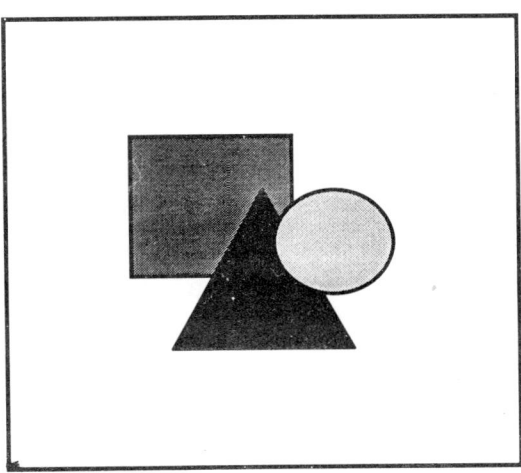

Fig. 2 :   The reliability of evidence Pyramid

Efforts to pool results from separate studies are not new. In his account on the preventive effect of serum inoculations against enteric fever, statistician Karl Pearson was in 1904 probably the first researcher reporting the use of formal techniques to combine data from different samples. The rationale put forward by Pearson for pooling studies is still one of the main reasons for undertaking meta-analysis today:

"Many of the groups ... are far too small to allow of any definite opinion being formed at all, having regard to the size of the probable error involved."

The term meta-analysis was coined in 1976 by the psychologist Glass.

The a prior definition of eligibility criteria for studies to be included and a comprehensive search for such studies are central to high quality meta-analysis.

As with criteria for including and excluding patients in clinical studies, eligibility criteria have to be defined for the data to be included. Criteria relate to the quality of trials and to the combinability of treatments, patients, outcomes, and lengths of follow up. Ideally, the investigator should consider including only controlled trials with proper randomization of patients that report on all initially included patients according 'to the intention to treat principle' and with an objective, preferably blinded, outcome assessment.

Assessing the quality of a study can be a subjective process, however, especially since the information reported is often inadequate for this purpose. It is therefore preferable to define only basic inclusion criteria and to perform a thorough sensitivity analysis.

The graphical display of results from individual studies on a common scale is an important intermediate step, which allows a visual examination of the degree of heterogeneity between studies

Different statistical methods exist for combining the data, but there is no single "correct" method

A thorough sensitivity analysis is essential to assess the robustness of combined estimates to different assumptions and inclusion criteria

Meta-analysis should be as carefully planned as any other research project, with a detailed written protocol being prepared in advance.

Surprising as it may appear but:

Meta-analyses of the same issue may reach opposite conclusions, as shown by assessments of low molecular weight heparin in the prevention of perioperative thrombosis[56, 57] and of second line antirheumatic drugs in the treatment of rheumatoid arthritis.[58, 59]

Meta-analysis should be seen as structuring the processes through which a thorough review of previous research is carried out. The issues of completeness and combinability of evidence, which need to be considered in any review, are made explicit. Was it sensible to have combined the individual trials that comprise the meta-analysis? How robust is the result to changes in assumptions? Does the conclusion reached make clinical and pathophysiological sense? Finally, has the analysis contributed to the process of making rational decisions about the management of patients?

An Example to illustrate

For a baseline risk of 1% a year, the absolute risk difference shows that two deaths are prevented per 1,000 patients treated. This corresponds to 500 patients (1 divided by 0.002) treated for one year to prevent one death. Conversely, if the risk is above 10%, less than 50 patients have to be treated to prevent one death. Many clinicians would probably decide not to treat patients at very low risk, given the large number of patients that have to be exposed to some of the adverse effects of Treatment e.g. Radiotherapy to delay one recurrence by six months. Appraising the number needed to treat from a patient's estimated risk without treatment and the relative risk reduction with treatment is a helpful aid when making a decision in an individual patient. A nomogram that facilitates calculation of the number needed to treat at the bedside has recently been published.[60]

TRIAL IDENTIFICATION

The most important step in a systematic review is trial identification. Every attempt should be made to ensure that as many as possible of the potentially eligible trials are found. This is especially important because of publication bias: trials with positive results are more likely to be published. As a consequence, the results of published and unpublished trials could be systematically different. Unless all trials are sought, regardless of their publication status, the review may contain a biased set of studies. Then, regardless of how the data are handled, a meta-analysis may be mathematically accurate, but clinically unreliable.

Before including trials for analysis it must be ascertained as to what is the degree of reliability of evidence. The following criteria are generally accepted.

RELIABILITY OF EVIDENCE

Class I: Evidence provided by one or more well-designed, randomized, controlled clinical trials.

Class II: Evidence provided by one or more well-designed clinical studies such as case control or cohort studies.

Class III: Evidence provided by expert opinions, nonrandomized historical controls, or case reports of one or more.

The present Meta-Analysis of low grade Astrocytomas was taken up with the objective of finding a statiscal model, primarily, for management of these tumours and in addition to collate general prevalence, distribution, anatomical and radiological data for analysis.

Literature reviewed for collecting the data:

Year———————————————1956 to 1998

Language————————————English

Publications———————————— Full text including on line Journals

The total number of publications selected for inclusion were 65 in number and were initially selected based on the Title of "Low grade Astrocytoma or Low grade Glioma" without first going through the contents of the article so as to eliminate selection bias

Total Number of studies Identified ————————65

Total Number of studies selected————————19

Total Number of Patients evaluated————————1869

As many as 46 studies could not be used for data analysis as there was no homogeneity of data whatsoever and the whatever statistics were provided were in percentages from which actual numbers could not be calculated. In these series the treatment V/S outcome protocols were not there for individuals or for any selected groups. This resulted in a loss of 3394 patients from being analysed.

The infirmity of Data with regard to each publication is given in INDEX 1.

It was not totally unexpected that a large number of these studies did not contain all the relevant data required to fill in the master chart but what was surprising was that none of the publications conformed to the standards of Class I evidence.

None of the publications conformed to Class II evidence either. There was only one study which came near to be labeled as Class II evidence study and that to for some of the parameters only.

Initially the following **inclusion criteria** were identified in the protocol stage for Supratentorial non optic pathway low grade tumors of astrocyte series.

Uniformity of histopathology pattern
Different treatment modalities for similar tumors
Informative treatment protocols
Out come statistics available for each group
Location of the tumor available
Age sex distribution available

The **exclusion criteria** were:

Those series which included chiasmal hypothalamic tumors under the generic umbrella of low grade gliomas without providing separate statistics.

Patients of oligodendrogliomas

Those series which have have included tumors like PXA, SEGA, gangliogliomas, along with low grade astrocytomas without a proper information regarding individual numbers.

Even in all the studies that have been included there was no uniformity of the Histopathology grading systems. To overcome this it was decided to convert the histopatholgy grading to a uniform pattern of **W.H.O. Criteria using the following conversion format.**

COMPARATIVE GRADING

WHO DESIGNATION	WHO GRADE	KERNOHANS	ANNE-MAYO
PILOCYTIC ASTROCYTOMA	1	1	EXCLUDED
ASTROCYTOMA	2	1, 2	1, 2
ANAPLASTIC ASTROCYTOMA	3	2, 3	3

There was no study which was randomized, nor was there any controlled study it was not possible to evaluate this meta analysis by the most frequently used statistical tools i.e. Odds ratio and the Relative risk parameters.

However one must understand these basic tools of statistical evaluation so that in planning any future study one must at least be able to submit the data for evaluation by these basic parameters.

ODDS AND ODDS RATIO

The odds is the number of patients who fulfil the criteria for a given endpoint divided by the number of patients who do not. For example, the odds of diarrhoea during treatment with an antibiotic in a group of 10 patients may be 4 to 6 (4 with diarrhoea divided by 6 without, 0.66); in a control group the odds may be 1 to 9 (0.11) (a bookmaker would refer to this as 9 to 1). The odds ratio of treatment to control group would be 6 (0.66d0.11).

RISK AND RELATIVE RISK

The risk is the number of patients who fulfil the criteria for a given end point divided by the total number of patients. In the example above the risks would be 4 in 10 in the treatment group and 1 in 10 in the control group, giving a risk ratio, or relative risk, of 4 (0.4 divided by 0.1).

For Meta-Analysis of low grade Astrocytomas only 19 studies could be short listed which had some of the common parameters which could be analysed to some degree. Here it would not be out of place to mention that all statistical results are based on Class III evidence

For the present analysis only the following parameters could be taken in to consideration as these were the only common factors in most of the publications selected.

Various parameters taken in to account in this study were: **(Table 1)**

A   Age

B   Sex Ratio
C   Signs & Symptoms
D   Location
E   Treatment Protocols (Surgery: Total, Partial or Biopsy with or without Radiotherapy)
F   Histological Grades (Conversion to WHO)
G   Survival Analysis
    (Based on Histology & Treatment protocols.)

Even amongst these limited parameters there was no parameter that was common to all the studies except the histological grades. While Age distribution was

Table 1 : *Shows the details of the 19 studies selected and also the 11 publications from which the survival data has been analyzed.*

Author	Journal	No. of Patients	Year
Pierre Janny et al	Cancer	58	April 1994
John T Fazekas	*Radiation Oncology Bio Phy*	*68*	*July 1977*
Yuta shibamoto, et al	Cancer Vol. 72 No. 1	119	July 1993
HJ Steiger, et al	*Acta Neurochir 106:99-105*	*50*	*1990*
Bruce M McCormac et al	Neurosurgery	53	Oct. 1992
Joseph Piepmeier et al	*Neurosurgery Vol. 38 No. 5*	*55*	*May 1996*
Dimitrios C.Nikas et al	Neurosurg Focus 4 [4]: Article 4	175	1998
Joseph M Piepmeier et al	J.N.S. 67: 177-181	60	1987
Clinton A Medbery III et al	*J Radiat Oncol Bio Phy*	*50*	*Oct. 1988*
Edward R Laws, Jr et al	J.N.S. 61: 665-73	461	1984
Edward G Shaw et al	*J.N.S. 70: 853-61*	*167*	*1989*
Antonto Nicolato et al	Surg Neurol 44:208-23	76	1995
P. Bahary et al	Journal of Neuro Oncology 27: 177-77	63	1996
Lucio Palma et al	J.N.S 62: 811-15	51	1985
Gary B Clark et al	Cancer 56: 1128-33	30	1985
R Miralbell et al	Radiotherapy & Oncology 27: 112-16	49	1993
Jacques H Philppon et al	Neurosurgery Vol. 32 No. 4	179	1993
D Afra, et al	*Acta Neurochir 81: 90-93*	*119*	*1986*
Harmon J Eyre et al	J.N.S. 78: 909-14	60	1993

*The Highlighted text shows the 11 studies from which data regarding treatment and survival could be analysed.

present in 17 studies the sex distribution was mentioned in 16 of the studies. Only 11 studies had treatment and survival data. (**Table 2**)

Table 2 : *Shows the details of the data collected from different series.*

	N		Mean		Median	Mode	Std. Deviation	Minimum	Maximum	Sum
	Valid	Missing								
	Statistic	Statistic	Statistic	Std. Error	Statistic	Statistic	Statistic	Statistic	Statistic	Statistic
ptno	19	0	98.37	22.75	60.00	50[a]	99.19	30	461	1869
ageMean	17	2	32.3724	2.2114	34.5000	18.00[a]	9.1179	14.00	50.00	550.33
sexM	16	3	43.63	5.57	38.50	28[a]	22.26	16	97	698
sexF	16	3	29.63	5.53	22.50	14[a]	22.13	9	91	474
histo1	10	9	49.50	15.13	32.50	8	47.85	6	123	495
histo2	17	2	80.65	17.50	60.00	60[a]	72.17	30	338	1371
Seizures	11	8	68.45	24.67	42.00	36	81.82	13	305	753
Headache	7	12	20.57	7.50	16.00	3[a]	19.84	3	63	144
Vomiting	4	15	25.50	12.80	24.50	2[a]	25.59	2	51	102
Papilledema	7	12	32.29	15.83	9.00	5	41.88	5	101	226
FND	6	13	55.50	49.32	8.00	2[a]	120.80	2	302	333
Altered Sensorium	2	17	17.00	13.00	17.00	4[a]	18.38	4	30	34
Cognitive Deficit	3	16	25.67	19.68	7.00	5[a]	34.08	5	65	77
Visual Deicit	5	14	22.20	14.27	2.00	2	31.91	2	75	111
Motor Deficit	6	13	29.00	19.00	7.50	3[a]	46.54	3	122	174
Sensory Deficit	1	18	50.00			50		50	50	50
Frontal	13	6	38.54	14.04	18.00	9	50.62	7	190	501
Temporal	13	6	40.54	13.75	24.00	24[a]	49.57	6	192	527
Parietal	13	6	17.85	4.69	11.00	10	16.92	9	71	232
Occipital	8	11	4.50	1.02	3.50	3	2.88	1	9	36
Deep Seated	9	10	15.78	3.84	12.00	3[a]	11.53	3	36	142
> 1 Lobe	7	12	15.00	4.49	13.00	4	11.87	4	35	105
Superficial	3	16	40.67	31.69	11.00	7[a]	54.88	7	104	122

a. Multiple modes exist. The smallest value is shown

From the foregoing table it is evident that the data available in the series with the most information is at best very sketchy.

Results of Analysis
Age distribution

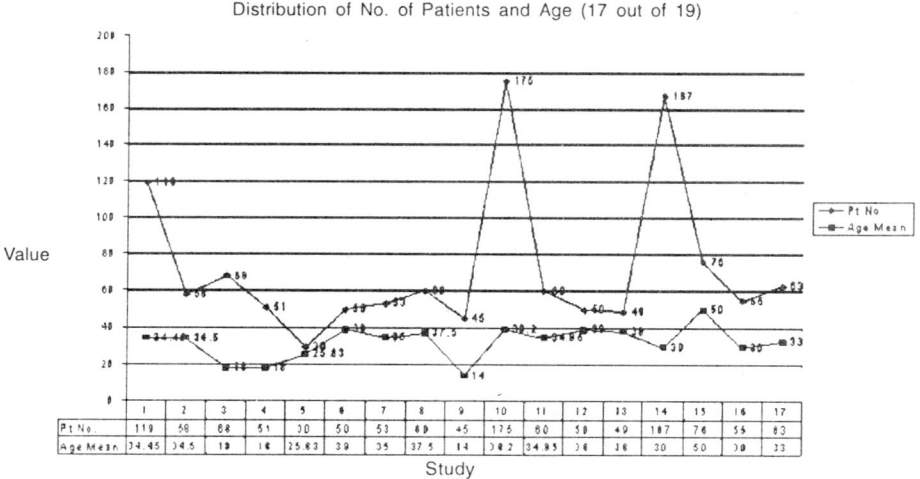

Mean age of the analyzed patient population was 33.4 yr. The patients were in between fourteen to fifty years of age The age distribution is shown in the graph.

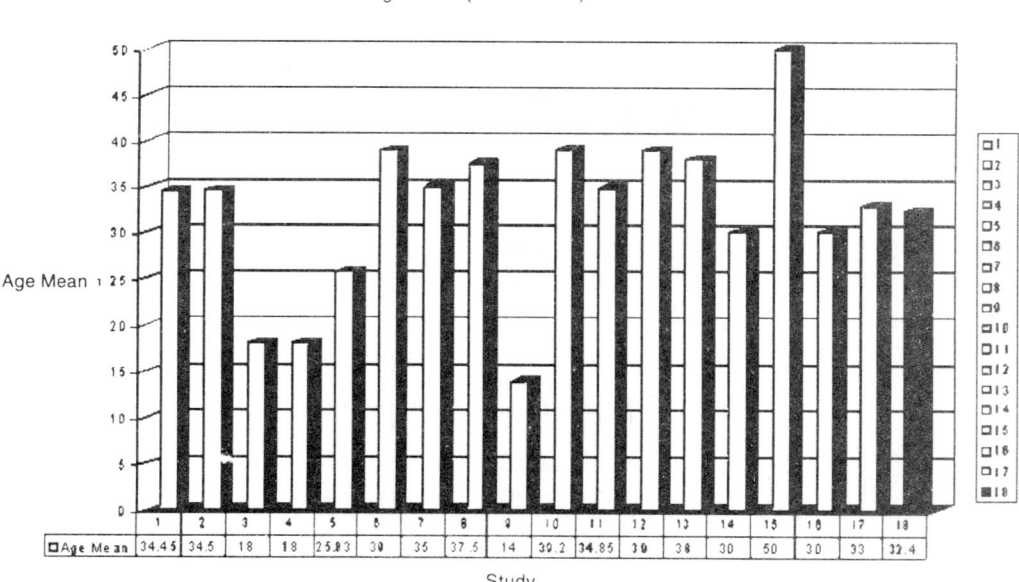

Sex Distribution

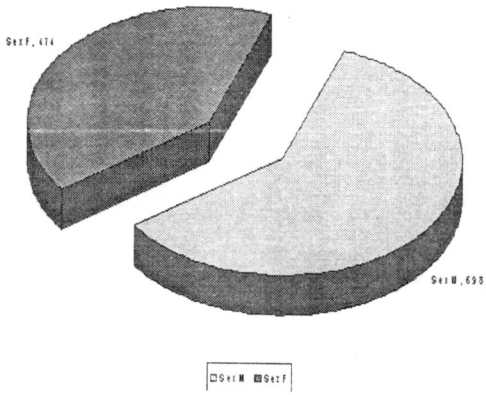

Clinical Features

Out of the nineteen studies signs and symptoms were mentioned only in eleven studies. Of all the signs and symptoms, Seizures (37%) Focal Neurological Deficit (17%) and Paplliooedema (11%) were the most common presentations though Motor deficit (9 %), Visual Defects (6%), Vomiting (5%), Headache (7%), Cognitive and Language Deficit (4%), Sensory Deficit (2%) and Altered Sensorium (2%) were also reported in different studies.

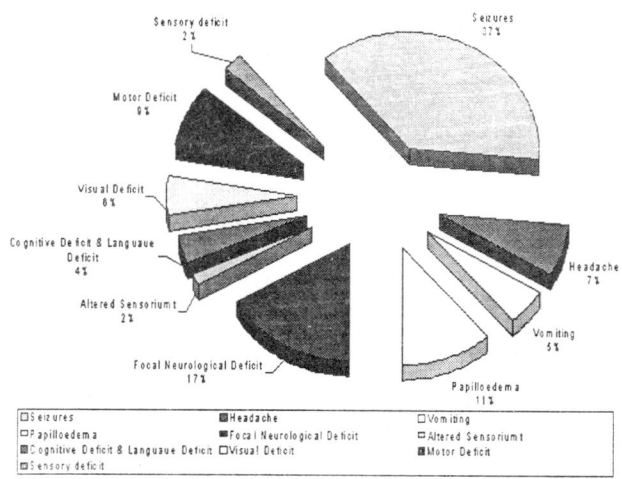

Location of Tumors

The location of tumor could be found only in fifteen of the nineteen studies. Tumors were most commonly located in the Temporal Lobe (32%) followed by

Frontal Lobe (30%), Parietal Lobe (14%), Deep Seated (9%), Superficial (7%), More than one lobe (6%) and Occipital lobe (2%).

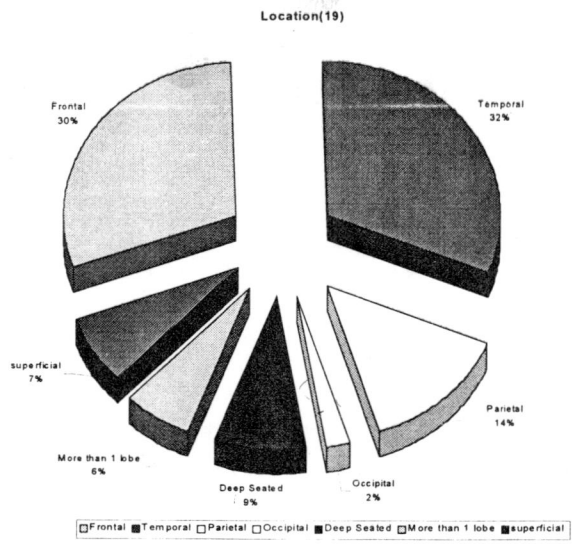

Histology Grading according to WHO or comparable Protocols was possible in all 19 series

GrII—73%   GrI——23%

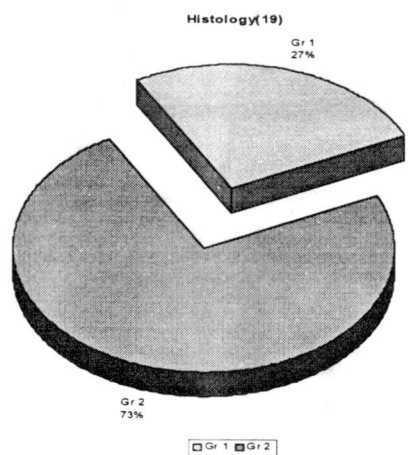

Treatment: Only eleven studies were finally selected for treatment analysis as the data for survival analysis, which could give an insight in to the appropriate management protocol to be followed, were present in these studies only. Detailed description of the treatment protocol as analysed from the eleven studies is mentioned in the table below.

Sample Size	Total Resection	Partial resection	Only Biopsy	Total Resection with Radiotherapy Py	Partial Resection Radiotherapy	Biopsy with Radiotherapy Py
68	18	5	0	4	40	1
51	39	0	-	6	6	-
50	26	0	14	6	4	-
50	0	0	0	8	30	12
49	0	0	0	16	24	9
179	29	26	6	16	69	33
461	44	356	-	61	0	-
167	14	14	0	19	92	28
76	4	0	-	13	59	-
55	28	6	.	3	18	-
63	10	10	0	4	24	15

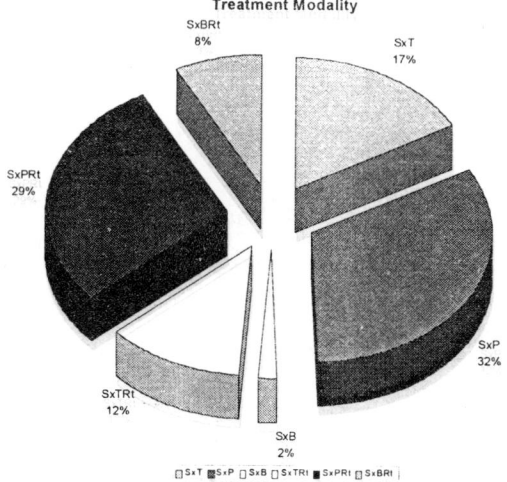

Partial Resection (32%) was found to be the most common mode of treatment followed by Partial Resection with Radiotherapy (29%), Total Resection (17%), Total Resection with Radiotherapy (12%), Biopsy With Radiotherapy (8%) and Only Biopsy (2%).

Survival Analysis

Sl.No	Parameters	No. of Study	N5	N10	n.5	n.10	Survival (5)%	Survival (10)%
1	Total Resection	6	133		104		78.19	
2	Partial Resection	4	387		170		44	
3	Biopsy Only	2	6		6		44.5	
4	Total Resection with Radiotherapy	3	24		12		50	
5	Partial Resection With Radiotherapy	3	40		17		42.5	
6	Histo1	5	177	38	95	19	53.67	50
7	Hsito2	6	812	297	374	63	46.05	21.35

After the Survival Analysis it was noted that best five year survival (78.9%) was seen in the group which had undergone Total Surgical Resection followed by Total Resection with Radiotherapy (50%), Partial Resection (44%), and Partial Resection with Radiotherapy (42.5%). Better five year survival evident in the Group which had undergone only biopsy is statistically insignificant because of the very small size of the Biopsy group as compared to other groups. Five and Ten year survival in the patients with WHO Gr.1 Low Grade Gliomas was 53.67% and 50% respectively and for the WHO Gr.2 Low Grade Gliomas it was found to be 46.05% and 21.35 % respectively.

P Value Analysis

In the following p value chart some short forms have been used while analyzing the data. Their expanded forms are given below.

SxT—Total Resection   SxTRt—Total Resection with Radiotherapy
SxP—Partial Resection   SxPRt – Partial Resection with Radiotherapy
SxB—Biopsy   SxBRt—Biopsy with Radiotherapy

P Value Chart

Variables	Parameters			p. Value
1 VS 2	SxT	VS	SxP	0.0000
1 VS 3	SxT	VS	SxB	0.0610
1 VS 4	SxT	VS	SxTRt	0.0047
1 VS 5	SxT	VS	SxPRt	0.0000
2 VS 3	SxP	VS	SxB	0.4805
2 VS 4	SxP	VS	SxTRt	0.2831
2 VS 5	SxP	VS	SxPRt	0.4278
3 VS 4	SxB	VS	SxTRt	0.8113
3 VS 5	SxB	VS	SxPRt	0.8718
4 VS 5	SxTRt	VS	SxPRt	0.2941
6 VS 7(5y)	Histo.1	VS	Histo2	0.0000
6 VS 7 (10y)	Histo.1	VS	Histo. 2	0.0269

Note:- Variables taken in this column are those from the survival analysis table.

Discussion

From all the available data it is evident that all the studies with the exception of two were not designed to evaluate a complete overview of low grade Astrocytomas.

There was no homogeneity of data from one study to another and indeed, in some cases, with in one study itself.

The inherent contradictions are obvious when the results are evaluated

The absolute figures for comparison between the Total Surgery and the Total Surgery with Radiotherapy group were first calculated as the percentage of total cases in the respective studies and then the figure arrived at was compared. This was done to eliminate the sample size conflict arising from different studies.

It is difficult to understand as to why should the patients who had undergone a total resection fare significantly better than those who had in addition to total resection had received Radiation treatment.

Since the data for comparison has originated from different studies it must be assumed that the conflict could have arisen from any of the two factors viz.

Firstly it is possible that with in the patient population, who received R.T. in addition to surgery, the numbers of grade II tumours must have been more as compared to grade I tumours. (There was no data with regard to this as to which of the patients with GrII and how many with GrI were subjected to Radiation treatment)

The other possibility is that only those patients underwent Radiations who had less than a complete resection and the surgeon decided to give the additional protection to the patient by Irradiation.

This bias could have been avoided if any study had been planned prospectively by assigning patients to one or the other treatment plan randomly and that such a study had been done either at a single place or as a part of a multicentric trials.

It would also be helpful if the histopathology of the specimens was evaluated by independent observers using a uniform classification method so that a uniformity of outcome on that account be evaluated. Even today the histopathology reports that we get for our patients are not in any uniform grading system but dependent upon which person or which centre is evaluating the specimens.

It would not be out of place to infer from personal experience that the objective evidence of total removal is post operative radiological evaluation as we have on occasions found residual tumour on MRI/CT even when per operative there was reasonable surety of total removal.

In a number of studies there is no mention whether the pilocytic astrocytomas have been included as Grl tumours or not. If in some studies these have been included as low grade astrocytomas then it is obvious that those series will show a better outcome than the others.

The other important inferences that can be drawn from this analysis is that in all cases irrespective of the other modalities used it is the completeness of resection of the low grade tumour that determines the length of the survival of the patient. Therefore every attempt must be made to remove the tumour completely either at the first operation itself or even the second time if the post op radiology reveals a residual tumour or if the first time a bit of the tumour was left in the belief that it might be a more malignant variety and the histopathology proves it otherwise.

The results between partial surgery and the biopsy group were not much different (radiation treatment being given in both situations)

There is very little mention about the time interval between the first symptom and detection of the tumour and between the detection and surgery. None of the tumours were subjected for genetic identification as to weather these tumours were likely to progress to Anaplastic tumours and represented an intermediate state or these tumour did not have any such gene abnormalities.

It is inferred that with the given set of publications we can not come to any definite conclusions and the answers regarding the entity of "low grade astrocytomas" and a numbers of questions remain largely unanswered.

It is suggested that a prospective randomized multicentric study be undertaken regarding this entity and then conclusions be drawn fro it.

Firstly we must define the entry level for the cases i.e.

Those patients who radiologically have a non enhancing diffuse lesion in the brain can be randomized for evaluation. If possible the histochemical markers for mitotic activity as evidenced by MR techniques should be taken in to consideration.

Those patients not assigned initially for study but found to have a low grade tumour on histopathology must also be included and the reasons why they were not suspected to be low grade tumours must be clearly illustrated.

All cases must be defined to be low grade tumours by a uniform histopathological grading classification and sequence. The histopathology must be evaluated by at least two different evaluators from different centres and in case of any conflict that can not be satisfactorily resolved the case should be removed from the protocol.

The inclusion criteria must be very strict and at the outset it must exclude all pilocytic tumours, tumours with oligodendriogliomatous component etc. so the overall statistics is not contaminated by these distinct entities which have been clubbed as low grade brain tumours.

If possible we should have a genetic analysis to evaluate if the tumour is one of those which has got specific gene damage which would make it in an intermediate stage of transition to anaplastic tumours and should be followed up more closely.

The histopathology feature should be uniformly low grade at least from three different sites of the tumour.

The patient must have a post operative radiological evaluation to ascertain the completeness of the surgical resection.

Those with incomplete resections and those with complete resections should again be randomized separately for further treatment.

At least a 5 Yr.follow up is required before making any conclusions and all patients lost to follow up should should be excluded.

All patients who constitute a part of the study should have special discharge cards so that

a) They are accorded a separate priority for follow up thus making them more compliant to regular revaluations.

b) These patients can be recognized by any other treating facility, as patients under trial, and that they can keep a record accordingly to be forwarded to any of the participating trial centers to record the followup.

The radiological and histopathology data should be posted on a common Web Site for some time so that if there are any different perspectives of the concerened fraternity the same can be considered.

We must conclude by saying that while this would go down as a failed meta-analysis yet it has served to make us aware that there is no use of publishing series after series of patient data which do not stand the scrutiny of rigorous statistical evaluation.

In the words of David Nylor, a well known personality in the field of statiscal analysis, "It is better to have Meta-Analyzed and failed than not to have Meta-Analyzed at all."

REFERENCES:

1. Barth R, Soloway A, Goodman J, et al: Boron Neutron Capture Therapy of Brain Tumors: An Emerging Therapeutic Modality. *Neurosurgery* Vol. 44, No. 3, 1999.
2. Gutin P, Posner J: Neuro-oncology: Diagnosis and Management of Cerebral Gliomas— Past, Present,

and Future. *Neurosurgery* Vol. 47, No. 1, 2000.
3. Kida Y, Kobayashi T, Mori Y: Gamma Knife radiosurgery for low-grade astrocytomas: results of long-term follow up. *J Neurosurg* (Suppl 3) 2000;93:42-46.
4. Palma L, Guideti B: Cystic Pilocytic astrocytomas of the cerebral hemispheres. Surgical experience with 51 cases and long-term results. *J Neurosurg* 1985;62:811-15.
5. Clark G, Henry J, McKeever P: Cerebral Pilocytic Astrocytoma. *Cancer* Vol. 56, 1985;1129-33.
6. Afra D, Muller W, Slowik F, et al: Supratentorial Lobar pilocytic Astrocytomas: report of 45 operated cases, Including 9 recurrences. *Acta Neurochirurgia* (wien) Vol. 81, 1986;90-93.
7. Phillipon J, Clemenceau S, Fauchon F, et al: Supratentorial Low Grade Astrocytoma in adults, *Neurosurgery* Vol. 32. 1993.
8. McCormack B, Miller D, Budzilovich G, et al: Treatment & Survival of Low-Grade Astrocytoma in Adults. *Neurosurgery* Vol.31, No. 4. 1992.
9. Mirabell R, Balart J, Molet J, et al: Radiotherapy for supratentorial low grade gliomas: Result and prognostic factors with special focus on tumor volume parameters. *Radiotherapy and oncology.* Vol.-27, 1993;112-16.
10. Eyre H, Crowley J, Townsend J, et al: A Randomized trial of Radiotherapy Vs Radiotherapy Plus CCNU for incompletely ressected Low-Grade gliomas: a south west Oncology Study Group. *J Neurosurg* Vol. 78: 1993; 909-14.
11. Piepmeier J. Observations on the current treatment of low-grade astrocytic tumors of the cerebral hemispheres. *J Neurosurg* 1987;67:177-81.
12. Nikas D, Bello L, Zamani A, et al: Neurosurgical Considerations in supratentorial low-grade Gliomas: Experience with 175 patients. *Neurosurg Focus.* 4(4): Article 4. 1998
13. Piepmeier J, Christopher S, Spencer D, et al: Variations in the Nature History & Survival of patients with Supratentorial Low Grade Astrocytomas. Vol. 38, No. 5. 1996.
14. Laws E, Taylor W, Clifton M, et al: Neurosurgical Management of Low grade Astrocytoma of the cerebral Hemispheres. *J Neurosurg* 1984;61:665-73.
15. Fazekas J. Treatment of Grade 1 & 2 Brain Astrocytomas. The Role Of Radiotherapy. *J Radiation Oncology Bio Phys* Vol. 2 1977; 661-66.
16. Fine H, Keith B, Dear G, et al: Meta-Analysis of Radiation Therapy with and without Adjuvant Chemotherapy for Malignant Gliomas in Adults
17. Shibamoto Y, Kitakabu Y, Takahashi M, et al: Supratentorial Low Grade Astrocytoma: Corelation of Computed Tomography Findings with Effect of Radiation Therapy and Prognostic Variable. *Cancer* Vol. 72 No. 1, 1993.
18. Steiger H, Markwalder R, Seiler R, et al: Early Prognosis of Supratentorial Grade 2 Astrocytoma in Adults Patients After Resection or Stereotactic Biopsy. An Analysis of 50 Cases Operated On Between 1984 and 1988. *Acta NeuroChirurgia* 1990;106:99-105.
19. Shaw E, Dupport-Dumas C, Scheithauer B, et al: Radiation therapy the in management of Low Grade Supratentorial Astrocytoma. *J Neurosurg* 1983;70:853-61.
20. Medbery C, Straus K, Steinberg S, et al: Low Grade Astrocytoma: Treatment Results and Prognostic Variables. *Int J Radiation Oncology Bio Phy* Vol. 15, 1998;837-43.
21. Sumer T, Freeman A, Cohen M, et al: Chemotherapy in Recurrent Noncystic Low-Grade Astrocytomas of the Cerebrum in Children. *J surgical Oncology* 1978;10:45-54.
22. Elvidge A: Long-Term Survival in the Astrocytoma Series.
23. Garcia D, Fulling K, Marks J: The Value of Radiation Therapy in Addition to Surgery for Astrocytomas of the Adults Cerebrum. *Cancer* 1985;55:919-97.
24. Lindegaard K, Mork S, Eide G, et al: Statistical analysis of clinicopathological features, radiotherapy, and survival in 170 cases of Oligodendroglioma. *J Neurosurg* Vol. 67. 1987.
25. Morantz R: Radiation Therapy in the Treatment of Cerebral Astrocytoma. *Neurosurg* Vol. 20, No. 6, 1987.
26. Rey J, Bello M, Campos J, et al: Chromosomal Composition of a Series of 22 Human Low-Grade Gliomas. *Cancer- Genet Cytogenet* 1987;29:223-37.

27. North C, North R, Epstein J, et al: Low-Grade Cerebral Astrocytomas: Survival and Quality of Life After Radiation Therapy. *Cancer* 1990
28. Shaw E, Dupport-Dumas C, Scheithauer B, et al: Postoperative Radiotherapy of supratentorial Low-Grade Gliomas. *Int J Radiation Onco* Vol. 16, No. 3, 1989
29. Shaw E: Low-Grade Gliomas: To Treat or Not to Treat? A Radiation Oncologist's View Point. *Arch Neurol* Vol. 47. 1990.
30. Cairncross J, Laperriere N: Low- Grade Glioma: To Treat or Not to Treat? *Arch Neurol* Vol. 46. 1989.
31. Recht L, Lew R, Smith T: Suspected Low-grade Glioma: Is Deferring Treatment safe? *Annals Of Neurology* Vol. 31. 1992.
32. Low Grade Gliomas : When To Treat? *Annals Of Neurology* Vol. 31, No. 4. 1992.
33. Whitton A.C, Bloom H.J.G: Low Grade Glioma Of the Cerebral Hemispheres in adults–A Retrospective Analysis of 88 cases. *Int J Radiation Oncology Bio Phy* Vol. 80. No. 4. 1990.
34. Westergaard L, Gjerris F, Klinken L: Prognostic Parameters in Benign Astrocytomas. *Acta NeuroChirurgica* 1993;123:1-7.
35. Ito S, Chandler K, Prados M, et al: Proloferative Potential and Prognostic evaluation of Low-Grade astrocytomas. *J Neuro-Oncology*. 1994;19:1-9.
36. Piepmeier J, Fried I, Makuch R. Low grade Astrocytoma May Arise From Different Astrocyte Lineages. Vol. 33, No. 4. 1993.
37. Berger M, Deliganis A, Dobbins J, et al: The Effect of Extent Of Resection on Recurrence in Patients with Low Grade Cerebral Hemisphere Gliomas. *Cancer* Vol. 74, No. 6. 1994.
38. Shaw E, Scheithauer B, Fallon J, et al: Mixed Oligoastrocytomas: A Survival and prognostic Factor Analysis. Vol. 34, No. 4, 1994.
39. Packer R, Sutton L, Patel K, et al: Seizure Control following tumor surgery for Childhood cortical Low- Grade gliomas. *J Neurosurg* 1994;80:998-1003.
40. Kross J, Pieterman H, Eden C, et al: Oligodendroglioma: The Rotterdam-Dijkzigt Experience. Vol. 34, No. 6. 1994.
41. Franzini A, Leocata F, Cajola L, et al: Low-Grade Glial Tumors in Basal Ganglia and Thalamus: Natural History and Biological Reappraisal. Vol. 35 No. 5. 1994.
42. Cellie P, Nofrone I, Palma L, et al: Cerebral Oligodendrogliomas: Prognostic factors and Life History. Vol. 33 No. 6. 1994.
43. Dirks P, Jay V, Becker L, et al: Develpoment of Anaplastic Changes in Low-Grade Astrocytomas of Childhood. Vol. 34 No. 1. 1994.
44. Ahmadi J, Savabi F, Apuzzo M, et al: Magnetic Resonance Imaging and Quantitative Analysis of Intracranial Cystic Lesions: Surgical Implication. Vol. 35, No. 2, 1994.
45. Pollack I, claassen D, Janosky J, et al: Low Grade Gliomas of the cerebral hemispheres in children: an analysis of 71 cases. *J Neurosurg* Vol. 82. 1995.
46. Glasser R, Rojiani A, Mickle J, et al: Dealyed Occurrence of cerebellar pleomorphic xanthoastrocytoma after supratentorial pleomorphic xanthoastrocytoma removal. *J Neurosurg* 1995;82:116-18.
47. Gajjar A, Bhargava R, Jenkins J, et al: Low-Grade astrocytoma with neuraxis dissemination at diagnosis. *J Neurosurg* 1995;83:67-71.
48. Karim A, Maat B, Hatlevoli R, et al: A Randomized Trial on Dose-Response in Radiation Therapy Of Low-grade cerebral Glioma: EORTC Study 22844. *Int J Of Radiation Oncology Bio Phy* Vol. 36, No. 3 1996;549-56.
49. Kreth F, Faist M, Warnke P, et al: Interstitial Radiosurgery of Low-grade Gliomas. *J Neurosurg* 1995;82: 418-29.
50. Negendank W, Sauter R, Brown T, et al: Proton magnetic resonance spectroscopy in patients with glial tumors: a multicenter study. *J Neurosurg* 1996;84:449-58.
51. Packer R, Ater J, Allen J, et al: Carboplatin and Vincristine Chemotherapy for children with newly diagnosed progressive low-grade gliomas. *J Neurosurg* 1997;86:747-54.
52. Decaestecker C, Salmon I, Dewitte O, et al: Nearest-neighbour Classification for identification of

aggressive versus nonaggressive low-grade astrocytic tumors by means of image cytometry-generated variables. *J Neurosurg* 1997;86:532-37.

53. Louw D, Bose R, Sima A, et al: Evidence of a High Free radical state in Low-grade Astrocytomas. *Neurosurg.* Vol. 41 NO. 5, 1999.
54. Lang F, Yung W.K, Sawaya R, et al: Adenovirus-mediated p53 Gene Therapy for Human Gliomas. Vol. 45 No. 5. 1999.
55. Schulder M, Maldjian J, Liu W, et al: Functional image-guided surgery of intracranial tumors located in or near the sensorimotor cortex. *J Neurosurg* 1996;89:412-18.
56. Nurmohamed MT, Rosendaal FR, Bueller HR, Dekker E, Hommes DW, Vandenbroucke JP, et al: Low-molecular-weight heparin versus standard heparin in general and orthopaedic surgery: a meta-analysis. *Lancet* 1992;340:152-6.
57. Leizorovicz A, Haugh MC, Chapuis F-R, Samama MM, Boissel J-P. Low molecular weight heparin in prevention of perioperative thrombosis. BMJ 1992;305:913-20.
58. Felson DT, Anderson JJ, Meenan RF: The comparative efficacy and toxicity of second-line drugs in rheumatoid arthritis. Arthritis Rheum 1990;33:1449-61.
59. Götzsche P C, Podenphant J, Olesen M, Halberg P: Meta-analysis of second-line antirheumatic drugs: sample size bias and uncertain benefit. *J Clin Epidemiol* 1992;45:587-94.
60. Chatellier G, Zaplet al E, Lemaitre D, Menard J, Degoulet P: The number needed to treat: a clinically useful nomogram in its proper context. *BMJ* 1996;312:426-9.

23
Surgical Approaches to Brain Stem Gliomas

VS Mehta, PS Chandra

INTRODUCTION

Brain stem gliomas are highly heterogenous, both in their clinical manifestation and in the pathology. It commonly occurs in children, comprising about 20% of all brain tumors although it may occur in adults also.[1] The term "brain stem glioma" is an imprecise description as it suggests that all such tumors behave in the same manner. More recently, it has been appreciated that brain stem tumors are not a homogeneous group with regard to their clinical, pathohistological, or biological features and that their prognosis may be directly related to tumor type and location.[2-6] The management is also quite varied ranging from direct radio/chemotherapy for diffuse lesions to open surgery in focal / exophytic tumors. The nature of surgery performed is a very important prognostic factor. Radical surgery for focal tumors may result in cure but can result in serious morbidity if not performed by an experienced surgeon. Pool[2] as early as 1968, reported useful long-term survival following surgical and x-ray treatment for verified brainstem gliomas. He suggested that if a brainstem tumor is suspected, an exploratory craniotomy is generally advisable; otherwise a benign lesion might be overlooked. The advent and the use of the operating microscope, the Cavitron (Valley Lab, Ontario, Canada), laser or tissue vaporizer (Japan Medical Dynamic Marketing Inc) and intra-operative evoked potentials and brain mapping has led to a more aggressive surgical approach to lesions in the brain stem.[7-9]

A number of classifications have been provided based on location, imaging, and histopathologies but, most of them obscure the fact that there exist basically two categories of brain stem tumors: (1) focal, discrete, sometimes exophytic, lesions associated with a favorable prognosis; (2) classic diffusely infiltrative lesions known for their relentless growth, resistance to radiotherapy and chemotherapy and with a bleak prognosis. Many authors[2-10] believe that the former mostly consist of pilocytic astrocytomas and the latter belong to a family of fibrillary astrocytomas, anaplastic astrocytomas, and glioblastoma multiforme. The focal tumors are particularly amenable to total excision, and carry a good prognosis provided the surgery has been performed by an experienced surgeon who has had enough exposure in operating such lesions. The following article apart from being a review of management of brain stem tumors also describes a series of 30 radically operated focal brain stem tumors.

PATIENT AND METHODS

This is a prospective study, which mostly includes focal intrinsic brain stem tumors operated by the senior author. All cases of exophytic and diffuse tumors were not excluded from this study. A total of 30 cases were operated between Feb 1999 and August 2001. The ages ranged from 4-58 years with a mean of 27 years and a median of 20 years. The maximum number of patients were between 21-40 years (11). The male to female ratio was 3:1. The clinical features included cerebellar signs (18), motor weakness (17-hemiparesis: 9, quadriparesis: 8), and headache in 12 patients. 12 cases had ocular involvement. Other cranial nerves involved included lower cranial palsy (13), facial palsy (9), facial hypoesthesia (5), and hearing deficit (2). One patient presented with altered sensorium due to acute hydrocephalus. Following a shunt, he improved and was later taken up for definitive surgery. MR imaging was performed in all cases. Contrast imaging was performed in 18 cases. This showed solid tumors in 21, partially cystic in 6 and cystic tumors in 3 cases. Ventriculomegaly was present in 8 cases. Tumor was located in pons in 12, midbrain in 8 (dorsal 6; ventrolateral 2) and medulla in 10 cases. All cases underwent a radical, total or near-total excision **(Figs. 1-2)**. The most common approach was midline suboccipital craniectomy used in 24 cases. Other approaches included retromastoid suboccipital approach (3), subtemporal-transtentorial (2) and supracerebellar-transtentorial (1).

RESULTS

Histopathology revealed a total of 16 astrocytomas (pilocytic:5, grade II:7, grade III:4). The other pathologies included cavernomas (6), hemangioblastoma (4), ependymoma (2), neurofibroma (1) and inflammatory pathology suggestive of tuberculosis in 1 case. Complications included CSF leak (3), menigotis (1), lower cranial palsy (2), chest infection (1), epidural hematomas (2- both re-explored), and mutism (1). All these cases subsequently improved. One patient died on the 6th post operative day due to aspiration. The follow up ranged from 1-26 months with an average of 17 months. A total of 22 patients had improvement in neurological features or remained in status quo ante. Improvement was for most of neurological features like motor weakness, cerebellar signs and even lower cranial palsies. One patient had complete improvement of his ocular palsy when seen at 18 months of follow-up **(Fig. 2)**. 7 patients had worsening in their neurological features.

DISCUSSION AND LITERATURE REVIEW

Defining brain stem tumors

In the pre CT scan era, and the early CT scan era, all tumors between the thalamus and the cervical spinal cord were called brain stem gliomas and were regarded as one entity. The traditional treatment of these tumors consisted of irradiation, often preceded by a biopsy, and sometimes followed by adjunctive chemotherapy.[11] This had met with little overall success. In part, this is because previous reports vary in their definition of "brain stem" tumors.[12,1-6] Some authors regard lesions of the midbrain, pons, and medulla oblongata all as brain stem lesions, whereas others make a distinction between the thalamic and midbrain lesions on the one hand and pons and medulla lesions on the other. In addition, many studies include patients from the era before CT scans,[12-14] as well as cases of brain stem tumors that were not histologically proven.[12-14]

Classification:

A number of classifications are available. Brain stem gliomas have been classified by site, imaging, and pathology.

Location: The terms applied include midbrain tumor,[15-17] tectal tumor,[17-19] pontine glioma,[20,21] focal medullary tumor[22] and cerviomedullary tumor.[21,22]

Imaging: diffuse gliomas,[22,23] intrinsic gliomas,[24] pencil gliomas,[25] exophytic tumors[21-23] (dorsal; ventral and lateral), focal and cystic tumors.[24]

Histopathology: low grade or benign tumors (pilocytic and grade II) and high grade tumors (WHO GRADES III and IV).[20-24]

Popularly a mixture of all categories are used and the types commonly used in clinical practice include focal (solid or cystic), diffuse, exophytic (ventral, lateral, or dorsal), tectal plate, and cervicomedullary.

Currently the most commonly used classification in clinical practice is based on imaging and location. **This includes focal (solid or cystic), diffuse, exophytic (ventral, lateral, or dorsal), tectal plate, and cervicomedullary.**

However, it is to be remembered that ultimately there are two categories of tumors: (1) **focal,** discrete sometimes exophytic lesions associated with a favorable prognosis, (2) **classic, diffusely** infiltrative lesions known for their relentless growth, resistance to radiotherapy, and bleak prognosis.[10,12] *It is the former category of tumors that need to be tackled aggressively.*

Epstein[7,8] and most of other authors[9-12] have defined focal tumors as "well circumscribed masses with a diameter <2 cm". Anything larger than this comes under the category of diffuse tumors. We have however, included even tumors >2 cm within the focal tumor category, provided there was no diffuse involvement of the brainstem. We feel that the definition of 'diffuse' should be included only for those tumors involving the entire brainstem axis homogenously where one cannot differentiate between normal brainstem and the tumor. In the authors' series a total of 30 cases were operated between Feb 1999 and August 2001. The ages ranged from 4-58 years with a mean of 27 years and a median of 20 years. The maximum number of patients were between 21-40 years (11). The male to female ratio was 3:1.

Pathology:

Most astrocytomas of the brain-stem are infiltrative tumors of fibrillary type, similar to diffuse cerebral astrocytomas. Macroscopically, they are usually characterized by a symmetrical enlargement of the pons (diffuse hypertrophy). The expanding structure encroaches posteriorly and superiorly upon the fourth ventricle. Occasionally, there may be anterior enlargement. The medulla is often spared. Sometimes, however it originates from medulla and may spread to the upper cervical region.[42] Histologically, there is a diffuse replacement of nervous tissue by small and large astrocytic cells, stellate, pilocytic, and gemistocytic either randomly dispersed or arranged in smaller or larger groups. This morphologic diversity reflects the cellular heterogeneity even in absence of anaplasia. In some cases there may be frank features of anaplasia, necrosis and endothelial proliferation as seen in glioblastoma multiformes.[42] Depending on the macroscopic appearance, the brain stem gliomas have been divided into the following categories.[42-44]

1. **Diffuse**: This is a 'stereotype' of brainstem tumors that have been recognized since the beginning of neurosurgery. These neoplasms present with a short history and present with multiple cranial nerve palsies along with involvement of long tracts. MR imaging is virtually diagnostic of these lesions. The tumor is invariably malignant and the only form of therapy advisable is radiotherapy or chemotherapy. Surgery is not indicated. Even stereotactic biopsy is not advised as the procedure may cause additional morbidity, and the biopsy may not contain representative tissue.

2. **Focal**: These more commonly involve the medulla and are often associated with a relatively long clinical history. Neurological examination usually reveals focal deficits i.e. either a sixth or seventh nerve palsy. These neoplasms are commonly low-grade astrocytomas and are amenable for surgical excision.

3. **Cervicomedullary tumors**: These are invariably low-grade astrocytomas and gangliogliomas and amenable to radical excision.

4. **Cystic tumors**: These are usually low-grade pilocytic astrocytomas and highly amenable for surgical excision.

5. **Exophytic tumors**: These tumors are characterized by exophytic growth either ventrally, laterally into the cerebellopontine angle or dorsally into the fourth

ventricle. Of these, the dorsally exophytic tumors have the best prognosis, are amenable for radical excision and are associated with long-term neurological recovery. The other variants have a poor prognosis.

6. **Tectal plate gliomas**: These are uncommon in the pediatric age group. They have been reported to be one of the commonest causes for hydrocephalus in the western literature. The main presenting feature is hydrocephalus. These are diagnosed on MR imaging with gadolinium contrast and are managed by performing a ventriculoperitoneal shunt.[19] These may be simply followed up and if they increase in size, may be subjected to radiotherapy. Radiosurgery has been also used recently. We have performed endoscopic stenting at the aqueduct along with simultaneous biopsy in two cases. Both the cases are doing well at follow-up. Endoscopic third ventriculostomy or aqueductal stenting may be a better alternative to a shunt surgery.

Histopathology in our series revealed a total of 16 astrocytomas (pilocytic: 5, grade II: 7, grade III: 4). The other pathologies included cavernomas (6), hemangioblastoma (4), ependymoma (2) neurofibroma (1) and inflammatory pathology suggestive of tuberculosis in 1 case.

Clinical features:

Albright et al[12] stated that approximately 70% of children with diencephalic tumors live 5 years after diagnosis, whereas only 30% of children with brain stem gliomas survive for that long. However, it is now widely recognized that brain stem tumors are a very heterogeneous group with regard to their clinical, histopathological, and biological features, and it is essential that they be therefore regarded as distinct entities.

Vomiting is a common symptom and is usually due to involvement of area postrema and the other medullary nuclei by the tumor. Ataxia is usually due to involvement of cerebral peduncles. Motor weakness due to involvement of pyramidal tracts is common. Extra-occular movement palsy is characterized by squint usually noticed by parents or school teachers. Additional signs include nystagmus, skew deviation, and internuclear ophthalmoplegia.[7-10] The latter sign is a hallmark of intrinsic brainstem involvement characterized by failure to adduct the eye on the side of the lesion and presence of nystagmus on abducting the eye on the opposite side. This occurs due to the involvement of medial longitudinal fasciculus. Other signs of cranial nerve involvement include facial weakness, dysarthria, and swallowing difficulties. Features of raised intracranial pressure may occur due to obstruction of CSF pathways in dorsally exophytic tumors. Focal intrinsic tumors from the tectum can cause obstruction early in the course of illness.[19] The following features are characteristic of an intrinsic brain-stem neoplasm: (1) Internuclear ophthalmoplegia, (2) Horner's syndrome, (3) cranial nerve involvement with crossed motor involvement e.g. ipsilateral third nerve paresis with contralateral hemiparesis (Weber's syndrome), (4) Combination of certain cranial nerve nuclei e.g. 6th and 7th cranial involvement on one side indicates an intrinsic pathology.

Patients with tumors at the *cervicomedullary junction* exhibit lower cranial nerve dysfunction and pyramidal tract signs, similar to those occurring in the upper cervical spinal cord. In these patients, radical excision is possible with little morbidity, showing a usually very protracted clinical course, and histopathologically, the vast majority consists of a low-grade glioma.[7-10]

The *intrinsic tumors of the pons* are infiltrative in nature and are mostly malignant. Symptoms at the time of admission are usually cranial nerve palsy, pyramidal tract signs, and ataxia. It is characteristic for these tumors to progress to an advanced stage without the development of increased intracranial pressure.[26] The course of the disease is steadily progressive, and most patients die within 2 years, even after radiation therapy. Epstein[7,8,16] refers to this group of tumors as "*diffuse*" and strongly advises against surgical intervention because this in no way alters the ultimate outcome. Even biopsy is not recommended by many authors, as it may not show the representative tissue due to the heterogenous nature of the tumor and poses a high risk to the patient. However, in authors experience if a focal tumor in pons is associated paucity of neurological signs, it may be benefited from radical surgery.

However it is important to differentiate these tumors from diffuse brain stem swelling which may occur in neurofibromatosis Type 1 (NF-1).[30-33] Among patients with NF-1, histological confirmation, when available, is consistent with low-grade glioma in the majority of cases. These two entities, diffuse PG in patients without NF-1 and diffuse brain stem enlargement in children with NF-1, are different in their natural histories and may represent different pathophysiological processes. A recent study has demonstrated the use of MR spectroscopy in differentiating the two entities. The MRS neuronal marker NA peak, was preserved in patients with NF-1 but significantly decreased in diffuse gliomas. They hypothesized that this finding reflects preservation of brain stem neuronal elements in the NF-1 group, consistent with their minor or absent symptoms of brain stem involvement. In contrast, patients with diffuse gliomas show significantly decreased NA and have major neurological deficits attributable to neuronal damage from their brain stem tumors.

A subgroup of benign brain stem gliomas, identified by Hoffman et al[4,27] is the *dorsally exophytic* transependymal benign brain stem glioma. The duration of the onset of symptoms is longer in this group, and these children often have hydrocephalus (75%). Ataxia is also common, whereas cranial nerve deficit and long tract signs are relatively rare. These tumors are very amenable to surgical resection, frequently requiring no further therapy.[28,29]

Focal midbrain astrocytomas form a special entity. Fifty percent of all tumors occur predominantly in the tectal region, giving rise to an obstructive hydrocephalus in 83%.[6-9] Most patients with hydrocephalus exhibit headaches, vomiting, and florid papilledema as a result of increased intracranial pressure. The other 50% of focal midbrain tumors occur predominantly in the tegmentum and cause either long tract signs, due to compression of the cerebral peduncles, or cranial nerve deficits and signs and symptoms referable to the eyes (up-gaze paresis, Perinaud's syndrome), caused by pressure on the nuclei of the oculomotor nerves and their connecting pathways.[6,-9, 21, 24, 25] In our series a total of 8 patients had tumors in the midbrain, of which 6 were dorsally located and 2 ventrolaterally. Two tumors were extending into thalamus.

Neuroradiology:

The advent of the MRI scan, with its readily available sagittal and coronal planes of imaging, has facilitated accurate localization of pathological lesions within the brain stem. Tumors should be assessed for location, signal intensity, focality, extent of infiltration, degree of brainstem enlargement, presence of cyst, necrosis or hemorrhage, or an exophytic component. Following this the morphology should be categorized whether it comes under focal, diffuse, cervicomedullary or exophytic. The MR imaging should be always correlated with the clinical history. Finally it is important to identify the benign subgroup of brainstem gliomas. Epstein et al[7-9] have retrospectively analyzed 88 cases of brainstem gliomas and have proposed that the more benign tumors tend to displace rather than infiltrate the secondary structures like pia, fibre tracts and the ependyma, the characteristics of which may be identified on good quality MRI's.

MR imaging was performed in all our cases. Contrast imaging was performed in 18 cases. This showed solid tumors in 21 cases, partially cystic in 6 and cystic tumors in 3 cases. Ventriculomegaly was present in 8 cases. Tumor was located in pons in 12 cases, midbrain in 8 (dorsal 6; ventrolateral 2) and medulla in 10 cases.

A retrospective review of all of our preoperative radiographs, made it possible to appreciate certain characteristic radiological features. All tumors in the midbrain were focal lesions, with a predilection for either the tectal plate or the tegmentum. Upward extension to the thalamus was present in 2 of the patients and downward extension to the pons was present in 4 of the patients. The edge of the tumor was always sharply defined, and most lesions were more or less round or spherical. Hydrocephalus was present in 8 patients.

Mapping of the cranial nerve nuclei:

Because the rhomboid fossa and the brain stem are densely packed with nerve nuclei and neural tracts,[34] surgery in this area carries a substantial risk of new or increased neurological deficits. For the same reason, neurosurgeons have long considered lesions in this "no-man's land" at least among the most difficult to manage and at most inoperable. However, new diagnostic and surgical techniques have reduced the risk of neurological deficit resulting from such surgery, and surgeries are being performed more frequently to treat intrinsic brain stem lesions.[34-36] Recent anatomic studies have identified safer approaches to the brain stem via the fourth ventricle,[37] but because disease often distorts the anatomy, landmarks may be difficult or impossible to identify with the aid of the operating microscope alone. Thus, electrophysiological techniques need to be developed to help distinguish the neural structures so that the surgeons can avoid manipulating or injuring those structures. Two kinds of monitoring methods are commonly used.

Intraoperative motor nucleus mapping: During motor mapping in the brain stem, the electrical stimulation of the hypoglossal or the vagal triangle elicits CMAPs from the tongue muscles and the soft palate, respectively. Stimulating the facial colliculus, which contains facial and abducens nerve fibers, elicits CMAPs simultaneously from the lateral rectus and facial muscles. Stimulating other ocular motor pathways elicits CMAPs from corresponding extraocular muscles.

Continuous intraoperative EMG monitoring: Because muscles amplify the electrical activity in the individual motor units, continuous monitoring of the EMG activity from certain muscles of the head can provide information about the status of the respective cranial motor nerves. Use of this technique has been described for the monitoring of the motor function of the facial nerve,[39, 40] and other motor cranial nerve nuclei.[41]

The prognostic value of these manipulation-evoked discharges is based on empirical information, rather than on the results of controlled studies or on a detailed understanding of the pathophysiological mechanisms involved. However, for practical purposes, the continuous recording of the EMG activity seems to provide a reliable way for the surgeon to identify the exact location of various cranial nerve nuclei to plan incision on the brainstem. For dorsally located lesions, we usually use electrodes connected to the orbicularis oris, orbicualris occuli and the tongue muscles with either intermittent or continuous stimuli. A single stimuli of 0.1-1 mA or a continuos stimuli of 10 Hz (50-400 mS) is generally used.

Management of diffuse tumors:

The diffuse tumors carry the worst prognosis and the survival is usually not more than 18 months after diagnosis. Parts of the tumor may start off as a fibrillary astrocytoma but by the time of death, usually the entire tumor is either a malignant astrocytoma or a glioblastoma multiforme. Thus a stereotactic biopsy is totally unreliable for diagnosis, while the imaging characteristics are almost diagnostic of this dreadful condition.[7-10] These patients may be directly subjected to radiotherapy. The type of radiotherapy is undergoing a change. The standard therapy has been 55 Gy to the tumor area with weekly dose of 800-1000 cGy. In hyper-fractionated therapy, higher doses up to 76 Gy may be tolerated and has improved short-term survival without obvious toxicity.[7-12, 44]

Surgical approaches:

Surgery is indicated for focal solid or cystic tumors, cervicomedullary tumors, and exophytic (particularly dorsal) tumors.[7, 9,16, 46] In a retrospective study of 111 cases from our institute, surgery was found to be definitely beneficial in selected cases.[46] The patients need to be followed up with regular neuroimaging. The patients with focal cystic tumors may undergo decompression of the cyst, which is fairly an easy procedure. However, focal radiotherapy may be required to prevent recollection of the cyst. In patients with dorsally exophytic tumors, decompression may be carried out till the tumor is shaved off flush with the floor of fourth ventricle. It is advisable not to remove a thin carpet of tumor if it is not coming off easily to prevent serious morbidity. A large number of patients with low-grade astrocytomas and gangliogliomas will have a good outcome.[26-28] Patients with tectal plate tumors may be subjected to a shunting procedure followed by direct radiotherapy or radiosurgery.[18, 19] Recently we have performed endoscopic third ventriculostomy and stenting in 2 cases of tectal plate gliomas along with simultaneous biopsy during the endoscopic procedure. This has an advantage of performing a biopsy and a shunting procedure at the same sitting.

All the cases in the present series were focal intrinsic brainstem tumors, and present the greatest surgical challenge. Of highest importance before surgery is to plan out the position and surgical approach.[16] The author prefers a sitting position for all tumors in dorsal midbrain and upper pons for a midline suboccipital approach. Prone is preferable in small children and the elderly and in patients with gross hydrocephalus and lower pons and medullary tumors. A lateral CP angle approach may be preferred for patients with ventrolateral tumors of the pons and medulla. Once the floor of the fourth ventricle is exposed (or the lateral surface of brainstem), it should be carefully inspected under high magnification for any abnormal "bulges" or any "discoloration". A mapping of the floor should be then performed as per technique described earlier. We have found direct stimulation of cranial nerve nuclei quite useful to map out the eloquent areas. The floor of the fourth ventricle is usually screened first with a low intensity bipolar. Following this a stimulation of 0.1 mA single or continuous stimulation with 10 Hz frequencies at 50-400 mS is given. The waveforms may then be recorded with electrodes from orbiculi occuli, oris and the tongue muscles. We have not found somatosensory evoked potentials or auditory evoked potentials very useful. This helps in localizing the 6th and the lower cranial nerve nuclei. If no exophytic component is seen, then one of the "safe areas"[38] should be used to make an incision after mapping. This usually includes the dorsolateral sulcus lateral to the cranial nerve nuclei. Otherwise the exophytic component should be tackled first. The incision should be as small as possible, longitudinal, nearest to the tumor surface and away from the nuclei. The author prefers to use the tissue vaporizer, as there is minimum heat dissipation to the surrounding structures even when applied close to the cranial nerve nuclei. The cyst aspiration if any, should be performed before removal of the solid component. Use of micro-tip ultrasonic aspirators and irrigating non-stick bipolars are very useful if available. The tumor rim should be left behind if no distinct plane is seen.[7, 9-11,16, 24, 26, 28, 34] All the patients should be started on methyleprednisolone just before giving incision on the brainstem and this should be continued for 24 hours in doses of 30 mg/kg for the first hour followed by 5.4 mg/kg/hour for the next 24 hours. All patients should be electively ventilated with sedation for 24 hours.

All cases in the author's series underwent a radical excision (total or near-total) (**Figs. 1-2**). The most common approach was midline suboccipital craniectomy used in 24 cases. Other approaches included retromastoid suboccipital approach (3), subtemporal-transtentorial (2) and supracerebellar-transtentorial for a tectal plate glioma (1).

Complications included CSF leak (3), meningitis (1), lower cranial palsy (2), chest infection (1), epidural hematomas (2-both re-explored), and mutism (1). All these cases subsequently improved. One patient died on the 6th postoperative day due to aspiration. The follow up ranged from 1-26 months with an average of 17 months. A total of 22 patients had improvement in neurological features or remained status quo. Improvement was for most of neurological features like motor weakness, cerebellar signs and even lower cranial palsies. One patient

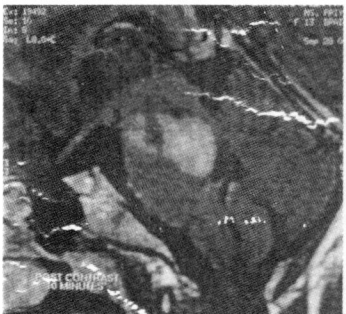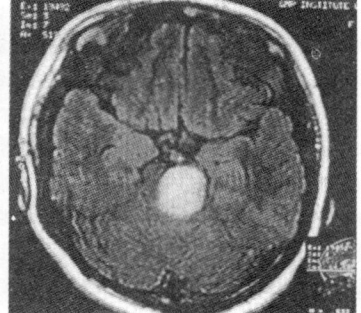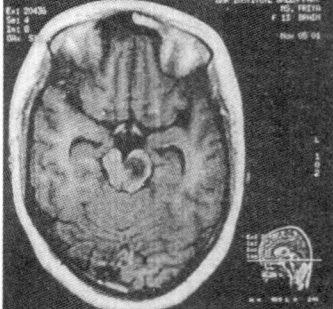

Figs. 1 : MR image of a large pontine glioma (a) saggital section with delayed contrast enhancement, (b) axial section and (c) following a total excision

had complete improvement of his ocular palsy when seen at 18 months of follow-up (**Fig. 2**). 7 patients had worsening in their neurological features.

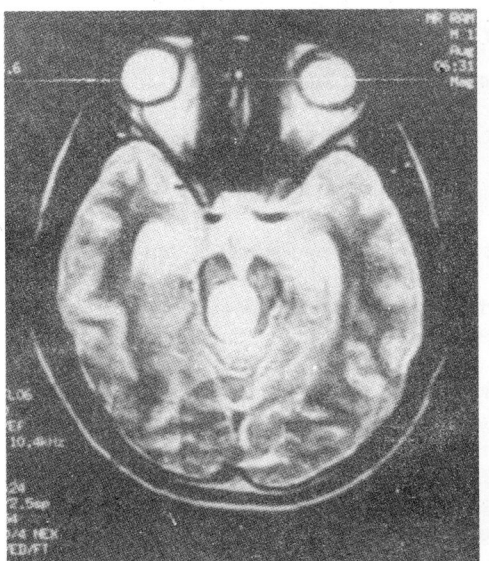 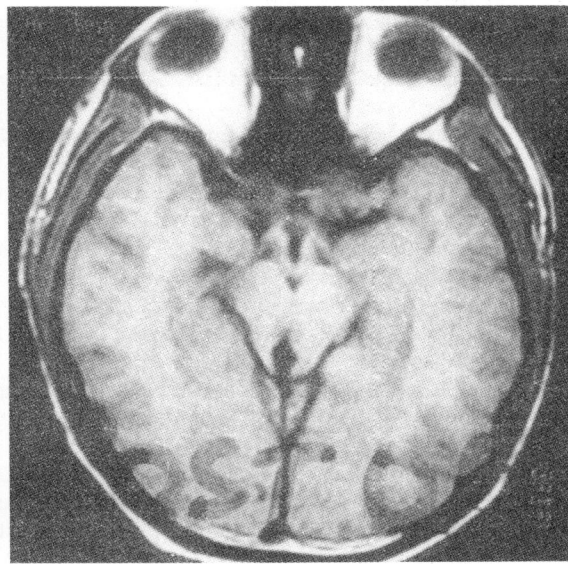

Figs. 2: (a) MR imaging with gad enhancement showing large focal solid midbrain tumor, (b) total excision following surgery

CONCLUSIONS:

The identification of various subgroups of brain stem tumors has led to more rational treatment strategies and allows for a more accurate assessment of prognosis. The focal tumors with paucity of neurological signs are a distinct subgroup of brain stem tumors. Most of them are low-grade astrocytomas and are amenable to radical surgical resection, but should be performed by an experienced surgeon in a set-up with adequate intra-operative and post-operative facilities. Surgical intervention in this subgroup of brain stem tumors is associated with an excellent long-term prognosis.

REFERENCES:

1. Bruno L, Schut L: Survey of pediatric brain tumors, in American Association of Neurological Surgeons (eds): Pediatric Neurosurgery. Surgery of the Developing Nervous System. New York, Grune & Stratton, 1982;361-65.
2. Albright AL: Midline intra-axial tumors, in Deutsch M (ed): Management of Childhood Brain Tumors. Boston, Kluwer Academic Publishers, 1990;401-09.
3. Artigas J, Ferszt R, Brock M, Kazner E, Cervos-Navarro J: The relevance of pathological diagnosis for therapy and outcome of brain stem gliomas. *Acta Neurochir Suppl* 1988;42:166-69.

4. Hoffman HJ, Becker L, Craven MA: A clinically and pathologically distinct group of benign brain stem gliomas. *Neurosurgery* 1980;7:243-48.
5. Nishio S, Fukui M, Tateishi J: Brain stem gliomas: A clinicopathological analysis of 23 histologically proven cases. *J Neurooncol* 1988;6:245-50.
6. Reigel DH: Brain-stem tumors during childhood, in Section of the American Association of Neurosurgery: Pediatric Neurosurgery. Surgery of the Developing Nervous System. New York, Grune & Stratton, 1982;409-17.
7. Epstein F, McCleary EL: Intrinsic brain-stem tumors of childhood: Surgical indications. *J Neurosurg* 1986;64:11-15.
8. Epstein F: Intrinsic brainstem tumors of childhood, in Homburger F (ed): Progress in Experimental Tumor Research. Basel, *Karger*, 1987;160-69.
9. Heffez DS, Zinreich SJ, Long DM: Surgical resection of intrinsic brain stem lesions: An overview. *Neurosurgery* 1990;27:789-98.
10. Fisher PG, Breiter SN, Carson BS, et al: A clinicopathological reappraisal of brain stem tumor classification. *Cancer.* 2000;89:1560-76.
11. Matson DD: Neurosurgery of Infancy and Childhood. Springfield, Illinois, Bannerstone House, 1969.
12. Albright AL, Guthkelch AN, Packer RJ, Price RA, Rourke LB: Prognostic factors in pediatric brain-stem gliomas. *J Neurosurg* 1986;65:751-55.
13. Eifel PJ, Cassady JR, Belli JA: Radiation therapy of tumors of the brainstem and midbrain in children: Experience of the Joint Center for Radiation Therapy and Children's Hospital Medical Center (1971-1981). *Int J Radiat Oncol Biol Phys* 1987;13:847-52.
14. Grigsby PW, Garcia DM, Simpson JR, Fineberg BB, Schwartz HG: Prognostic factors and results of therapy for adult thalamic and brainstem tumors. *Cancer* 1989;63:2124-29.
15. Barkovich JA, Kricher J, Kun LE, et al: Brain stem gliomas: a classification system based on magnetic resonance imaging. *Pediatr Neurosurg* 1991;16:73-83.
16. Epstein F, Constantini S: Practical decisions in the treatment of pediatric brain tumors. *Pediatr Neurosurg* 1996;24:24-34.
17. Fischbein NJ, Prados MD, Wara W, et al: Radiologic classification of brain stem tumors: correlation of magnetic resonance imaging with clinical outcome. *Pediatr Neurosurg* 1996;24:9-23.
18. Boydston WR, Stanford RA, Muhlbauer MS, et al: Gliomas of the tectum and periaqueductal region of the mesencephalon. *Pediatr Neurosurg* 1992;17:234-38.
19. Pollack IF, Pang D, Albright AL: The long-term outcome in children with late onset aqueductal stenosis resulting from benign intrinsic tectal tumors. *J Neurosurg* 1994;80:681-88.
20. Lassman LP, Arjoma VE: Pontine gliomas of childhood. *Lancet* 1967;1:913-15.
21. Broniscer A, Gajjar A, Bhargava R, et al: Brain stem involvement in children with neurofibromatosis type I: role of magnetic resonance imaging and spectroscopy in the distinction from diffuse pontine gliomas. *Neurosurgery* 1997;40:331-38.
22. Epstein FJ: farmer JP: Brain stem growth patterns. *J Neurosurg* 1993;78:408-12.
23. Freeman CR, Farmer JP: Pediatric brain-stem gliomas: a review. *Int J Radiat Oncol Biol Phys* 1998;40:265-71.
24. Epstein F, Wiscoff JH: Intrinsic brainstem tumors in childhood: surgical indications. *J Neurooncol* 1988;6:309-17.
25. Sanford RA, Bebin J, Smith WR: Pencil gliomas of the aqueduct of sylvius. *J Neurosurg* 1982;57:690-96.
26. Lassman LP: Tumors of the pons and medulla oblongata, in Amador LV (ed): Brain Tumors in the Young. Springfield, Illinois, Charles C Thomas, 1983;546-64.
27. Hoffman HJ: Benign brainstem gliomas in children, in Homburger F (ed): Progress in Experimental Tumor Research. Basel, *Karger*, 1987;154-59.
28. Koos WTH, Miller MH: Intracranial Tumors of Infants and Children. St. Louis: CV Mosby Company, 1971;148-49, 221-22.

29. Kelly PJ: Computer-assisted stereotaxic laser microsurgery, in Apuzzo MLJ (ed): Surgery of the Third Ventricle. Baltimore, Williams & Wilkins, 1987;811-28.
30. Broniscer A, Gajjar A, Bhargava R, et al: Brain Stem Involvement in Children with Neurofibromatosis Type 1: Role of Magnetic Resonance Imaging and Spectroscopy in the Distinction from Diffuse Pontine Glioma. *Neurosurgery* 1997;40(2):331-34.
31. Cohen ME, Duffner PK, Heffner RR, Lacey DJ, Brecher M: Prognostic factors in brain stem gliomas. *Neurology* 1986;36:602-05.
32. Ilgren EB, Kinnier-Wilson LM, Stiller CA: Gliomas in neurofibromatosis: A series of 89 cases with evidence for enhanced malignancy in associated cerebellar astrocytomas. *Pathol Annu* 1985;20:331-58.
33. Milstein JM, Geyer JR, Berger MS, Bleyer WA: Favorable prognosis for brain stem gliomas in neurofibromatosis. *J Neurooncol* 1989;7:367-71.
34. Bricolo A, Turazzi S, Cristophori L, Talacchi A: Direct surgery for brainstem tumors. *Acta Neurochir (Suppl) Wien* 1991;53:148-58.
35. Daube JR: Intraoperative monitoring of motor cranial nerves, in Schramm J, Møller AR (eds): Intraoperative Neurophysiologic Monitoring in Neurosurgery. Berlin, Springer Verlag, 1991;246-67.
36. Møller AR: Intraoperative monitoring of cranial motor nerves, in Møller AR (ed): Evoked Potentials in Intraoperative Monitoring. Baltimore, Williams & Wilkins, 1988;99-120.
37. Strauss C, Romstöck J, Nimsky C, Fahlbusch R: Intraoperative identification of motor areas of the rhomboid fossa using direct stimulation. *J Neurosurg* 1993;79:393-99.
38. Kyoshima K, Kobayashi S, Gibo H, Kuroyanagi T: A study of safe entry zones via the floor of the fourth ventricle for brain-stem lesions. *J Neurosurg* 1993;78:987-93.
39. Harner S, Daube J, Ebersold M, Beatty C: Improved preservation of facial nerve function with use of electrical monitoring during removal of acoustic neuromas. *Mayo Clin Proc* 1987;62:92-102.
40. Kartush JM: Electroneurography and intraoperative facial monitoring in contemporary neurotology. *Otolaryngol Head Neck Surg* 1989;101:496-503.

24

Brain Stem Gliomas: Stereotactic Biopsy and Radiotherapy

RK Moorthy, V Rajshekhar

INTRODUCTION

Brain stem gliomas are a group of tumours that have been well described and studied since the routine use of magnetic resonance imaging in diagnostic neuroradiology. They have been classified into several groups based on the location of lesion and the enhancement pattern by Stroink et al.[1] Epstein et al have grouped intrinsic brain stem gliomas into Cervicomedullary, Diffuse lesions and Focal lesions based on MR findings.[2] The Children's cancer group study has advocated the use of MRI findings in making an accurate diagnosis in a suspected case of brain stem gliomas.[3] The differential diagnosis of lesions in the brain stem is presented in **(Table I)**. The differential diagnosis of brain stem lesions encompass a wide variety from brain stem encephalitis to arteriovenous malformations thus making histological confirmation prior to treatment mandatory in certain select situations.

Table 1 : *Differential Diagnosis of Brain Stem Lesions Based on Radiologic Appearance*

Focal
    Brain stem gliomas
    Ependymomas
    Non glial primary neoplasms
    Metastasis
    Tuberculomas
    Cysticercal granulomas
    Pyogenic abscess
    Vascular malformations

Diffuse
    Brain stem gliomas
    Brain stem encephalitis
    Demyelination

INDICATIONS FOR STEREOTACTIC BIOPSY

Epstein and Mcleary have advocated radical surgery for focal intrinsic and cervicomedullary lesions.[2] In their experience, microscopic pathology of gliomas of the brain stem is not homogeneous. They have also speculated that in diffuse lesions, what appears to be oedema may be infiltrating lesions. All the patients with diffuse lesions in this series had malignant gliomas (astrocytoma grade III or IV). The Children's cancer group study showed that biopsy of brain stem masses in children always showed astrocytic tumours in histopathology while non glial tumours were more often seen in adults. Prior to the availability of MRI, biopsy was done for all brain stem masses in children to obtain histological confirmation and for prognostication. However, this study showed that **diffuse non enhancing lesions in the brain stem** on an MRI in children might be treated empirically without histological confirmation.[3] However, **focal enhancing masses** or dorsally exophytic tumours require histological confirmation. Of these, stereotactic biopsy is indicated in focal enhancing masses in children. However, **in adults all lesions in the brain stem** irrespective of their diffuse or focal nature require histological confirmation and hence a stereotactic biopsy is indicated in these lesions.

Apart from histological verification, stereotaxy can be used therapeutically to obtain symptomatic relief. Many lesions in the brain stem have a cystic component, aspiration of which provides resolution of symptoms. In addition, radioisotopes have been instilled into cavities of gliomas after aspiration of cyst contents.[4-6]

CONTRAINDICATIONS

While specific indications have evolved over the years for stereotactic biopsy of brain stem lesions, there are certain situations where a stereotactic biopsy should not be performed. These include dorsally exophytic brain stem tumours, cervicomedullary gliomas and focal midbrain gliomas. Diffuse non enhancing brain stem lesions on MRI in children with clinical features of an intrinsic brain stem mass do not warrant a histopathological confirmation prior to radiation therapy.[2,3] While stereotactic biopsy is not suited for dorsally exophytic tumours and cervicomedullary gliomas, the benign nature of focal midbrain lesions make follow up a safe and optimal treatment option in this condition.

TECHNICAL DETAILS[7-10]

The Brown-Roberts-Wells (BRW) or Cosman-Roberts-Wells (CRW) stereotactic systems are used to obtain the biopsies. The head ring should be fixed as low as possible on the head to ensure optimal access to the target in the brain stem. The head ring may have to be angulated in the anteroposterior direction to make this possible. For a transcerebellar approach, the anterior part of the ring should be placed posterior. This 180^0 rotation is as the slots for attaching the Mayfield head clamp are posterior and these will be needed to support the head in the prone position. The localiser ring with the nine fiducial markers made of carbon-fibre is then fixed to the head ring and a contrast CT scan is done. The head of the patient is not required to be secured to the CT couch. The x and y coordinates of the nine fiducial markers and the target on an appropriate CT slice are obtained from the computer on the scan machine and these are fed into the SCS1 computer to obtain the AP, lateral and vertical coordinates of the target. The rest of the procedure depends on the approach being used.

For the trancerebellar approach, a CRW system is preferred as the clearance of its arc system from the posterior aspect of the head is much greater than in the BRW system. The cranial entry point is made using a twist drill craniostomy or a burr hole. In our institution, a 1.2 mm diameter probe and a Gildenberg's (1x2mm) cup forceps are used to obtain tissue from a solid tumour. The same probe and a hand held syringe is used to gently aspirate any cystic component.

ACCESS TO THE BRAIN STEM

The brain stem can be accessed stereotactically through two approaches- transfrontal and transcerebellar. The transfrontal approach[5, 6] allows a trajectory parallel to and through the axis of the brain stem. The transcerebellar approach described by Abernathy et al[8] provides access to the pons through the middle cerebellar peduncle.

In both these approaches, the corridor remains within the brain parenchyma. The most important principle in stereotaxy is that one should not transgress a second pial plane. Violation of this second pial plane would lead to entry into the basal cisterns with resultant increased risk of bleed from the vessels situated therein. Stereotactic biopsies of the brain stem may be performed under local anaesthesia in cooperative adults via the transfrontal route. The transcerebellar approach, however required general anaesthesia. General anaesthesia is also indicated in children and in uncooperative adults.

TRANSFRONTAL ROUTE

The transfrontal approach[4, 7, 10] requires the patient to be in supine position under local anaesthesia. A cranial entry point just anterior to or posterior to the coronal suture **(Fig. 1)** about 3 cm lateral to the midline is chosen. A right frontal entry is preferred for all masses, except those that are lateralised left of midline. Through a twist drill craniostomy or a burr hole, the stereotactic probe is passed. The probe can be directed through a trajectory parallel to and through the entire axis of the brain stem. The probe passes through the cerebral white matter, body of the lateral ventricle and the thalamus into the brain stem. The basilar artery, posterior cerebral artery, vertebral artery and the internal cerebral veins should be avoided and the interpeduncular cistern should not be transgressed. The

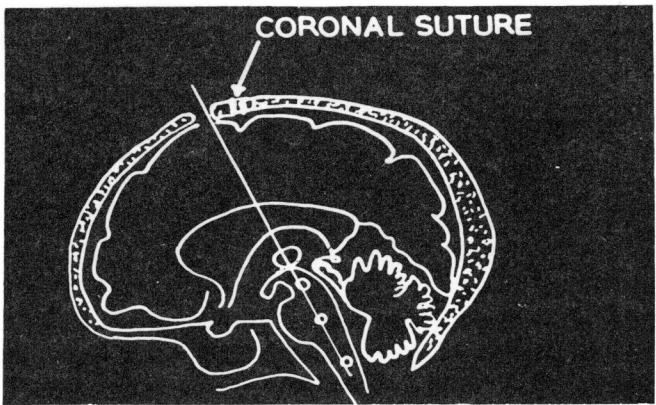

Fig. 1 : Diagram showing the precoronal entry site and the trajectory along the axis of the brain stem in the transfrontal route.

entry point and trajectory should be planned as not to enter the tentorium at its hiatus. The transfrontal approach allows tissue sampling at various depths along the axis of the brain stem in an infiltrative lesion. This decreases the error that may occur due to sampling.

TRANSCEREBELLAR ROUTE

The transcerebellar route requires the patient to be in prone position.[7-9] The anterior portion of the frame is positioned low on the posterior cervical region and the posterior portion of the frame is positioned at the level of the nose with the vertical support pins in the supraorbital region. A cranial entry point is chosen in the suboccipital region below the level of the transverse sinus and an appropriate distance lateral to the midline. Access to the pons is through the avascular plane within the middle cerebellar peduncle (**Fig. 2**). This infratentorial

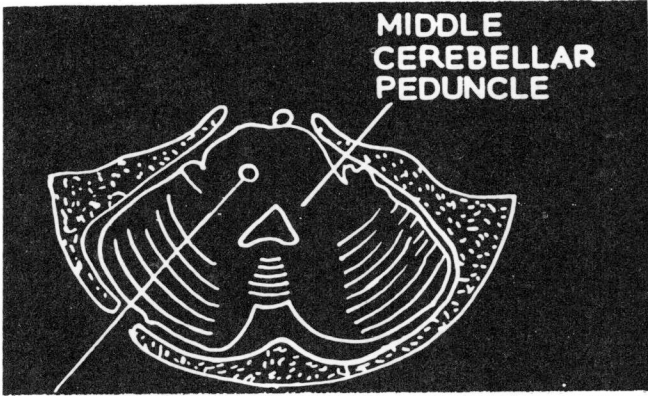

Fig. 2 : Diagram showing an axial section through the posterior fossa demonstrating the trajectory along the middle cerebellar peduncle to the pons in the transcerebellar route.

route to the cerebellum and the pons avoids the ventricular system and crosses only the cerebellar pial plane. This decreases the possibility of haemorrhage and neurological deficits postoperatively. The limitation of this approach is that it cannot be used to access lesions in the midbrain and the medulla.

All patients undergoing stereotactic biopsy of the brain stem are administered steroids from 24 hours prior to the procedure. A non-contrast CT scan is done four to six hours after the procedure to confirm the target site, look for evidence of any haematoma at the biopsy site or worsening of brain stem oedema or hydrocephalus.

COMPLICATIONS

The most feared complications in any biopsy of the brain stem would be haemorrhage and worsening of neurological deficits. These are the main complications encountered with stereotactic biopsy also. A risk-benefit analysis of computerized tomography-guided stereotactic surgery for brain stem masses (including lesions other than gliomas) from our institution showed no procedure related mortality and transient worsening of neurological deficits in 5.6% patients.[11] A review of the 158 cases of brain stem gliomas done from our institution from 1987-2000 has shown 1.3% mortality and 2.6% permanent morbidity. Coffey et al who first reported stereotactic surgery for mass lesions of the brain stem also had no morbidity or mortality.[10] The series of Abernathy et al also had no morbidity or mortality.[8]

BENEFITS OF STEREOTACTIC SURGERY

A review of literature on brain stem gliomas suggests that most patients – especially in the paediatric age group – present with significant and disabling neurological deficits. The poor general condition makes them a poor candidate for prolonged anaesthesia that would be required for open biopsy or radical resection. Most authors[1-3, 11, 12] agree that an open procedure should not be done to merely obtain a biopsy as the same can be obtained using stereotactic procedures that are safer and reliable. Stroink et al observed that the chance of a negative result with open biopsy was more likely in diffuse masses of the brain stem.[1] While Abernathy et al[8] and Kratimenos et al[13] find good correlation between the histological grade of the tumour on stereotactic biopsy and the survival of patients with brain stem gliomas, other authors (tissue obtained by stereotactic biopsy as well as open biopsy) have found no such correlation as these lesions are known to be heterogeneous.[1, 2, 11, 12] Moreover, stereotactic surgery has a yield of 94% to 100%[4, 10, 11, 13, 14] and cystic lesions can be safely aspirated for immediate amelioration of various symptoms.[4-6] Instillation of radioisotopes and placement of Ommaya reservoir for long term aspiration of the cysts have also been done.[14] Thus stereotactic surgery is a safe management option for brain stem gliomas.

RADIATION THERAPY FOR BRAIN STEM GLIOMAS

There are several large series of patients who have been treated with radiation therapy for brain stem gliomas. In our institution, these patients are treated with conventional external beam fractionated radiation with a cumulative dose of 56 to 60 Gy. Albright et al have recommended a dose of 40 to 60 Gy of radiation.[15] There have been several studies on hyperfractionated radiotherapy for brain stem gliomas but none have shown definitive improvement in survival. Increasing the dose of radiation or addition of chemotherapeutic agents during radiotherapy has not been shown to be of conclusive benefits and is not recommended.[16,17]

Stereotactic radiotherapy has been used as a boost for focal, discrete brain stem gliomas. Kondziolka et al have recommended the use of stereotactic radiosurgery as an adjuvant approach in patients with small volume; well circumscribed brain stem glioblastoma and they reported no complications among patients with anaplastic astrocytoma or glioblastoma.[18]

Brachytherapy or interstitial radiation therapy with implantation of radioactive sources into the brain stem lesion is another novel modality of therapy. A study by Mundinger et al showed a definite improvement in survival with implantation of I^{125} radionuclide into well circumscribed lesions less than 40mm in radius.[19] Implantation of Ir^{192} did not improve outcome significantly. The advantage of this procedure was that it could be done at the same sitting as the stereotactic biopsy. Local radionecrosis caused by radionuclides achieves a kinetically significant reduction of tumour cellular burden. In contrast to external beam radiation, an interstitially implanted radioisotope delivers radiation continuously at very low dose rates, thus minimising unwanted side effects like nausea, vomiting and hair loss usually seen after cranial irradiation.

RESULTS

Selvapandian et al have followed up 49 children and 18 adults with brain stem gliomas who underwent stereotactic biopsy followed by radiation therapy. At a mean follow up of 11 months, 21 of the children were alive and at a mean follow up of 25 months, 15 of the adults were alive. Analysis of data from our institution has shown no correlation between the tumour grade and survival in children while the tumour grade was of prognostic relevance in adults. Adults had a statistically significant longer duration of survival. The Kaplan-Meier analysis showed median survival of less than 10 months among children and median survival of more than 60 months among adults.[20] Albright et al have reported a median survival time of 1.1 years among children with brain stem gliomas with the tumour grade having on effect on the prognosis.[15] Another group has shown a survival range of 13.2 months to 26 months in children with brain stem gliomas treated with external beam radiation therapy of 40 to 60Gy.[21] The Children's Cancer group review of 119 cases of brain stem glioma in children showed very poor survival overall and duration of symptoms more than 1 month was associated with better survival. Thus further research in tumour biology and newer therapeutic modalities would be required to improve outcome for children with brain stem gliomas.[17]

CONCLUSIONS

Brain stem gliomas are a heterogeneous group of tumours that have a dismal prognosis in children with better survival rates in adults. Stereotactic biopsy is indicated for all tumours in adults as tumour grade has a definite prognostic value in our experience. Children with non gadolinium enhancing diffuse brain stem lesions can be empirically treated with radiation therapy and do not require histological confirmation of the astrocytoma. Stereotactic biopsy is a safe procedure in the diagnosis of brain stem gliomas with lesser morbidity than open surgeries for these lesions.

REFERENCES:

1. Stroink AR, Hoffman HJ, Hendrick EB, Humphreys RP: Diagnosis and management of pediatric brain-stem gliomas. *J Neurosurg* 1986;65:745-50.
2. Epstein F, Mcleary EL: Intrinsic brain-stem tumours of childhood : surgical considerations. *J Neurosurg* 1986;64:11-15.
3. Albright LA, Packer RJ, Zimmerman R, Rorke LB, Boyett J, Hammond DJ: Magnetic resonance scans should replace biopsies for the diagnosis of diffuse brain stem gliomas: A report from the Children's Cancer Group. *Neurosurgery* 1993; 33:1026-30.
4. Hood TW, Gebarski SS, Mckeever PE, Venes JL: Stereotactic biopsy of intrinsic lesions of the brain stem. *J Neurosurg* 1986;65:172-76.
5. Hood TW, Mckeever PE: Stereotactic management of cystic gliomas of the brain stem. *Neurosurgery* 1989;24:373-78.
6. Hood, et al: *Neurosurgery* 1989;65:172-176.
7. Rajshekhar V: Image guided stereotactic management of brain stem masses. *Progress in Clin Neurosciences* 1995;10:329-36.
8. Abernathy CD, Camacho A, Kelly PJ: Stereotactic suboccipital transcerebellar biopsy of pontine mass lesions. *J Neurosurg* 1989;70:195-200.
9. Guthrie BL, Steinberg GK, Adler JR: Posterior fossa stereotaxic biopsy using the Brown-Roberts-Wells stereotaxic system. *J Neurosurg* 1989;70:649-52.
10. Coffey R, Lunsford LD: Stereotactic surgery for mass lesions of the midbrain and pons. *Neurosurgery* 1985;17:12-18.
11. Rajshekhar V, Chandy MJ: Computerized tomography-guided stereotactic surgery for brain stem masses: a risk-benefit analysis in 71 patients. *J Neurosurg* 1995;82:976-81.
12. Albright AL: Tumours of the pons. *Neurosurg Clin North Am* 1993;4:529-36.
13. Kratimenos GP, Thomas DGT: The role of image-directed biopsy in the diagnosis and management of brain-stem lesions. *Br J Neurosurg* 1993;7:155-64.
14. Giovanini MA, Mickie JP: Long-term access to cystic brain stem lesions using the Ommaya reservoir: technical case report. *Neurosurgery* 1996;39:404-07.
15. Albright AL, Prie PA, Guthkelch AN: Brain stem gliomas of children. *Cancer* 1983;52:2313-19.
16. Freeman CR, Kepner J, Kun LE, Sanford RA, Kadota R, Mandell L, Friedman H: A detrimental effect

17. of a combined chemotherapy-radiotherapy approach in children with diffuse intrinsic brain stem gliomas? *Int J Radiat Oncol Biol Phys* 2000;47:561-64.
17. Kaplan AM, Albright AL, Zimmerman RA, Rorke LB, Li H, Boyett JM, Finlay JL, Wara WM, Packer RJ: Brain stem gliomas in children. A Children's Cancer group review of 119 cases. *Paediatr Neurosurg* 1996;24:185-92.
18. Kondziolka D, Lunsford DL, Flickinger JC: Intraparenchymal brain stem radiosurgery. *Neurosurg Clin North Am* 1993;4:469-80.
19. Mundinger F, Braus DF, Kraus JK: Long term outcome of 89 low grade brain stem gliomas after interstitial RT. *J Neurosurg* 1991;75:740-46.
20. Selvapandian S, Rajshekhar V, Chandy MJ: Brainstem glioma: Comparative study of clinico-radiological presentation, pathology and outcome in children and adults. *Acta Neurochirurgica (Wein)* 1999;141: 721-27.
21. Berger MS, Edwards MSB, La Masters D: Paediatric brain stem tumours. Radiographic, pathological and clinical course. *Neurosurgery* 1983;12:298-302.

25

Brain Stem Gliomas: Radiotherapy without Biopsy

NJ Laperriere

Pediatric brain stem tumours account for approximately 10-20% of central nervous system tumours in children. There are several different classification systems for brain stem tumours in children, but Freeman et al classified them as follows: focal 5-10%, dorsal exophytic 10-20%, cervicomedullary 5-10%, and diffuse pontine gliomas 60-75%.[1] Fisher and colleagues reviewed 76 consecutive cases registered at the Johns Hopkins Medical Institutions from 1980 to 1997 and reduced the classification to a 2 tier system: fibrillary astrocytomas which were associated with central pontine location and a 1 year survival of 23% and pilocytic astrocytomas which were associated with location outside the central pons and dorsal exophytic growth and a 5 year survival of 95%.[2] Extrapolating from these 2 review articles and others, it is reasonable to conclude that diffuse pontine gliomas are fibrillary astrocytomas associated with a very poor outcome and all other brain stem tumours are very likely to be pilocytic astrocytomas associated with a favourable survival.

At the outset, I am recommending that diffuse pontine gliomas not be biopsied at diagnosis, but that all other brain stem gliomas will benefit from some surgical intervention, either of the primary tumour itself, cyst resection or aspiration, or for ventriculoperitoneal shunting.

Characteristics of focal tumours include limited size (usually < 2 cm), well circumscribed lesions with no infiltration or edema, occasionally cystic components, and usually involving the midbrain or medulla.[1] Tectal region tumours are often associated with hydrocephalus. These tumours are typically pilocytic astrocytomas and are associated with a favourable outcome with a long natural history of indolent lesions in many instances. Focal tumour management typically includes the following options: observation, shunting or 3rd ventriculostomy, decompression of a cystic component, resection in some instances, and radiation therapy only in the setting of progression in surgically inaccessible lesions. In that circumstance, radiation delivered would be 54 Gy in 30 fractions via conformal techniques. On occasion, intracavitary irradiation for recurrent cystic tumours has been a very appropriate and successful approach in highly selected cases.[3,4]

Doral exophytic tumour arise from the floor of the 4th ventricle. They typically present as large neoplasms filling the 4th ventricle and are typically pilocytic astrocytomas. Surgical resection is the treatment of choice, and radiotherapy is only considered in cases of recurrent unresectable cases. In two recently reported series, 12/16 and 7/10 patients treated with initial surgery only had not recurred after median intervals of 113 and 26 month intervals respectively.[5, 6]

Cervicomedullary tumours arise from the cervical cord and grow up into the medulla posteriorly. Most of these tumours are low grade well circumscribed tumours and gross total resection is achievable in up to 75% of cases, and routine post-operative radiation therapy is reserved for recurrent disease activity in selected cases.[7, 8] A small proportion of tumours in this location will be high grade infiltrative tumours, and in this instance a gross total resection is not possible, and post-operative radiation therapy is indicated post-operatively as part of initial management.

Diffuse pontine gliomas account for approximately 60-75% of all brain tumours. They typically are centered on the pons and cause diffuse enlargement of the pons and can be associated with axial and/or exophytic in approximately two thirds of cases.[9] **Figure 1** is a an example of a diffuse pontine glioma. Tumours in this location are unresectable, and the only a biopsy can be done for tumours in this location. Biopsies are generally not done in most centres because the imag-

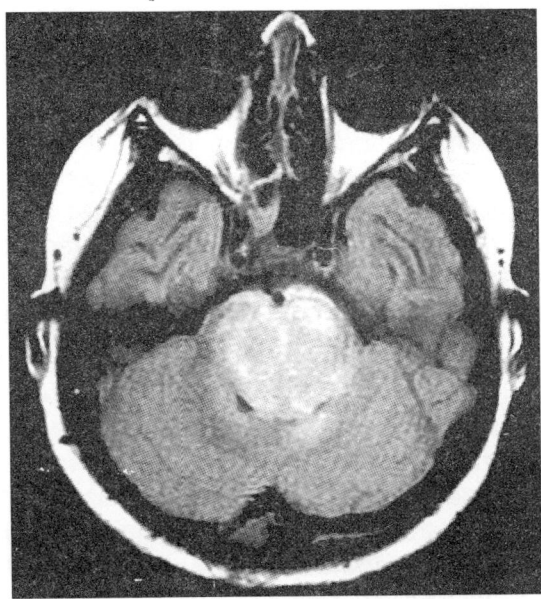

Fig. 1 : An MRI flair examination demonstrating diffuse involvement of the pons by an infiltrative neoplasm with anterior extension and partially surrouding the basilar artery.

ing characteristics combined with the clinical scenario is highly characteristic for this tumour. As well, a small biopsy may not be representative of the whole lesion, and the result is unlikely to influence subsequent management. To exemplify the difficulties with a small biopsy from a larger tumour, Jackson et al from M. D. Anderson reviewed a consecutive series of 82 patients both a stereotactic biopsy and a subsequent resection in patients with primary intra-axial tumours of the brain.[10] Diagnoses differed in 40/82 (49%) of cases, and most discrepancies were related to the grade of the tumour. The surgical specimen upgraded the tumour in all instances. Cartmill et al biopsied a series of 18 patients in diffuse pontine gliomas and found the following distribution of pathologies: 8 glioblastomas, 5 anaplastic astrocytomas, 5 low grade fibrillary astrocytomas.[11] Five of these 18 biopsies were associated with transient neurologic complications. Most studies were biopsies were done demonstrated that biopsy results were not of prognostic value.[12-14] Standard therapy consists of radiation therapy to a dose of 54 Gy in 30 fractions via a lateral opposed parallel pair, and approximately 75% of children will improve clinically. However, early recurrence is the rule, and median survival is of the order of 6 to 9 months, with survival rates at 2 years generally < 20%. (ref 21 freeman) Adverse prognostic factors include short duration of clinical symptoms, larger tumours and pontine enlargement, poorly circumscribed tumour on imaging, and ring enhancement suggestive of a malignant glioma.[9, 15-19]

In summary, for pilocytic astrocytomas of the brain stem, surgery is a major therapeutic modality, and radiation is reserved for recurrent unresectable tumours. For the commoner diffuse pontine gliomas, radiation therapy without a biopsy is advocated because the diagnosis is rarely in doubt with the typical constellation of clinical and imaging characteristics, and biopsies may not be representative of the actual tumour present. In addition, biopsies may be associated with additional neurologic morbidity which is unnecessary under the circumstances, and the results of the biopsies will not alter the therapeutic approach. It is possible that in the future that biopsies may be useful for the molecular characterization of genetic abnormalities in these neoplasms for either research purposes or hopefully for the use of some yet to be discovered therapy that would be directed at the specific genetic profile of that individual tumour.

Brain stem gliomas are far less common in adults, but they are all fibrillary astrocytomas with a variable natural history in keeping with whether the tumour is either low grade or high grade. Generally MRI appearances can differentiate this latter issue by the presence or absence of enhancement, and biopsies can be associated with significant neurologic sequelae. There remain unresectable tumours in this location, and standard therapy consists of radiation therapy only. Currently the standard of care is to not biopsy these lesions, but as previously mentioned, biopsies may be indicated at some future time for molecular characterization of these neoplasms that may influence yet to be available therapy.

REFERENCES:

1. Freeman CR, Farmer JP: Pediatric brain stem gliomas: a review. *Int J Radiat Oncol Biol Phys* 1998;40(2):265-71.
2. Fisher PG, Breiter SN, Carson BS, Wharam MD, Williams JA, Weingart JD, et al: A clinicopathologic reappraisal of brain stem tumor classification. Identification of pilocytic astrocytoma and fibrillary astrocytoma as distinct entities. *Cancer* 2000;89(7):1569-76.
3. Hood TW, Shapiro B, Taren JA: Treatment of cystic astrocytomas with intracavitary phosphorus 32. *Acta Neurochir* Suppl (Wien) 1987;39(3):34-37.
4. Hood TW, McKeever PE: Stereotactic management of cystic gliomas of the brain stem. *Neurosurgery* 1989;24(3):373-78.
5. Khatib ZA, Heideman RL, Kovnar EH, Langston JA, Sanford RA, Douglas EC, et al: Predominance of pilocytic histology in dorsally exophytic brain stem tumors. *Pediatr Neurosurg* 1994;20(1):2-10.
6. Pollack IF, Hoffman HJ, Humphreys RP, Becker L: The long-term outcome after surgical treatment of dorsally exophytic brain-stem gliomas. *J Neurosurg* 1993;78(6):859-63.
7. Bricolo A, Turazzi S, Cristofori L, Talacchi A: Direct surgery for brainstem tumours. *Acta Neurochir* Suppl (Wien) 1991;53:148-58.
8. Epstein F, McCleary EL: Intrinsic brain-stem tumors of childhood: surgical indications. *J Neurosurg* 1986;64(1):11-15.
9. Barkovich AJ, Krischer J, Kun LE, Packer R, Zimmerman RA, Freeman CR, et al: Brain stem gliomas: a classification system based on magnetic resonance imaging. *Pediatr Neurosurg* 1990;16(2):73-83.
10. Jackson RJ, Fuller GN, Abi-Said D, Lang FF, Gokaslan ZL, Shi WM, et al: Limitations of stereotactic biopsy in the initial management of gliomas. *Neuro-oncol* 2001;3(3):193-200.
11. Cartmill M, Punt J: Diffuse brain stem glioma. A review of stereotactic biopsies. *Childs Nerv Syst* 1999;15(5):235-37; discussion 238.
12. Berger MS, Edwards MS, LaMasters D, Davis RL, Wilson CB: Pediatric brain stem tumors: radiographic, pathological, and clinical correlations. *Neurosurgery* 1983;12(3):298-302.
13. Albright AL, Packer RJ, Zimmerman R, Rorke LB, Boyett J, Hammond GD: Magnetic resonance scans should replace biopsies for the diagnosis of diffuse brain stem gliomas: a report from the Children's Cancer Group. *Neurosurgery* 1993;33(6):1026-9; discussion 1029-30.
14. Kratimenos GP, Thomas DG: The role of image-directed biopsy in the diagnosis and management of brainstem lesions. *Br J Neurosurg* 1993;7(2):155-64.
15. Albright AL, Guthkelch AN, Packer RJ, Price RA, Rourke LB: Prognostic factors in pediatric brain-stem gliomas. *J Neurosurg* 1986;65(6):751-55.
16. Grigsby PW, Thomas PR, Schwartz HG, Fineberg B. Irradiation of primary thalamic and brainstem tumors in a pediatric population. A 33-year experience. *Cancer* 1987;60(12):2901-06.
17. Grigsby PW, Thomas PR, Schwartz HG, Fineberg BB: Multivariate analysis of prognostic factors in pediatric and adult thalamic and brainstem tumors. *Int J Radiat Oncol Biol Phys* 1989;16(3):649-55.
18. Halperin EC, Wehn SM, Scott JW, Djang W, Oakes WJ, Friedman HS: Selection of a management strategy for pediatric brainstem tumors. *Med Pediatr Oncol* 1989;17(2):117-26.
19. Sanford RA, Freeman CR, Burger P, Cohen ME: Prognostic criteria for experimental protocols in pediatric brainstem gliomas. *Surg Neurol* 1988;30(4):276-80.

26

Perioperative Management of Patients with Brainstem Glioma

B Mohanty, H H Dash

Gliomas of the brainstem constitute between 10% and 20% of all childhood central nervous system (CNS) tumors. Brainstem gliomas may occur at any time of life. However, approximately three fourth of patients are less than 20 years of age at the time of diagnosis. The peak prevalence of these tumors is in the latter half of the first decade of life. There is no race or sex predilection.[1] The brainstem gliomas were once regarded as a universally fatal disorder not amenable to any form of surgical therapy. But the above concept has recently been challenged. Remarkable changes in the field of imaging, excellent perioperative monitoring, appropriate anesthetic technique, use of ultrasonic suction aspirator and microscope during surgery, and above all, intense post-operative care in the neurosurgical intensive care unit (Neuro-ICU) have changed the whole complexion of management of brainstem gliomas over the years.

MANAGEMENT

Many years ago 'Matson' stated "regardless of specific histology, brainstem gliomas must be classified as malignant tumors since their location itself renders them inoperable".[2] However, currently the treatment of brainstem gliomas can be surgical, non-surgical (radiotherapy, chemotherapy), or a combination of both. Prognosis of patients with cervicomedullary masses, and possibly for those with tectal masses, is relatively good. Poor outcome is expected for patients with diffuse intrinsic masses. Children with tectal lesions, presenting with hydrocephalus, probably should be observed initially, before any intervention is undertaken. Steriotactic biopsy of the lesions would provide useful information for surgical intervention or radiotherapy. Children with diffuse brainstem glioma rarely survive after conventional therapy. There is little evidence that more aggressive modes of treatment (hyperfractionated radiotherapy, pre or post-radiation chemotherapy, and aggressive attempts at surgical resection) increase the rate of disease free survival. For patients with cervico-medullary lesions, aggressive attempts at surgical resection are, probably, indicated. Before going into the details of anesthetic management one should consider anatomy and physiology relevant to brainstem gliomas.

ANATOMICAL AND PHYSIOLOGIC CONSIDERATIONS

Because of its anatomical location, neurosurgical intervention for brainstem glioma is considered as posterior fossa surgery. The brainstem is housed in the posterior cranial fossa which, in fact, lies in proximity to large venous sinuses (transverse, sigmoid, and torcula) and its postero- inferior limits, the occipital squama, is traversed by emissary veins. The presence of these non-collapsible venous structures partly explains the higher incidence of venous air embolism (VAE) during craniotomy done in sitting position than during surgery in the same position on the high cervical spine.[3] The pons and medulla contain the major motor and sensory pathways, the primary respiratory and cardiovascular centers, and the lower cranial nerve nuclei. Patients with brainstem tumors may have decreased level of consciousness, increased sensitivity to sedative medication, depressed respiration, and impaired airway protective reflexes. These factors should carefully be considered throughout the perioperative period.[4] Proper clinical examination in the preoperative period helps a great deal in planning the intraoperative anesthetic management as well as the postoperative care.

PREOPERATIVE ASSESSMENT

During preoperative evaluation of patients with brainstem glioma the anesthesiologist should focus on cardiovascular, pulmonary and airway assessment, because these are the determinant of the choice of patient position during surgery. It is important to assess and correct intra-vascular volume deficits to minimize the chances of hypotension during induction of anesthesia or during positioning for surgery. History of hypotension, cardiovascular diseases, cerebrovascular insufficiency, or previous carotid endarterectomy, may suggest altered limits of cerebral autoregulation, impaired cerebral perfusion, or abnormal baroreceptor function.[5, 6] This in fact makes the patient more vulnerable to cerebral ischemia during systemic hypotension. Gradual and slow changes in the position and adequate volume infusion prior to the induction of anesthesia, prevent these unwarranted situations. The presence of a patent foramen ovale increases considerably the chances of paradoxical air embolism, and is, therefore, a contraindication to the sitting position.[7, 8] Excessive cerebro spinal fluid (CSF) drainage can occur in the sitting position in patients with ventricular shunts and can result in tissue displacement or intracranial bleeding. In such situations, consideration must be given to temporary occlusion of the shunt during surgery. Antibiotic prophylaxis is required for the patients with ventricular shunt in situ, since they are at risk of infection. Impaired consciousness and airway protective reflexes may have allowed "silent" aspiration to occur. Patients with brainstem lesion may suffer from sleep apnea, aspiration, dysphagia, recurrent pneumonia, and difficulties with coughing or clearing of secretions, due to the pathology affecting the lower cranial nerves. There may be the possibility of significant muscular weakness and potential respiratory compromise. Hence close respiratory monitoring is of paramount importance. Neurosurgical patients with previous craniotomies may have pseudoankylosis and limited mouth opening.[9] Excessive neck flexion in the sitting position after muscle paralysis can lead to spinal cord ischemia and quadriparesis; therefore, the range of neck movement should carefully be observed.[10] Sensitivity to general anesthesia can be estimated from the response to the sedative if they were used for preoperative neuroradiologic procedures. Prolonged somnolence after sedatives usually indicates a decrease in intraoperative anesthetic requirement. Intraoperative evoked potential monitoring is usually planned. Hence it is important to have preoperative baseline studies done prior to the surgery. Alternatively these studies may be done just before induction of anesthesia. For the patients presenting with respiratory muscle weakness it is prudent to get baseline negative inspiratory force measurement and arterial blood gas analysis done to guide the postoperative airway management.

ANESTHETIC MANAGEMENT

No one anesthetic technique can be advocated for brainstem glioma surgery. It should be properly planned and executed to achieve a preset goal by taking account of any pre-existing disease status, presenting pathology, as well as limitations of specific operation or positioning. Before embarking on any anesthetic technique the anesthesiologist must have a thorough knowledge about the anesthetic considerations so that goals can be fixed. This in-fact helps in achieving an uneventful intraoperative period.

Anesthetic considerations

As has already been discussed, brainstem gliomas often occur at a very tender age, hence it is the duty of the anesthesiologist to be aware of the anesthetic requirements for a pediatric patient. All possible steps should be taken to maintain the core temperature of the patient, since younger patients are at a greater risk of getting hypothermic. Abnormal position poses unique challenge for its potential capability to cause circulatory disturbances, VAE, musculo-skeletal, and nerve injury. Increased ICP is a late feature in patients with brainstem glioma as per international literature.[1] But vast majority of Indian patients seek late medical advice, which makes raised ICP almost a presenting symptom. Hence preoperative shunt placement to control the ICP is usually carried out. Hemodynamic disturbances could be due to inappropriate brainstem manipulation or excessive hemorrhage. Patients with brainstem tumors have decreased level of consciousness, increased sensitivity to sedatives, depressed respiration, and less reactive airway protective reflexes, which all point to a need for postoperative ventilation.

Anesthetic goals

1. Maintain proper homeostasis, preserve cerebral perfusion and avoid increase in ICP.

2. Provide a conducive environment to carry out intraoperative neurophysiologic monitoring.
3. Prompt detection and effective management of intraoperative complications.
4. Preservation of cardiovascular and respiratory responsiveness to the surgical stimuli.
5. Try to achieve faster postoperative recovery from anesthesia.

Premedication

Premedication for the patients of brainstem glioma surgery are no different from the premedication protocols of any other intracranial surgery. Narcotics premedication is avoided for the resultant hypoventilation and CO_2 retention, which may increase ICP. Oral benzodiazepines, 60 to 90 minutes before anesthesia are effective in reducing anxiety with no significant effects on ICP. Chronic antihypertensive medications, corticosteroids, and antibiotics are routinely continued till the morning of surgery. Oral atropine or glycopyrrolate are usually advised for pediatric patients for their antisialogogue action. In adult patients, parenteral atropine or glycopyrrolate can be prescribed, along with promethazine (0.5mg/kg) for its mild sedation and antiemetic action.

MONITORING

The goals of monitoring during brainstem glioma surgery are to assure adequate central nervous system (CNS) perfusion, maintain cardiovascular and respiratory stability, and detect and treat possible air embolism. **Table 1** lists the monitors used, regardless of patient position.[5] Specialized monitoring for brainstem glioma surgery includes monitoring for VAE, nerve tract injury and electro-physiologic activity.

Table 1 : *Monitors for Brainstem Tumor Surgery*

Preinduction/Induction:	Postinduction:
Five-lead electrocardiogram	Central venous (Right atrial, pulmonary artery) cather
Blood pressure monitoring	Precordial Doppler probe
Pulse oxymetry	Esophageal stethoscope
Precordial stethoscope	Esophageal or nasopharyngeal temperature probe
Capnography	$EtCO_2$, EtN_2 monitoring
Electrophysiologic monitoring	Trans-esophageal ECHO cardiography

Monitoring for VAE

The VAE can be detected by changes in the precordial Doppler sound, and end tidal carbon dioxide ($EtCO_2$). Trans-esophageal echocardiography (TEE) is the most sensitive method for the detection of VAE.[3] Its advantages over the precordial Doppler are that TEE can demonstrate the presence of patent foramen ovale, help in satisfactory location of the right arterial catheter for air aspiration, and detect air in the heart chambers. Vegetations in heart chambers and valvular lesions can also be demonstrated by TEE.

Neurophysiologic monitoring

Intraoperative neuro-physiological monitoring (IONM) represents the application of electro-physiological techniques to detect changes in the functional state of the nervous system consistent with ischemia or injury. Displacement of surface electrodes, effects of anesthetic drug, interference by electrical equipments, and absence of reliable baseline data can diminish the sensitivity, reliability, and usefulness of IONM. Hence it is necessary that the depth of anesthesia or intensity of neuro-muscular block be titrated to facilitate such monitoring. Since the cranial nerves are in close proximity to the brainstem, stimulation of these nerves during surgery can give rise to cardiovascular and respiratory reflexes, for instance, tachycardia and hypertension from 5th nerve, bradycardia and arrhythmia from 10th nerve, and cough reflex from 9th and 10th nerve. To prevent potential injury to the brainstem and preserve functions of the cranial nerves, monitoring of somatosensory-evoked potential (SSEP), brainstem auditory-evoked potential (BAEP), and evoked electromyogram (EMG) of 5th, 7th, 11th, and 12th cranial nerves are frequently employed.[11] Recording of EMG requires normal neuromuscular function, which in turn precludes the administration of higher dose of inhaled anesthetics and omission of muscle relaxants. In comparison to SSEP, BAEP monitoring is much more resistant to the effect of anesthetics. Anesthetics decrease the cortical SSEP amplitude and increase the latency. The combination of nitrous oxide and volatile anesthetic agents tends to have greater depressant effect than an equipotent administration of a single agent. Both volatile and intravenous anesthetics increase the latency of SSEP and BAEP. The amplitude of the response may be unaffected by intravenous agents (i.e., propofol) or even br increased by ketamine and etomidate. Ultrasonic Doppler study of the basilar artery can provide excellent information regarding morbidity and mortality.

Central venous cannulation

Central venous cannulation such as internal jugular vein (IJV) or subclavian vein, can be achieved with 95% success rate following normal anatomical landmarks. The IJV cannulation is not preferred at our center for its potential risk of kinking during neck flexion. The incidence of serious complications like pneumothorax and hemothorax following subclavian vein cannulation are reported to be 3% to

5%. To avoid such complications we prefer to have central venous access through the cannulation of the anti-cubital veinsr. The successful placement of the catheter tip at the junction of superior venacava and right atrium for venous pressure monitoring, and retrieval of air in the event of VAE, through the anti-cubital vein cannulation is reported to be more than 80%.[12]

Induction of anesthesia

Goals during this stage should be to preserve cerebral perfusion and avoid increase in ICP. Induction of general anesthesia for brainstem glioma surgery is usually accomplished with thiopental and propofol. In the presence of hypotension induction with etomidate is advisable. In small children induction of anesthesia can be carried out with spontaneous respiration through a mask by using either halothane or isoflurane. Sevoflurane is a faster acting inhalational agent which is preferred now a days. Hyperventilation during the cries helps in maintaining the ICP during the induction of anesthesia. Though there is a controversy regarding the use of succinylcholine for intubation in neuroanesthesia, it can be safely used following pretreatment with metocurine, if there is no pre-existing neurologic deficit.[13] Kovarik et al, have reported that patients with neurologic disease do not manifest an intracranial response to succinylcholine.[14] In our center, we use non-depolarizing muscle relaxant for intubation (rocuronium–0.6 mg/kg or vecuronium–0.1 mg/kg). Intravenous fentanyl (2 to 3 mgm/kg) is used to provide intraoperative analgesia. Pre-existing and functioning ventriculo-peritoneal shunts (VP shunts) in majority of patients, act as a deterrent against rise in ICP. Before endotracheal intubation, lidocaine 1.5 mg/kg, and thiopentone 1-2 mg/kg, are given intravenously to counter the hemodynamic responses to laryngoscopy and intubation thereby reducing any increase in ICP. A flexometallic or reinforced armored endotracheal tube may be used to prevent kinking during positioning. Oral airway is best avoided to minimize potential venous obstruction to the tongue and neck. A small soft bite block may be used during facial nerve monitoring as the patient tends to bite down when stimulated. Arterial and central venous cannulations are usually performed after induction of anesthesia. Direct arterial blood pressure monitoring (by cannulating either radial artery or dorsalis pedis or posterior tibial artery) established before induction of anesthesia allows tighter control of blood pressure.

Maintenance of anesthesia

The maintenance of anesthesia should be tailored to the needs of the patient and requirement of the surgical procedure or position. The target should be a "relaxed brain" to reduce retraction pressure and to maintain adequate cerebral perfusion in a hemodynamically stable patient. Muscle paralysis is usually maintained with a non-depolarizing muscle relaxant, using peripheral nerve stimulator as a guide. Mannitol 1 to 2 gm/kg, with or without furosemide (0.1–0.4 mg/kg), is administered intravenously to reduce the brain bulk. It may be deferred in patients with functioning shunt and while using sitting position. Intraoperative use of methyl prednisolone is controversial.

Inhalational versus Intravenous anesthetics

Opinion is divided over whether to use inhalational anesthetics or intravenous agents or a combination of both, in-order to maximally benefit the patient. Inhalational agents like desflurane and sevoflurane with low blood gas partition coefficient provide a rapid onset and offset of action. Sevoflurane, in particular, is good for induction of anesthesia in pediatric age group. These agents can maintain better anesthetic depth, which in turn obviate the need for nitrous oxide. However, total intravenous anesthesia is an alternative to the use of inhalational agents. The newer sedative-hypnotics, narcotics, and muscle relaxants with short elimination half-lives are the drugs to be chosen for use.[15, 16] Intravenous anesthetic agents, such as thiopentone and propofol, are reported to provide neuroprotection.[17] Inhalation anesthesia can interfere with electrophysiological monitoring and cause systemic hypotension. Remifentanil or alfentanil can provide excellent intraoperative analgesia. But its termination of infusion after surgery can lead to postoperative pain due to its rapid elimination. Hence it would be prudent to take a median path broad and use a based balanced technique.[3] In such techniques one can supplement low concentrations of inhalational agents to increase the depth of anesthesia maintained by intravenous agents or vice versa. This, in fact, minimizes the risk of systemic hypotension, increases in ICP, or interference with the electrophysiological monitoring.

Preservation of Cardiovascular Responsiveness

The second issue is the preservation of cardiovascular responsiveness to surgical manipulation of brainstem structures. It can be achieved by avoiding the use of anticholinergic drugs or long acting beta-adrenergic blockers.

Use of Nitrous Oxide

A third issue is the use of Nitrous oxide (N_2O) in case where the risk of VAE is high. However, it is clinically observed that the use of 50% N_2O does not significantly alter the incidence and severity of VAE, provided precordial Doppler monitoring is used and N_2O is discontinued as soon as an air embolus is detected.[18] Nitrous oxide has the potential to increase cerebral metabolic rate, cerebral blood flow, ICP, and is also associated with increased postoperative nausea and vomiting (PONV).[19, 20] It has also been blamed in the etiopathogenesis of tension pneumocephalus. Though there is lack of strong evidence against use of N_2O, but, yet some advanced centers defer using N_2O totally in neuroanesthesia practice.[21]

Controlled ventilation versus spontaneous ventilation

Spontaneous breathing may have merits as a sensitive monitor of the brainstem well being.[22] However, less specific cardiovascular signs are equally sensitive. Effectiveness of hypocapnia in reducing intracranial blood volume is undisputed, which makes controlled ventilation superior over the spontaneous ventilation. Moreover spontaneous ventilation is limited to the sitting position, and in this it abates VAE. More superior monitors (electrophysiologic monitoring) to determine brainstem integrity are a good choice to eliminate the use of spontaneous ventilation.

Intravenous fluid

Both overhydration and fluid restriction are avoided. The necessity of blood transfusion is only felt when the hematocrit falls below 30%. The filling pressure and periodic electrolyte determination are the guide to fluid and electrolyte replacement. Solutions containing glucose should be avoided since hyperglycemia can aggravate ischemic insult to the brain. Isotonic saline is preferred. In small children serum glucose monitoring helps in averting hypoglycemia.

POSITIONING

The position of a patient during surgery for brainstem glioma has important repercussions on intracranial hemodynamics and homoeostasis. The position of the patient is made to permit proper anatomical orientation and visualization with least tissue retraction. The surgery for brainstem glioma is usually undertaken either in sitting position or in prone position or in semi prone ('park bench') position.

Sitting Position

The sitting position provides optimum access to midline lesions, improves cerebral venous decompression, lowers intracranial pressure (ICP), promotes gravity assisted drainage of blood and CSF, and provides a comfortable visualization of the surgical site to the surgeon.[23] It also grants better access for the anesthesiologist to the airway. The disadvantages of this position include hypotension, venous air embolism (VAE), paradoxical air embolism (PAE), pneumocephalus, and quadriparesis from extreme neck flexion. In this position, the essential monitoring includes a capnograph and a precordial Doppler to assist in the early detection of VAE. This position is avoided in children less than 3 yrs of age and in case of patients with a known intracardiac shunt, because of greater risk of PAE. Other recognized complications of the sitting position include peripheral nerve palsies, macroglossia, and intracranial hemorrhage in patients with shunt due to excessive CSF drainage. Relative contraindications to the sitting position include known pulmonary arterio-venous malformations, severe hypovolemia, gross hydrocephalus, and high vascular lesions. Sitting position is safe for patients with beta-adrenergic blockade.[24]

Prone Position

The prone position is an alternative to the sitting position. Though it is associated with lower incidence of VAE,[25] the elevated head position above the level of heart does not eliminate the risk of air embolism completely. Care must be taken to ensure free abdominal and chest wall movements to avoid both impairment of respiration and increase abdominal pressure leading to an increase in venous pressure and bleeding. Enough padding must be provided to prevent patient's face, eye, genitalia and other pressure points from getting compressed or bruised. Lower limbs of the patient should be flexed at the hips, knees and ankles to avoid venous pooling in the lower extremities.

Park-bench or Semiprone Position

The park-bench position (semi prone) is a modification of the lateral position with the patient's back elevated to 45 degrees. In this position in the event of VAE the patient can be made more horizontal, which is a clear advantage over sitting position.[26]

Pin fixation

Pin fixation of the head during different positions, can result in tachycardia and hypertensive response. This can be prevented by infiltration of a local anesthetic agent at the site of pin application, regional nerve block of the scalp, or by intravenous administration of a narcotic.[27,28]

INTRAOPERATIVE PROBLEMS

A plethora of intraoperative complications may manifest during brainstem glioma surgery. Prompt detection through continuous monitoring and effective management helps in combating these myriad problems.

Hypotension

Hypotension is a very common complication seen during brainstem glioma surgery. It can be attributed to the sitting position, VAE, existing hypovolemia, or use of higher concentration of volatile anesthetics. Gradual positioning of the patient, maintenance of adequate intravascular volume and judicious use of volatile anesthetic is answer to this problem. Manipulation of the brainstem structures and hemorrhage can also give rise to hypotension, in which event it is important to inform the surgeon so that he can stop manipulating. Use of inotropes to increase the blood pressure is usually not necessary. However, titrated dose of inotropes like mephenteramine, dopamine, or dobutamine may be administered to treat refractory hypotension after correcting intravascular volume deficit.

Venous air embolism (VAE):

Venous air embolism (VAE) is an inherent problem of posterior cranial fossa surgery performed in sitting position. Slow insidious air entrainment usually occurs through the open venous sinuses. Very rarely it may manifest as large bolus air embolism occluding the pulmonary artery thereby resulting in cardiovascular collapse and cardiac arrest. But, slow and insidious air entrainment may manifest in the form of tachycardia, hypotension, with or without 'millwheel' murmur. Immediate detection and prompt management is of paramount importance so as to prevent any mortality and morbidity. Ultrasonic Doppler and end-tidal carbon dioxide ($EtCO_2$) has been recommended for prompt detection as well as for resolution of VAE. Transesophageal echocardiography (TEE), no doubt is highly sensitive but it can also detect paradoxical air embolism (PAE). However, the main constraint is its exorbitant cost and it can not predict the complete resolution of VAE. Best way of management of VAE is to discontinue N_2O and institute ventilation with 100% O_2 keeping the total gas flow unaltered. At the same time inform the surgeon so that he will irrigate the wound with saline or keep the wound pressed with saline soaked gauze to prevent further air entrainment. Appropriate measures should be taken to counter the hemodynamic problems. Simultaneously one may try to aspirate air through the central vein and if need arises the neck veins may be kept compressed intermittently (maximum for 30 seconds) to prevent further air entrainment.

Hypertension

Hypertension is often encountered during decompression of a brainstem glioma. It could be attributed to lack of analgesia, lighter plane of anesthesia, or manipulation of the vaso motor center. Best way to tackle such a situation is to inform the surgeon so that he stops excising or fiddling with the vasomotor center. Anesthesiologist should also ensure proper anesthesia and analgesia. Antihypertensive therapy is seldom required to tide over the situation. However, short acting betablockers (esmolol, labetolol) may occasionally be considered.

Arrhythmias

Both brady-arrhythmia and tachy-arrhythmia are encountered during the manipulations in and around brainstem structure. Surgeon should be cautioned immediately to move away from the brainstem to prevent injury to the vital area. Persistent brady-arrhythmia can be treated with intravenous administration of atropine (5-10 mgm/kg). Arrhythmia in the form of ventricular ectopics or bigemini is seen during moderate hypothermia. Hence it is prudent for the anesthesiologist to administer warm fluids and apply convective air device to maintain the core temperature of the patient.[29] Lignocaine 1 mg/kg is useful in controlling the ventricular events. Anesthesiologist should ensure adequate intravascular volume, analgesia and depth of anesthesia when tachy-arrhythmia is a problem. Electrolyte imbalance can lead to all kind of arrhythmias. Hence ensuring fluid and electrolyte balance is the first step to counter arrhythmias.

Bleeding

Bleeding during the surgical procedure is a common problem. Use of cell saver minimizes the blood loss and need for intraoperative blood transfusion. Preoperative clotting profile of the patient should be within the normal limits. Risk and benefit of heterologous blood and blood product transfusion should be carefully weighed against each other.

Neurological injury

It is very difficult to determine the intraoperative neurological injury. However, intraoperative neurophysiologic monitoring and cardiovascular responsiveness to brainstem manipulation can provide sufficient warnings to prevent any neurological injury.

EMERGENCE FROM ANESTHESIA

The possibility of immediate postoperative extubation depends mainly on the nature and extent of surgery. Extensive brainstem manipulation can cause postoperative brainstem edema which may jeopardize the respiration and cardiovascular function in the postoperative period. A possible brainstem compression or hematoma in that region can produce persistent hypertensive reaction in a previously normotensive patient during the postoperative period. Direct injury to the respiratory centre in the brainstem can also impair the respiratory drive. Airway obstruction may manifest postoperatively due to macroglossia, partial damage to the vagus nerve, or extreme flexion of the cervical spine. In all the above situations a secured airway should be maintained until the patient is awake, with intact protective airway reflexes.[1] So it is prudent to provide ventilatory support to the patient postoperatively until the return of airway reflexes.

Extubation

Hypertensive response to extubation can be managed effectively by the use of either beta adrenergic blocking drugs (Esmolol, Labetolol), or intravenous verapamil (0.1 mg/kg) or diltiazem (0.2 mg/kg). A strategy should be evolved to prevent the patient from coughing and straining on the endotracheal tube, as such reactions can increase the ICP.[3] Local anesthetics such as intravenous lignocaine can be of good use in this situation by blunting the cough reflex but, one should be careful about its central sedative effects.

Postoperative complications

Surgical complications such as acute hydrocephalus, intracranial hemorrhage, or acute pneumocephalus can cause slow emergence. In such situations it is imperative to maintain the airway and support ventilation till the patient recovers fully.

Postoperative nausea and vomiting (PONV) following craniotomy is a well-established complication. Use of ondansetron provides prophylaxis against vomiting.[30] Other antiemetics also can be of great help in controlling PONV. Convulsion in the postoperative period, though rare, may occur in 5 to 10% of the patients due to hematoma or tension pneumocephalus. Patients of brainstem glioma may need postoperative tracheostomy to maintain the airway for postoperative ventilation in presence of lower cranial nerve involvement or severe chest infection.

POSTOPRATIVE ANALGESIA

Hypertension, hyperventilation, and hyperglycemia associated with postoperative pain can adversely affect neurosurgical outcome. Since majority of the patients of brainstem glioma remain on ventilator in the postoperative period, it becomes necessary to provide them adequate sedation and analgesia, without influencing postoperative neurological assessment. Propofol can be used in such situations to provide conscious sedation without interfering in the neurological examination

Narcotics such as morphine or codeine can be used for postoperative analgesia in neurosugical patients with no significant effects on cerebral hemodynamics. Codeine can cause nausea and vomiting. Narcotics with short elimination half-lives, such as alfentanil, or remifentanil, may also be used. Mu-receptor agonist-antagonists such as tramadol hydrochloride, cause less sedation and are therefore, better suited for postoperative period.

Regional techniques such as skin infiltration or greater or lesser occipital nerve blocks can be of good use without any systemic side effect. Ropivacaine is preferred over bupivacaine due to less cardiac toxicity.

CONCLUSION

There have been very few studies done on anesthetic management of brainstem glioma surgery. Most information remains anecdotal and is based largely on small case series. But, depending on the available information it is concluded that, meticulous preoperative assessment, proper institution of anesthetic technique, continuous monitoring of the vital functions, VAE and neurological wellbeing and postoperative care of the airway and ventilation helps in salvaging majority of the patients.

REFERENCES:

1. Jeam-Pierre F, Jose LM, Carolyn RF, et al: Brainstem Gliomas. A 10-year institutional review. *Pediatr Neurosurg* 2001;34:206-14.
2. Matson DD: Neurosurgery of infancy and childhood, 2nd edition. Springfield, IL: *Thomas,* 1969; 469-77.
3. Joshi S, Dash HH, Ornstein E: Anesthetic considerations for posterior fossa surgery. *Current Opinion in Anesthesiology* 1997;10:321-26.

4. Artru AA, Cucchiara RF, Messick JM: Cardiorespiratory and cranial nerve sequellae of surgical procedures involving the posterior fossa. *Anesthesiology* 1980;52:83.
5. Smith DS, Irene Osborn: Posterior Fossa: Anesthetic considerations. in Anesthesia and Neurosurgery, 4th edn. Edited by Cottrell JE, Smith DS. St. Louis, Missouri: Mosby, 2001;335.
6. Wade JG, Larson PC, Hickey RF, et al: Effect of carotid endarterectomy on carotid chemoreceptor and baroreceptor function in man. *New Eng J Med* 1970;282:823-29.
7. Papadopoulos G, Kuhly P, Brock M, et al: Venous and paradoxical air embolism in the sitting position. A prospective study with transoesophageal echocardiography. *Acta Neurochir* (Wein) 1994;126:140-43.
8. Cucchiara RF, Seward JB, Nishimura RA, et al: Identification of patent foramen ovale during sitting position craniotomy by transesophageal echocardiography with positive airway pressure. *Anesthesiology* 1985;63:107-09.
9. Kawaguchi M, Sakamoto T, Furuya H, et al: Pseudoankylosis of the mandible after supratentorial craniotomy. *Anesth Analg* 1996;83:731-34.
10. Spiekerman BF, Stone DJ, Bogdonoff DL, Yemen TA: Airway management in neuroanesthesia. *Can J Anaesth* 1996;43:820-34.
11. Solan TB: Evoked potential monitoring. *Int Anesthesiol Clin* 1996;34:109-36.
12. Bithal PK, Dash HH, Vishnoi N: Comparative study on proper placement of central venous catheters with and without stillete. *Indian J Med Res* 1991;[B] 94:238-40.
13. Stirt JA, Grosslight KR, Bedford RF, et al: 'Defasciculation' with metocurine prevents succinylcholine-induced increase in intracranial pressure. *Anesthesiology* 1987;67:50-53.
14. Kovarik DW, Mayberg TS, Lam AM, et al: Succinylcholine does not change intracranial pressure, cerebral blood flow velocity, or the electroencephalogram in patients with neurologic injury. *Anesth Analg* 1994;78:469-73.
15. Warner DS, Hindman BJ, Todd MM, et al: Intracranial pressure and hemodynamic effects of remifentanil versus alfentanil in patients undergoing supratentorial craniotomy. *Anesth Analg* 1996;83:348-53.
16. Liu J, Singh H, White PF: Electroencephalographic bispectral index correlates with intraoperative recall and depth of propofol induced sedation. *Anesth Analg* 1997;84:185-89.
17. Hans P, Deby-Dupont G, Vrijens B, et al: Effect of propofol on in vitro lipid peroxidation induced by different free radical generating systems: a comparison with vitamine E. *J Neurosurg Anesth* 1996;8:154-58.
18. Losasso TJ, Muzzi DA, Ditz NM, Cucchiara RF: Fifty percent nitrous oxide does not increase the risk of venous air embolism in neurosurgical patients operated upon in the sitting position. *Anesthesiology* 1992;77:21-30.
19. Hartung J: Twenty-four of twenty-seven studies show a greater incidence of emesis associated with nitrous oxide than with alternative anesthetics. *Anesth Analg* 1996;83:114-16.
20. Divatia JV, Vaidya JS, Badwe RA, Haawaldar RW: Omission of nitrous oxide during anesthesia reduces the incidence of postoperative nausea and vomiting. *Anesthesiology* 1996;85:1055-62.
21. Dash HH: Tension pneumocephalus in sitting position. *Ind J Anaesth* 1996;44:345-51.
22. Abraham M, Singh D, Singh AK, et al: Initial experience with spontaneous ventilation monitoring for posterior fossa surgery involving the brainstem. *J Anaesth Clin Pharmacol* 1999;15:253-59.
23. Porter JM, Pidgeon C, Cunningham AJ: The sitting position in neurosurgery: a critical appraisal. *Br J Anaesth* 1999;82:117-28.
24. Morisson SC, Kumana CR, Rudnick KV, et al: Selective and non-selective beta adrenoreceptor blocked in hypertension. Resposes to change in posture, cold and exercise. *Circulation* 1982;65:1171-77.
25. Black S, Ockert DB, Oliver WC, et al: Outcome following posterior fossa craniotomy in patients in the sitting or horizontal positions. *Anesthesiology* 1988;69:49-56.
26. Calliauw L, Van Aken J, Rolly G, et al: The position of the patient during neurosurgical procedures on the posterior fossa. *Acta Neurochir (Wien)* 1987;85:154-58.
27. Pinosky ML, Fishman RL, Reeves ST, et al: The effect of bupivacaine skull block on the hemodynamic response to craniotomy. *Anesth Analg* 1996;83:1256-61.

28. Bithal PK, Joshi S, Dash HH, Sarmah H: Cardiovascular changes during scalp pin fixation. *J Anaesth Clin Pharmacol* 1993;9:143-45.
29. Mogera C, Dash HH, Chaturvedi A, et al: Control of body temperature with forced-air warming system during neurosurgery. *J Anaesth Clin Pharmacol* 1997;13:207-12.
30. Kathirvel S, Dash HH, Bhatia A, et al: Effect of prophylactic ondansetron on postoperative nausea and vomiting after elective craniotomy. *J Neurosurg Anesthesiol* 2001;13:207-12.

27
Biological Therapies of Human Astrocytomas

G Zadeh, A Guha

CONTENTS

1. BACKGROUND OF ADULT GLIOMAS
1.1: Epidemiology
1.2: Classification
1.3: Current Treatment and Natural History

2. MOLECULAR GENETICS OF ADULT ASTROCYTOMAS
2.1: Overall Molecular Pathogenesis
2.2: Aberrations in cell cycle regulatory pathway
2.3: Aberrations in growth factors & growth factor receptors
2.4: Regulators of astrocytoma tumor angiogenesis
2.5: Regulators of astrocytoma invasion and cytoskeletan

3. ABERRANT SIGNAL TRANSDUCTION PATHWAYS IN ASTROCYTOMA
3.1: p21-ras:MAPK
3.2: PI 3'Kinase:Akt/PKB
3.3: PKC mediated pathways
3.4: Apoptotic pathways

4. TARGETED ONCOGENIC SIGNALING PATHWAYS IN ASTROCYTOMAS
4.1: Receptor tyrosine kinase inhibitors
4.2: Ras pathway inhibitors
4.3: Protein Kinase C inhibitors
4.4: Anti-angiogenic therapy
4.5: Anti-invasion therapies

5. FUTURE DIRECTIONS

1. BACKGROUND OF ADULT GLIOMAS

1.1: Epidemiology

Tumors of the central nervous system (CNS) can be categorized as primary or secondary. Metastatic or secondary tumors of the CNS are increasing in frequency, as we are better able to control local disease of common human cancers resulting in longer life expectancy. Incidence of primary CNS tumors are also increasing for reasons which are currently unclear, with approximately 30,000 new cases diagnosed yearly in North America according to the Central Brain Tumor Registry of the United States in 1998.[1] This represents about 4% of all cancer related deaths in adults, and the most common pediatric cancer, second only to leukemias. The types, location and molecular pathogenesis of pediatric CNS tumors are for the most part different from those in adults, with this article restricting its comments to adult tumors. More than 50% of all primary CNS tumors arise from glial cells,[2,3] which are further characterized according to their presumed cell of origin, giving rise to astrocytomas, oligodendroglioma, ependymomas, and choroid plexus papillomas.

Amongst gliomas, astrocytomas are the most common by far, with the most malignant variety termed a glioblastoma multiforme (GBM), which unfortunately accounts for more than 50% of all gliomas in adults, with a mean age of onset of 62 years.[1] There is a slight predominance of males in all three astrocytoma grades (male:female ratio 1.2-1.5:1), with no proven pre-disposing environmental factors. Co-relationships with trauma, radiation exposure and cell-phone use have been speculated. The vast majority of astrocytomas occur sporadically, with germline pre-disposing syndromes such as Neurofibromatosis-1 (NF-1), Li-Frauemeni, Turcott, and Tuberous Sclerosis contributing to less than 5% of all newly diagnosed astrocytomas. However, these syndromes have helped to shed much light into the molecular pathogenesis of the more common sporadic astrocytomas.

1.2: Classification

Astrocytomas are graded according to the World Health Organization (WHO) into four increasing grades of malignancy which usually progress from lower to higher grades[4] **(Fig. 1)**. Pilocytic astrocytomas (WHO grade 1) are infiltrating tumors usually found in children and young adults, and are characterized by a mild increase in the number of astrocytes and cellular atypia. They usually occur sporadically and are rarely associated with germline pre-disposition syndromes such as NF-1. Since these tumors usually have a long indolent behavior without progression to higher grades of malignancy, they often do not require any intervention other than close radiological follow-up after the diagnosis is made. Low grade astrocytomas (LGA=WHO grade II) are diffusely infiltrating tumors with moderate cellularity, mild nuclear atypia, and rare or absent mitotic figures. Anaplastic astrocytomas (AA=WHO grade III) are diffusely infiltrating tumors with increased cellularity, nuclear atypia, mitotic activity and characteristic endothelial hyper-proliferation. Finally, glioblastoma multiforme (GBM=WHO grade IV) are characterized by all the microscopic features of AA, in addition to macroscopic and microscopic heterogeneity, glomeruloid microvascular proliferation, and regional and geographic necrosis.[4]

1.3: Current Treatment and Natural History

Grade is the most important prognosticator of astrocytomas,[5] with the mean life expectancy of patients with GBMs = 9-12 mths, AAs = 18 mths, LGAs = 5 yrs.[6] As mentioned, LGAs usually progress to a GBM, with a mean time to transformation of 4-8 years.[7,8] Patients who present with LGA are typically one to two decades younger than those who at presentation harbor a GBM.[1,9] In addition to grade, other important positive prognosticators include younger age and a good functional state at the time of diagnosis (referred to as the Karnofsky Performance Score: KPS). Molecular and immunohistochemical (IHC) markers of proliferation such as the Ki67 labeling index, although increases with grade, has not proven to be an independent prognosticator beyond the main variables mentioned above. Recent evidence from our lab examining expression of mutant Epidermal Growth Factor Receptor (EGFRvIII) with RT-PCR, western blots and EGFRvIII specific IHC expression, demonstrates that this common molecular aberration found in GBMs is an independent negative prognosticator, especially in patients less than 50 years of age.[10]

Unfortunately, other than radiation, very little has altered the ultimate prognosis of astrocytomas, especially for GBMs. Aggressive surgical resection, while in certain cases improves the functional level of the patient and allows for a definitive diagnosis to be made, it does little to alter survival. External beam radiation, as evaluated in a landmark randomized controlled trial in 1978,[11] has increased survival in GBMs from its historical levels of 4 months to the current 9-12 months.[12] Chemotherapy has added little to the median survival of GBM patients, which is most likely related to tumor cell resistance and heterogeneity. Chemotherapeutic agents that have been evaluated are many, including mithramycin[13] and BCNU (1,3-bis(2-chloroethyl)-1-nitrosurea; carmustine), with the latter nitrosurea family perhaps demonstrating little and unpredictable efficacy in combination with surgery and radiation in GBMs (36 weeks median survival with BCNU vs. 34.5 weeks without BCNU).[11] Recent attempts to utilize chemotherapy in GBMs include local delivery in the tumor bed after surgery to overcome the blood brain barrier (BBB) and systemic toxicity associated with slow release polymers.[14] In addition, a second-generation alkylating agent, Temozolomide, has demonstrated some promise in early clinical trials and awaits larger clinical verification.[15,16] Compared to GBMs, patients with AAs often do respond to chemotherapy[17] and therefore they are recommended to receive chemotherapy in addition to surgery and external beam radiation. The discovery that loss of yet to be identified gene(s) on chromosomes #1p and #19q confers chemosensitivity and improved prognosis in oligodendrogliomas,[18] represents a recent major positive change in the management of gliomas. Whether such chemosensitivity resides in a sub-population of the much more common astrocytic lineage tumors, is an exciting but yet unknown possibility.

2. MOLECULAR GENETICS OF ADULT ASTROCYTOMAS

2.1: Overall molecular pathogenesis

The most common adult astrocytoma is the GBM, which may develop de-novo and are termed primary GBMs, or they may progress from lower grade astrocytomas and are termed secondary GBMs, **(Fig. 1)**. Whether all GBMs arise from pro-

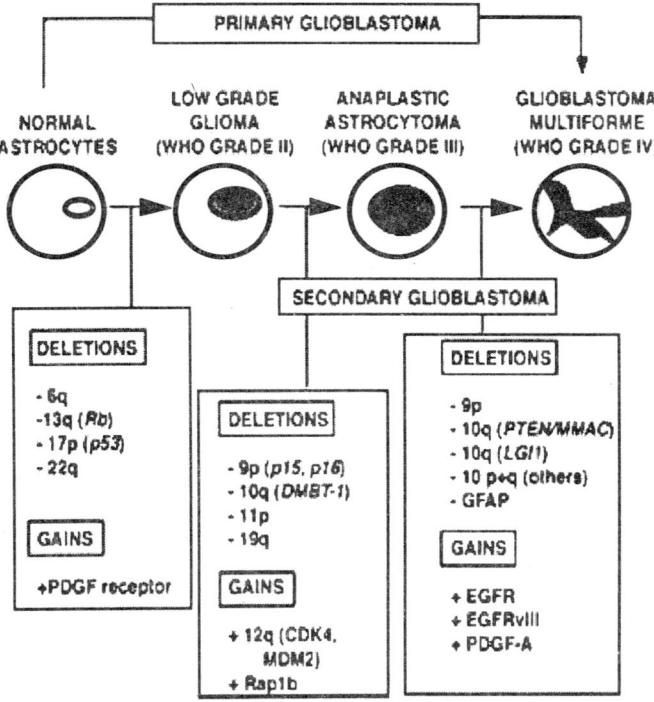

Fig. 1: Molecular pathogenesis of varying grades of human astrocytomas with associated gain and loss of function genetic alterations. At least two molecular progression profiles have been proposed to the most malignant grade IV of astrocytoma also known as a glioblastoma multiforme (GBM) : A- Primary- Usually arising de novo in the elderly without presence of p53, Rb and other genetic alterations; B- Secondary- More prevalent in the younger adult with progression from lower grades of astrocytomas to a GBM. The hallmarks of a GBM, which is the most common grade is loss of chromosome #10q (PTEN) and gain/amplification of chromosome #7p (wild type and mutant EGFR).

gression, with the lower grade astrocytoma remaining clinically silent and hence presenting only as a primary GBMs, is of debate. However, molecular characterization suggests that these pathologically heterogenous tumors are also molecularly heterogenous, with at least two if not more molecular pathways leading to development of a GBM. Similar to other human malignancies, astrocytomas are characterized by the stepwise accumulation of gain-of-function mutations or amplification (oncogenes) and of loss-of-function mutations or deletions (tumor suppressor genes).[19, 20] The molecular signatures that seem to be more specific to GBMs, as compared to lower grade astrocytomas, are loss of parts or all of chromosome#10 and amplifications involving chromosome#7. On chromosome#10, loss of PTEN/MMAC1 expression by mutations, LOH, and alternative mechanisms of gene silencing have been documented in most GBMs, with yet other tumor suppressor genes to be identified.[21, 22] Amplifications of chromsome#7 involve EGFR in a third of GBM patients, with a large percentage harboring various mutations of EGFR such as EGFRvIII, as described in further detail below.[23, 24]

Additional common genetic alterations in astrocytomas include those that alter the p53-MDM2-p19/ARF[25-26] and the Rb-p16-CDK4 cell cycle regulatory pathways[27-29] as discussed below. In addition to EGFR, other aberrantly regulated growth factors/receptors and signaling pathways are involved in promoting the proliferation of astrocytoma cells, recruitment of tumor vessels, invasion, apoptosis etc. These include over-expression of Platelet Derived Growth Factor Receptors (PDGFR) and their cognate ligands.[30, 31, 32] Transforming Growth Factor (TGF), Insulin-like Growth Factor (IGF), etc.[33, 34] Angiogenic specific growth factors and their cognate receptors that are aberrantly regulated in high grade astrocytomas and contribute to the growth of these highly vascularized tumors include Vascular Endothelial Growth Factor (VEGF),[35, 36] Fibroblast Growth Factor[37, 38] and Angiopoietins.[39, 40]

2.2: Aberrations in cell cycle regulatory pathway

Like majority of human cancers, perturbations in both the p53 and Rb regulated cell cycle regulatory pathways, are present in human astrocytomas. p53 protein is a transcription factor that can inhibit cell cycle progression or induce apoptosis in response to stress or DNA damage, and inactivation of p53 attenuates both of these cellular responses.[41, 42] Loss of heterozygosity (LOH) of chromosome#17p is found in 30-40% of all astrocytomas grades.[44] Approximately one third of all astrocytomas with #17p loss have p53 mutations with 25% in GBMs, 34% in AA and 30% in LGAs.[45] Most mutations are missense mutations found on the conserved domains of Exon 5-8, with no clear studies reporting brain specific mutations, except one study that identified a preponderance of Exon 4 mutations in GBMs.[46] There is good evidence that a second tumor suppressor gene on chromosome#17p13.3 exists, as a large number of tumors with 17p LOH do not have corresponding p53 mutations.[47] LOH of 17p or p53 mutations are rarely found in GBMS with EGFR amplification. This confirms the presence of two clinical categories of GBM: 1-primary or de-novo GBM with EGFR amplification and no p53 mutation and 2-secondary or progressive tumors with frequent p53 mutation and no EGFR amplification.[48,49]

The MDM2 oncogene is an important negative regulator of p53, which acts in a feedback loop to limit the action of p53[50] both by inhibiting its trans ADDIN activating activity and by catalyzing its destruction.[51,52] Less than 5% of astrocytomas demonstrate MDM2 amplification and none of these tumors have p53 mutations.[53] Furthermore, 50% of GBMs have over-expressed MDM2 without gene amplification.[54] Although over expressed in a large percentage of GBMs, the functional role of MDM2 in glioma tumorigenesis remains unknown. Loss of p19 expression is also prevalent in astrocytomas and seen in 30% of GBMs.[53] Since p19 is a negative regulator of MDM2, this provides another mechanism of aberrant regulation of the p53 mediated cell cycle regulatory pathways in astrocytomas.

Similar to the p53 pathway, loss of either pRb or other regulators of the pRb mediated cell cycle regulatory pathway is also prevalent in the majority of astrocytomas. The p16/cdk4/cyclinD/Rb cell-cycle regulatory pathway is integral in G1 to S phase transition. Inactivation of p16 through homozygous deletion of CDKN2A gene occurs most commonly in 24% of AA and 33% of GBMs.[27,53] Rare point mutations of CDKN2A, or transcriptional silencing due to CDKN2A promoter methylation, may also inactivate or down-regulate p16 in GBMs (55,56). p16 deletion is also associated with a high proliferative index as demonstrated by Ki67 staining.[57] LOH of Rb or point mutation of the Rb gene, occurs in 30%-40% of GBMs,[27] while amplification or over-expression of cdk4 is found in 10-20% of GBMs.[26] Primary GBMs demonstrate a higher rate of p16 deletion compared to secondary GBMs, whereas pRb LOH and CDK4 amplification occurs with similar frequency.[58]

2.3: Aberrations in growth factors & growth factor receptors

Receptor Protein Tyrosine Kinases (RPTKs) have clearly been linked to the pathogenesis of astrocytomas, with the two major classes being the platelet-derived growth factor receptor (PDGFR) and the epidermal growth factor receptor (EGFR). Human astrocytomas express high amounts of both the PDGF-A and -B ligands and together with their cognate receptors represent a paracrine and/or autocrine stimulatory action.[30] Amplification or mutation of either PDGF ligand or receptor genes is rare, though increased expression is common. PDGFR-α is over-expressed in 24% of human astrocytomas, and probably is a fairly early event in the pathogenesis of astrocytomas since it is expressed in all grades of astrocytomas compared to normal glial cells.[30] PDGFR-β is over-expressed in the higher astrocytoma grades, and often associated and postulated to be causally related to the characteristic florid tumor associated vasculature.[30,59] Over expression of ligands increases with AA and GBMs[31,60] suggesting that the "closing" of the autocrine/

paracrine loop is important in the pathogenesis of higher-grade astrocytomas. The functional relevance of these loops in the growth of astrocytomas has been tested with several approaches, including neutralizing antibodies, small molecule inhibitors and dominant-negative mutants.[32] These encouraging pre-clinical data have led to clinical trials; targeting PDGF mediated stimulation in astrocytomas, as discussed below.

EGFR, located on chromosome 7p11-p12,[61, 62] is over-expressed late in the pathogenesis of astrocytomas.[63] Unlike PDGFR, over-expression of EGFR is as a result of gene amplification and/or rearrangement that produces both wild type and several mutant types of EGFR, some of which are truncated and constitutively active. Amplification of EGFR is seen in only 3% of LGAs, 7% of AA, but 40%-50% of GBMs.[53, 63, 64] The most common mutant EGFR in astrocytomas is the receptor variably known as EGFRvIII or p140EGFR, being expressed in at least 40% of GBMs that have EGFR amplification.[65, 66] Although there are various intragenic deletions that occur in Exons 2-7 of EGFR, which result in expression of EGFRvIII, coercive splicing results in an identical truncated protein in these cases. The mature coercively spliced mRNA encoding the mutant receptor lacks the 801 bases encoding amino acids #6-273 of the wild type EGFRs extracellular domain.[24, 66] The resulting EGFRvIII truncated protein is constitutively phosphorylated[66, 67] and confers an in vivo and in vitro growth advantage.[67] Our recent studies demonstrate that the cohort of GBMs, especially those patients younger than 50 years of age, expressing EGFRvIII have a worse prognosis.[68] At a molecular level this may be a result of the EGFRvIII conferring increased proliferative signals to the astrocytoma cells, along with recent results demonstrating increased protease activity and expression of angiogenic molecules such as Vascular Endothelial Growth Factor (VEGF).[69] Strategies to target EGFR and its mutants are of interest in treatment of GBMs, with early clinical trials initiated.

In addition to PDGF and EGF, other RPTKs have also been implicated in promoting the growth of astrocytomas. These include the TGFβ family, expressed in AAs and GBMs, but not in LGAs and normal brain.[70] In contrast to its usual physiological role of growth inhibition, TGFβ is mitogenic to astrocytoma cell lines.[71] This may be due to either an up-regulation of the TGFβ receptors, induction of other growth factors by TGFβ, or a dysregulation of the downstream signaling pathways in higher-grade astrocytomas. For example, both PDGF ligands are induced in astrocytoma cell lines by activation of TGFβ,[72] along with EGFR,[73] and other angiogenic factors such as VEGF and b-FGF.[74]

2.4: Regulators of astrocytoma tumor angiogenesis

Malignant astrocytomas are one of the most vascularized of all human cancers. The tumor-induced vessels in addition to being numerous are also abnormal, in that they do not maintain the blood-brain-barrier (BBB) leading to peri-tumoral edema. In addition, they often lack a normal capillary bed leading to shunting and often intra-tumoral hemorrhage. Like in other solid cancers, anti-angiogenic therapy, either alone or often in conjunction with radiation or chemotherapy, is an area of intense interest in astrocytomas. Several angiogenic cytokines have been implicated in the tumor-induced neo-angiogenesis, but most factors such as PDGF, FGFs, TGFβ have pleotropic effects, in addition to their contribution to angiogenesis. However, VEGF and Angiopoietins are two angiogenic specific growth factor families, with aberrant expression in astrocytomas. VEGF is highly expressed by GBM cells and is principally induced by tumor hypoxia and aberrant cytokine expression by astrocytoma cells such as PDGF, EGF etc (36,69). Expression of VEGFRs, especially VEGFR2, by the tumor endothelium is up-regulated secondary to the hypoxia, and increased cytokines by astrocytoma cells. Antibodies against VEGF and VEGFRs, antisense strategies and small molecule inhibitors have demonstrated encouraging pre-clinical activity, leading to early clinical trials targeting VEGF.

Similar to VEGF, Angiopoietins are specific for angiogenesis in that their receptors are only found in endothelial cells.[75, 76] Tie2, the RPTK that is activated by Ang-1 and inhibited by Ang-2, is over-expressed and phosphorylated in GBMs.[39] Activation of Tie2 is however not an endothelial cell mitogen, unlike VEGFR2, but is postulated to be involved in vessel maturation. The functional role of Angiopoietins in tumor angiogenesis and whether it does interact with VEGF and other angiogenic molecules are under current study in several tumor types, including astrocytomas. Much pre-clinical work is pending before therapeutic strategies targeting Angiopoietins and/or its receptors, come to the clinical arena.

2.5: Regulators of astrocytoma invasion and cytoskeleton

One of the main reasons why malignant gliomas remain incurable by local therapies such as surgery or radiation is their highly infiltrative and invasive nature. The central process of invasion is degradation of the extra-cellular matrix (ECM) by proteolytic enzymes expressed by tumor cells. Matrix metalloproteases (MMPs, including collagenases, stromelysins and gelatinases) and serine proteases (including urokinase-type plasminogen activator, uPA, and its receptor, uPAR) play a fundamental role in this process. An imbalance between expression and/or activity of MMPs and their endogenous tissue inhibitors (TIMPs), is in-part responsible for tumor cell invasion. This is similar to the balance of pro-angiogenic factors and endogenous anti-angiogenic factors that regulate the "angiogenic switch".[77] In fact, the factors that regulate invasion are an integral and vital part of the angiogenesis cascade.

There is a positive correlation between tumor malignancy and level of MMP-2, -9 and -12 expression in astrocytomas.[78, 79, 80] MMP-2 and -9 have created additional interest in GBMs due to their co-localization around proliferating blood vessels, suggesting a role in both angiogenesis and tumor invasion.[78, 79] Angiogenic factors directly regulate MMP expression, such as VEGF mediated induction of MMP-1, -3, and -9 in vascular smooth muscle cells.[81] This would be required to breakdown the ECM allowing not only tumor cell invasion, but also sprouting of new blood vessels. The endogenous negative tissue regulators of MMPs or TIMPs are also important regulators of astrocytoma invasion. The reports on TIMP-1 and TIMP-2 levels in astrocytomas remains inconclusive, with most of the earlier studies demonstrating a decreased level with increasing glioma grade, whereas recent studies have shown an increase in TIMP-1 in AAs and GBMs compared to LGA and normal brain.[79, 82] Pre-clinical investigations with over or under expression of TIMPs using cell culture and transgenic models may be of use in helping decipher which of the TIMPs are of functional relevance in astrocytoma invasion. Exogenous inhibitors of MMPs, known as metalloprotease inhibitors, have shown promise in both pre-clinical and limited clinical trials in human malignant astrocytomas.

3. ABERRANT SIGNAL TRANSDUCTION PATHWAYS IN ASTROCYTOMA

3.1: p21-ras MAPK

Neuroectodermal derived tumors, such as astrocytomas, do not harbor activating p21-ras mutations, the most common oncogene found in 25-30% of all human cancers.[83] However, we have demonstrated that GBMs and their derived cell lines have high levels of p21-ras activation, which is functionally relevant in proliferation, angiogenesis and radiation resistance.[69, 84] Increased p21-ras activation was also a feature of LGAs, compared to normal brain, suggesting these tumors may also be candidates for agents targeting this pathway, such as farnesyl transferase inhibitors (FTIs), as discussed later. Activated Ras-GTP leads to stimulation of several downstream effector pathways; the most recognized of which is MAPKinase. Activated MAPK appears to follow a somewhat different profile across astrocytoma grades than Ras-GTP. Although highest in GBMs, MAPK activity levels were not elevated in the LGAs, where Ras-GTP levels were elevated compared to normal brain.[85] This suggests that different downstream effector pathways of p21-ras are being employed in GBMs and LGAs. In addition to proliferation, MAPK signaling may also regulate angiogenic signals, as evidenced by our results on VEGF transcriptional regulation under normoxic and hypoxic conditions.[86] Although inhibition of MAPK activation by MEK (MAPKK) inhibitors do show anti-proliferative effects on astrocytoma cells in vitro, their effects in vivo remains to be determined.

3.2: PI 3'Kinase-Akt/PKB

The PI3'Kinase pathway and its main downstream effector Akt/PKB is a major signaling pathway that can be activated in both a p21-ras dependent and independent manner. This pathway is of relevance in astrocytomas, since it is activated in astrocytoma cell lines.[87, 88, 89] Additionally, PTEN/MMAC1, which is the most common tumor suppressor gene whose expression is lost in GBMs, is a negative regulator of PI3'K activity. PTEN is located on chromosome#10q23 and its expression is lost in many human tumors, including GBMs.[21, 22] PTEN functions as a dual specific protein and lipid phosphatase, with its lipid phosphatase function thought to be critical in its anti-tumorgenic effects. PTEN is an inositol phospholipid phosphatase, dephosphorylating the 3' phosphate of two main PI3'K products, PtdInsP3 and PtdInsP2, thereby acting as a negative regulator of PI3'K activity.[88, 90] Unchecked PI3'K activity by mutation, deletion, or gene inactivation of PTEN results in accumulation of a major downstream substrate of PI3'K: activated Akt/PKB. It in turn promotes cell survival, proliferation and cytoskeletal organization.[91, 92]

PTEN/MMAC1 mutations are absent in LGAs, rare in AAs, and detected in 25-30% of GBMs.[93] Secondary GBMs progressing from LGAs (Fig. 1) have a 4% frequency of PTEN mutations, compared to about a 32% mutational rate detected in the primary or de-novo GBMs.[53] In GBMs, about 18% had both mutations of PTEN and amplification of EGFR, making these two the most tightly correlated molecular signatures of these tumors. PTEN protein expression is lost in over 70-95% of GBMs, suggesting that in addition to mutations and associated LOH, gene inactivation is also a major mechanism contributing to loss of PTEN function in these tumors.[88, 95, 96] The functional relevance of aberrant PI3'K-Akt/PKB activity and loss of PTEN in GBM cell lines, has been demonstrated by reintroduction of wild type PTEN, resulting in arrest of the GBM cells in G1.[21, 88, 89, 94] It should be noted that investigations to identify additional tumor suppressor genes on chromosome#10 in GBMs are ongoing, since multiple areas on both arms of the chromosome are known to be lost in GBMs.[97] For example, DMBT is a putative gene on chromosome#10q25.3-26.1, which is homozygously deleted in a subset of GBMs and also in lower grade astrocytomas.[98]

3.3: PKC mediated pathways

Protein Kinase-C (PKC) is a large family of phospholipid-dependent serine/threonine kinases that are involved in a variety of signal transduction pathways. The various isoforms of PKC differ in their enzymatic properties, tissue expression and intracellular localization, and consist of an N-terminal regulatory and a C-terminal kinase domain. PKC activation can be induced by increased intracellular calcium, anionic phospholipids, diacylglycerol (DAG), or tumor promoting phorbol esters.[99] PKC is expressed at high levels in fetal brain, and is important in normal proliferation and differentiation of fetal but not adult glial cells.[100,101] Malignant astrocytomas express high levels of PKC compared to normal adult glial cells, and approach those of fetal astrocytes, perhaps as a result of de-differentiation.[102] Certain isozymes appear to be particularly over-expressed such as α-PKC, with anti-sense constructs directed against the α-PKC resulting in reversion of the malignant phenotype.[103, 104] However, it should be noted that the PKC isoforms that are relevant in astrocytomas, their functional role, and alterations with tumor grade is not fully clear, with one study demonstrating an inverse relation of PKC activity with astrocytoma grade.[105] Nevertheless, due to encouraging pre-clinical data with inhibitors of PKC, clinical trials with tamoxifen (a non-specific PKC inhibitor, with a long track record in management of breast cancer, and known to cross the BBB) has been undertaken as discussed later. It is hoped that some of the encouraging results may be further exploited with more specific PKC isoform inhibitors in the future.

3.4: Apoptotic pathways

The signaling pathways that regulate apoptosis are of relevance in astrocytomas, as in other tumors. Fas ligand, and tumor necrosis factor can induce apoptosis through activation of death receptors such as Fas. Fas is highly expressed in astrocytoma cell lines and tumor specimens, with a direct correlation between levels and grade of tumor. There is also regional variation, with increased expression in areas of necrosis, leading to apoptotic bodies surrounding these regions in GBMs. Increased expression of Bcl-2 has been reported in GBMs, but not in areas of necrosis. This would suggest that increased Bcl-2 expression is to counter balance the apoptotic effects of Fas and Fas ligand expression in GBMs.[107] It is hoped that our increased understanding of the regulation of these cell death pathways in astrocytomas will allow us to favorably manipulate the balance towards increased apoptosis in these tumors.

4. TARGETED ONCOGENIC SIGNALING PATHWAYS TARGETED IN ASTROCYTOMAS

4.1: Receptor tyrosine kinase inhibitors

EGFRs and PDGFRs are of particular interest in astrocytomas, as discussed above. A variety of approaches have been taken to inhibit these RPTKs, including generation of neutralizing antibodies directed against the extracellular domain of full-length wild type EGFR[108] and antisense therapy against the extracellular domain.[109] As astrocytoma cells expressing the truncated EGFRvIII confer a growth advantage, toxin-labeled antibodies directed against the novel glycine splice site of EGFRvIII may prove useful in blocking the mitogenic advantage conferred by this oncoprotein.[110] Additional trials in GBMs with antibody dependent strategies targeting EGFRvIII is about to be initiated. However, the clinical utility of these approaches may depend in part on the cellular localization of EGFRvIII, as in certain cells EGFRvIII appears to be primarily localized to a subcellular location, rather than being on the cell surface.[111] However, examination of GBM specimens does support that a significant portion of EGFRvIII is accessible to antibodies on the cell surface.[111]

Small molecule inhibitors of RPTKs and signaling pathways have the theoretical advantage of penetrating the BBB and reaching the infiltrating astrocytoma cells. These two challenges may be a major obstacle of macromolecule therapies such as antibodies and gene based therapies. Naturally occurring fungal derived tyrosine kinase inhibitors have shown promise, but have the major disadvantage of being non-specific.[112,113,114] Synthetic tyrosine kinase inhibitors or tyrphostins, based on combinatorial chemistry knowledge of signaling molecules and modifications of natural inhibitors, have shown remarkable specificity. For example, quinazolines such as AG-1478 and PD153035 are potent and specific inhibitors of EGFR, with an IC50 in the nanomolar range for PD153035.[115] The quinoxaline AG-1296, on the other hand, is a specific inhibitor of PDGFR.[116] The 4,5-dianilinophthalimides are specific inhibitors of both EGFR and ErbB2, but do not inhibit PDGFR.[117] Similarly SU-101 was developed from a drug screen and was shown to have efficacy and specificity against PDGFRs. Promising pre-clinical results in a variety of human tumor types in immunocompromised mice including astrocytomas led to a Phase1[118] followed by a Phase2/3 trial, which has recently been completed and unfortunately no benefit was demonstrated. Other inhibitors against this RPTK family are currently undergoing pre-clinical evaluation. Although the specific mode of action of these tyrosine kinase inhibitors is not known in all cases, most appear to be competitive inhibitors of the ATP binding site on the RPTKs.

Early-generation tyrphostins were shown to inhibit EGF-tyrosine phosphorylation in astrocytoma cell lines in a dose-dependent fashion, which correlates with its potency as an anti-proliferative agent in these cell lines.[119] The EGFR specific tyrphostin AG-1478 has recently been evaluated on U87MG astrocytoma cells and U87MG cells transfected to express EGFRvIII.[120] Both cells were growth-inhibited by the agent, but U87:EGFRvIII cells were considerably more sensitive to the drug's effect. Despite such promising results in tissue culture, early results in nude mice injected with the human epidermoid cancer cell line A431 and treated with the tyrphostin PD153035 have not shown a significant beneficial effect.[121]

ZD-1839 (Iressa) is a quinazoline derivative that selectively inhibits the EGFR tyrosine kinase and is under clinical development in cancer patients. The anti-proliferative activity of ZD-1839 alone or in combination with cytotoxic drugs differing in mechanism(s) of action, such as cisplatin, carboplatin, doxorubicin, etoposide, etc, was evaluated in-vitro, demonstrating anti-proliferative and apoptotic effects.[122] Also, the anti-tumor effect of ZD1839 is accompanied by inhibition in the production of autocrine and paracrine growth factors that sustain autonomous local growth and angiogenesis, therefore this effect can be potentiated by the combined treatment with certain cytotoxic drugs, such as paclitaxel. Promising in vivo results provided a rationale for its clinical evaluation in combination with cytotoxic drugs. Presently phase 3 clinical trials are being carried out in non-small cell lung cancer and other tumor types are being considered, including phase 2 trials in astrocytomas.[122, 123]

4.2: p21-ras pathway inhibitors

There has been great interest in developing pharmacological agents that inhibit p21-ras, since oncogenic mutations are seen in a wide variety of human cancers.[83] p21-ras pathway inhibition has focussed on farnesyl transferase inhibitors (FTIs), which blocks the post-translational modification of p21-ras. This post-translational modification is a three or four-step process necessary for p21-ras to be recruited to the inner cell membrane and become activated **(Fig. 2)**. The first and most critical step involves the addition of a 15-carbon isoprenoid, farnesyl pyrophosphate (FPP) through a thioester bond at the cysteine residue of the CAAX (cysteine - aliphatic amino acid - aliphatic amino acid - other amino acid) box at the extreme C-terminus of p21-ras.[124] It needs to be pointed out that 1 in 200 cellular proteins undergo farnesylation,[125] in addition to p21-ras and other members of the Ras superfamily such as Rap2 proteins and RhoB.[106,126] Indeed, there is some evidence and debate that much of the anti-tumor effects of FTIs

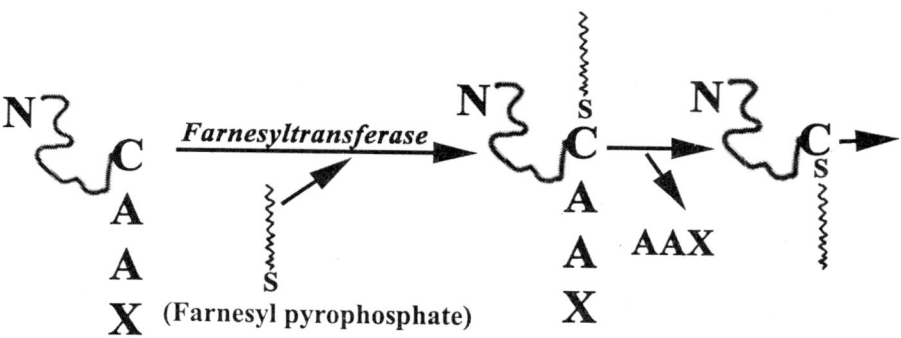

Fig. 2 : Processing of p21-Ras requires addition of isoprenyl moieties to the C-terminal CAAX box, allowing p21-Ras-GDP to bind to the inner cell membrane, where it can undergo guanine-nucleotide exchange by interaction with signalling molecules generated by activated receptors to become activated Ras-GTP. Ha-Ras is obligatorily farnesylated (15 carbon isoprenyl group added by farnesyl transferase from the donor farnesyl pyrophosphate), hence most sensitive to farnesyl transferase inhibitors (FTIs). K-Ras and to a letter extent N-Ras, can be geranylated if farnesylation is inhibited, hence more resistant to FTIs. Astrocytomas, although not harboring oncogenic p21-Ras mutations, do have elevated Ras-GTP levels and are susceptible to growth inhibition by FTIs, which can be predicted by isoform specific p21-Ras activity assay.

are not related to its action on p21-ras at all, but rather on the alteration of RhoB activity.[127] Nevertheless, over the last few years, there has been much interest in creating FTIs in order to target p21-ras signaling in human cancers. Developed agents include tetrapeptide analogs of the CAAX motif, benzodiazepine peptidomimetic CAAX inhibitors, FPP analogs, and bisubstrate analogs, which inhibit both FPP and the CAAX motif.

FTIs have been evaluated for their ability to inhibit the progression of a large number of human tumor cell lines. L-739, 749 (Merck Pharmaceuticals) was able to inhibit the proliferation of approximately 70% of a wide variety of cell lines in anchorage-independent (soft agar) assays.[128,129,131] Our own experiments with the FTIs have demonstrated that the proliferation of a wide variety of GBM cell lines was inhibited in an anchorage-dependent growth assay in a dose-dependent manner.[131] At low doses the effect was reversible with cessation of the drug, while at higher doses it led to cell kill. Anti-proliferative effects were a result of block at both the G1-S and G2-M checkpoints and increased apoptosis. In addition, GBM cell lines transfected to express the mutant EGFRvIII were more sensitive to the FTIs than the parental cells, suggesting that this subgroup of patients who have a poorer prognosis perhaps would benefit most from these agents. Furthermore, we have also demonstrated that FTIs also possess anti-angiogenic effects, by reducing the amount of VEGF secreted by astrocytoma cells, with the effect being most notable under hypoxic conditions.[132,133] Recently we have demonstrated that indeed FTIs do have significant anti-tumorgenic effects in a GBM subcutaneous explant model in mice, with both anti-proliferative and anti-angiogenic effects.[85] In addition, FTIs may be helpful in astrocytomas by radio-sensitization of the tumors, since recent data from several labs demonstrate that p21-ras activity confers radio-resistance.[134] With these and other pre-clinical data, demonstrating that GBMs may also be a rationale candidate for FTI therapy though it does not have any oncogenic p21-ras mutations, a Phase 1/2 clinical trial has been proposed and hopefully will start shortly.

Recognizing the large number of farnesylated proteins and potential side effects, FTIs do not appear to demonstrate adverse effects on normal cells or in animals at doses, which are potent inhibitors of tumor cell mitogenesis.[128] FTIs have however been shown to suffer from another common problem with conventional chemotherapy, namely that of drug resistance. It has become evident that these inhibitors are much more potent at inhibiting tumors with oncogenic H-Ras mutations than those tumors with K-Ras or N-Ras mutations.[135] Farnesylation occurs in all isoforms of p21-ras, though there is isoform specificity, with H-ras being fully dependent on it for proper activation. In contrast, K-Ras and more specifically K4B-Ras that are the main oncogenic isoform mutations seen in human cancers are relatively more resistant to FTIs.[136] The explanation for this differential sensitivity is that K-Ras and N-Ras (to a lesser extent), but not H-Ras, can undergo geranyl-geranylation, when farnesylation is inhibited.[136] We have recently been able to predict which GBMs would be more sensitive to FTIs, by measuring isoform specific p21-ras activity in the cell lines and GBM explants.[85] This measurement can be undertaken on the operative specimens to select the most likely sensitive tumors, as part of future clinical trials.

4.3: Protein Kinase C inhibitors

Pharmacological inhibitors of the PKC pathway have led to promising initial results in astrocytomas. A variety of non-specific PKC inhibitors (polymyxin B and tamoxifen) and specific PKC inhibitors (staurosporine) have been shown to reduce astrocytoma cell proliferation.[137, 138] Tamoxifen caused a dose dependant inhibition of proliferation in astrocytoma cell lines through an estrogen receptor blockade mechanism.[139, 140] Earlier clinical trials using the non-specific PKC inhibitor tamoxifen have been disappointing; despite an established acceptable side effect profile, this drug has shown only minor benefit in patients with astrocytomas.[141] However, more recent studies have reported that 25% of patients had a 50% decrease in tumor volume on magnetic resonance imaging and 19% had stabilization of their disease.[142] 7-hydroxystaurosporine (UCN-01) is more selective for PKC inhibition and produces marked reduction in proliferation of in vitro astrocytoma cell lines and in vivo mouse models.[143] Phase 2 clinical trials of UCN-01, has been approved for adult recurrent malignant gliomas. One major hurdle in the use of these inhibitors involves the isozyme specificity of each agent. We are only now beginning to understand what roles the individual PKC isozymes play in individual cells, and it is likely that optimum pharmacological inhibition will depend on both an understanding of isozyme functions and on pharmacological agents which specifically inhibit certain isozymes while not inhibiting others.

4.4: Anti-angiogenic therapy

Anti-angiogenic therapy has certain appealing features outside of the fact that GBMs are highly vascularized tumors. These therapies have limited drug resistance, since endothelial cells do not posses the genetic heterogeneity or instability of cancer cells.[144] Anti-angiogenic agents employ a cytostatic strategy and therefore end points of therapy include progression free survival and not tumor shrinkage, which may be highly desirable in a large percentage of astrocytomas, especially LGAs. In recent years, multiple molecular mediators and inhibitors of angiogenesis have emerged and have been tried in various clinical trials. VEGF inhibitors are identified as promising agents in glioma therapy, with various studies demonstrating in vitro and in vivo effects of targeting VEGF. Anti-sense VEGF constructs inhibit tumor formation in animal experiments of C6 rat gliomas and xenografts of malignant gliomas.[145, 146] Dominant negative inhibition of VEGFR2 decreases vascularization and tumor growth of C6 rat glioma.[147] Small molecule inhibitors of VEGFR2, or related members of the split kinase RPTK family are also of promise in a variety of cancers, though they have not been examined in GBMs. In addition to direct inhibition of VEGF or VEGFR, biological targets against mitogenic receptors or signaling pathways such as EGFR, PDGFR or p21-ras in astrocytomas may have as a major component of their anti-tumor effects, an anti-angiogenic mechanisms of inhibiting the VEGF pathway.[69]

Non-angiogenic receptor mediated endothelial cell inhibitors are also a category of angiogenic therapies that have undergone clinical testing in astrocytomas. Thalidomide is best known in this category and well recognized through its teratogenic effects in humans. Thalidomide was assessed in phase 2 trials against malignant gliomas in a small group of patients and demonstrated minimal radiographic response.[148, 149, 150] Efficacy of thalidomide together with radiation therapy is currently being studied. Angiostatin and endostatin are endogenous anti-angiogenic proteins that are also found in many solid tumors including astrocytomas. Angiostatin has undergone pre-clinical studies in gliomas[148, 151] and is currently being evaluated in a clinical trial on GBM patients. Endostatin is being studied against various solid tumors but not yet in gliomas.

4.5: Anti-invasion therapies

MMPs break down the ECM, inducing invasion of gliomas and endothelial cells. Targeting this pathway theoretically inhibits tumorgenesis by preventing invasion and tumor vascularization. Endogenous MMP inhibitors or TIMPs, have limited tissue penetrance and poor oral absorption, hence they are yet not practical therapeutic strategies. Exogenous pharmaceutical compounds that inhibit MMPs are Marimastat, and Prinomastat. Marimastat acts against all major classes of MMPs and is currently studied in a double-blinded placebo-controlled trial in GBMs in patients receiving surgery and/or radiotherapy. In contrast, Prinomastat inhibits MMPs-2,-9,-13 and -14. In animal studies it has shown promising results in gliomas and is currently being evaluated in clinical trials as a phase 1 trial for GBMs.

5. FUTURE DIRECTIONS

Delivery of biological therapies against relevant oncogenic signaling pathways in astrocytomas located in the brain poses unique challenges that are different from other organ sites. First, is consideration of toxicity, which can have devastating consequences not only in focal but more global neurological functions such as intellect, emotion and memory. The lessons learned from the consequence of whole brain radiation, especially in the developing nervous system must be kept in mind. Second, is the heterogeneity both at a molecular and pathological level in astrocytomas, and one that differs between and within a tumor. Therefore, it would be ideal but not realistic, to have a molecular profile of each individual astrocytoma being treated, though we know regional variations may be an insurmountable barrier. Perhaps non-invasive imaging methods will allow us to determine the molecular profile of these tumors, and guide our therapies in the future. Third, is the issue of the BBB, which although broken in the vascularized and necrotic regions of GBMs, is still intact in the infiltrating edge where tumor recurrence occurs after conventional surgery and radiation. Local delivery with pumps and convection based delivery systems, breaking the BBB with hyper-osmolar or pharmaceutical strategies, intravascular delivery, biodegradable slow release wafers are just some of the innovative ideas being tried in astrocytomas to improve local delivery. These local delivery methods must balance the requirements for sufficient delivery of the biological compound to the infiltrating astrocytoma cells with long-term efficacy. Due to some of these challenges related to the profile of astrocytomas, and some attributes of growing in the brain, it is most unlikely that any of the novel biologicals will provide a significant cure by themselves. It is hoped that small incremental improvements in survival may be made by these biological therapies in combination with each other, and with current conventional therapies, without increasing the morbidity to the patient.

REFERENCES:

1. Central Brain Tumor Registry of the United States. 1997 Annual Report. Chicago, IL:CBTRUS,1998.
2. Schoenberg BS, Christine BW, Whisnant JP: The descriptive epidemiology of primary intracranial neoplasms: the Connecticut experience. *Am J Epidemiol* 1976;104:499-510.
3. Zimmerman HM: The ten most common types of brain tumor. *Semin Roentgenol* 1971;6:48-58.
4. Kleihues P, Cavenee WK: World Health Organization Classification of Tumours: Pathology and Genetics of Tumours of the Nervous System. Lyon: IARC Press, 2000.
5. Kleihues P, Burger P, Scheithauer B: Histological Typing of Tumors of the Nervous System. Berlin: Springer-Verlag, 1991.
6. Mahaley MS, Jr., Mettlin C, Natarajan N, Laws ER, Jr., Peace BB: National survey of patterns of care for brain-tumor patients. *J Neurosurg* 1989;71:826-36.
7. Recht LD, Lew R, Smith TW: Suspected low-grade glioma: is deferring treatment safe? *Ann Neurol* 1992;31:431-36.
8. Watanabe K, Sato K, Biernat W, et al: Incidence and timing of p53 mutations during astrocytoma progression in patients with multiple biopsies. *Clin Cancer Res* 1997;3:523-30.
9. Davis FG, Freels S, Grutsch J, Barlas S, Brem S: Survival rates in patients with primary malignant brain tumors stratified by patient age and tumor histological type: an analysis based on Surveillance, Epidemiology, and End Results (SEER) data, 1973-1991. *J Neurosurg* 1998;88:1-10.
10. Feldkamp MM, Lala P, Lau N, Roncari L, Guha A: Expression of activated Epidermal Growth Factor Receptors, Ras-Guanosine triphosphate, and Mitogen-activated Protein Kinase in human glioblastoma multiforme specimens. *Neurosurgery* 1999;45:1442-53.
11. Walker MD, Hunt WE, MacCarty CS, et al: Evaluation of BCNU and/or radiotherapy in the treatment of anaplastic gliomas: a cooperative clinical trial. *J Neurosurg* 1978;49:333-43.
12. Salcman M: Survival in glioblastoma: historical persepctive. *Neurosurgery* 1980;7:435-39.
13. Jelsma R, Bucy PC: The treatment of glioblastoma multiforme of the brain. *J Neurosurg* 1967;27:388-400.
14. Brem H, Piantadosi S, Burger PC, et al: Placebo-controlled trial of safety and efficacy of intraoperative controlled delivery by biodegradable polymers of chemotherapy for recurrent gliomas. The Polymer-brain Tumor Treatment Group. *Lancet* 1995;345:1008-12.

15. Janinis J, Efstathiou E, Panopoulos C, et al: Phase II study of temozolomide in patients with relapsing high grade glioma and poor performance status. Med *Oncol* 2000;17:106-10.
16. Yung WK: Temozolomide in malignant gliomas. Semin Oncol 2000;27:27-34.
17. Levin VA, Silver P, Hannigan J, et al: Superiority of post-radiotherapy adjuvant chemotherapy with CCNU, procarbazine, and vincristine (PCV) over BCNU for anaplastic gliomas: NCOG 6G61 final report. *Int J Radiat Oncol Biol Phys* 1990;18:321-24.
18. Cairncross JG, Ueki K, Zlatescu MC, et al: Specific genetic predictors of chemotherapeutic response and survival in patients with anaplastic oligodendrogliomas. *J Natl Cancer Inst* 1998;90:1473-79.
19. Kleihues P: Subsets of glioblastoma: clinical and histological vs. genetic typing. *Brain Pathol* 1998;8:667-68.
20. Louis DN, Gusella JF: A tiger behind many doors: multiple genetic pathways to malignant glioma. *Trends Genet* 1995;11:412-15.
21. Li L, Ernsting BR, Wishart MJ, Lohse DL, Dixon JE: A family of putative tumor suppressors is structurally and functionally conserved in humans and yeast. *J Biol Chem* 1997;272:29403-06.
22. Steck PA, Pershouse MA, Jasser SA, et al: Identification of a candidate tumour suppressor gene, MMAC1, at chromosome 10q23.3 that is mutated in multiple advanced cancers. *Nat Genet* 1997;15:356-62.
23. Libermann TA, Razon N, Bartal AD, Yarden Y, Schlessinger J, Soreq H: Expression of epidermal growth factor receptors in human brain tumors. *Cancer Res* 1984;44:753-60.
24. Yamazaki H, Fukui Y, Ueyama Y, et al: Amplification of the structurally and functionally altered epidermal growth factor receptor gene (c-erbB) in human brain tumors. *Mol Cell Biol* 1988;8:1816-20.
25. Louis DN: The p53 gene and protein in human brain tumors. *J Neuropathol Exp Neurol* 1994;53:11-21.
26. Reifenberger G, Reifenberger J, Ichimura K, Meltzer PS, Collins VP: Amplification of multiple genes from chromosomal region 12q13-14 in human malignant gliomas: preliminary mapping of the amplicons shows preferential involvement of CDK4, SAS, and MDM2. *Cancer Res* 1994;54:4299-303.
27. Ichimura K, Schmidt EE, Goike HM, Collins VP: Human glioblastomas with no alterations of the CDKN2A (p16INK4A, MTS1) and CDK4 genes have frequent mutations of the retinoblastoma gene. *Oncogene* 1996;13:1065-72.
28. Ueki K, Ono Y, Henson JW, Efird JT, von Deimling A, Louis DN: CDKN2/p16 or RB alterations occur in the majority of glioblastomas and are inversely correlated. *Cancer Res* 1996;56:150-53.
29. Nishikawa R, Furnari FB, Lin H, et al: Loss of P16INK4 expression is frequent in high grade gliomas. *Cancer Res* 1995;55:1941-45.
30. Guha A, Dashner K, Black PM, Wagner JA, Stiles CD: Expression of PDGF and PDGF receptors in human astrocytoma operation specimens supports the existence of an autocrine loop. *Int J Cancer* 1995;60:168-73.
31. Guha A, Glowacka D, Carroll R, Dashner K, Black PM, Stiles CD: Expression of platelet derived growth factor and platelet derived growth factor receptor mRNA in a glioblastoma from a patient with Li-Fraumeni syndrome. *J Neurol Neurosurg Psychiatry* 1995;58:711-14.
32. Shamah SM, Stiles CD, Guha A: Dominant-negative mutants of platelet-derived growth factor revert the transformed phenotype of human astrocytoma cells. *Mol Cell Biol* 1993;13:7203-12.
33. Antoniades HN, Galanopoulos T, Neville-Golden J, Maxwell M: Expression of insulin-like growth factors I and II and their receptor mRNAs in primary human astrocytomas and meningiomas;in vivo studies using in situ hybridization and immunocytochemistry. *Int J Cancer* 1992;50:215-22.
34. Trojan J, Johnson TR, Rudin SD, Ilan J, Tykocinski ML: Treatment and prevention of rat glioblastoma by immunogenic C6 cells expressing antisense insulin-like growth factor I RNA. *Science* 1993;259:94-97.
35. Millauer B, Shawver LK, Plate KH, Risau W, Ullrich A: Glioblastoma growth inhibited in vivo by a dominant-negative Flk-1 mutant. *Nature* 1994;367:576-79.

36. Plate KH, Breier G, Welch HA, Risau W: Vascular Endothelial Growth Factor is a potential tumour angiogenesis factor in human gliomas in vivo. *Nature* 1992;359:845-48.
37. Libermann TA, Friesel R, al. e: An Angiogenic Growth Factor is Expressed in Human Glioma Cells. *EMBO J* 1987;6:1627-32.
38. Morrison RS: Supression of Basic Fibroblastic Growth Factor by Antisense Oligodeoxynucleotides Inhibits the Growth of Transformed Human Astrocytes. *J of Biological Chemistry* 1991;266:728-34.
39. Ding H, Roncari L, Wu X, et al: Expression and hypoxic regulation of angiopoietins in human astrocytomas. *Neuro-oncol* 2001;3:1-10.
40. Stratmann A, Risau W, Plate KH: Cell type-specific expression of angiopoietin-1 and angiopoietin-2 suggests a role in glioblastoma angiogenesis. *Am J Pathol* 1998;153:1459-66.
41. Ko LJ, Prives C: p53: puzzle and paradigm. *Genes Dev* 1996;10:1054-72.
42. Levine AJ: p53, the cellular gatekeeper for growth and division. *Cell* 1997;88:323-31.
43. Giaccia AJ, Kastan MB: The complexity of p53 modulation: emerging patterns from divergent signals. *Genes Dev* 1998;12:2973-83.
44. el-Azouzi M, Chung RY, Farmer GE, et al: Loss of distinct regions on the short arm of chromosome 17 associated with tumorigenesis of human astrocytomas. *Proc Natl Acad Sci US A* 1989;86:7186-90.
45. Fulci G, Ishii N, Van Meir EG: p53 and brain tumors: from gene mutations to gene therapy. *Brain Pathol* 1998;8:599-613.
46. Li Y, Millikan RC, Carozza S, et al: p53 mutations in malignant gliomas. *Cancer Epidemiol Biomarkers Prev* 1998;7:303-8.
47. Chattopadhyay P, Rathore A, Mathur M, Sarkar C, Mahapatra AK, Sinha S: Loss of heterozygosity of a locus on 17p13.3, independent of p53, is associated with higher grades of astrocytic tumours. *Oncogene* 1997;15:871-74.
48. Watanabe K, Tachibana O, Sata K, Yonekawa Y, Kleihues P, Ohgaki H: Overexpression of the EGF receptor and p53 mutations are mutually exclusive in the evolution of primary and secondary glioblastomas. *Brain Pathol* 1996;6:217-23;discussion 23-24.
49. Louis DN: A molecular genetic model of astrocytoma histopathology. *Brain Pathol* 1997;7:755-64.
50. Wu X, Bayle JH, Olson D, Levine AJ: The p53-mdm-2 autoregulatory feedback loop. *Genes Dev* 1993;7:1126-32.
51. Haupt Y, Maya R, Kazaz A, Oren M: Mdm2 promotes the rapid degradation of p53. *Nature* 1997;387:296-99.
52. Kubbutat MH, Jones SN, Vousden KH: Regulation of p53 stability by Mdm2. *Nature* 1997;387:299-303.
53. Rasheed BK, Wiltshire RN, Bigner SH, Bigner DD: Molecular pathogenesis of malignant gliomas. *Curr Opin Oncol* 1999;11:162-67.
54. Newcomb EW, Cohen H, Lee SR, et al: Survival of patients with glioblastoma multiforme is not influenced by altered expression of p16, p53, EGFR, MDM2 or Bcl-2 genes [see comment]. *Brain Pathol* 1998;8:655-67.
55. Costello JF, Berger MS, Huang HS, Cavenee WK: Silencing of p16/CDKN2 expression in human gliomas by methylation and chromatin condensation. *Cancer Res* 1996;56:2405-10.
56. Merlo A, Herman JG, Mao L, et al: 5' CpG island methylation is associated with transcriptional silencing of the tumour suppressor p16/CDKN2/MTS1 in human cancers. *Nat Med* 1995;1:686-92.
57. Ono Y, Tamiya T, Ichikawa T, et al: Malignant astrocytomas with homozygous CDKN2/p16 gene deletions nave higher Ki-67 proliferation indices. *J Neuropathol Exp Neurol* 1996;55:1026-31.
58. Biernat W, Tohma Y, Yonekawa Y, Kleihues P, Ohgaki H: Alterations of cell cycle regulatory genes in primary (de novo) and secondary glioblastomas. *Acta Neuropathol (Berl)* 1997;94:303-09.
59. Hermansson M, Nister M, Betsholtz C, Heldin CH, Westermark B, Funa K: Endothelial cell hyperplasia in human glioblastoma: coexpression of mRNA for platelet-derived growth factor (PDGF) B chain

and PDGF receptor suggests autocrine growth stimulation. *Proc Natl Acad Sci US A* 1988;85:7748-52.
60. Guha A: Platelet Derived Growth Factor: A General Review with Emphasis on Astrocytomas. *Pediatric Neurosurgery* 1992;17:14-20.
61. Haley J, Whittle N, Bennet P, Kinchington D, Ullrich A, Waterfield M: The human EGF receptor gene: structure of the 110 kb locus and identification of sequences regulating its transcription. *Oncogene Res* 1987;1:375-96.
62. Kondo I, Shimizu N: Mapping of the human gene for epidermal growth factor receptor (EGFR) on the p13 leads to q22 region of chromosome 7. *Cytogenet Cell Genet* 1983;35:9-14.
63. Louis DN, Gusella JF: A tiger behind many doors: multiple genetic pathways to malignant glioma. *Trends Genet* 1995;11:412-15.
64. Wong AJ, Bigner SH, Bigner DD, Kinzler KW, Hamilton SR, Vogelstein B: Increased expression of the epidermal growth factor receptor gene in malignant gliomas is invariably associated with gene amplification. *Proc Natl Acad Sci U S A* 1987;84:6899-903.
65. Steck PA, Lee P, Hung MC, Yung WK: Expression of an altered epidermal growth factor receptor by human glioblastoma cells. *Cancer Res* 1988;48:5433-39.
66. Ekstrand AJ, Longo N, Hamid ML, et al: Functional characterization of an EGF receptor with a truncated extracellular domain expressed in glioblastomas with EGFR gene amplification. *Oncogene* 1994;9:2313-20.
67. Nishikawa R, Ji XD, Harmon RC, et al: A mutant epidermal growth factor receptor common in human glioma confers enhanced tumorigenicity. *Proc Natl Acad Sci U S A* 1994;91:7727-31.
68. Feldkamp MM, Lala P, Lau N, Roncari L, Guha A: Expression of activated epidermal growth factor receptors, Ras-guanosine triphosphate, and mitogen-activated protein kinase in human glioblastoma multiforme specimens. *Neurosurgery* 1999;45:1442-53.
69. Feldkamp MM, Lau N, Rak J, Kerbel RS, Guha A: Normoxic and hypoxic regulation of vascular endothelial growth factor (VEGF) by astrocytoma cells is mediated by Ras. *Int J Cancer* 1999;81:118-24.
70. Yamada N, Kato M, Yamashita H, et al: Enhanced expression of transforming growth factor-beta and its type-I and type-II receptors in human glioblastoma. *Int J Cancer* 1995;62:386-92.
71. Horst HA, Scheithauer BW, Kelly PJ, Kovach JS: Distribution of transforming growth factor-beta 1 in human astrocytomas. *Hum Pathol* 1992;23:1284-88.
72. Jennings MT, Hart CE, Commers PA, et al: Transforming growth factor beta as a potential tumor progression factor among hyperdiploid glioblastoma cultures: evidence for the role of platelet-derived growth factor. *J Neurooncol* 1997;31:233-54.
73. Battegay EJ, Raines EW, Seifert RA, Bowen-Pope DF, Ross R: TGF-beta induces bimodal proliferation of connective tissue cells via complex control of an autocrine PDGF loop. *Cell* 1990;63:515-24.
74. Mandriota SJ, Pepper MS: Vascular endothelial growth factor-induced in vitro angiogenesis and plasminogen activator expression are dependent on endogenous basic fibroblast growth factor. *J Cell Sci* 1997;110:2293-302.
75. Suri C, Jones PF, Patan S, et al: Requisite role of angiopoietin-1, a ligand for the TIE2 receptor, during embryonic angiogenesis. *Cell* 1996;87:1171-80.
76. Suri C, McClain J, Thurston G, et al: Increased vascularization in mice overexpressing angiopoietin-1. *Science* 1998;282:468-71.
77. Folkman J: The role of angiogenesis in tumor growth. *Semin Cancer Biol* 1992;3:65-71.
78. Forsyth PA, Wong H, Laing TD, et al: Gelatinase-A (MMP-2), gelatinase-B (MMP-9) and membrane type matrix metalloproteinase-1 (MT1-MMP) are involved in different aspects of the pathophysiology of malignant gliomas. *Br J Cancer* 1999;79:1828-35.
79. Kachra Z, Beaulieu E, Delbecchi L, et al: Expression of matrix metalloproteinases and their inhibitors in human brain tumors. *Clin Exp Metastasis* 1999;17:555-66.
80. Wagner S, Stegen C, Bouterfa H, et al: Expression of matrix metalloproteinases in human glioma cell lines in the presence of IL-10. *J Neurooncol* 1998;40:113-22.

81. Webb KE, Henney AM, Anglin S, Humphries SE, McEwan JR: Expression of matrix metalloproteinases and their inhibitor TIMP-1 in the rat carotid artery after balloon injury. *Arterioscler Thromb Vasc Biol* 1997;17:1837-44.
82. Lampert K, Machein U, Machein MR, Conca W, Peter HH, Volk B: Expression of matrix metalloproteinases and their tissue inhibitors in human brain tumors. *Am J Pathol* 1998;153:429-37.
83. Bos JL: p21ras: an oncoprotein functioning in growth factor-induced signal transduction. *Eur J Cancer* 1995;31A:1051-54.
84. Guha A, Feldkamp MM, Lau N, Boss G, Pawson A: Proliferation of human malignant astrocytomas is dependent on Ras activation. *Oncogene* 1997;15:2755-65.
85. Feldkamp MM, Lau N, Roncari L, Guha A: Isotype-specific Ras.GTP-levels predict the efficacy of farnesyl transferase inhibitors against human astrocytomas regardless of Ras mutational status. *Cancer Res* 2001;61:4425-31.
86. Favata MF, Horiuchi KY, Manos EJ, et al: Identification of a novel inhibitor of mitogen-activated protein kinase kinase. *J Biol Chem* 1998;273:18623-32.
87. Haas-Kogan D, Shalev N, Wong M, Mills G, Yount G, Stokoe D: Protein kinase B (PKB/Akt) activity is elevated in glioblastoma cells due to mutation of the tumor suppressor PTEN/MMAC. *Curr Biol* 1998;8:1195-98.
88. Furnari FB, Lin H, Huang HS, Cavenee WK: Growth suppression of glioma cells by PTEN requires a functional phosphatase catalytic domain. *Proc Natl Acad Sci U S A* 1997;94:12479-84.
89. Sun H, Lesche R, Li DM, et al: PTEN modulates cell cycle progression and cell survival by regulating phosphatidylinositol 3,4,5,-trisphosphate and Akt/protein kinase B signaling pathway. *Proc Natl Acad Sci U S A* 1999;96:6199-204.
90. Cantley LC, Neel BG: New insights into tumor suppression: PTEN suppresses tumor formation by restraining the phosphoinositide 3-kinase/AKT pathway. *Proc Natl Acad Sci U S A* 1999;96:4240-45.
91. Franke TF, Yang SI, Chan TO, et al: The protein kinase encoded by the Akt proto-oncogene is a target of the PDGF-activated phosphatidylinositol 3-kinase. *Cell* 1995;81:727-36.
92. Stambolic V, Suzuki A, de la Pompa JL, et al: Negative regulation of PKB/Akt-dependent cell survival by the tumor suppressor PTEN. *Cell* 1998;95:29-39.
93. Steck PA, Lin H, Langford LA, et al: Functional and molecular analyses of 10q deletions in human gliomas. *Genes Chromosomes Cancer* 1999;24:135-43.
94. Myers MP, Pass I, Batty IH, et al: The lipid phosphatase activity of PTEN is critical for its tumor supressor function. *Proc Natl Acad Sci U S A* 1998;95:13513-18.
95. Maher EA, Furnari FB, Bachoo RM, et al: Malignant glioma: genetics and biology of a grave matter. *Genes Dev* 2001;15:1311-33.
96. von Deimling A, von Ammon K, Schoenfeld D, Wiestler OD, Seizinger BR, Louis DN: Subsets of glioblastoma multiforme defined by molecular genetic analysis. *Brain Pathol* 1993;3:19-26.
97. Ichimura K, Schmidt EE, Miyakawa A, Goike HM, Collins VP: Distinct patterns of deletion on 10p and 10q suggest involvement of multiple tumor suppressor genes in the development of astrocytic gliomas of different malignancy grades. *Genes Chromosomes Cancer* 1998;22:9-15.
98. Mollenhauer J, Wiemann S, Scheurlen W, et al: DMBT1, a new member of the SRCR superfamily, on chromosome 10q25.3-26.1 is deleted in malignant brain tumours. *Nat Genet* 1997;17:32-39.
99. Nishizuka Y: Intracellular signaling by hydrolysis of phopholipids and activation of protein kinase C. *Science* 1992;258:607-14.
100. Honegger P: Protein kinase C-activating tumor promoters enhance the differentiation of astrocytes in aggregrating fetal brain cell cultures. *J Neurochem* 1986;46:1561-66.
101. Bhat NR: Role of protein kinase C in glial cell proliferation. *J Neurosci Res* 1989;22:20-27.
102. Couldwell WT, Antel JP, Yong VW: Protein kinase C activity correlates with the growth rate of malignant gliomas: Part II. Effects of glioma mitogens and modulators of protein kinase C. *Neurosurgery* 1992;31:717-24;discussion 724.
103. Xiao H, Goldthwait DA, Mapstone T: The identification of four protein kinase C isoforms in human glioblastoma cell lines: PKC alpha, gamma, epsilon, and zeta. *J Neurosurg* 1994;81:734-40.

104. Ahmad S, Mineta T, Martuza RL, Glazer RI: Antisense expression of protein kinase C alpha inhibits the growth and tumorigenicity of human glioblastoma cells. *Neurosurgery* 1994;35:904-8;discussion 908-09.
105. Benzil DL, Finkelstein SD, Epstein MH, Finch PW: Expression pattern of alpha-protein kinase C in human astrocytomas indicates a role in malignant progression. *Cancer Res* 1992;52:2951-56.
106. Weller M, Frei K, Groscurth P, Krammer PH, Yonekawa Y, Fontana A: Anti-Fas/APO-1 antibody-mediated apoptosis of cultured human glioma cells. Induction and modulation of sensitivity by cytokines. *J Clin Invest* 1994;94:954-64.
107. Newcomb EW, Bhalla SK, Parrish CL, Hayes RL, Cohen H, Miller DC: bcl-2 protein expression in astrocytomas in relation to patient survival and p53 gene status. *Acta Neuropathol* (Berl)1997; 94:369-75.
108. Jannot CB, Beerli RR, Mason S, Gullick WJ, Hynes NE: Intracellular expression of a single-chain antibody directed to the EGFR leads to growth inhibition of tumor cells. *Oncogene* 1996;13:275-82.
109. De Giovanni C, Landuzzi L, Frabetti F, et al: Antisense epidermal growth factor receptor transfection impairs the proliferative ability of human rhabdomyosarcoma cells. *Cancer Res* 1996;56:3898-901.
110. Reist CJ, Archer GE, Kurpad SN, et al: Tumor-specific anti-epidermal growth factor receptor variant III monoclonal antibodies: use of the tyramine-cellobiose radioiodination method enhances cellular retention and uptake in tumor xenografts. *Cancer Res* 1995;55:4375-82.
111. Ekstrand AJ, Liu L, He J, et al: Altered subcellular location of an activated and tumour-associated epidermal growth factor receptor. *Oncogene* 1995;10:1455-60.
112. Graziani Y, Chayoth R, Karny N, Feldman B, Levy J: Regulation of protein kinases activity by quercetin in Ehrlich ascites tumor cells. *Biochim Biophys Acta* 1982;714:415-21.
113. Akiyama T, Ishida J, Nakagawa S, et al: Genistein, a specific inhibitor of tyrosine-specific protein kinases. *J Biol Chem* 1987;262:5592-95.
114. Onoda T, Isshiki K, Takeuchi T, Tatsuta K, Umezawa K: Inhibition of tyrosine kinase and epidermal growth factor receptor internalization by lavendustin A methyl ester in cultured A431 cells. *Drugs Exp Clin Res* 1990;16:249-53.
115. Fry DW, Kraker AJ, McMichael A, et al: A specific inhibitor of the epidermal growth factor tyrosine kinase. *Science* 1994;265:1093-95.
116. Levitzki A, Gazit A: Tyrosine kinase inhibition: an approach to drug development. *Science* 1995; 267:1782-88.
117. Buchdunger E, Trinks U, Mett H, et al: 4,5-Dianilinophthalimide: a protein-tyrosine kinase inhibitor with selectivity for the epidermal growth factor receptor signal transduction pathway and potent in vivo antitumor activity. *Proc Natl Acad Sci U S A* 1994;91:2334-38.
118. Mason W, Malkin M, Lieberman F, Cropp G, Hannah A: Pharmacokinetics of SU101, a novel signal transduction inhibitor, in patients with recurrent malignant glioma. *Proc Am Assoc Cancer Res* 1996;37:166.
119. Miyaji K, Tani E, Shindo H, Nakano A, Tokunaga T: Effect of tyrphostin on cell growth and tyrosine kinase activity of epidermal growth factor receptor in human gliomas. *J Neurosurg* 1994;81:411-19.
120. Han Y, Caday CG, Nanda A, Cavenee WK, Huang HJ: Tyrphostin AG 1478 preferentially inhibits human glioma cells expressing truncated rather than wild-type epidermal growth factor receptors. *Cancer Res* 1996;56:3859-61.
121. Hook KE, Kunkel MW, Elliott WL, Howard CT, Leopold WR: Epidermal growth factor receptor tyrosine kinase in A431 xenografts: inhibition by PD 153035 (3-(3-bromoanilino)-6,7-dimethoxyquinazoline). *Proc Am Assoc Cancer Res* 1995;36:434.
122. Sirotnak FM, Zakowski MF, Miller VA, Scher HI, Kris MG: Efficacy of cytotoxic agents against human tumor xenografts is markedly enhanced by coadministration of ZD1839 (Iressa), an inhibitor of EGFR tyrosine kinase. *Clin Cancer Res* 2000;6:4885-92.
123. Ciardiello F, Caputo R, Bianco R, et al: Inhibition of growth factor production and angiogenesis in human cancer cells by ZD1839 (Iressa), a selective epidermal growth factor receptor tyrosine kinase inhibitor. *Clin Cancer Res* 2001;7:1459-65.

124. Hancock JF, Magee AI, Childs JE, Marshall CJ: All ras proteins are polyisoprenylated but only some are palmitoylated. *Cell* 1989;57:1167-77.
125. Marshall CJ: Protein prenylation: a mediator of protein-protein interactions. *Science* 1993; 259:1865-66.
126. Farrell FX, Yamamoto K, Lapetina EG: Prenyl group identification of rap2 proteins: a ras superfamily member other than ras that is farnesylated. *Biochem J* 1993;289:349-55.
127. Lebowitz PF, Davide JP, Prendergast GC: Evidence that farnesyltransferase inhibitors suppress Ras transformation by interfering with Rho activity. *Mol Cell Biol* 1995;15:6613-22.
128. Yan N, Ricca C, Fletcher J, Glover T, Seizinger BR, Manne V: Farnesyltransferase inhibitors block the neurofibromatosis type I (NF1) malignant phenotype. Cancer Res 1995;55:3569-75.
129. Sepp-Lorenzino L, Ma Z, Rands E, et al: A peptidomimetic inhibitor of farnesyl:protein transferase blocks the anchorage-dependent and -independent growth of human tumor cell lines. *Cancer Res* 1995;55:5302-09.
130. Kohl NE, Omer CA, Conner MW, et al: Inhibition of farnesyltransferase induces regression of mammary and salivary carcinomas in ras transgenic mice. *Nat Med* 1995;1:792-97.
131. Feldkamp MM, Lau N, Roncari L, Guha A: Isotype-specific RasòGTP-levels predict the efficacy of farnesyl transferase inhibitors against human astrocytomas regardless of Ras mutational status. Submitted 2000.
132. Feldkamp MM, Lau N, Guha A: Growth inhibition of astrocytoma cells by farnesyl transferase inhibitors is mediated by a combination of anti-proliferative, pro-apoptotic, and anti-angiogenic effects. *Oncogene* 1999;18:7514-26.
133. Feldkamp MM, Lau N, Rak J, Kerbel RS, Guha A: Normoxic and hypoxic regulation of Vascular Endothelial Growth Factor (VEGF) by astrocytoma cells is mediated by Ras. *Int J Cancer* 1999; 81:118-24.
134. Kokunai T, Urui S, Tomita H, Tamaki N: Overcoming of radioresistance in human gliomas by p21WAF1/CIP1 antisense oligonucleotide. *J Neurooncol* 2001;51:111-19.
135. James G, Goldstein JL, Brown MS: Resistance of K-RasBV12 proteins to farnesyltransferase inhibitors in Rat1 cells. *Proc Natl Acad Sci U S A* 1996;93:4454-58.
136. Lerner EC, Qian Y, Blaskovich MA, et al: Ras CAAX peptidomimetic FTI-277 selectively blocks oncogenic Ras signaling by inducing cytoplasmic accumulation of inactive Ras-Raf complexes. *J Biol Chem* 1995;270:26802-6.
137. Pollack IF, Randall MS, Kristofik MP, Kelly RH, Selker RG, Vertosick Jr FT: Response of malignant glioma cell lines to activation and inhibition of protein kinase C-mediated pathways. *J Neurosurg* 1990;73:98-105.
138. Couldwell WT, Hinton DR, Law RE: Protein kinase C and growth regulation in malignant gliomas. *Neurosurgery* 1994;35:1184-86.
139. Pollack IF, Randall MS, Kristofik MP, Kelly RH, Selker R, Vertosick Jr FT: Effect of tamoxifen on DNA synthesis and proliferation of human malignant glioma lines in vitro. *Cancer Res* 1990;50:7134-38.
140. Baltuch GH, Couldwell WT, Villemure J-G, Yong VW: Protein kinase C inhibitors suppress cell growth in established and low-passage glioma cell lines. A comparison between staurosporine and tamoxifen. *Neurosurgery* 1993;33:495-501.
141. Couldwell WT, Weiss MH, DeGiorgio CM, et al: Clinical and radiographic response in a minority of patients with recurrent malignant gliomas treated with high-dose tamoxifen. *Neurosurgery* 1993;32:485-9; discussion 489-90.
142. Couldwell WT, Hinton DR, Surnock AA, et al: Treatment of recurrent malignant gliomas with chronic oral high-dose tamoxifen. *Clin Cancer Res* 1996;2:619-22.
143. Pollack IF, Kawecki S, Lazo JS: Blocking of glioma proliferation in vitro and in vivo and potentiating the effects of BCNU and cisplatin: UCN-01, a selective protein kinase C inhibitor. *J Neurosurg* 1996;84:1024-32.
144. Folkman J: Angiogenesis and angiogenesis inhibition: an overview. *Exs* 1997;79:1-8.

145. Cheng SY, Huang HJ, Nagane M, et al: Suppression of glioblastoma angiogenicity and tumorigenicity by inhibition of endogenous expression of vascular endothelial growth factor. *Proc Natl Acad Sci U S A* 1996;93:8502-07.
146. Saleh M, Stacker SA, Wilks AF: Inhibition of growth of C6 glioma cells in vivo by expression of antisense vascular endothelial growth factor sequence. *Cancer Res* 1996;56:393-401.
147. Millauer B, Shawver LK, Plate KH, Risau W, Ullrich A: Glioblastoma growth inhibited in vivo by a dominant-negative Flk-1 mutant. *Nature* 1994;367:576-79.
148. Puduvalli VK, Sawaya R: Antiangiogenesis — therapeutic strategies and clinical implications for brain tumors. *J Neurooncol* 2000;50:189-200.
149. Fine HA, Figg WD, Jaeckle K, et al: Phase II trial of the antiangiogenic agent thalidomide in patients with recurrent high-grade gliomas. *J Clin Oncol* 2000;18:708-15.
150. Cha S, Knopp EA, Johnson G, et al: Dynamic contrast-enhanced T2-weighted MR imaging of recurrent malignant gliomas treated with thalidomide and carboplatin. *AJNR Am J Neuroradiol* 2000;21:881-90.
151. Meneses PI, Abrey LE, Hajjar KA, et al: Simplified production of a recombinant human angiostatin derivative that suppresses intracerebral glial tumor growth. *Clin Cancer Res* 1999;5:3689-94.

28

Immunomodulation of Gliomas

S Chaudhuri, S Sarkar, S Chaudhuri

INTRODUCTION

Tumors of the central nervous system (CNS) occupy a unique position within the spectrum of human cancers, and represent often 'difficult to treat' neoplasm. Nevertheless, the conventional therapeutic approaches such as surgery, radio – and chemotherapy cause a generalised immunosuppression in addition to the lesion in the brain.[1-3] Therefore, a new therapeutic approach with a compatible tolerance profile are needed for the treatment of brain tumors. Immunotherapy may offer such a form of therapy and agents called biological response modifiers (BRMs) are gradually occupying an interesting position as immunotherapeutic adjunct.[4]

Sheep red blood cells (SRBC), long been used as a classical antigen, have been shown to exert an immunopotentiating and antitumor property in experimental animals.[5-6] The active component of SRBC which is responsible for such effects was found to be a cell surface glycoprotein molecule, known as 'T11 target structure' (T11TS) or sheep form of LFA3 or more currently CD58 that binds with the T11 (CD2) molecules of the lymphocyte.[7-10] Although some of its immunological role have been discussed, surprisingly no reports on disease relevance or function of such molecule in intact animals have yet been made.[11]

In the present course of investigations, attempts have been made to isolate and purify the T11TS molecule from the SRBC membrane and to investigate its effects on rats with experimentally induced brain tumor.[12-15] Histological studies have been made to ascertain the tumor growth following ENU administration and tumor inhibitory and/or abrogatory effect of the purified T11TS (CD58) molecule following administration, which has also been investigated through cytokinetic studies. Simultaneous studies towards time-dose response, CMI functions and flow cytometric analysis have been made to identify active involvement of the immunocytes across the blood brain barrier as well as the peripheral immune system. Modulation of 'CD2 molecules' on the microglial cells has been investigated through rosette and rosette inhibition and their surface ultrastructural characteristics. Moreover, receptor analysis studies with flow cytometric scanning has also been performed on purified microglial cells to study their activation state, if any, with T11TS administration. Scope for successful trials of T11TS as sheep form of LFA3 for anti-tumor immunostimulatory therapeutic agent has been sought for.

MATERIALS AND METHODS

Preparation of glycopeptide (T11TS) from sheep red blood cell

The method followed steps slightly modified from that of Kitao et al.[16] Briefly, 1ml volume of packed sheep red blood cells have been incubated for one hour at 37°C in the presence of trypsin-phosphate buffer (100mg/ml). The red tinted supernatant was removed and then treated with one-quarter volume of 25% trichloroacetic acid to precipitate the non-specific proteins. The clear supernatant was obtained by centrifugation, neutralized with NaOH, and dialyzed against distilled water. Since the glycoprotien is acidic in nature, it was separated from neutral peptides by ion exchange chromatography on a DEAE-Cellulose column (1.5 X 8 cm) previously equilibrated with 0.05M formate buffer, pH 6.8. The acidic glycopeptide was then eluted with a five chamber gradient system containing 1 ml each of (1) water, (2) 0.05M formic acid (3) 0.2M formic acid (4) 0.4M formic acid and (5) 0.4M formic acid in 0.3M sodium chloride. Fractions of each elute (3 ml) were collected and analyzed for absorbance at 280 nm as well as for sialic acid. Finally, rosette inhibition assay was performed with variable concentrations of glycopeptide. Protein estimation of the elutes were determined by Lowry's method,[17] and also the absorbance emission of different elutes were determined at 280 nm.

Animals

Healthy rats of Druckrey strain of both sexes (strain originally supplied by Central Drug Laboratory, Kolkata, India and maintained subsequently in our laboratory) formed the materials of the whole investigations. They were grouped as follows: (I) Normal untreated controls (N), (II) Normal-T11TS: Normal animals receiving (1 ml, i.p.) of T11TS fraction, (III) ENU: Animals received 80 mg/kg body wt. of ethyl nitrosourea (ENU) 7-10 days after birth (i.p.). Group (IV), (V) and (VI) received 1 ml, 2 ml and 3 ml of T11TS fraction respectively at the end of 7th month of ENU administration and assigned as ET1, ET2 and ET3 respectively. Group (VII) consisted of rats injected with 0.5 ml of 7% SRBC (PCV/Vol.). The animals, comprising of 20 in each group were weaned at 30 days of age and housed separately in groups of five animals in isolated cages. All animals were fed autoclaved Hind-Lever pellet and water *ad libitum*. Rats were examined daily and weighed weekly throughout the experimental period. Maintenance and animal experiment procedure strictly followed "Principles of laboratory animal care" (NIH) and also local "ethical" regulations.

ENU-administration

ENU was freshly prepared by dissolving 10 mg/ml in sterile saline and adjusting the pH to 4.5 with crystalline ascorbic acid. ENU was injected intraperitoneally (i.p.) to 100 animals with an acute dose of 80 mg/kg body weight in the first week after birth.[12-15]

Rosette formation

Lymphocytes were separated from spleen cell suspension on a percoll density gradient elution method,[5,18] 0.25 ml of $3 - 4 \times 10^6$ lymphocytes were mixed with 0.25 ml of 1% (PCV/saline volume) sheep erythrocytes (SRBC) and incubated at 37°C for 15 minutes. Following brief centrifugation, the preparation was kept at 4°C overnight. Number of rosettes formed were counted per 200 lymphocytes and expressed as rosette %. *Microglia cells* were separated from adherent brain cells on a Nycodenz gradient (1.068 specific gravity) (Nycomed, Oslo, Norway) and rosette study was done in the similar way as above.[19]

Rosette inhibition assay

Selection of the elute by maximum CD2 ligand binding assay was performed with each elute fraction (I – V). 0.25 ml of $3 - 4 \times 10^6$ / ml *lymphocytes* (splenic/blood) were mixed with 0.5 ml, 1 ml and 1.5 ml portions of glycopeptide fractions in different tubes. They were then incubated at 37°C for 15, 30 and 60 minutes, washed off twice with PBS, and finally, pellets were suspended in 0.25 ml of media (RPMI-1640) without FBS (fetal bovine serum). The treated lymphocytes were then subjected to rosetting technique described as above. The elute fraction forming minimum number of rosettes was considered as the richest source of T11TS. Similarly rosette inhibition assay was performed with the *microglial cells*.

Administration of T11TS fraction

The fraction of choice was that having the highest absorbence peak, and also minimum rosette forming capacity as determined by rosette inhibition assay. The dose of injected volume was calibrated from the time-dose responsiveness of rosette inhibition assay. The protein content of the elutes has been determined and also a relation with the body weight was established. The first dose 1 ml of T11TS (i.p.) from the third elute fraction (EFIII) was followed by second booster dose on the sixth day and third booster dose on the 12th day, making a dose schedule of 1 ml, 2 ml, and 3 ml to the group IV, V and VI animals respectively.

Survival studies

Of all the animals prepared as above, observations were made to account for the total number of days survived by individual animals and the mean survival time in each group were determined. Further, progressive neurologic signs and weight loss were taken into account in selecting the animals for the tumor development study at the initial phase. Moreover, histological studies of brain of such animals were made to ascertain the study.

Histological Studies

Portions of brain tissues from respective group of animals were prepared for routine histological studies: tissues were fixed in 10% formal-buffer overnight and finally dehydrated and embedded in paraffin through histokinet processing. Sections were cut at 5µ thickness and finally stained with routine haematoxiline / eosine.

Cell surface immunofluorescence and fluorescence activated cell sorting

Splenic lymphocytes separated on a percoll gradient[18] and *microglial cells* separated on a Nycodenz gradient were washed twice with PBS and final volume adjusted to 1×10^7 cells, incubated in dark with FITC conjugated CD25 monoclonal antibody (5µ) (Beckton Dickinson, USA). After 30 minutes of incubation the cells were washed thrice with PBS, and analysed on FACS calibur (Beckton Dickinson, USA) (Argon Laser, excitation at 488 nm, 515 band pan filter). Whole brain cells were also tagged with anti-CD25-FITC conjugate and in both the preparations isotype control (IgGκa) was maintained. In total 10,000 events were acquired and data were analysed using Cell Quest software (Beckton Dickinson, USA). In a separate set of experiment, in order to assess the infiltration of the lymphocytes in the brain compartment, whole brain cells were again tagged with anti-CD4-FITC conjugate and isotype control (IgGκa) was also maintained. The cells were gated with the leukogate software and positive controls of lymphocytes as well as microglial cell markers were maintained.

Cytokinetic study of brain tumors

The whole brain homogenate was prepared as described before from each group of animals, namely the normal (N), the ENU treated (E) and those having both ENU and T11TS (Ist dose) (ET1). The cells from different groups as prepared in a homogeneous suspension were counted in the order of 1×10^5 cells/ml and then incubated with 10 µg of HO-33342 fluorochrome dye for 24 hours in PBS. Following several washes the cells were resuspended in PBS and fluorescence activity of the respective cell mass measured in a spectrofluorimeter (Hitachi, Japan.[20]

Functional Assay for Cellular Immune responses

Lymphocytes, macrophage, PMN, seperated from the spleen and microglial cells from brain tissue of animals of different groups were subjected for cellular immune response studies.

    a) Rosette Formation: Lymphocytes isolated from spleen cell suspension by percoll density gradient and the microglial cells seperated from the brain tissue were assayed for spontaneous E-rosette formation by methods described as above.

b) **Lymphocyte Mediated Cytotoxicity(CTL Assay):** A newer approach to this method has been adopted using a fluorochrome dye Hoechst assay was performed by maintaining an effector – target ratio at 10:1 through an incubation (37°C, 4% CO_2 – air environment) period of 18 hours. Fluorochrome released as per target lysis measured in a spectrofluorimeter (Hitachi, Tokyo) provided an index of cytotoxic efficacy of effectors.[20-21]

c) **PMN – mediated phagocytosis:** Splenic tissues was obtained from group I to VII animal groups. Single cell suspension were prepared by teasing splenic tissues with forceps, followed by percoll density gradient centrifugation. 3 ml of cell suspensions was layered on 5 ml of percoll gradient (density 1.089) and centrifuged for 20 minutes at 800g. The PMN layer was then removed from the interface, washed thrice with PBS and finally suspended in 1 ml of media. Then neutrophills (PMNs) were allowed to react with target tumor cells in presence of Nitroblue tetrazolium (NBT) for 18 hours. Reduction of yellow NBT to blue formazan indicated the extent of phagocytic burst by effectors concerned.[22]

d) **Macrophage (Mϕ) mediated phagocytosis:** Single cell suspension was prepared from the splenic tissue as described above. The cell suspensions were incubated in plastic petridish (Corning, USA) in a CO_2 incubator for 30 minutes for adherence. The non-adherent cells were then washed off and Mϕ were separated by washing thrice with PBS-EDTA. The cells were then washed thrice with PBS and suspended in 1 ml of media. The macrophages were then allowed to react with the target tumor cells in presence of NT as described above.[23]

Inhibition Studies with elute fraction III

Lymphocytes, PMN and macrophages (purified, as described before) from the normal splenic tissue (group I) as well as microglial cells from normal brain were incubated each separately with 0.5 ml of elute fraction EF-III for 60 minutes. Cells were washed thrice with PBS and suspended in 1 ml of media (RPMI-1640). Each type of cells were then used as effector cells in functional assays for immune parameter such as CTL, PMN mediated phagocytosis.

SRBC control study

For each immunological parameters effects of whole SRBC (7% PCV/Vol, 0.5 ml) have been evaluated along with.

Preparation of cells for SEM study

The cells thus obtained were prepared for SEM study as described below. They were fixed in 2.5% gluteraldehyde overnight at 4°C followed by washing in PBS and dehydration in graded alcohol (50% onwards) and finally brought to 100% acetone. The cells thus prepared were then coated with gold for 2 mins. (35 ma, 1KV) in a diode sputtering system and viewed under SEM ((Jeol JSM 5200) at 25KV beam voltage. Each specimen was scanned for 64 sec. and photographed at direct magnification at 1500X to 7500X.

RESULTS

Protein Content of elutes

The elutes from the DEAE-cellulose column, were collected as a series of fraction EF I, EF II, EF III, EF IV and EF V and analysed for absorbance at 280nm which showed a higher O.D. value for fraction EF III (0.346) compared to those for another fractions **(Table 1)**. This indicates the total extent of protein content from the fraction EF III and it further shows that EF III is a high density protein. To estimate the protein content qualitatively from this fraction, Lowry's method[15] was performed and the amount determined was found to be 50 μg/ml. Consequently, the amount of T11TS administered per Kg body weight was calculated to be 0.41 mg/ Kg body weight which represent the content in 1 ml.

Table 1 : *Spectrophotometric reading (O.D. at 280 nm) of different elute fractions (EF) isolated from SRBC membrane. EFIII showed the highest O.D.*

Elute Fractions	$O.D._{280}$
EFI	0.001
EFII	0.153
EFIII	0.346
EFIV	0.319
EFV	0.131

Spontaneous E-Rosette Formation

Lymphocytes separated from the splenic tissue of the normal animal which formed rosettes with sheep erythrocytes, counted as number of rosettes per 200 lymphocytes and was found to be 19% ± 4.31 while in case of *microglia* the value was found to be 21.33±3.01%.

Rosette Inhibition Assay

Rosette inhibition assay carried out for the identification of T11TS–glycoprotein part from the elute fractions (EF I to EF V), and the minimum number of rosette forming elute was determined as the chosen elute fraction. Although rosette inhibition was found in all the fractions, EF III exhibited complete inhibition **(Table 2)**.

Table 2 : *Represent "rostte inhibition assay" by incubating EFI to EFV in vitro with lymphocytes for 60 minutes. EFIII showed complete inhibition of rosette formation and was considered the fraction of choice.*

Fraction	Rosette/ 200 lymphocytes	% Rosettes
Normal Control	38	19%
EFI	2	1%
EFII	7	3.5%
EFIII	0	0%
EFIV	9	4.5%
EFV	11	5.5%

Optimum Time and Dose Requirement for the Glycopeptide Activity

The optimum time for rosette inhibition tested with EF III at 15, 30 and 60 minutes intervals showed its maximum efficacy at 60 minute. And the optimum concentration, as determined through rosette inhibition assay, tested with EF III 0.5 ml, 1 ml and 1.5 ml, showed same level of inhibition with all the doses (complete inhibition). For optimisation 1 ml was considered to be the chosen dose.

Survival studies

The effect on survival following administration of SRBC in rats with both acute and fractionated of ENU treated groups were studied. The average survival for the normal control animal was found to be 705±35 days(n= 10). In contrast, the average survival value for the ENU treated animals was significantly ($p < 0.001$) reduced to 190±30 days. The administration of 1st, 2nd and 3rd booster doses of 1ml. of purified T11TS fraction into ENU –treated animals elevated the mean survival period to 640±25 days, 675± 20days and 700±30 days respectively. SRBC administration in such animals was found to be 650±35 days. The values were compared and found to be significantly greater than that of ENU group **(Fig. 1)**

HISTOLOGICAL EVIDENCES

Histological studies show normal glial cell populations in white matter of cerebral cortex **(Fig. 2a)** with few astrocytes oligodendrocytes and a few neurons. The effect of ENU showed grade 4 oligodendroglioma with mitotic figure, giant cells and absence of intercellular spacing **(Fig. 2b)**. Effect of first dose of T11TS fraction showed reduced glial cell population with enlarged nucleie due to increased permeability of nuclear membrane to water. Most importantly reduction in the number of oligodendroglioma cells with spongiosis and appototic figures,

SURVIVAL DATA OF ENU INDUCED DRUCKRAY RATS FOLLOWING T11TS ADMINISTRATION

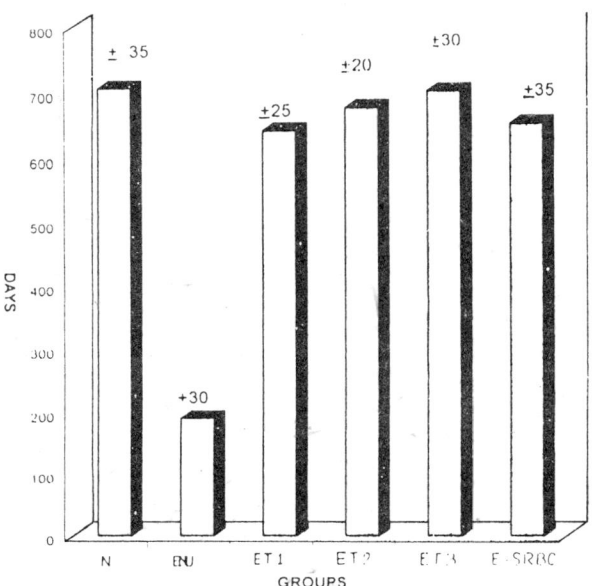

Fig. 1 : Survival Date of ENU induced Druckray compared with normal (N) and T11TS treated ENU rates in 3 sucessive doees denoted as (ET1), (ET2), and (ET3) and also data shows END—SRBC treated rats. Better survival noted in (ET3) rats.

lymphocyte infiltration and margination have been observed with this dose **(Fig. 2c)**. The hypocellularity with oedema and degenerative changes and presence of lymphocytes have been observed following administration of second dose of T11TS fraction **(Fig. 2d)**. The third dose of T11Ts fraction shows return to the normal glial features with gliosis. Evidence of necrosis and calcification was noted **(Fig. 2e)**. Finally the effect of SRBC has been observed through reduced oligodendroglial cell population with gliosis nearly same that of first dose of T11TS fraction **(Fig. 2f)**.

Hyperkinetic malignant cells of brain are regulated by T11TS

Compared to the healthy normal control, the high fluorescent spectra (six times) of ENU treated rat brain cells presented a hyperkinetic malignant feature of brain cells concerned **(Fig. 3)**. Administration of T11TS *in vitro* was able to limit the cytokinetic process of the ENU treated brain cells at normal physiology.

IL-2R (CD25) as an indicator for T-Cell Activation

The effects of T11TS fraction on **CD25** *expression on splenic T lymphocytes* in

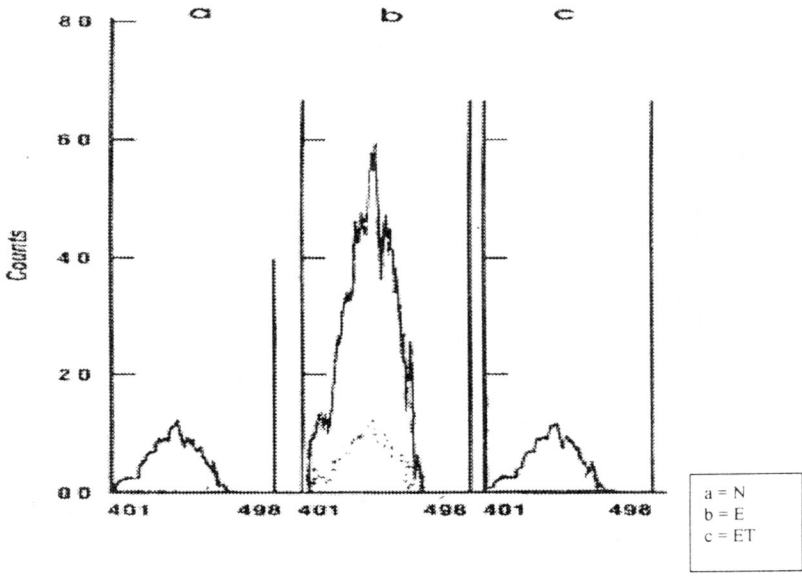

Fig. 3 : Fluorescent spectral analysis of brain cell kinetics in normal (N), following ENU mediated tumor indusction (8 months) (E) and after T11TS administration in ENU (E) animals. The hyperkinetic "E" cells were brought back to near normal level with T11TS (ET).

animals with experimentally induced tumor (E) (with ENU) revealed interesting results. The CD25 (IL-2R) expression on splenic T cells of normal group of animals (Gr. I) was 2.66%. The lymphocytes of the group III animals (ENU-group) showed a marked decrease in IL-2R expression in comparison to the normal ones. The group IV (ET1), i.e. the tumor bearing animals that received a single booster dose of 1 ml. of T11TS-fraction, showed a highly significant increase in IL-2R expression on splenic lymphocytes (97.06%) whereas group V and VI (ET2, ET3) were found to be poorly responding (3.46% and 1.34% respectively) **[Figs. 4(a–g)]**.

FLOWCYTOMETRIC ANALYSIS OF BRAIN INFILTRATED LYMPHOCYTES AND MICROGLIAL CELLS OF THE BRAIN

The search for **CD25** expression in whole brain cells further revealed activation of a population of **CD25+** *lymphocytes* which were gated with the leukogate software and positive controls of lymphocytes were maintained with lymphocyte markers. In the normal animals though the lymphocyte population in brain was small, 79.26% showed CD25 [**Fig. 5(a)**] positivity. In ENU treated animals though the infiltrated lymphocytes were greater, activation was greatly reduced (10.69%) [**Fig. 5(b)**]. But T11TS treatment in ENU animals with the 1st, 2nd and 3rd booster

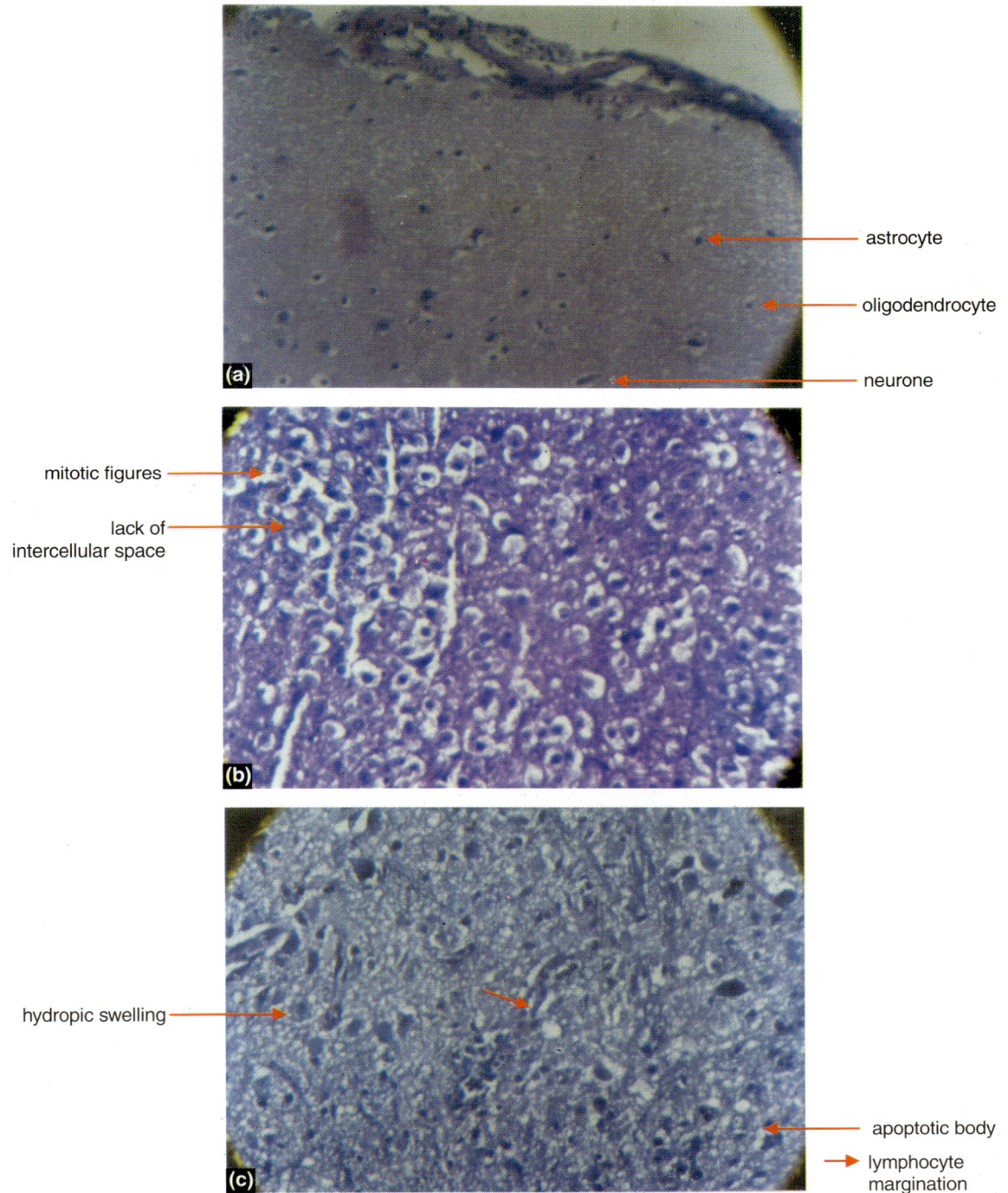

Figs.2: Hitological features of normal(N) (a) and ENU-treated brain (ENU) (b) and following administration of 1st (c), 2nd (d), and 3rd (e) booster doses of T11TS administration in ENU-treated animals, (f) section showing the effect of sheep red blood cells in ENU-treated animals.

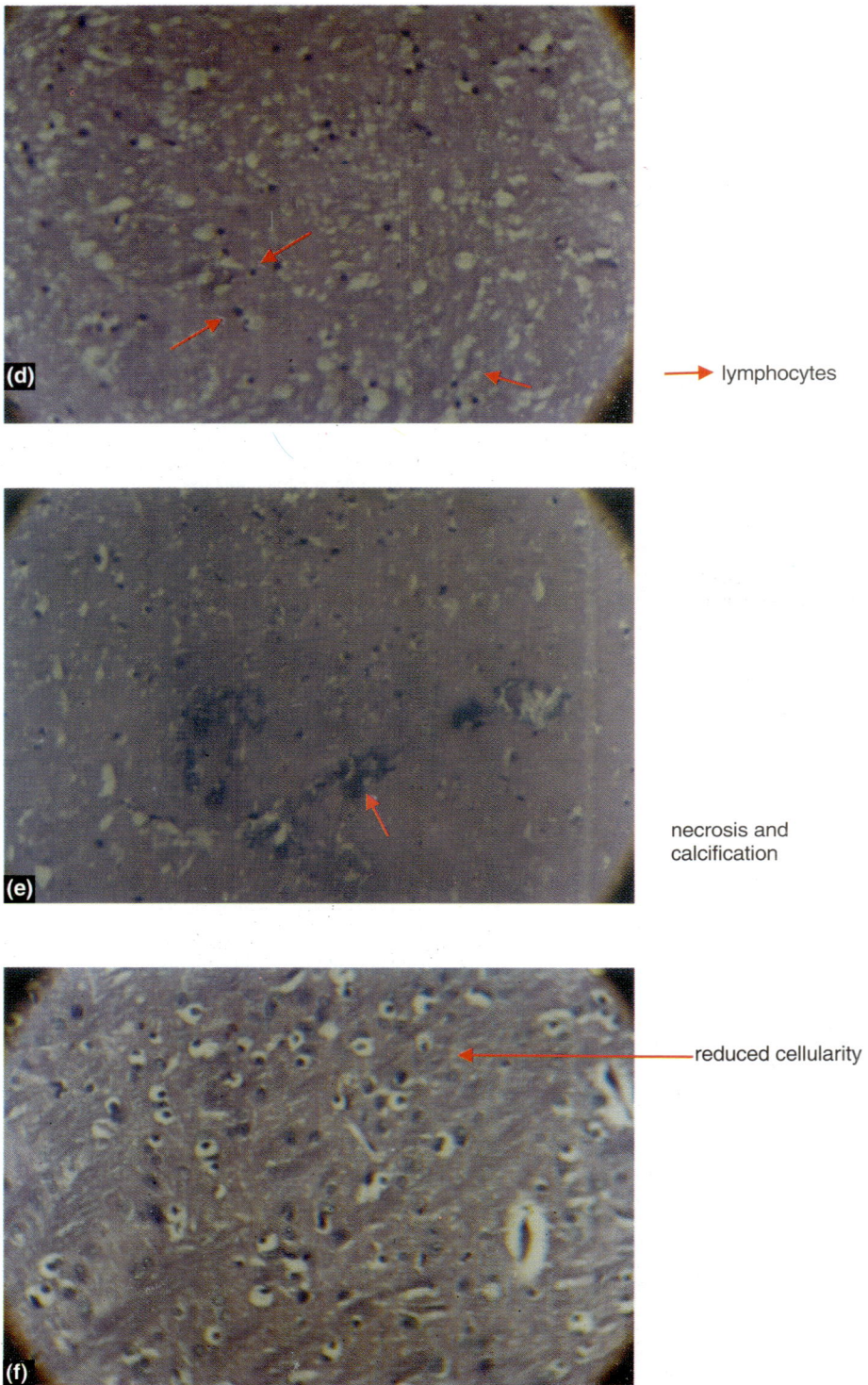

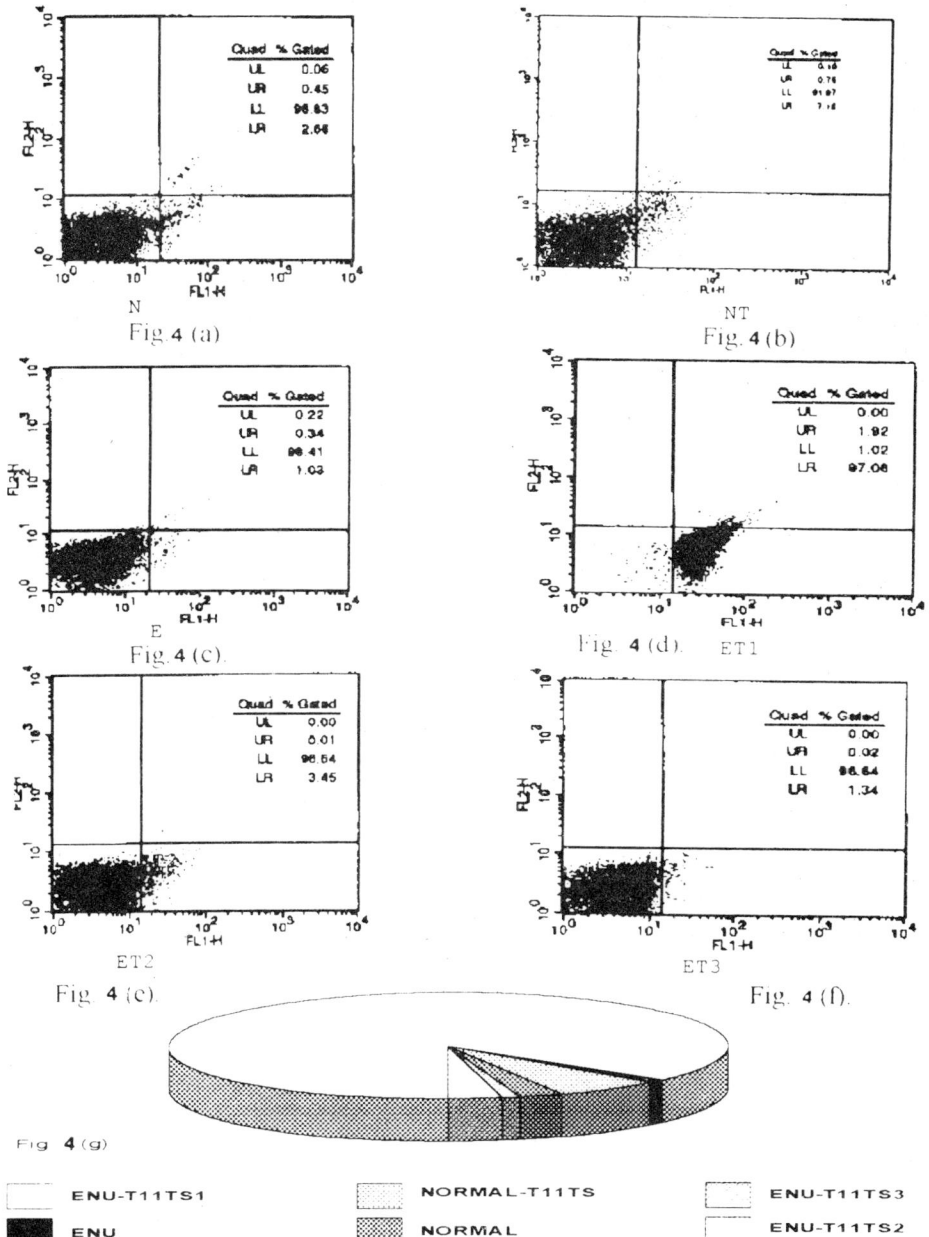

Figs. 4 : Effects of T11TS administration (ET1-ET3) on IL-3R (CD25) expression on splenic lymphocytes (a) N, (b) NT, (c) E, (d) ET1, (e) ET2, (f) ET3): cells with ET1 showed maximum CD25 expression as revealed from flowcytometric analysis (FACS-Calibur, Beckton—Dickinson, USA)

Fig. 4(g) : Pie-chart showing percentage distribution of CD25 expression on splenic lymphocites at different T11TS booster doses.

doses showed increasing activation or CD25 positivity of the lymphocytes, compared to the ENU group [Figs. 5(c–e)]

Flowcytometric analysis of the *infiltrated lymphocytes* in the brain compartment showed 1.6% infiltration in the normal untreated control; of which 76.61% was found to be **CD4+ lymphocytes** and 25.81% CD4– cell types. In ENU treated animals, the infiltration was greater than the normal untreated control (4.18%), whereas CD4+ lymphocytes was 62.28% and 36.72% was CD4– cell types. Following 1st dose of T11TS treatment in ENU animals, the infiltration was found to be increased to 6.35% and was greater than that of normal and ENU animals, whereas CD4+ lymphocytes in the brain was 34.32% and CD4– cell type was 65.25%. The administration of 2nd booster dose of T11TS fraction in ENU treated

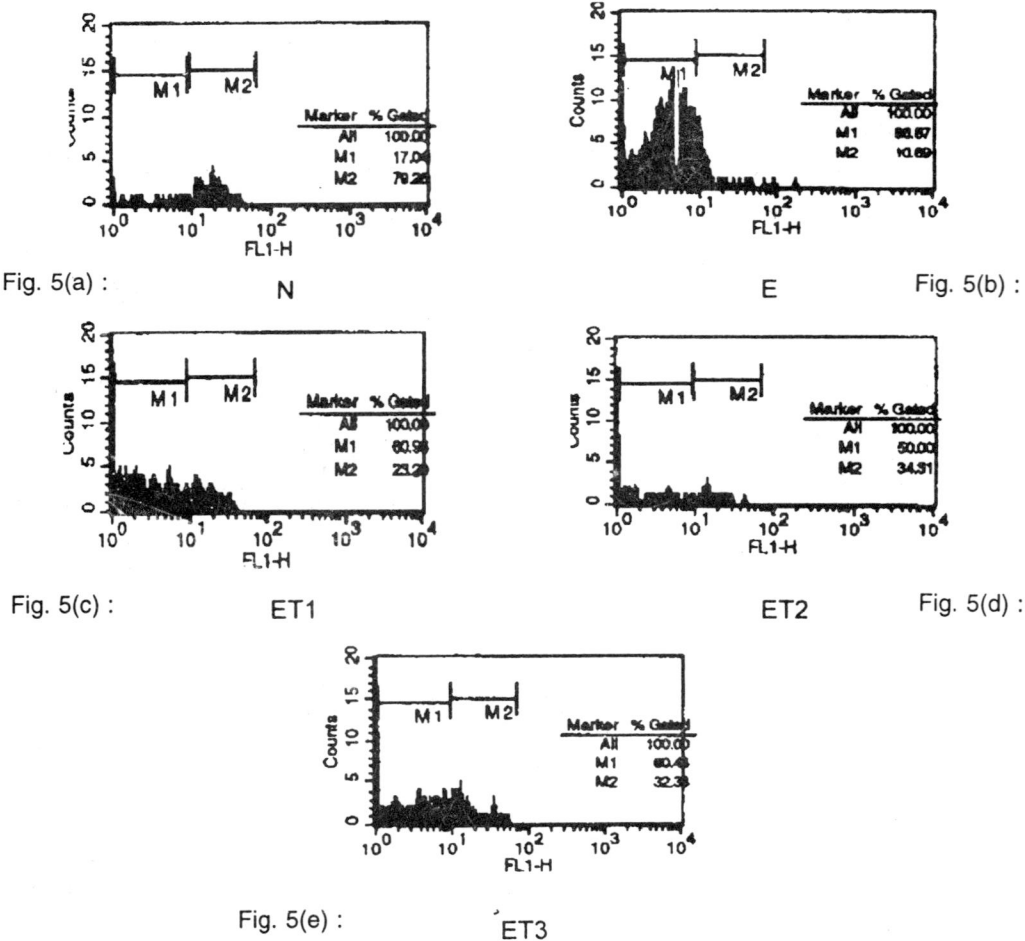

Figs. 5 : T11TS administraction *in vivo* showing activation of brain infiltrated lymphocytes (5a)-N, (5b)-E, (5c)-ET1, (5d)-ET2, (5e)-ET3).

animals showed the highest degree of infiltration among all the experimental groups which was 13.53% with 30.13% CD4+ lymphocytes and 68.58% CD4– cells. No further enhancement of infiltration was observed following 3rd booster dose of T11TS fraction in ENU treated animals, rather it was reduced almost to the normal level with 1.38%, in which 13.66% cells were found to be CD4+ and 86.34% were CD4– lymphocytes **[Figs. 6(a-e)]**.

Exciting results shown by the IL-2R (CD25) expression on splenic and brain infiltrated lymphocytes following T11TS administration in tumor bearing animals provided impetus to search further for the IL-2R **(CD25)** *expression on the brain microglial cells*—the chief immunomodulatory cells of the brain. The microglial cells seperated from the normal brain tissue showed 1.52% CD25 expression. In

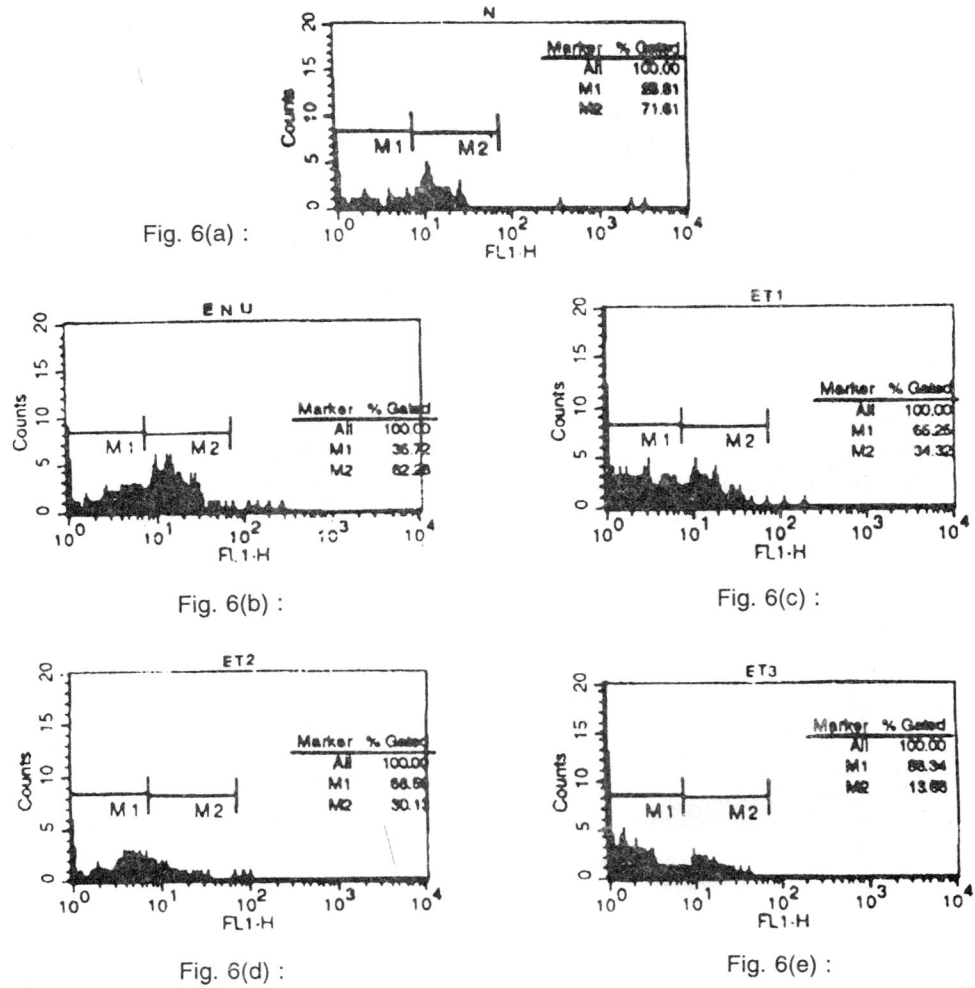

Fig. 6(a) :

Fig. 6(b) :

Fig. 6(c) :

Fig. 6(d) :

Fig. 6(e) :

Figs. 6 : (a–e) Flowcytometric analysis of brain infiltrated lymphocytes.

ENU treated animals expression was found to be 9.05%, which was significantly greater than the normal untreated control. Further upregulation of CD25 (IL-2R) expression was noted followig 1st and 2nd booster doses of 1 ml of T11TS fraction in ENU treated animals. The 2nd booster dose demonstrated the highest receptor (CD25) expression on microglial cells (33.36%) which was greater than the normal untreated control and ENU-treated group. However, no further upregulation of CD25 receptor was observed with the third booster dose, and was 2nd booster dose **(Figs. 7(a-e)]**.

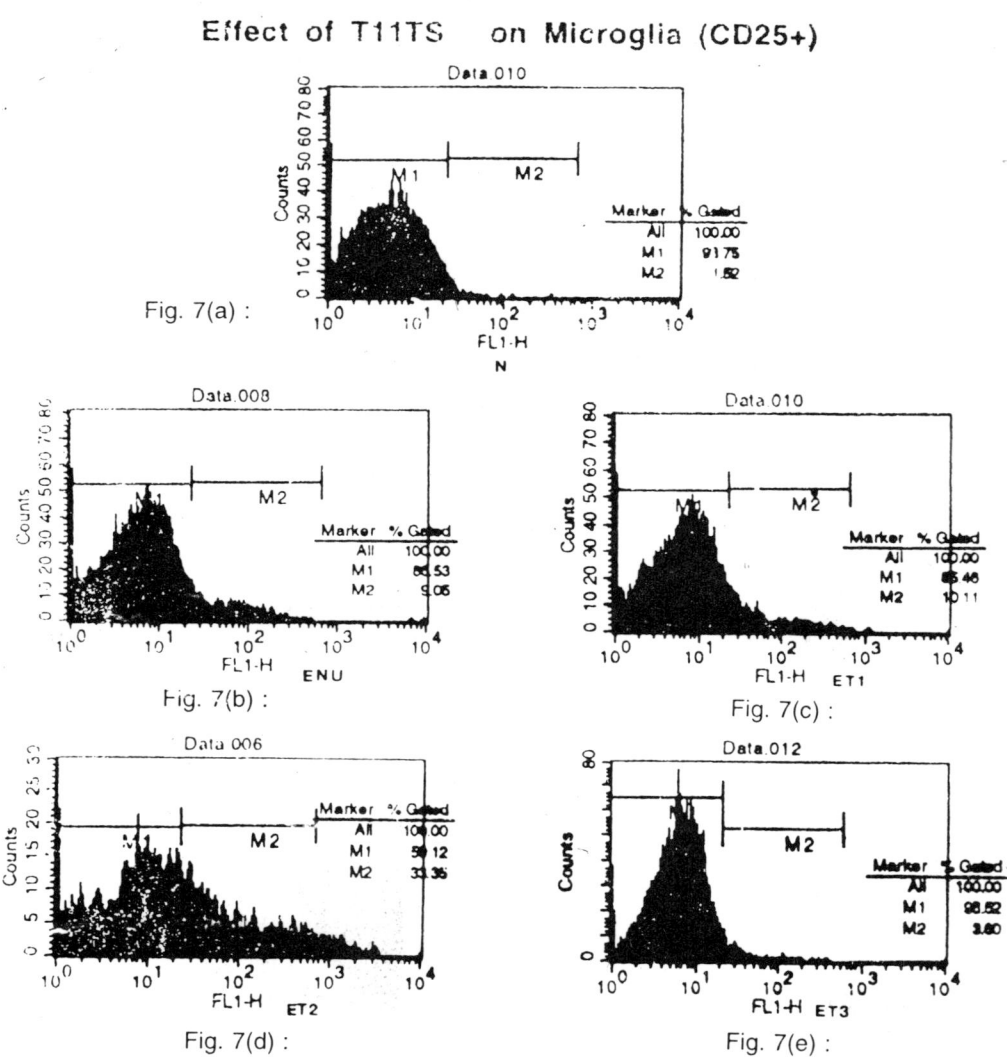

Figs. 7 : (a–e) Effects of T11TS administration (ET1-ET3) on CD25 (IL-2R) expression on microglial cells of the brain. Cells with ET2 showed maximum CD25 expression as revealed from flowcytometric analysis.

IMMUNE PROFILE OF T11TS TREATED ANIMAL

a) E-Rosette Formation:

Rosetting with Lymphocytes: The normal animal group (Gr.I) showed its rosette forming capacity to be 19% ± 4.31 and in T11TS injected normal animals it was 24% ± 3.92. The ENU animal groups (Gr.III) showed significant decrease in rosette forming capacity, which is 2% ± 1.03 ($p < 0.001$). The group IV, ENU-T11TS (ET1), showed significant increase in rosetting capacity from that of the ENU-control ($p < 0.001$) and was nearly the normal value. The rosetting capacity observed in group V animals, receiving the 2nd booster dose of T11TS (ET2) was 42.75% ± 7.21 which was significantly higher ($p<<0.001$) than the ENU group as well as normal control and was almost double to that found in group IV. Although the splenic lymphocyte of group VI animals showed greater rosetting capacity than Gr. I and Gr. II but it was somewhat lower than group V. On the other hand, the rosetting capacity following administration of 0.5 ml. of 7% SRBC – ENU animals was shown to be 20% ± 2.84 **(Fig. 8)**.

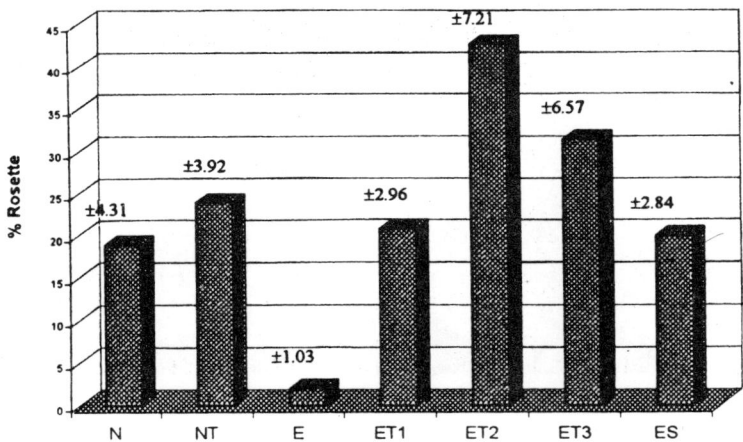

Fig. 8: *In vivo* effects of T11TS administration on rosette forming lymphocytes of different groups of rats. Second booster dose (ET2) produced maximum rosette forming efficacy in splenic lymphocytes of ENU treated animals.

Rosetting with Microglia (MG): separated on Nycodenz gradient, from normal untreated rat brain demonstrated rosette forming capacity with a value of 21.33±3.01%. When 1 ml. Of T11TS fraction was administered into such animals (i.p.), the number of rosette forming microglia was found to be increased to 78.33±1.04%. The number of rosette forming microglia was significantly reduced ($p< 0.001$) in ENU treated tumor bearing animals (3.50 ± 0.50%), in comparison to normal untreated control as well as normal T11TS treated animals. Interestingly, the number of rosetting microglia was significantly (p<0.001) increased to 96.16±1.25% in ENU group of animals when 1st booster dose of T11TS was administered, and was significantly greater than the ENU and normal control group. However, number of rosetting microglia was gradually reduced following administration of 2nd and 3rd booster dose of T11TS fraction in ENU treated animals with values of 12.83 ±0.76% and 4.83±1.04% respectively. In addition, *in vitro* assay, T11TS fraction directed against the microglial cells showed complete inhibition of rosette formation **(Fig. 9)**.

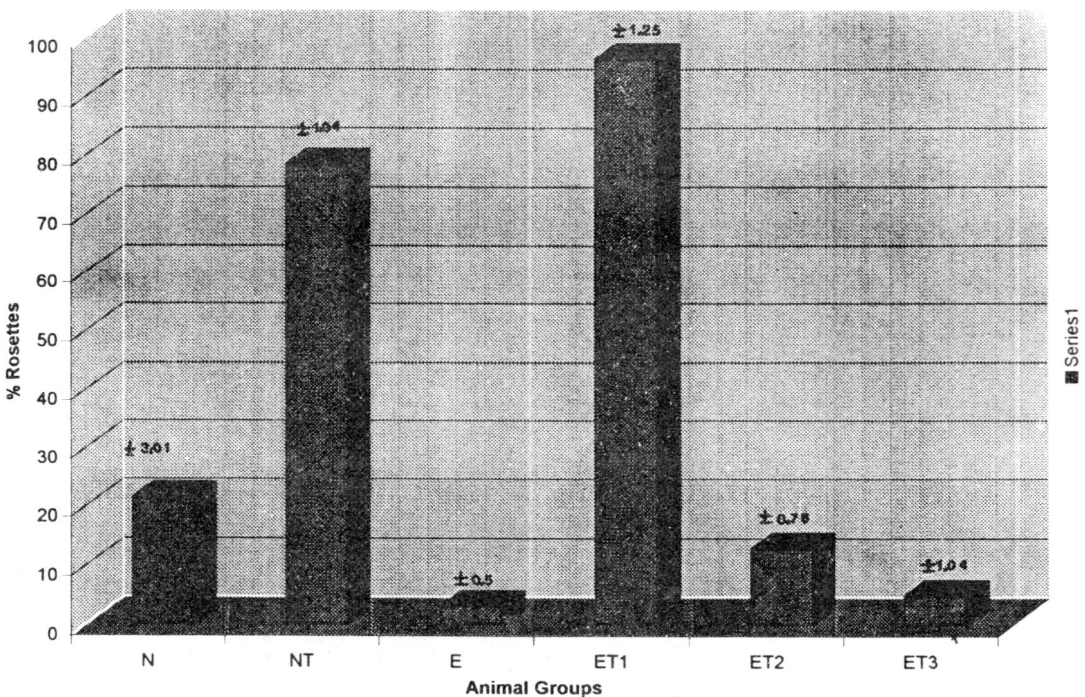

Fig. 9 : *In vivo* effects of T11TS administration on rosette forming microglia of different groups of rats. 1st dose (ETI) showed maximum rosette forming efficacy in brain microglia of ENU–treated animals.

SEM study with *microglial cells* with and without the application of T11TS in normal and different experimental groups revealed interesting features: Microglial cells forming rosettes with SRBC, sometimes super rosettes with clear anchorage. with SRBC have been observed. T11TS administration in ENU-treated groups showed rosetting microglia(s) (activated microglial cells), some of them liberating secretory molecules, extending large projections, pseudopodias **(Figs. 10(a)-c)**.

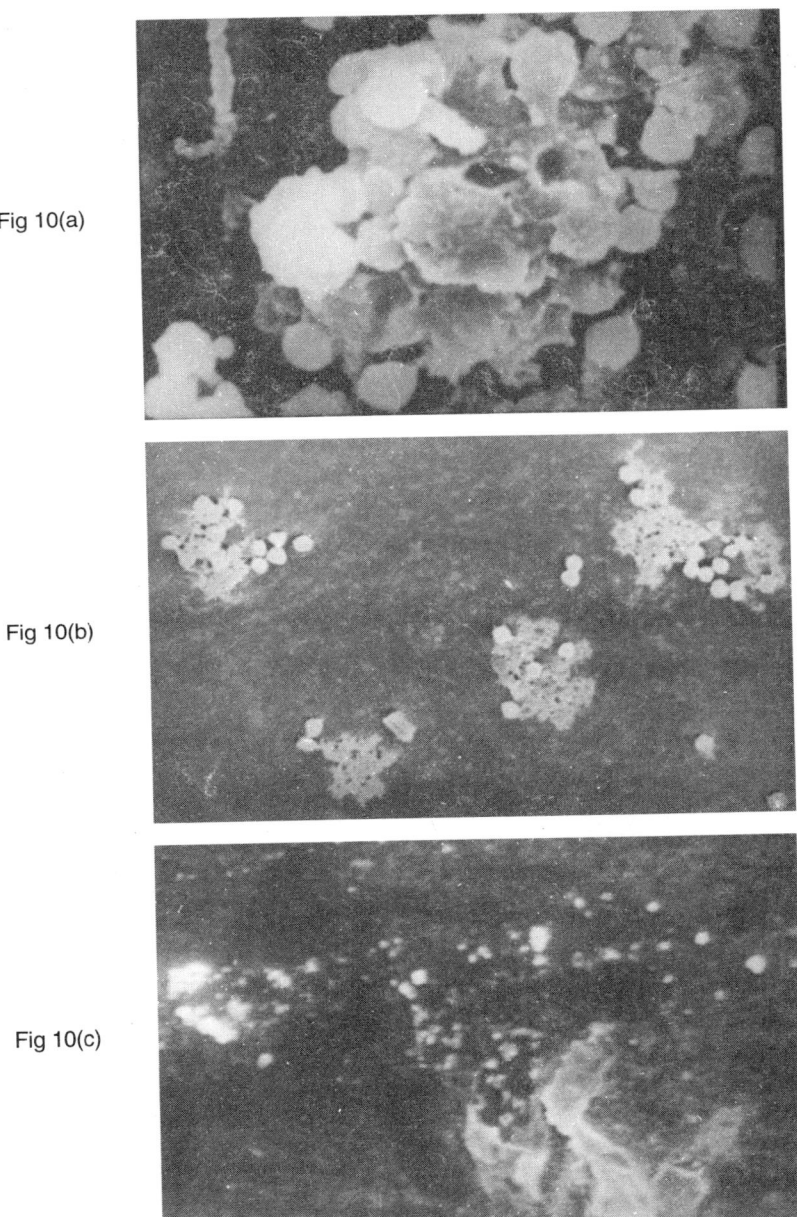

Figs. 10 : (a) **SEM study** with *microglial cells* with and without the application T11TS in normal and different experimental groups revealed interesting features : Microglial cells forming rosettes with SRBC10(a), sometimes super rosettes with clear anchorage with SRBC have been observed (b) T11TS administration in ENU-treated groups showed rosetting microglia(s) (activated microglial cells), some of them liberating secretory molecules (c), extending large projections, pseudopodias.

b) **Enhancement of cell mediated cytotoxic activity as observed with elute fraction III:** *In vitro* functional assay for CTL activity with EF III (incubated, washed lymphocytes) has been found to be 6% ± 0.62 **(Table 3)**. Cytotoxic activity of lymphocytes of normal group of animals was 30% ± 3.62 which decreased significantly to 7% ± 1.69 (p<0.001) with ENU administration *in vivo*. Revival of the value to 24% ± 2.69 was observed after 1ml of T11TS – fraction administrated to the ENU – group (ET1). Significant increase in CTL activity has also been observed in ET2 and ET3 groups but it was more (p<0.001) pronounced in the ET3 (Gr. IV) (51.5% ± 6.94) **(Fig. 11)**. ENU–SRBC group, on the other hand, yielded a value of 22% ± 3.92.

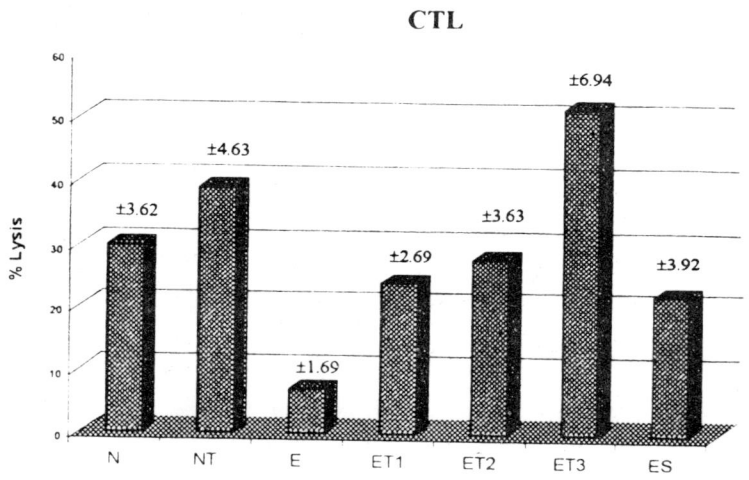

Fig. 11 : *In-vivo* effects of T11TS on cytotoxic efficacy of splenic lymphocytes. Third booster dose (ET3) was found to induce CTL at its maximum ENU induced animals.

c) **Phagocytic activity of the polymorphonuclear neutrophil has been observed in T11TS administered animal :** the phagocytic activity of PMN was significantly reduced in ENU induced tumor group (Gr. III animals), as revealed from reduced colour density of the phagosomes (0.016 ± 0.001) whereas it was 0.03 ± 0.002 in normal control group **(Fig. 12)**. The normal group receiving T11TS showed an elevation of phagocytic activity (0.053 ± 0.003). Further, it was significantly increased with the single booster dose of T11TS fraction in ET1 animals, (0.033 ± 0.002) which restored the value back to its normal level. The tumor bearing animals that received the second booster dose of 1 ml of T11TS fraction (ET2) showed still significantly higher phagocytic activity (0.065 ± 0.006), (p<<0.001). No further improvement of phagocytic activity was observed with third booster dose of T11TS – fraction (ET3) **(Fig. 12)**. *In vitro* effects of T11TS showed functional inhibition for phagocytosis of PMN **(Table 3)** PMN of ENU – SRBC group also showed significant increase of phagocytic activity (0.043± 0.004).

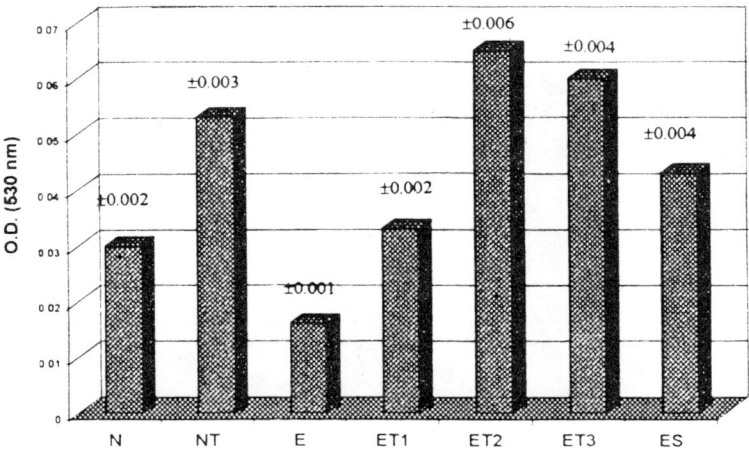

Fig. 12 : Effects of T11TS administered *in vivo* on PMN mediated phagocytosis. The most significant effect was observed with ET1 with highest phagocytic burst by PMN in tumor bearing animals.

d) **Induction of macrophage-mediated phagocytosis observed in T11TS treated animal:** The phagocytic activity of normal macrophages (Mφ) as determined through NBT reduction assay (O.D. of reduced phagosome) was found to be 0.037 ± 0.002 **(Fig. 13)** ENU treated animals showed significant decline ($p<0.001$) in phagocytic activity (0.021 ± 0.002). The activity was increased very significantly ($p<<0.001$) in T11TS treated normal (Gr. II) animals. The tumor bearing animal group that received a single booster dose of 1ml of T11TS fraction (ET 1) exhibited significantly greater phagocytic activity than the ENU induced tumor-control 0.055 ± 0.003 **(Fig. 13)**. Other doses were mostly ineffective. ENU-SRBC group also showed moderately improved phagocytic activity (0.048 ± 0.003). T11TS *in vitro* showed complete inhibition of phagocytosis by Mφ **(Table 3)**.

Table 3 : In vitro *effects of T11TS on the functional efficacy of immunocyte of normal rats: Rosetting and cytotoxic capacity of lymphocytes and phagocytic efficacy of PMN and Mφ were all inhibited drastically.*

Parameters	Normal	In vitro T11TS
Rosette (%)	19%	0%
CTL(%)	30%	6±62%
PMN (O.D.)	0.030±0.002	—
Mφ (O.D.)	0.037±0.002	--

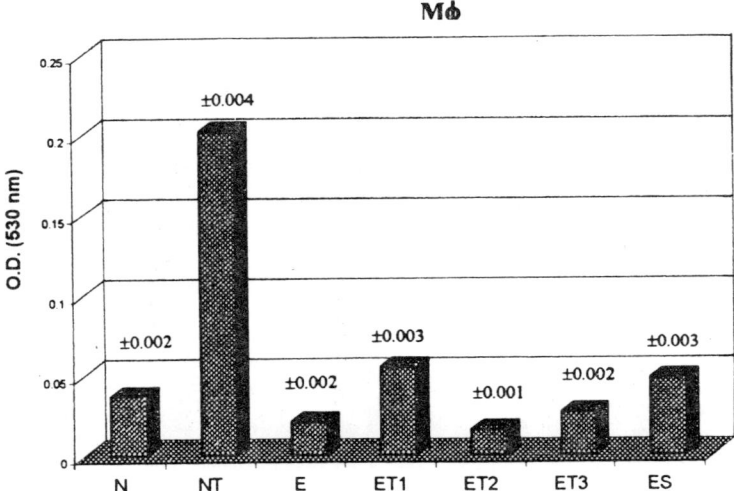

Fig. 13 : *In vivo* effects of T11TS on macrophage mediated phagocytosis in different groups of rats. ET1 showed the most significant stimulatiion of Mf for tumor phagocytosis under the event.

DISCUSSION

The results obtained from the present study showed that a minor glycoprotein of sheep red blood cell membrane termed T11TS or the sheep form of LFA3[7-10, 24] or more currently CD58 having non-reducing terminal residues of N-linked sugar chain and involved in rosette formation with lymphocytes,[25] exhibited profound immunostimulatory and tumor inhibitory effects when administered in the animals concerned.

The elutes obtained through DEAE-cellulose column and further subjected to functional assays like rosette inhibition to identify the glycoprotein T11TS, showed that the elute fraction III (EF III) exhibited the complete rosette inhibition in the assay. During the process of isolation, glycopeptides were released from the intact sheep erythrocytes through trypsin treatment. Splenic lymphocytes were used to react with sheep erythrocytes to form rosettes wherein rosetteing capacity was abolished after trypsin treatment.[16] This indicated that the T-cell recognising site on sheep erythrocytes were released by trypsin treatment. Among the elute fractions used, EF III showed the maximum rosette inhibition in comparison to other fractions. The rationale for such inhibition lies in the fact that the fraction that binds with the T11 receptor (CD2) of the lymphocytes, inhibit the rosette formation to occur. Thus, elute fraction III could be determined as the possible T11TS fraction of choice. The amount of protein present in the elute fraction III, as determined through Lowry's method showed that very small amount of protein in the elute fraction III, which has been identified as T11TS through functional assay. In addition, the results have also shown that a dose of 1 ml of T11TS fraction with an incubation period of 1 hour with the effectors (10^7) are good enough for the efficacy as determined through rosette inhibition assay (**Table 2**). No further selection of time dose were necessitated, since receptors (T11/ CD2) seemed to get saturated with those selected amounts.

The first effective outcome of T11TS administration (0.41 mg/ Kg body weight) in ENU treated animals was the abrogation or inhibition of brain tumor induction in rats. This has long been proved that N-ethyl-Nitrosourea (ENU) is the most potent of neuro carcinogens and produce tumors with morphological and biological similarities to naturally occurring neural neoplasms in man and animals.[12] Administration of 1 ml. of purified T11Ts fraction in ENU-treated animals revealed some interesting results: total survival period significantly increased, tumor induction was delayed or absent over the period of one year after ENU administration as observed macroscopically / microscopically. The results indicate that administration of T11TS can develop effective anti tumoral immune response in tumor bearing animals resulting in a mean increase in survival period. Histological studies have provided evidence for neoplastic growth following five months of ENU administration. When compared to normal rat brain it clearly demonstrates grade IV oligodendroglioma with multiple mitotic figures and oligodendroglial cells in close apposition with no intercellular spaces. This is corroborated with the previous findings that the foci of oligodendrocytes have been reported in the cortex and basal ganglia in rats treated transplacentally with ENU[26] a dual pathogenesis for ENU-induced gliomas seems to be the most plausible explanation.[27] Oligodendroglioma thus develop from these early oligodendrocytic proliferation while all other gliomas owe their origins to the subependymal plate cells. In the present study, administration of T11TS fraction in ENU treated animals demonstrated anti-tumor activity in a dose dependent manner as evidenced through the reversion of neoplastic glial features to normal glial features. Infiltration of lymphocytes and margination of the endothelial lining in the brain is the hallmark of the first dose of T11TS administration in ENU animals. The hypocellularity observed in the first dose gives way to apoptotic and enlarged nuclei due to enhanced permeability of the nuclear membrane. Finally, the third dose leads to the total clearance of the neoplastic cells with necrosis and calcification. It seems likely that T11TS fraction has the capability to inhibit or block tumor proliferation not only through the potentiation of the immune network, but also by interfering with the mitotic rate of the malignant cells. It is assumed that it possibly do that by (1) inducing a cytostatic effect, (2) differentiation induction, (3) blocking DNA synthesis and growth factor interactions and (4) interaction on cellular oncogene expression. Recently molecular biology of ENU induced rat brain tumor identified a new oncogene, *neu* which is detectable by transfection into mouse NIH 3TC cells.[28-29] The data obtained from fluoroscopic analysis further directed the investigation towards the causative factors responsible for tumor inhibition (**Fig. 3**). The role of cellular immune components as direct killer of tumor cells was considered at the first instance. However, the role of T11TS in inhibiting brain tumor in a direct or cytokine induced manner cannot be ruled out under the event.

T cell growth and clonal expansion is a tightly regulated process requiring the growth factor IL-2, and hence the extent of the T cell proliferative response is determined by the concentration of IL-2 available to the cell and the level of IL-2R (CD25) expression on the cell.[30-31] CD25 is also an activation marker for T cells, so greater CD25 expression will also indicate the higher activation of T cells. Previously Fox et al,[32] demonstrated that activation of human thymocytes via the 50 Kd T11 sheep erythrocytes binding protein induces the expression of IL-2R both on T3$^+$ and T3$^-$ populations. Furthermore, Meuer et al.,[33] hinted at an alternative pathway of T-cell activation through 50 Kd T11 sheep erythrocyte receptor protein. In the present course of investigation FACS analysis has shown that very low level of IL-2R expression has been observed on splenic lymphocytes of tumor bearing animals (ENU – group) which can be attributed to the immune suppressive effect of tumor on lymphocytes. The significant enhancement of IL-2R expression on lymphocytes of tumor bearing animals that received a single booster dose of 1 ml of T11TS fraction is an important finding. It is possible that a single booster dose of T11TS fraction may generate a strong mitogenic stimulus through T11 (CD2) that closely simulates the view of Meuer and co-workers. It can further be pointed out that the specific interaction between the CD2 (T11) and T11TS (sheep form of LFA3 or sheep CD58) play important role via a CD2 mediated co-stimulatory pathway during T-cell activation and proliferation.[34,35] Hunig T et al, have demonstrated that an activation of T-cells through an alternative pathway was observed in vitro when a combination of anti-CD2 monoclonal antibodies bind to the distinct epitopes of CD2, such as mAbs against T11$_2$ + T11$_3$. It was further demonstrated that the binding of the cell surface molecule, T11TS to T11 (CD2) induces reactivity in resting T-cells to a mitogenic stimulus given by a mAb to the T11$_3$ determinant or by concentration of anti-T11$_{2+3}$. Thus, it was confirmed that one of the signals required for T cell activation through the alternative path way is provided by the interaction of CD2 with a naturally occuring complementary cell surface molecule. Further, the interaction between CD2 and T11TS can reverse T cell anergy[36,37] caused by tumorigenic inhibition. Hatikeyama et al,[38] and Bell et al,[39] demonstrated that such a strong mitogenic stimulus through T11 (CD2) associate *src*-like protein tyrosin kinase from the surface of the T lymphocytes. Thus the immediate bio-chemical events triggered at the cell surface link up with the transcription factor that control IL-2R expression[40,41] and consequently IL-2 titre. Thus in our study, the higher expression of IL-2R in group IV animals is well corroborated with the view of other workers. The gradual down-regulation of IL-2R expression in group V (ET2) and group VI (ET3), can be explained in terms of receptor saturation kinetics that most if not all T11 (CD2) antigens would have been *saturated* at the first booster dose of T11TS-fraction. Another possibility could be the repeated antigenic stimulation which may produce *anergy*.[37] Also, the possibility of the occurrence of *desensitization* cannot be ruled out.

Flowcytometric analysis (FACS) has also suggested the occurrence of *activated lymphocytes* in the brain compartment which corroborates the fact that activated but not the resting T-cells are not able to migrate through the blood brain barrier.[42-45] The diminished expression of IL-2R on tumor infiltrating lymphocytes could be due to the presence of some soluble blocking factor, such as IL2 blocking factor and/or some other inhibitory effects encountered in that microenvironment.[4, 46-47] T11TS in its successive booster doses could, however remove the microenvironmental interference to re-express CD25 (IL-2R).

Flowcytometric analysis of the *infiltrating lymphocytes* in the brain tissue demonstrated some interesting features. While enhancement of the total infiltration of the lymphocytes and the presence of *CD4 positivity* have been observed following 1st and 2nd booster doses of T11TS fraction in ENU treated animals, the gradual decrease in the number of CD4+ lymphocytes have noticed, whereas the significant increase of the CD4 negative cell types have been observed. Thus it can be assumed that the administration of T11TS fraction in tumor bearing animals can enhance the infiltration of lymphocytes into the brain compartment across the BBB in search of tumor antigen in order to execute immune reactivity. The significant increase in the number of CD4 negative cell types, in T11TS treated animals possibly involve CD8+, NK cells, etc. (data not shown).

In the present study, FACS analysis of the *microglial cells* (MG) froom the normal brain tissue showed CD25 (IL-2R) positivity, possibly on resident microglia. The increase in CD25 expression on microglial cells in ENU –treated animals and following 1st dose of T11TS administration in such animals explains its (MG) *activation* state, which is possibly due to the upregulation of CD2 (T11) molecules and is well corroborated with the findings of the microglial-rosette formation and SEM study. The most significant up regulation of the CD25 expression on the microglial cells was observed following 2nd booster dose of T11TS fraction in ENU-treated animals, assumed to be due to the effect of secretory molecules or cytokines liberated from the MG cells and acting on them (autocrine) and /or possible modulatory effect of cytokines from the increased number of brain infiltrating lymphocytes that occurred with the 2nd booster dose. Thus, the 2nd booster dose depicts the higher level of microglial cell activation. Down regulation of CD25 expression following administration of T11Tsfraction in ENU-treated animals can be attributed to the saturation and/or desesitization of the CD2 (T11) molecules. *In vitro* inhibition study showing the complete inhibition of the microglial-rosette formation with T11TS fraction further strengthened the findings of the CD25 expression on the microglial cells. This is the first report claiming the findings for the presence of CD25 and CD2 receptors on the microglial cells.

Apart from the immunopotentiating and anticancer properties of whole SRBC, T11TS when isolated produced still better effects in animals. Results obtained through immunological parameters suggest potentiation of cell-mediated immune response (CMI) in tumor bearing animals. Rosetting capacity of normal splenic lymphocyte was 19% ± 4.31 which was drastically reduced to 2% ± 1.03 in group III (ENU animals). The group IV, tumor bearing animals that received a single booster dose of 1 ml of T11TS fraction demonstrated greater rosetting capacity than any other group, indicating significant lymphocyte proliferation and activation both qualitatively and quantitatively. Indeed, up-regulation of CD2 (T11) molecules both in individual lymphocytes indicated by super rosetting and also quantitative lymphocyte poliferation activity was indicated by significant increase in rosette forming cells. Although the rosetting capacity of lymphocytes increased to 42.75% ± 7.21 with the second booster dose, the third booster dose could not increase the rosetting capacity further possibly due to the saturation of the CD2 molecules. *In vitro* incubation of lymphocytes with T11TS blocked the CD2 receptors with subsequent absence of rosette formation as revealed in the present study. In addition, immunoregulatory property of CNS –resident cells, particularly of *microglia* has been found to play a crucial role in brain tumor immunity. When modulated this cell can facilitate and amplify immune effector mechanisms in CNS. The interaction between the glial cells and lymphoid cells, specifically with CD4+ and CD8+T cells, are important constituents of a complex immunoregulatory system as envisaged by several workers.[44-45, 47] Administration of T11TS in ENU treated tumor bearing animals demonstrated significant enhancement of microglial rosette formation, which can be attributed to the possible upregulation of CD2 (T11) molecules on microglial cells as evidenced by **super rosetting** and **SEM** study. The observed inhibition with T11 TS fraction in the in vitro inhibition study further hints at the possible binding sites for T11TS fraction.

The extent of cytotoxic efficacy of lymphocytes, obtained through CTL assay, indicates a linear relationship with the dose of T11TS fraction to the cytotoxic efficacy of lymphocytes. Since T11(CD2) is a pan T cell marker, the administration of T11TS *in vivo* may thus involve the CD4+, CD8+ T cells as well as NK cells. The activation of CD4+ T cell through with T11TS provides means to recruit the other effector cells as well as such as Mϕ, PNM. Activation of CTL and NK function through the T11 sheep erythrocyte binding problem have also been demonstrated by other workers.[48-49] Our data strongly indicated the linear increase of cytotoxic efficacy of lymphocytes in tumor bearing animal with T11Ts of first, second and third booster dose correspondingly. *In vitro* functional inhibition study has shown that purified T11TS fraction directed against T11 (CD2) molecule inhibit cytotoxic T lymphocyte mediated killing. The accessory molecules such as CD2, LFA-1 function to increase adhesion between T cell and the APC or target cell.[50-51] It seems reasonable to argue that the binding of T11TS to CD2 *in vitro* thus inhibit the adhesion between tumor cell with the T cell and thereby hinders other T cell function including cytotoxicity, etc.

Results of phagocytic activity of polymorphonuclear neutrophil have been expressed in a dose-dependent manner. The phagocytic activity of PMN with the first booster dose of 1 ml of T11TS fraction (ET1) seems to be normocytic, the second booster dose (ET2) showed the highest stimulation and the third at nearly equal manner. It is possible that activated lymphocytes in the T11TS treated animal recruited PMNs indirectly through the cytokine network (neutrophil activitating factor). Our previous study[22] has shown that tumoricidal phagocytic activity of PMN is increased in SRBC treated animals. The reason for the absence of phagocytic activity of PMN in *in vitro* functional assay can be explained on the basis of a report on the presence of T11 (CD2) molecule on PMN surface.[52] The observed inhibition with T11TS fraction in the *in vitro* inhibition assay hints at the possible binding site for T11TS fraction on PMN. Whether nonspecific activity with the ingestion of T11TS occurs may pose a question.

Macrophages play a very important role in CMI response, from antigen presentation to effector function. Our data suggest lymphokine mediated activation of macrophages that might have occurred in a dose dependant manner as well. Significantly higher ($p<0.001$) and optimum activation was observed with the first booster dose of T11TS fraction in tumor bearing animals. As observed with other effector stimulation T11TS might have a modulatory effect on Mϕ through series of appropriate cytokines or directly interacting through cytoskeletal system of the cells concerned. Absence of Mϕ mediated phagocytosis in *in vitro* functional assay could be explained since macrophages have been reported to bear CD2 molecule on their surface.[53]

The data obtained on the above investigations suggested that T11 target structure (T11TS) of sheep erythrocytes or sheep form of LFA3 (CD58) as isolated and purified in our laboratory exhibited inhibitory role on experimentally induced brain tumor in rats by way of augmenting the CMI function primarily at the peripheral extent and subsequently activating immunocompetent cells infiltration across the blood brain barrier and also the *microglial* cells (MG). These findings, therefore, can be effective in establishing therapeutic module against cancer and any other immunodeficiency state of different diseased conditions.

REFERENCES:

1. Diengdoh JV, Booth AE: Post-irradiation necrosis of the temporal lobe presenting as a glioma. *J Neurosurg* 1976;44(6):732.
2. Dix AR, Brooks WH, Roszman TL, Morford LA: Immune defects observed in patients with primary malignant brain tumors. *J Neuroimmunol* 1999;100(1-2):216.
3. Nakagaki M, Brunhart G, Kempertl and Caveness WF: Monkey brain damage from radiation in the therapeutic range. *J Neurosurg* 1976;44:3.

4. Gillespie GY, Mahalay MS: Biological response modifier therapies for patients with malignant gliomas: In (G.T.Thomas) Neuro-Oncology. Churchill Livinstone, NewYork, London, 1995;242.
5. Basu Swapna, Chaudhuri S, Roy B: Surface ultrastructural changes of percoll separated splenic lymphocytes from syngeneic and allogenic mice with transplantable ascites fibrosarcoma tumor. *Indian J Exp Biol* 1982;20:227.
6. Roy RU, Sarkar S, Duttachaudhuri M, Dutta SK, Chaudhuri Swapna, Chaudhuri S: Antigen presenting capacity of mononuclear phagocytes in experimental leukemia and in patients with hematological malignancies, in India. *J Hematol & Blood Transf* 1997;15 (1):33.
7. Hunig T: The cell surface molecule recognized by the erythrocyte receptor of T ly mphocyte. *J Exp Med* 1985;162:890.
8. Hunig T, Mitracht R, Tiefenthaler G, Kohler C, Miyasaka M: The cell surface molecule binding to the 'erythrocyte receptor' of T-lymphocytes. Cellular distribution, purification to homogeneity and biochemical properties. *Eur J Immunol* 1986;16(12):1615.
9. Hunig T, Tiefenthaler G, Mitracht R, Kholer C, Lottspeich F, Meuer S: The erythrocyte receptor of T-lymphocytes and T-11 target structure (T11TS): Complementary cel interaction molecules involved in T-cell activation. *Behring Inst Mitt* 1987;81:31.
10. Giegerich GW, Hein WR, Miyasaka M, Tifenthaler G, Hunig T: Restricted expression of CD2 among subsets of sheep thymocytes and T lymphocytes. *Immunology* 1989;66(3):354.
11. Springer TA, Dustin ML, Kishimoto TK, Marlin SD: The lymphocyte function – associated LFA-1, CD2 and LFA-3 molecules: Cell adhesion receptors of the immune system. *Annu Rev Immunol* 1987;5:223.
12. Druckrey H, Ivancovic S, Preussman R: Teratogenic and carcinogic effects in the offspring after single injection of ethylnitrosourea to pregnant rats. *Nature* 1966;210:1378.
13. Koestner A, Swenberg JA, Wechsler W: Transplacental production with ethylnitrosourea of neoplasms of the nervous system in Sprague-Dawley rats. *A M J Pathol* 1971;63:37.
14. Lantos PL: Chemical induction of tumors in the nervous system: In (GT Thomas) Neuro-Oncology. Churchill Livingstone, New York, London 1993;85.
15. Lantos PL, VandenBerg SR, Paul Kleihues: Tumors of the nervous system. In Greenfield's Neuropathology, Vol. 2, Edition 6th. Eds. David I Graham & Peter Lantos, Arnold, London, Sydney, Auckland, 1997; pp. 583.
16. Kitao T, Takeshita M, Hattori K: Studies on glycopeptide released by trypsin from sheep erythrocytes. *J Immunol* 1976;117:310.
17. Lowry OH, Rosebrough NJ, Farr AL, Randell RJ: Protein measurement with the Folin-Phenol reagent. *J Biol Chem* 1951;193:265.
18. Raha SK, Dey SK, Roy SK, Chaudhuri S, Chakraborty SL: Antitumor activity of L-Asparaginase from *Cylindrocarpon obstutisporum* MB10 and its effects on the immune system. *Biochem Int* 1990;21(6):1001.
19. Chaudhuri Swapna, Sarkar S, Roy RU, Dutta S, Dutta SK, and Chaudhuri S: Microglia – The APC of the brain, possess immunocompetent receptors and the excitatory-inhibitory potential to protein kinase activity. *The Immunologist Suppl.* 1998;1:145.
20. Chaudhuri Swapna, Ganguly S, Ghosh SN, Dutta S, Sarkar S, Begum Z, Roy RU, Bhattacharya MK, Dutta SK and Chaudhuri S: Immunological insurgence during intracranial tumor development: Cellular immunity in astrocytoma patients. *J Physiol & Appl Sc* 2000;54(3):118.
21. Law Sujata, Maiti D, Palit Aparna, Majumder D, Basu K, Chaudhuri Swapna, Chaudhuri S: Facilitation of functional compartmentalization of bone marrow cells in leukemia mice by biological response modifiers: an immunotherapeutic approach. *Immunology Letters* 2001;76:145.
22. Chaudhuri Swapna, Sinha A, Sengupta A and Chaudhuri S: Sheep erythrocytes provide metabolic triggers for tumor phagocytosis in polymorphonuclear neutrophils: A possible mechanism of tumor inhibition in mice. *Biochem Int* 1991;23 (20):231.
23. Hudson L, Hay FC: Cell dynamics in vivo. In: Practical Immunology. Blackwell Scientific Publications. Oxford, London, Edinburg, Boston, Melbourne, 1989;182.

24. Sun ZJ, Dotsch V, Kim M, Li J, Reinherz EL, Wagner G: Functional glycan-free adhesion domain of human cell surface receptor CD58: design, production and NMR studies. *EMBO* 1999;18:2941.
25. Ogasawara H, Kusuy K, Takasaki S: Role of terminal galactose residues in N-linked sugar chains of sheep erythrocyte membrane glycoproteins in rosette formation with T lymphocytes. *Immunol Lett* 1995;48:35.
26. Schiffer D, Grordana MT, Pezzotta S, Lenchner C, Paoletti P: Cerebral tumors induced by transplacental ENU: study of different tumoral stages, particularly of early proliferations. *Acta Neuropathologica (Berlin)* 1978;41:27.
27. Lantos PL: The fine structure of periventricular pleomorphic gliomas induced transplacentally by N-ethyl-N-nitrosourea in BDIX rats-with a note on their origin. *J Neurological Sciences* 1972;17:443.
28. Shih C, Padhya LC, Murry M', Weinberg RA: Transforming genes of carcinomas and neuroblastomas introduced into mouse fibroblast. *Nature* 1981;290:261.
29. Schubert D, Heinemann S, Carlis W, Tarikas H, Kimes B, Patric J, Steinbach JH, Culp W, Brandth BL: Clonal cell lines from the rat central nervous system. *Nature* 1974;249:224.
30. Smith KA: Interleukin-2. *Curr Opin Immunol* 1992;4:271.
31. Taniguchi T, Minami Y: The IL-2/ IL-2 receptor system: a current overview. *Cell* 1993;75:5.
32. Fox DA, Hussey RE, Fitzgerald KA, Bensussan D, Daley JF, Schlossman SF, Reinherz EL: Activation of human thymocytes via the 50 Kd T11 sheep erythrocytes binding pprotein induces the expression of IL.-2 receptors on both T3$^+$ and T3$^-$ populations. *J Immunol* 1985;134:330.
33. Meuer SC, Hussey RE, Fabbi M, Fox D, Acuto O, Fitzgerald KA, Hodgdon JC, Protentis JP, Schlossman SF, Reinherz EL: An alternative pathway of T-cell activation: a functional role for the 50 Kd T11 sheep erythrocyte receptor protein. *Cell* 1984;36:897.
34. Koyasu S, Lawton T, Novick D, Recny MA, Siliciano RF, Wallner BP, Reinherz EL: Role of interaction of CD2 molecule with lymphocyte function – associated antigen3 in T-cell recognition of normal antigen. *Proc Natl Acad Sci USA* 1990;87:2604.
35. Semnanirt, Nutman TB, Hochman P, Shaw S, van Seventer GA: Co-stimulation by purified intercellular adhesion molecule 1 and lymphocyte function- associated antigen3 induces distinct proliferation, cytokine and cell surface antigen profiles in human "naïve" and "memory" CD4$^+$ T-cells. *J Exp Med* 1994;180:2125.
36. Boussiotis VA, Freeman GJ, Griffin JD, Gray GS, Gribben JG, and Nadler LM: CD2 is involved in maintenance and reversal of human alloantigen-specific clonal anergy. *J Exp Med* 1994;180:1665.
37. Bell GM, Imboden JB: CD2 and regulation of T cell anergy. *J Immunol* 1995;155: 2805.
38. Hatikeyama M, Kono T, Kobayashi N, Kawahara A, Levin SD, Perlmutter RM: Interaction of the IL-2 receptor with the *src*-family kinase p^{56} lck: identification of novel intermolecular association *Science* 1991;252:1523.
39. Bell GM, Bolen JB, Imboden JB: Association of *src*-like protein tyrosine kinases with the CD2 cell surface molecule in rat T lymphocytes and natural killer cells. *Mol Cell Biol* 1992;12:5548.
40. Crabtree GR: Contingent genetic regulatory effects in T lymphocytes activation. *Science* 1989;243.
41. Ullman KS, Northrop JP, Verweij CL, Crabtree GR: Transmission of signals from the T lymphocyte antigen receptor to the gene responsible for cell proliferation and immune function: the missing link. *Annu Rev Immunol* 1990;8:421.
42. Weller RO, Engelhardt B, Phillips MJ: Lymphocytes targetting of the central nervous system: a review of afarent and efferent CNS -immune pathways. *Brain Pathol* 1996;6(3):275.
43. Hickey WF, Hsu BL, Kimura H: T-lymphocyte entry into the central nervous system. *J Neurosci Res* 1991;28:254.
44. Selmaj K: Pathophysiology of the blood-brain barrier In: Immunoneurology, Eds. Choffion M & Steinman L, Springer-Verlag, Berlin, Heidelberg, New York. 1996;175.
45. Bradle M: Immune control of the brain. In: Immunoneurology, Eds. Choffion M & Steinman L, Springer-Verlag, Amsterdam, New York. 1996;153.
46. Sido B, Otto G, Zimmermann R, Muller P, Meuer SG, Dengler TJ: Modulation of the CD2 receptor

and nondisruption of the CD2/ CD48 interaction is the principle action of CD2-mediated immunosuppression in the rat. *Cell Immunol* 1997;182:57.
47. Wekerle H, Livington C, Lassmann H, Meyermann R: Cellular Immune reactivity within the CNS. *Trends Neurosci* 1986;9:271.
48. Siliciano RF, Pratt JC, Schmidt RE, Ritz J, Reinherz EL: Activation of cytolytic T lymphocyte and natural killer cell function through the T11 sheep erythrocyte binding protein. *Nature* 1985;317:428.
49. Krensky AM, Robbins E, Springer TA, Burakoff SJ: LFA-1, LFA-2 and LFA-3 antigens are involved in CTL-target conjugation. *J Immunol* 1984;132:2180.
50. Dustin ML, Selvaraj, Mattaliano RJ and Springer TA: Purified lymphocyte function – associated antigen 3 binds to CD2 and mediates T lymphocyte adhesion. *J Exp Med* 1987;165:677.
51. Moingeon P, Chang HC, Wallner BP, Stebbins C, Frey AZ and Reinherz EL: CD2-mediated adhesion facilitates T lymphocyte antigen recognition function. *Nature* 1989;339:312.
52. Niiazova CP, Osipova SO, Badalova NS, Dekhkan-Khodzhaeva NA: Early and late E-rosette forming neutrophils in persistent lambliasis. *Med Parazitol (Mosk)* 1995;(1):43
53. Gutierrez M, Froster FI, Mc Conell SA, Cassidy JP, Pollock JM, Bryson DG: The detection of CD2+, CD4+ CD8+, and WC1+ T lymphocytes, B cells and macrophages in fixed and paraffin embedded bovine tissue using a range of antigen recovery and amplification techniques. *Vet Immunol Immunopathol* 1999;71(3-4):321.

29

Cortical Mapping and Functional MRI: Value in Surgical Planning

RP Tripathi, A Batra, S Khushu

INTRODUCTION

Magnetic resonance imaging (MRI) has seen a phenomenal progress in the last fifteen years. It has become the imaging modality of choice in various conditions due to its capability to delineate the structural anatomy at a high spatial resolution with exquisite soft tissue contrast. Multinuclear MR spectroscopy helps evaluate cellular metabolism quantifying several metabolites and is also now capable of providing a two-dimensional metabolite image that shows the spatial distribution of various metabolites in a particular area. In addition, MR is now capable of performing angiography, flow quantification, perfusion & diffusion studies, and magnetization transfer techniques which have provided new dimensions for exploration of brain function. However, the most fascinating development among all the advances is that of the cortical activation techniques, which enable one to map various functional areas of brain with a high degree of temporal and spatial resolution. These techniques have been foreseen to have tremendous impact in cognitive neuroscience and surgical planning of intracranial tumors. In contrast to functional MRI (fMRI) which is completely non-invasive and does not require any contrast administration, the routinely used brain function studies such as EEG lack the necessary spatial resolution and other techniques such as SPECT or PET require administration of radionuclides. FMRI as of now can be projected as a convenient technique for providing complimentary information to that available with other brain function studies.

FMRI PHYSICS AND TECHNIQUE

Animal experimental studies initially demonstrated that the level of cerebral blood oxygenation influences the signal intensity of $T2^*$-weighted gradient echo MR images. Oxygenated haemoglobin (oxyhemoglobin) is diamagnetic while the deoxygenated hemoglobin is paramagnetic and serves as an endogenous paramagnetic contrast agent. Deoxyhemoglobin causes dephasing of water protons in its direct vicinity, resulting in signal loss. The degree of signal loss depends on the absolute concentration of deoxyhemoglobin per voxel and scales with voxel size and gradient echo time. Conversely, an increased concentration of oxyhemoglobin or decrease in deoxyhemoglobin as a result of increased cerebral blood flow & volume can cause an increase in both the effective spin-spin relaxation time (T2*) and the corresponding signal intensity of a gradient echo image. As is clear the contrast generated is directly dependent on the blood oxygenation level. The technique is thus popularly known as the BOLD (Blood oxygenation level dependent) imaging.

FMRI is an indirect way of assessing the neuronal integrity. Neuronal activity results in change in the brain physiology resulting in increased oxygen consumption in the particular region. This in turn results in a disproportionately increased cerebral blood flow in the region causing an overcompensation of the fall in oxyhemoglobin. Reduced deoxyhemoglobin and increased oxyhemoglobin levels thus contribute to the fMRI signal which is subsequently processed and activation maps created[1] **(Fig. 1)**.

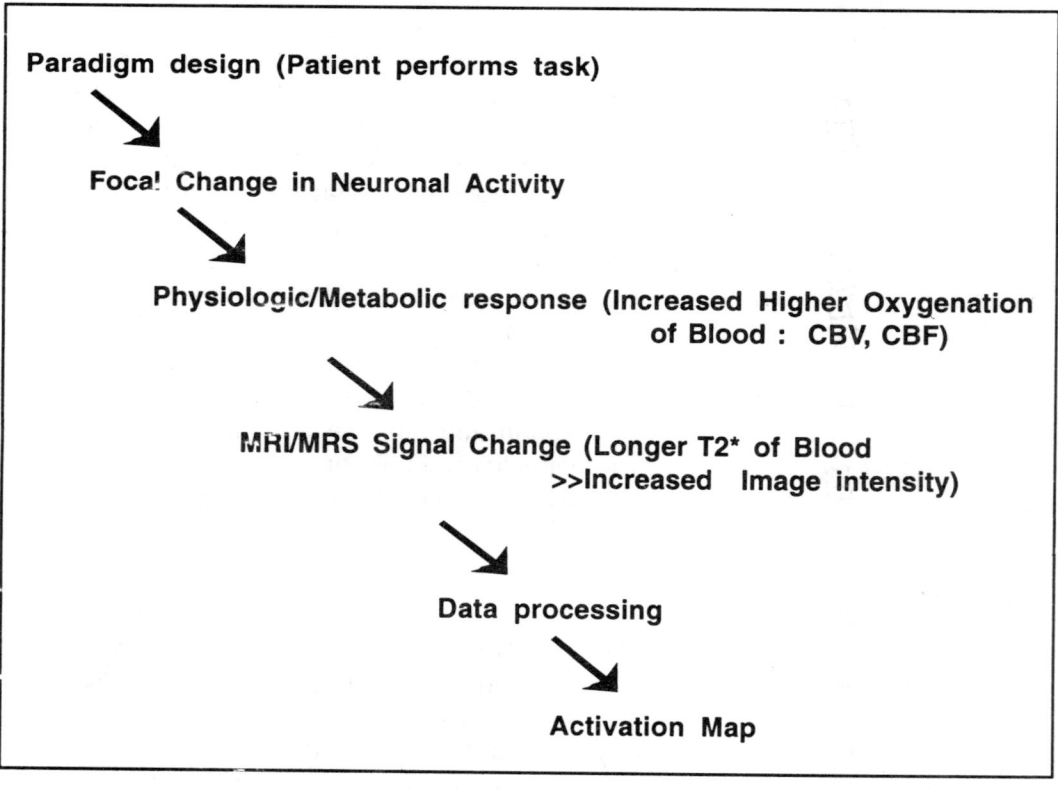

Fig. 1: Schematic representation of transformation of focal change of neuronal activity into activation map [1].

Besides the BOLD technique, fMRI can also be performed by using the Perfusion method based on the injection of contrast agent. However this method does not allow for frequent repetition and thus has not become popular.

In order to carry out fMRI studies, one requires a state of art MR equipment with fast switching gradients and sub second imaging techniques such as EPI (echo planar imaging). The scanner should have a minimum strength of 1.5 T. Advanced software capable of handling the large amount of data generated is now available to not only create maps but also perform statistical analysis. Areas of activity can be accurately localized so that adequate stereotactic guidance is possible on standardized images. This is of utmost help to the neurosurgeon and others involved in planning therapy.

Accurate and convenient paradigms need to be prepared in order to carry out fMRI. Among its various applications, fMRI has been used experimentally to map the sensorimotor cortex, visual cortex, primary auditory cortex, association areas & language regions. Besides, activation during higher cognitive processes, such as language tasks, motor learning and ideation or visual mental imagery have provided greater insights for the neuroscientists in identifying non-invasively the various and complex intracerebral neural networks and pathways.

CLINICAL APPLICATIONS

The preoperative identification of essential functional regions allows evaluation of both surgical feasibility and approach. Resulting clinical benefits include improved identification of candidates for successful surgery, improved outcome of these surgeries undertaken & reduced overall treatment cost. FMRI offers possibility of performing these cortical localization routinely and in existing rather than new instrumentation. FMRI can directly identify functional regions at high resolution and provide an anatomic basis to these localization since there is a natural correspondence of fMRI data with MRI structural images. The non-invasive nature of this technique allows repeated examinations of the same patient, so that it can also provide information on the post-surgical neural reorganization and recovery of function.[1] FMRI technique is based on hemodynamic responses and is thus likely to overlap in application with other modalities such as SPECT and PET. Given its higher spatial and temporal resolution, lack of ionizing radiation and lower cost (little or no additional cost to medical centres already having a clinical MRI unit), fMRI may become more widely used than these technologies for clinical functional brain mapping based on hemodynamic responses to neural activation.

Various clinical applications of this technique are becoming evident, the most direct being the presurgical mapping for patients with lesions near essential functional cortical areas esp. sensorimotor, visual, language and memory areas.[2] The principal goal of neurosurgery is to maximize lesion resection while minimizing neurologic deficit.[2] The location of various functional areas have been based on anatomic landmarks but distortion by pathology or individual variability make this approach inaccurate. Direct electrical cortical stimulation provides best accuracy for functional mapping but this requires a craniotomy and only a limited brain area can be mapped. FMRI is especially suited for preoperative assessment of surgical risk, planning and intraoperative navigation.[2] Studies evaluating spatial specificity of fMRI by comparing it with intraoperative electrocortical mapping have shown the intraoperative site & MR activation site within 20 mm in 100% correlation and within 10 mm in 87% of correlations.[3] Comparative PET and intraoperative electro-cortical stimulation studies have validated value of sensorimotor fMRI for presurgical planning.[4-6] FMRI has been used to map sensorimotor functions in patients with a variety of brain pathologies including tumors, arteriovenous and cavernous malformations and cortical dysplasias.

In patients having paresis with a brain tumor in proximity to the motor cortex, functional deterioration could be either due to the direct invasion of the functional area or due the mass effect. As cerebral tumors are known to spread beyond their apparent margins as seen on MRI, detection of cortical activity by fMRI at the apparent margin of the tumor suggests partial sparing of neurologic function by the tumor. Roux et al have shown that in patients with paresis having brain tumors invading the contralateral motor cortex the ipsilateral motor cortex is responsible for taking over and providing the residual function if present. Such vital information greatly influences therapeutic and surgical decisions before debulking so that regions with morphologic abnormality but intact function can be preserved.[7-8]

In the treatment of perisylvian lesions and medically intractable temporal lobe epilepsy, presurgical assessment regarding localization and lateralization of language function and capacity of contralateral lobe to support memory function is required.[2] The "Wada test" which consists of intracarotid injection of amobarbital is currently accepted as the gold standard for assessing lateralization of hemiphereic language dominance but it does not give information regarding localization. Studies have shown fMRI to be helpful in localizing of language areas and lateralization of hemispheric language dominance.[9-13] The major limiting factor that has prevented fMRI from replacing Wada test is the fact that Wada testing evaluates both language and memory function. However, in recent studies in patients with medial temporal lobe epilepsy, asymmetries in memory activation were detected on fMRI studies showing a promising role for fMRI in improving the overall presurgical evaluation in TLE patients.[14] Lateralization of the epileptic focus with respect that of the language and memory function helps in planning lobectomy.

FMRI can also be used for visualize epileptic focus with or without associated clinical seizure. Besides use of fMRI for the monitoring of functional recovery after stroke or head injury appears especially important for patients with hemiplegia or aphasia. Studies can be repeated any number of times and at the will of the patient during the recovery period without concerns about radiation. fMRI can also be used for neuropharmacological research by visualizing areas of the brain where receptors are activated or inhibited through effects on blood flow or oxygenation.

There are certain limitations of fMRI. Besides all the contraindications applicable to conventional MRI, fMRI also requires that the patient is highly cooperative lies immobilised for the period of the study to avoid misregistration artefacts, and actively participates in the activation task. Sophisticated software capable of handling the large amount of raw data generated is needed and is often required to be developed in-house. FMRI study has found to fail in 8-30% of patients because of motion, patient inability to co-operate or severe tumor induced damage to patients sensorimotor cortex.[3,7,9] Another serious limitation relates to the accuracy localization of the activated regions. At times the localization could be seen in large vessels rather than the actual sites of neural activation.

The ability to distinguish cystic or necrotic areas from regions of solid or extremely viscous tissue before surgery is of significant clinical use in designing the surgical approach to the lesion.[15] Routine MR imaging does not consistently distinguish cystic from solid gliotic alterations. ADC & diffusion anisotropy characterizations using diffusion weighted MRI have been used to distinguish tissue consistency as there is a higher diffusion coefficient in fluid than in solid tissue even in face of equivalent relaxation characteristics.[8]

Attempts to use DWI to distinguish peritumoral edema from spreading tumor mass are being made but since a tumor often extends beyond the visualised abnormality on routine images, this technique may have inherent limitations. Relative cerebral blood volume (rCBV) mapping using perfusion MRI may be more relevant way to localize the aggressive portions of a tumor.

Also rCBV maps in combination with cortical activation mapping can be performed to locate viable functioning cortex in the peritumoral region esp. in cases in which the tumor has displaced or destroyed normal anatomical landmarks.[7-8]

INMAS EXPERIENCE

At the NMR research centre, INMAS, we have now performed forty fMRI studies. Task paradigms for activation of the motor, speech and visual functional areas have been evolved. The response of neural activation in the motor cortex with increasing load has been demonstrated in normal volunteers. Eleven patients with brain tumors in close relation to the motor cortex (n=9) **(Fig. 2)** and speech areas (n=2) have also been evaluated. Two studies have been carried out in the post-operative period to demonstrate preservation of the motor area in

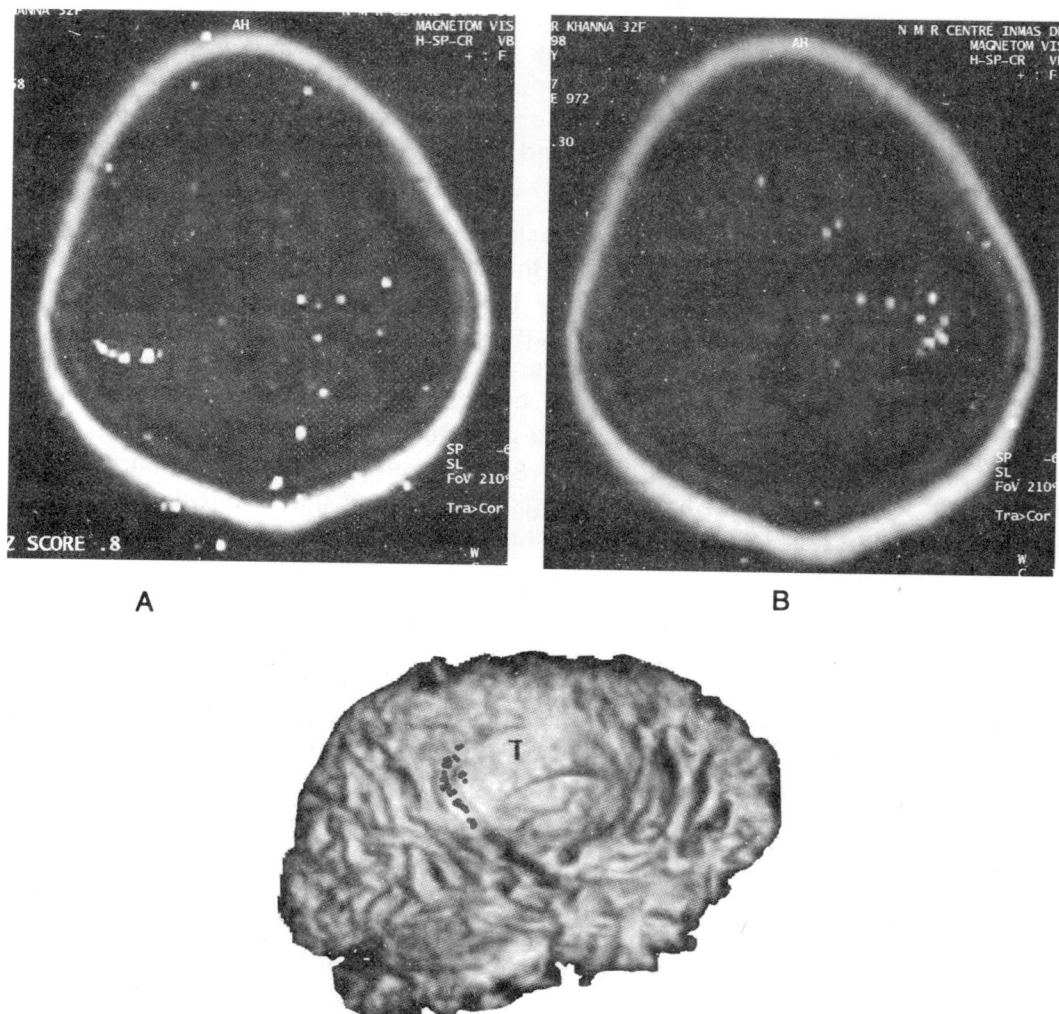

Fig. 2 : Motor cortex activation map in a case of right frontal cerebral glioma with grade IV power of the left hand with left finger tapping as the task paradigm: Note the posterior relation and displacement of the motor cortex. Information such as illustrated is vital for deciding the surgical approach to preserve the functional status of the patient.

these patients **(Fig. 3)**. FMRI studies of the motor cortex on hyperthyroid patients **(Fig. 4)** have also been done and bilateral activation of the motor cortices have been shown in these patients suggesting greater need recruitment of neural networks for the necessary tasks.

CONCLUSIONS

In conclusion, fMRI has tremendous potential in the evaluation of the normal as

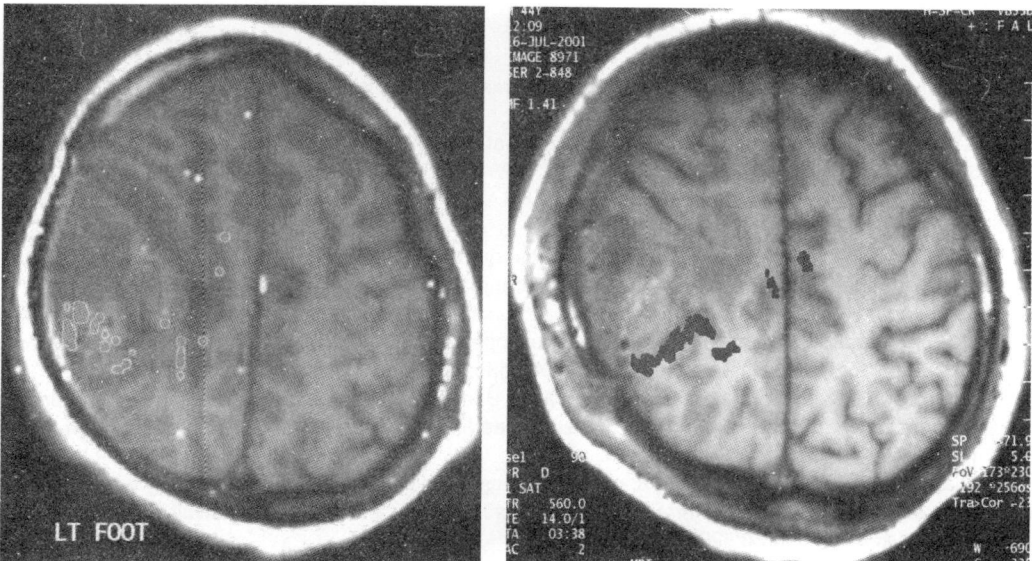

Fig. 3 : Right sided motor area activation in a case of crenral glioma. Pre-operative (A) and Post-operative (B) studies showing the distorted motor functional area posterior to the tumor (A) and preservation of the same in the post-operative study (B).

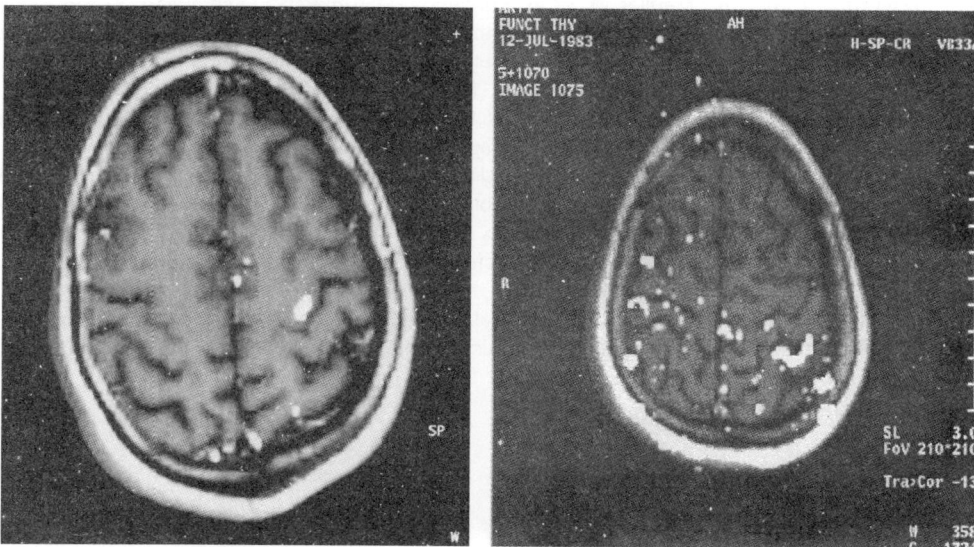

Fig. 4 : Bilateral motor area activation seen in an hyperthyroid patient (B) as compared to only left sided activation (A) in normal volunteer using the right index finger tapping paradigm.

well as the abnormal brain. Various clinical and research applications have been recognized and appear promising. Extensive scope still exists in exploration of the new technique for different aspects in brain surgery and medicine.

REFERENCES:

1. Frahm J: Magnetic resonance functional neuroimaging: New in sights into the human brain. *Current Science* 1999;76:735-43.
2. Castelijns JA, Lycklama GJ, Mukherji SK: Functional MRI: Background and clinical applications. Seminars in Ultrasound, CT & MRI 21, 2000;428-33.
3. Yetkin FZ, Ulmer JL, Meuller W, et al: Functional MR activation correlated with intra-opertive cortical mapping. *AJNR* 1997;18:1311-15.
4. Jack CR, Thompson RM, Butts RK, et al: Sensory motor cortex : correlation of presurgical mapping with functional MR imaging and invasive cortical mapping. *Radiology* 1994;190:85-92.
5. Buchbinder BR, Tiang Hj, Cosgrove GR, et al: Functional mapping of sensorimotor cortex: correlation between functional MRI, 0 – 15 PET and intra-operative cortical stimulation in individual subjects. Proceedings of 32nd Annual meeting of the American Society of Neuroradiology 1994;162.
6. Berkelbach vander Sprenkaal JW, Verheul J, de Boer RW, et al: Functional imaging of dominance in sensorimotor cortex of the human brain. Proceedings of 12th SMRM Annual meeting, New York 1993;1397.
7. Atlas SW, Howard RS, Maldjian J, et al: Functional Magnetic Resonance Imaging of the regional brain activity in patients with intracerebral gliomas : Findings and implications in clinical management. *Neuroradiology* 1996;38:329-38.
8. Sorenson AG, Rosen BR: Functional MRI of the brain in Magnetic Resonance Imaging of the brain and spine by Atlas S.W. (ed). Lippincott-Raven Publishers, Philadelphia 1996;1501-45.
9. Haughton VM, Turski P, Meyerband B, et al: the clinical applications of the functional MR imaging. *Neuroimaging Clin N Am* 1999;9:295-308.
10. Yetkin FZ, Ulmer JL, Meuller W, et al: Functional MR imaging assessment of the risk of post operative hemiperesis after excision of cerebeal tumors. *Instt L Neuroradiol* 1997;4:253-57.
11. Desmond JE, Sum JM, Wagner AD, et al: Functional MR measurement of language lateralization in Wada tested pateints. *Brain* 1995;118:1411-19.
12. Binder JR, Swanson SJ, Hammeke TA, et al: Determination of language dominance using functional MRI: A comparison with Wada test. *Neurology* 1996;46:978-84.
13. Hinki RM, Hu X, Kim T, et al: The use of multislice functional MRI during internal speech to demonstrate the lateralization of language function. Proceedings of 12th SMRM Annual meeting , New York, 1993;63.
14. Detre JA, Maccotta L, King D, et al: Functional MRI lateralization of memory in temporal lobe epilepsy. *Neurology* 1998;50:926-32.
15. Brunberg JA, Chenvert TL, Mekeever PE, et al: In vivo MR determination of water diffusion coefficients and diffusion anisotropy: Correlation with structural alteration in Gliomas of the cerebral hemispheres. *AJNR* 1995;16:361-71.

30

Role of Rapid Histopathological and Cytological Diagnostic Techniques

M Tatke

The role of the pathologist is in guiding the surgeon in making a clinically relevant intraoperative diagnosis which in most cases means whether a tumour is present[1] or not and the neuropathologist is facilitated by two diagnostic techniques available in response to an operating room consultation request by the neurosurgeon, namely histological frozen sections by means of the cryostat and cytological smear preparations.

Use of frozen section examination as an intraoperative diagnostic method is popular in North America[2] and this was facilitated by the development in the 6th decade of the last century of the cryostat-a cabinet cooled to −20°C to −30°C and enclosing a microtome knife.

On the other hand, utilization of smears for rapid intraoperative diagnosis is popular in the United Kingdom and was first introduced in the 1920's and 1930's.

An understanding of the advantages and limitations of these preparations is essential for effective communication between the neurosurgeon and neuropathologist during such a consultation, and equally important is the recognition that use of tissue for rapid diagnosis reduces the amount left for subsequent permanent paraffin embedded sections which are the gold standard for pathologic diagnosis. Hence the choice of technique should be judicious taking into consideration the sample size and type of tissue.

The information required of a neuropathologist on an intraoperative specimen is whether the tissue is adequate i.e., whether the tissue is viable and whether it will suffice for further ancillary techniques like electron microscopy, study of tumour biology, biochemical analysis and secondly whether it is representative of the lesion. For this it is essential that the neuropathologist be informed regarding the site of the biopsy, the neurosurgeon's operative findings and gross impressions, the radiological appearance of the lesion, the clinical differential diagnosis and whether the patient has received any prior chemo or radiotherapy, since these can alter the morphological appearances of the lesion.

On an adequate specimen, the neuropathologist is expected to opine whether the lesion is neoplastic or nonneoplastic and if the latter whether it is reactive or any specific pathology is seen.

The presence of reactive astrocytosis, perivascular lymphocytic cuffing, presence of foamy macrophages **(Fig. 1)** may be associated with CNS neoplasms especially those previously treated by surgery, chemo or radiotherapy, but their presence usually indicates a demyelinating process like demyelination, infection (abscess, encephalitis) or infarction. Granulomas of the epithelioid cell type go in favour of tuberculosis, and the foreign body type suggest a fungal infection. The fungal body when seen of course clinches the diagnosis.

If the lesion is neoplastic, one has to say whether it is primary to the CNS or metastatic, and if primary if benign or malignant and the cell type and grade (whether high or low grade).

ADVANTAGES OF SMEAR PREPARATIONS

1. The main advantage over frozen sections is the rapidity with which the smear preparations are made. From the time the tissue is received in the laboratory, it takes approximately 5 minutes for the smears to be made, fixed and stained while a frozen section preparation takes approximately 15 minutes and requires considerable technical expertise.

2. Preservation of cytological features–another important advantage of smears is the superior preservation of nuclear and cytoplasmic detail which usually are lost during freezing. This happens since the soft oedematous brain tissue is very susceptible to ice crystal artifact which mars the morphological details of the lesion. Nuclear details which can be appreciated in smears and helps in differentiation is the dense uniform chromatin in medulloblastomas **(Fig. 2)** and fine chromatin with prominent nucleoli in lymphomas. These are also seen in germinomas. Another area of differentiation is between ependymoma and astrocytoma the former may have nucleoli and the latter do not. This difference also comes in useful in differentiating oligodendrogliomas from central neurocytomas **(Fig. 3)** which have similar appearances in smears except that the latter have nucleolar prominence. Nuclear pseudoinclusions seen in meningiomas often help in differentiation, especially in CP angle tumours like schwannomas and meningiomas especially the fibroblastic variant. These can also be seen in glioblastomas and melanomas.

Cytoplasmic features of help are glial processes and fibrillary background seen in ependymomas and astrocytomas. These may be obscured in frozen sections. Vacuolated cytoplasm can be seen in chordomas **(Fig. 4)**, germinomas and metastatic carcinomas and this feature in the latter helps in differentiating from glioblastomas in some cases.

Cell types with interdigitating cell processes like astrocytomas, ependymomas or tumours with cell to cell junctions like meningiomas, craniopharyngiomas and metastatic carcinomas smear in cohesive sheets whereas those without like medulloblastomas, lymphomas, pituitary adenomas and oligodendrogliomas smear as individual cells.

The background in a smear may contribute to the diagnosis, eg. a necrotic background indicates a high grade tumour like glioblastoma, lymphoma, metastatic carcinoma or caseous necrosis in cases of tuberculosis.

Psammoma bodies indicate a meningioma, chordoma has a myxoid background and presence of Rosenthal fibres **(Fig. 5)** indicate a low grade tumour (pilocytic astrocytoma, ependymoma, and around hemangioblastomas and craniopharyngiomas). Endothelial hyperplasia indicates a high grade glial tumour usually glioblastomas **(Fig. 6)**.

3. Preservation of architecture–although in smears the tissue architecture is lost, it may be preserved in certain tumours like papillae in choroid plexus papillomas **(Fig. 7)**, whorls in meningiomas and canals in ependymomas **(Fig. 8)**.

4. Conservation of tissue- since very little tissue is required in preparing smears, tissue free from freezing artifact can be conserved for permanent preparations and electron microscopy, and is ideal for tissue obtained at stereotactic biopsy or open biopsies from brain stem and spinal cord.

5. Smear preparations help in infection control because viral infections, prion diseases, HIV or tuberculosis can contaminate the cryostat.

6. Delineation of tumour margins- Since smear preparations are made fast, this technique can be used to determine tumour margins.

DISADVANTAGES OF SMEAR PREPARATIONS

1. Firm tissue cannot be smeared eg. certain cases of TB, fungal infections neurofibromas, hemangiopericytomas, hemangiomas, angiofibromas.

2. Smear thickness cannot be controlled and hence it is difficult to comment upon cellularity and anaplasia.

3. Sampling error is possible – as only a small piece is required to make a smear, the tissue sent may not be adequately sampled. This is overcome by taking pieces from all parts of sample sent.

ADVANTAGES OF FROZEN SECTIONS

1. The main advantage is familiarity of pathologist with tissue appearances since this is similar to examination of paraffin sections.

2. Sampling error is overcome since tissue received is frozen and cut in its entirety.
3. Preservation of tissue architecture
4. Thickness of the tissue is controlled and therefore evaluation of cellularity and anaplasia is easier.

DISADVANTAGES OF FROZEN SECTIONS

These are akin to advantages of smear preparations ie take longer to prepare, more tissue is required, presence of freezing artifact, cryostat contamination by infectious material, and not suitable for examination of bony tissue.

BONY TISSUE EXAMINATION

Since this type of tissue cannot be subjected to smear preparation or cryostat sectioning, impression smears can be made from this type of material and is useful in diagnosis of osteolytic metastasis, intraosseous meningiomas and other bony lesions like eosinophilic granulomas and chordomas.

USE OF MICROWAVE TECHNOLOGY

This has tremendous potential in diagnostic neuropathology[4] involving almost all methodologies from tissue fixation to processing for light and electron microscopy, mmunocytochemistry and molecular neuropathology. The main advantage is the remarkable shortening of processing time in all procedures and thus facilitates rapid diagnosis.

Examples are shortening of time required for smear fixation from 2 minutes to 30 seconds, prior fixing of the unfixed tissue in the microwave oven and then cutting frozen sections in the cryostat- this improves the quality of the frozen sections remarkably.

Another important field is shortening of time taken for histochemical stains eg. reticulin from 45 minutes to 10 minutes. This stain is of considerable importance in differentiating brain tumours and thus an intraoperative diagnosis is possible.

In immunocytochemistry, the processing time is shortened from 3-4 hours or overnight to 20 minutes, with reduction in time taken for all steps like antigen retrieval in formalin fixed tissues, blocking of endogenous peroxidase, incubation with primary and secondary antibodies and visualization of reaction product.

However it is very important to optimize the procedure carefully since overheating in the oven can lead to destabilization of antigen antibody complexes and destruction of epitopes leading to nonspecific staining.

TISSUE PROCESSING

For smears all the unfixed tissue received has to be sampled adequately with

1mm pieces taken from all parts of the tissue sent, after which smears are made by smearing the tissue on the slide with the help of another slide held at right angles. The tissue should not be compressed excessively since this causes crush artifact. If the tissue is firm, it should be frozen and cut in a cryostat. After the smear is made it is fixed in 90% ethanol, stained with either 1% toluidine blue solution which gives excellent nuclear details or by the Hematoxylin and Eosin stain which allows comparison with paraffin sections and gives good cytoplasmic outlines.

It is essential to keep tissue for permanent paraffin sections.

If the tissue is firm, it should be cut in the cryostat after fixing in the microwave oven if possible, since this improves the quality of sections. The cryostat temperature is maintained at –20°C to –30°C, and frozen sections of 5-8 micrometer thickness are cut and stained with H.& E. As already mentioned the cryostat cut sections can be utilized for special stains like reticulin, PAS/Alcian Blue for mucins and immunocytochemistry, after fixing in the microwave.

We have reported upon 1054 intraoperative specimens by the smear technique from 1995-2000 and the smear diagnosis was compared with permanent paraffin section diagnosis. Out of these an accurate diagnosis (regarding tumour versus nontumour, primary versus metastatic tumours, benign versus malignant tumours, reactive tissue, specific nontumour lesions and type and grade of tumours) was made in 964 (91.5%) specimens **(Graph 1)**. Of these, 872 were tumours, 82 were nontumour lesions and 5 each were inadequate and nonrepresentative. These were confirmed on the paraffin sections prepared from the tissue sent for

(Graph 1)

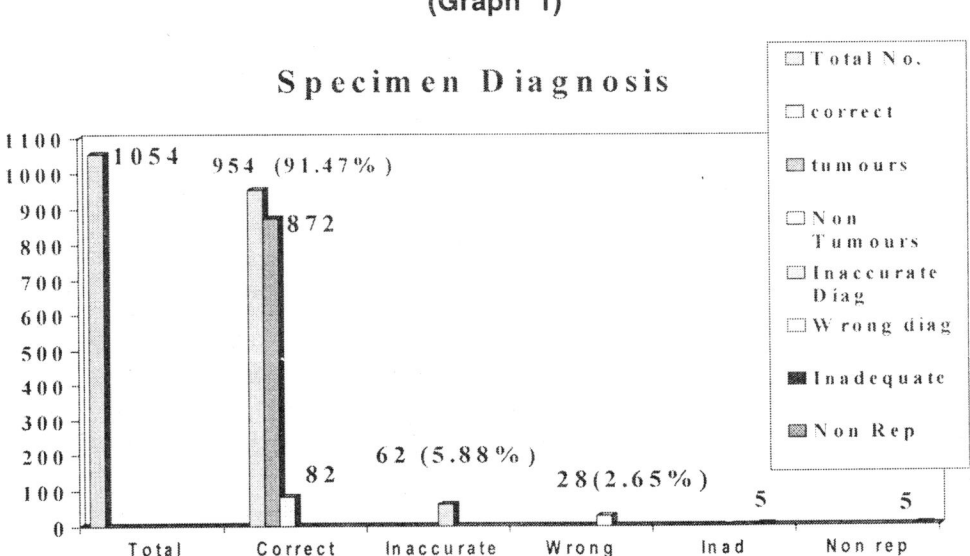

intraoperative diagnosis. The inadequate tissues were necrotic and subsequent samples sent were glioblastomas (2 cases), metastatic carcinoma, tuberculosis and gliosarcoma (1 case each).

The nonrepresentative tissues comprised the infiltrating part of tumours, again confirmed by paraffin sections. Subsequent specimens sent of these cases were anaplastic astrocytoma (2 cases), glioblastoma (2 cases) and one case of metastatic neuroendocrine tumour.

Of the 82 nontumour lesions 53 were tuberculosis (intracranial and spinal), 12 were normal cortex and white matter, 6 cases showed reactive changes around tumours or cysts, 4 cases of epidermoid cysts, 3 cases each of arachnoid cysts and nonspecific abscess and 1 case of infarction. **(Graph 2).**

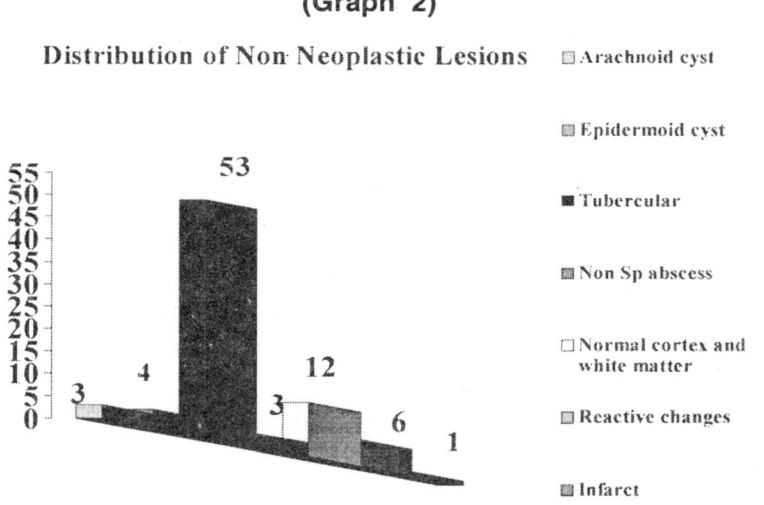

(Graph 2)
Distribution of Non Neoplastic Lesions

The 872 tumours diagnosed on smears are listed in **Table 1.**

In 62 cases (5.88%), an inaccurate diagnosis was made, i.e. grading was correct but the exact glial cell type was not recognized on smears. Of these 18 cases diagnosed as low grade gliomas on smears, 10 were astrocytomas, 5 were oligodendrogliomas and 1 case each of clear cell epenymoma, pilocytic astrocytoma and SEGA on paraffin sections.

9 cases were diagnosed as low grade astrocytomas on smears but all were mixed oligoastrocytomas. 33 high grade gliomas on smears were glioblastoma (22 cases), anaplastic oligodendroglioma (10 cases) and 1 case of gliosarcoma. **(Graph 3)**.

I case diagnosed as atypical meningioma was a clear cell meningioma, and one meningioma was actually a meningeal hemangiopericytoma.

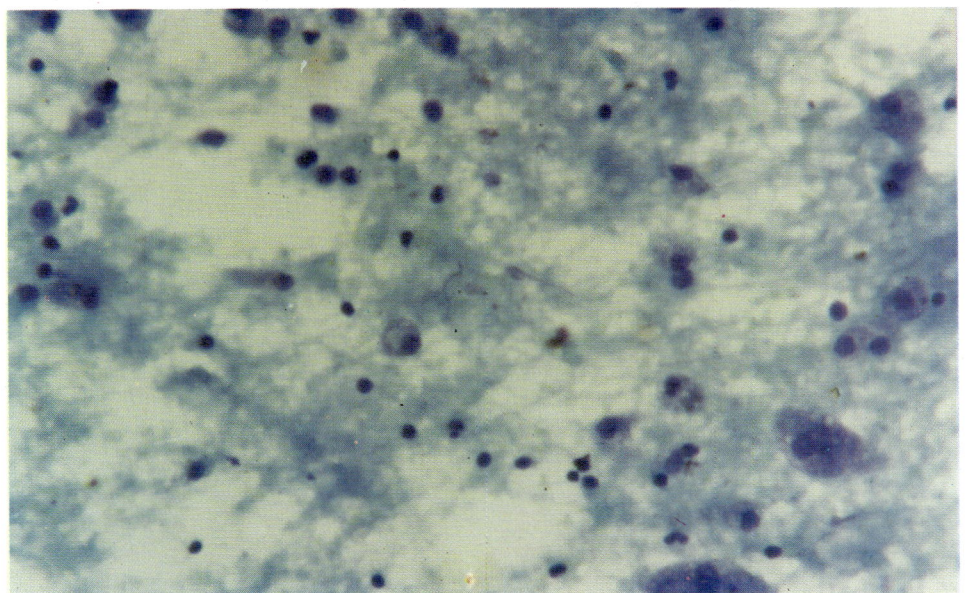

Fig. 1 : Smear showing foamy macrophages, inflammatory cells and oedema in white matter in a case of infarction

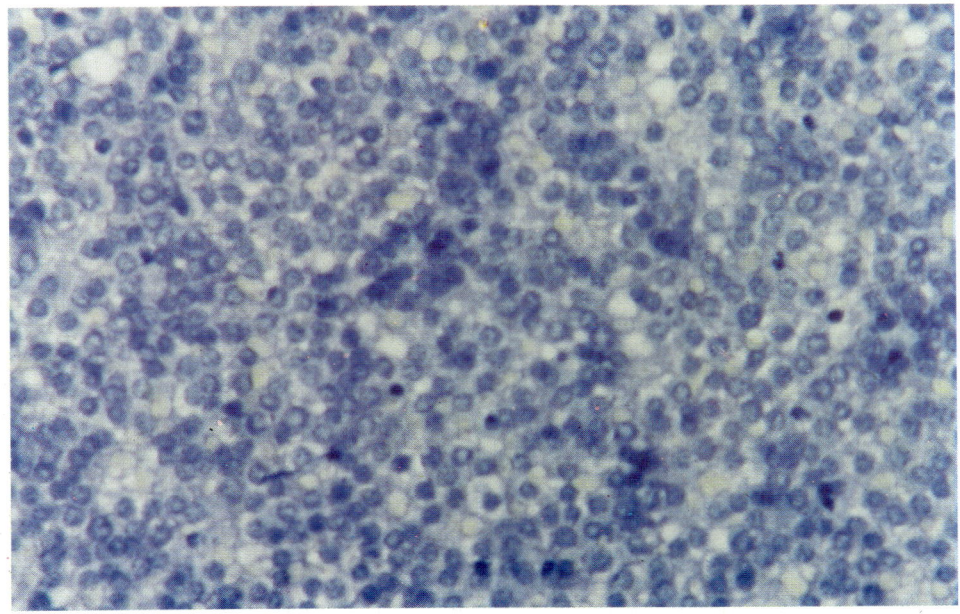

Fig. 2 : Cellular Smear from a case of medulloblastoma. Mitotic figures are seen, and the cells are smeared individually.

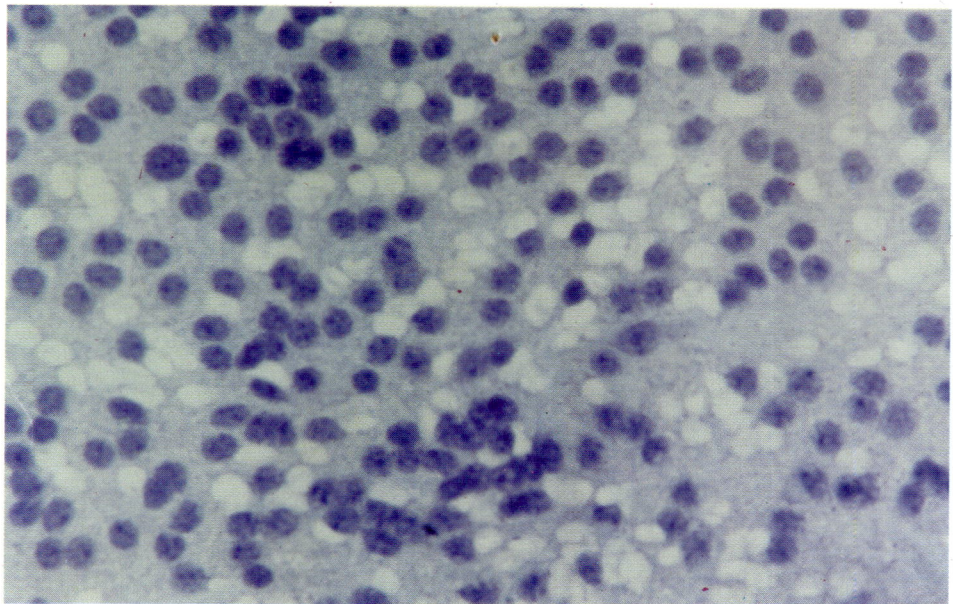

Fig. 3 : Smear from a case of central neurocytoma, comprising of monomorphic cells. The nuclei have 1-2 prominent nucleoli.

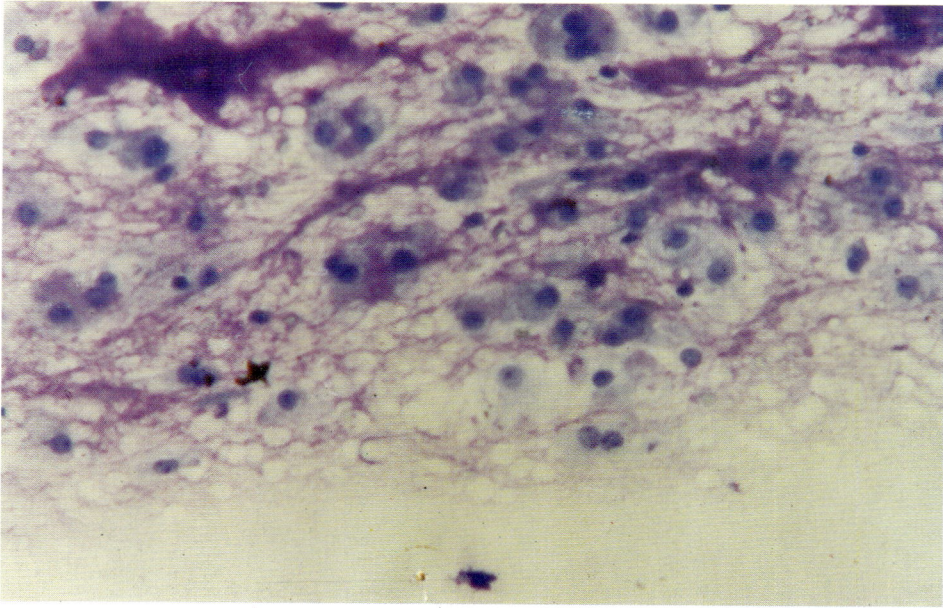

Fig. 4 : Smear from a chordoma showing vacuolated cytoplasm of the cells and a myxoid metachromatic background.

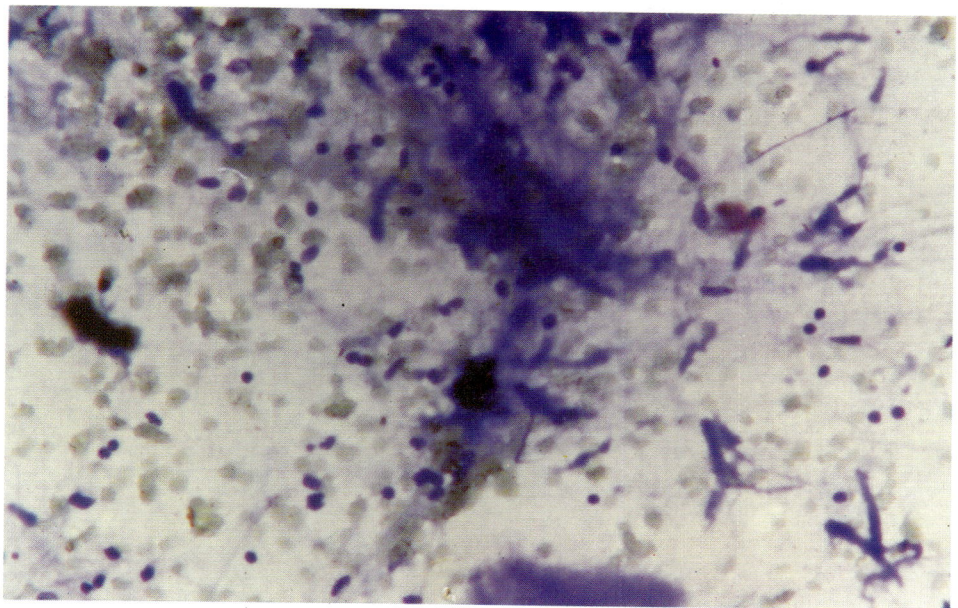

Fig. 5 : Rosenthal fibres seen around a hemangioblastoma.

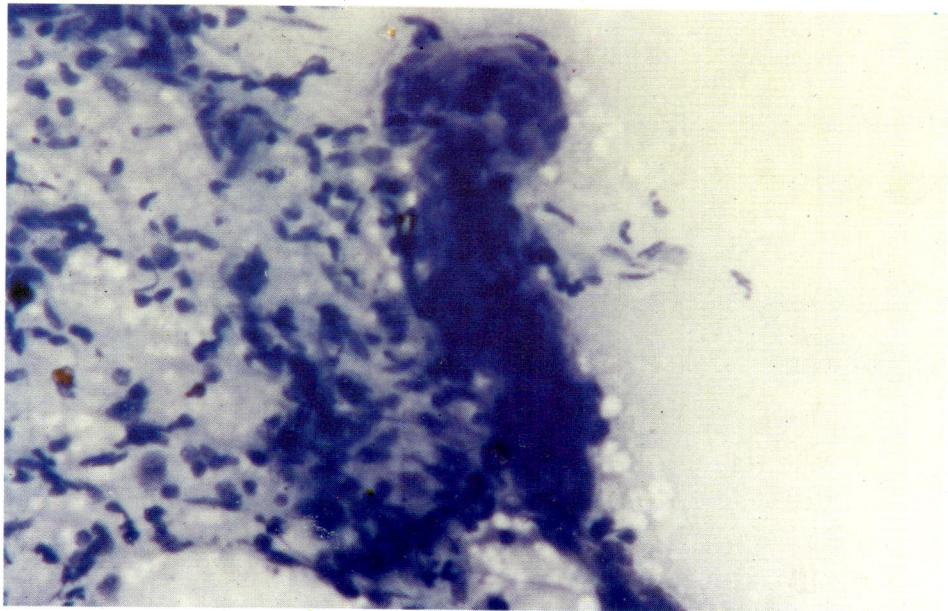

Fig. 6 : Endothelial hyperplasia in a glioblastoma.

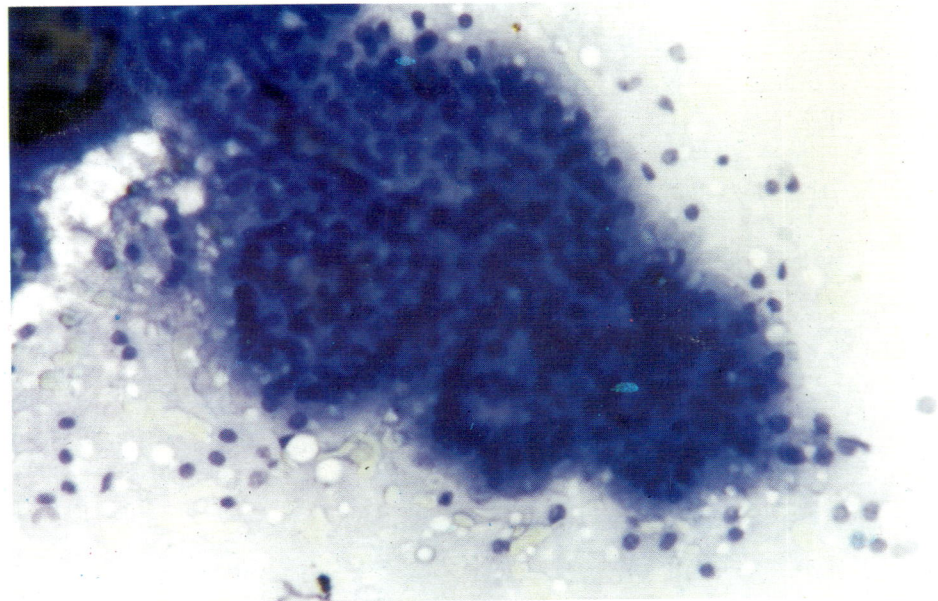

Fig. 7 : Papillary process and few individually smeared cells in a choroid plexus papilloma

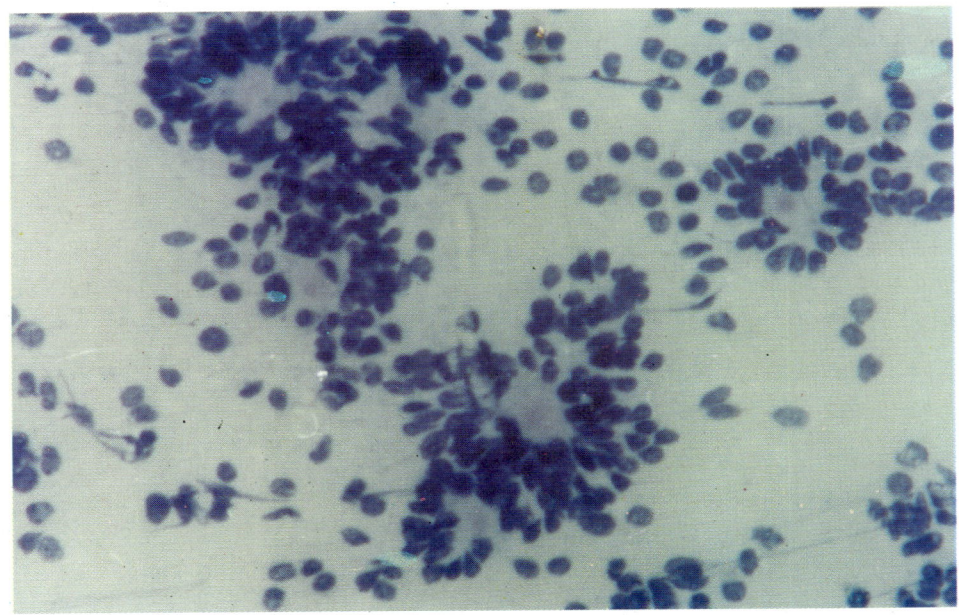

Fig. 8 : Ependymal canals in a 4 ventricular ependymoma

Table 1

Tumour	Count
Pilocytic astrocytoma	10
Low Grade Astocytoma	81
Anaplastic Astrocytoma	24
GBM	140
SEGA	6
Oligodendroglioma	23
Anaplatic Oligodendroglioma	12
Mixed tumours (Oligoastro)	3
(Oligoastro G III)	6
Ependymoma	32
Myxopapillary Ependymoma	3
CPP	5
CN	3
PNET	10
Medulloblastoma	43
Schwannoma	105
Neurofibroma	3
Meningioma	223
Hemangiopericytoma	8
NHL	5
Germinoma	3
Pituitary adenoma	92
Craniopharyngioma	8
Metastatic carcinoma	31
Plasmacytoma	7
Chordoma	4
Giant cell tumour	2

(Graph 3)

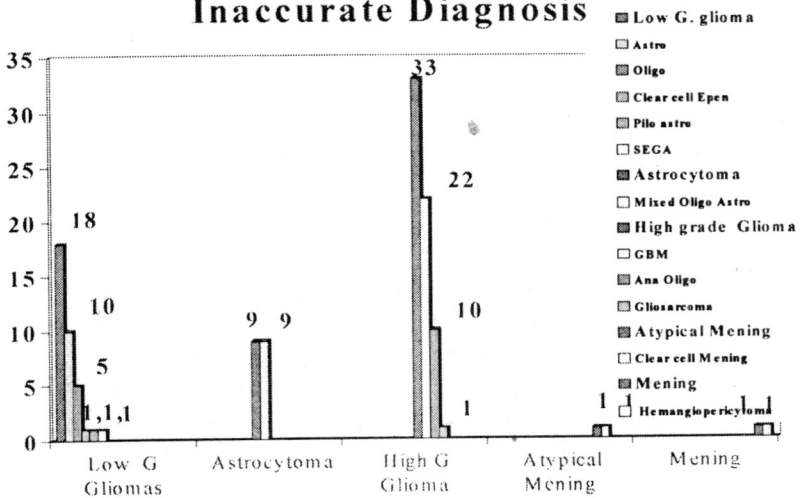

Inaccurate Diagnosis

The 28 cases (2.65%) wrongly dignosed on smears are listed in **Table 2,** along with the paraffin section diagnosis. The reasons for these were either firm tissue smeared due to lack of frozen section section facility or inadequate information or inadequate sampling.

The lesions found difficult to smear were some cases of tuberculosis, especially spinal extradural lesions, fungal infections, neurofibromas, some hemangiopericytomas, angiofibromas and hemangiomas, which were then cut in the cryostat.

Table 2

Smear Diagnosis	Paraffin Diagnosis
SEGA	Fibroblastic meningioma
Ependymoma	Medulloblastoma
Meningioma	Metastatic carcinoma
SCC ?Anaplastic meningioma	Sebaceous carcinoma
Hemangioblastoma	Oligo III
Astrocytoma G II	Oligo II
Tuberculosis	Non Specific Inflammation
Schwannoma	Meningioma
Pineal parenchymal tumour	Immature teratoma
Meningioma	Oligodendroglioma
Medulloblastoma	ATRT
Medulloblastoma	Ependymoma
Meningioma	Schwannoma
Medulloblastoma	anaplastic oligodendroglioma
Astrocytoma G II	Pilocytic Astrocytoma
Meningioma	Olfactory neuroblastoma
Pilocytic astrocytoma	Ependymoma
Oligodendroglioma GII	Astrocytoma G II
Ependymoma	Pilocytic astrocytoma
Papillary Ependymoma	Choroid plexus carcinoma
High grade glioma	Oligodendrogliom GII
? Meningioma,? Ependymoma	Anaplastic ependymoma
Meningioma	Craniopharyngioma
Arachnoid cyst	Rathke's cleft cyst
Granulomatous Inflammation	PML

In conclusion before making an intraoperative diagnosis, the pathologist should have the relevant clinical, radiological and operative details on the patient. The choice of technique used for this should depend upon the type of tissue ie. whether firm or soft, and provided one is familiar with the morphological appearances.The smear preparation has proved to be a reliable and fast technique for making an intraoperative diagnosis with results comparable to other reported studies. There are two studies reported from India which we are aware of, with diagnostic accuracy of 87% and 89.5%.

REFERENCES:

1. Scheithauer BW: CNS and Pituitary from Intraoperative Pathologic Diagnosis, Ed. Silva EG and Balfour Kraemer B 1987;167-220
2. Folkerth RD: Smears and frozen sections in the intraoperative diagnosis of central nervous system lesions. *Neurosurgery clinics of North America* 1994;5(1):1-18.
3. Ironside JW: Update on central nervous system cytopathology, II. Brain smear technique. *J Clin Patol* 1994;47:683-88.
4. Ainley CD, Ironside JW: Microwave technology in diagnostic neuropathology. *J Neurosc Metds* 1994;55:183-90.
5. Asha T, Shankar SK, Vasudev Rao T, Das S: Role of squash smear technique for rapid diagnosis of neurosurgical biopsies- a cytomorphological evaluation. *Ind J Pathol Microbiol* 1989;32:3:152-60.
6. Shah AB, Mazumdar GA, Chitale AR, Bhagwati SN: Squash smear preparation and frozen section in intraoperative diagnosis of CNS tumours. *Acta Cytol* 1998;42(5):1149-54.

31

Malignant Gliomas: Radiosensitizers

NJ Laperriere

There are many compounds that sensitize tissues to the effects of radiation therapy. However, we are only interested in compounds that differentially sensitize tumour tissue more so than normal tissues. The ideal radiation sensitizer will sensitize tumour and have no effects on normal tissue.

Hypoxic cell sensitizers were one of the first class of compounds investigated as a possible radiation sensitizer in malignant gliomas. The cell killing effect of radiation therapy has been well established to be significantly influenced by the presence or absence of oxygen. In a completely hypoxic environment, radiation dose must be doubled or tripled to achieve the same degree of cell killing in any given biologic system as when that same system is well oxygenated.[1] When a patient is breathing room air, his tissue oxygenation level will be approximately 155 mm Hg. Tissue oxygen levels start to be associated with increasing radiation resistance when oxygen levels fall below approximately 10 mm Hg.[1]

Figure 1 is a coronal MRI image of a patient with a glioblastoma multiforme. It

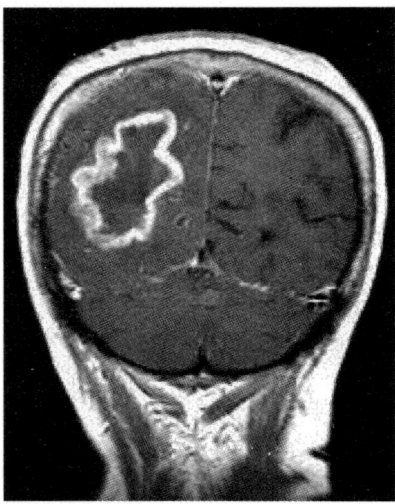

Fig. 1: Coronal MRI T1 weighted image with gadolinium demonstrating in a patient with a glioblastoma multiformes

has long been suspected that viable tumour cells that exist in a hypoxic environment surround the large necrotic central core. Increasing distance from a vascular capillary is associated with a decreasing level of oxygen as a result of oxygen consumption by the intervening cells, such that beyond 150 microns oxygen levels are so low that cells die.[1] However, just inside that 150-micron distance, there are viable cells existing in a zone of hypoxia that at a theoretical level are suspected of being increasingly resistant to radiation.[1] Rampling and colleagues have performed direct intra-operative oxygen concentration measurements in 10 patients with glioblastoma multiforme.[2] There were 2486 measurements performed. The median intra-tumoral oxygen concentration was 7.4 mm Hg, and 39.6% of measurements were < 2.5 mm Hg. These direct measurements in human tumours clearly demonstrated significant areas of hypoxia in patients with malignant gliomas.

Chang and colleagues performed a non-randomized study of the use of hyperbaric oxygen in patients with malignant gliomas.[3] Thirty-eight patients received radiation therapy with hyperbaric oxygen and 42 patients were treated with radiation only. Survival at 18 months was 28% in the hyperbaric oxygen group versus 10% for the non-hyperbaric oxygen group. Obviously, treating patients in a hyperbaric chamber is not a very feasible procedure for patients in general, and other means of sensitizing hypoxic tumours were sought.

The nitroimidazoles are a class of compounds that have a significant radiosensitizing effect on hypoxic tissues. The lead compounds of this group were metronidazole and misonidazole. In 1976 Urtasun et al published a randomized study in 31 patients where there was a delay in dying of 4.5 months for the radiation therapy with metronidazole versus for the radiation therapy only group.[4] There are several concerns about this study: the number of randomized patients was small, the radiation delivered in this study was an unusual fractionation (30 Gy/9 fractions, 3 fractions per week), and the median survival for the radiation only group was less than expected. Metronidazole was associated with significant toxicity, so subsequent studies utilized misonidazole, which in addition to being better tolerated clinically, was a more potent hypoxic cell sensitizer in the laboratory. **Table 1** displays all the subsequent published studies with nitroimidazoles in much larger numbers of patients with more conventionally fractionated radiation therapy, and all studies have been negative. Possible reasons for the failure of these studies may include any or all of the following: the intratumoral concentrations of misonidazole may not have been adequate as a result of dose-limiting neurotoxicity, reoxygenation may occur during the 5 to 6 weeks of daily fractionated radiotherapy to counter the effect of hypoxia, or hypoxia may not be a rate-limiting phenomenon in this disease.

The halogenated pyrimidines 5–bromodeoxyuridine (BUdR) and 5–iododeoxyuridine (IUdR) are similar to the normal DNA precursor thymidine, having a halogen

Table 1 : Randomized studies comparing sensitized radiotherapy with nitroimidazoles to radiotherapy alone in malignant glioma.

Study (Reference)	Hypoxic Drug Studied	Total # Patients	Median Survival (Months) Sensitizer	Median Survival (Months) Radiation	Overall Survival p value
Urtasun, 1976 [4]	Metronidazole	29	7	4	p<0.02
Bleehen, 1981 [22]	Misonidazole	38	9	7	n.s.
Urtasun, 1982 [23]	Metronidazole	36	5	6	n.s.
	Misonidazole	42	7	6	n.s.
Sack, 1982 [24]	Misonidazole	102	10	12	n.s.
EORTC, 1983 [25]	Misonidazole	163	11	12	n.s.
MRC, 1983 [26]	Misonidazole	384	8	9	n.s.
Stadler, 1984 [27]	Misonidazole	45	13.8	9.8	n.s.
Shin, 1985 [28]	Misonidazole	86	12	10	n.s.
Hatlevoll, 1985 [29]	Misonidazole	244	10	10	n.s.
Nelson, 1986 [30]	Misonidazole	146	11.5	12.5	n.s.
Okkan, 1988 [31]	Ornidazole	40	15	10	n.s.
Deutsch, 1989 [32]	Misonidazole	279	9	10	n.s.

Note: n.s., not statistically significant.

substituted in place of a methyl group. These compounds are incorporated into DNA in place of thymidine in a competitive fashion, which leads to an increased sensitivity of cells incorporating these compounds to the effects of radiation and ultraviolet light.[1] The rationale for using these compounds in the treatment of brain tumours is that mitotically active tumour cells are much more likely to incorporate these compounds than the slowly replicating glial and vascular cells in the normal brain.

Phillips et al reported an increase in median survival for anaplastic astrocytoma patients from 82 weeks in prior studies to 252 weeks in patients treated with radiation, BUdR, and chemotherapy.[5] There was no significant improvement seen with the use of BUdR for patients with glioblastoma.[5] As a result of this observation, the RTOG embarked on a randomized study for patients with anaplastic astrocytoma: 60 Gy in 30 fractions with and without BUdR, both arms followed by PVC chemotherapy.[6] The study was closed prematurely when the initial 190 patients were analysed. The one-year survival rate for radiotherapy, PVC, and BUdR was 68% versus 82% for radiotherapy plus PVC (one-sided p=0.96)(54). In a recent update, 4 year survival was 51% for both arms of the study, with the same conclusion that BUdR was not of benefit in this patient population.[7]

Tirapazamine is a lead drug of a class of compounds referred to as hypoxic cytotoxins.[8] These compounds are preferentially cytotoxic to hypoxic cells, and the differential hypoxic cytotoxicity ratio has been found to be in the range of 50-150 for a variety of human cell line tumours for this drug.[8-10] Because these compounds are significantly more toxic to hypoxic cells and not particularly toxic to well oxygenated cells, hypoxic cytotoxic agents would be complementary to the effects of radiation therapy which is selectively more toxic to well oxygenated cells. Del Rowe et al performed a phase II study of concurrent tirapazamine with conventionally fractionated radiation therapy in 124 patients with glioblastoma multiforme. Patients received 60 Gy/30 fractions over 6 weeks with tirapazamine 3 times per week for a total of 12 doses.[11] Median survival time was 10.8 months for the cohort who received tirapazamine at 159 mg/m^2, and 9.5 months for the cohort treated with the 260 mg/m^2 dosage. These results were not significantly different from the expected survival utilizing the RTOG recursive partitioning analysis based on the distribution of prognostic factors in these cohorts of patients on this study.

The ras protein is an important transducer of signals from growth factor receptors to downstream effector molecules, and cells transformed with the activated ras-oncogene are often more resistant to cell killing by radiation therapy.[12,13] To be active, ras needs to be located at the inner surface of the cell membrane, and this process is dependent on prenylation. Ras is prenylated by farnesyl-protein transferase. Farnesyltransferase inhibitors (FTIs) inhibit this latter process and inhibit the growth of human tumour cell lines in vitro.[14,15] Glioblastomas are not tumours that are associated with ras mutations, but are associated with a number of ligand-dependant and independent growth factors. Receptor induced ras activation is a common feature of glioblastomas, and laboratory studies have confirmed inhibition of growth of glioma models with FTIs.[16,17] Accordingly, the RTOG is about to embark on a phase II study of an FTI in 72 patients with glioblastoma. The FTI will be administered both during radiotherapy for its possible radiation senitization effects, and for a total of 18 months for its possible innate anti-tumour effects.

Epidermoid growth factor (EGFR) is activated by ligand binding to transmit mitogenic signals to the nucleus, and approximately one third of patients with glioblastomas are associated with EGFR gene amplification.[18,19] EGFR expressed at high levels have been associated with more aggressive tumours and poor prognosis and increasing radiation resistance.[20] Accordingly, antibodies that prevent the binding of ligand to EGFR could sensitize tumours to radiotherapy, and when utilized with fractionated radiation therapy over several weeks, could be useful to inhibit cellular proliferation during radiation therapy.[21] A number of monoclonal antibodies have been developed against EGFR, and the RTOG will soon be embarking on a phase II study of an EGFR inhibitor in a cohort of patients with glioblastoma.

In summary, there has been no benefit to the use of nitroimidazoles or halogenated pyrimidines in the treatment of malignant gliomas. The potential benefit of additional hypoxic cytotoxins, FTIs and EGFR inhibitors remain to be explored in future clinical studies. It is unlikely that any single strategy will be effective in a heterogeneous population of patients with glioblastoma multiforme, but the future therapy of these patients will likely evolve to individual tumour characterization at a molecular level and utilization of a battery of tumour specific radiosensitizers and biologic modifiers.

REFERENCES:

1. Hall EJ: Radiobiology for the Radiologist. 3 ed. New York: J.B. Lippincott Co.; 1988.
2. Rampling R, Cruickshank G, Lewis AD, Fitzsimmons SA: Workman P. Direct measurement of pO2 distribution and bioreductive enzymes in human malignant brain tumors. *Int J Radia Onco Bio Phys* 1994;29:427-31.
3. Chang CH: Hyperbaric oxygen and radiation therapy in the management of glioblastoma. *Natl Cancer Inst Monogr* 1977;46:163-69.
4. Urtasun R, Band P, Chapman JD, Feldstein ML, Mielke B, Fryer C: Radiation and high-dose metronidazole in supratentorial glioblastomas. *N Engl J Med* 1976;294(25):1364-67.
5. Phillips TL, Prados MD, Bodell WJ, Levin VA, Uhl V, Gutin PH: Rationale for and experience with clinical trials of halogenated pyrimidines in malignant gliomas: the UCSF/NCOG experience. In: Dewey WC, Edington M, Fry RJM, Hall EJ, Whitmore GF, eds.: Radiation Research: A Twentieth-Century Perspective. Volume II: Congress Proceedings. San Diego: *Academic Press*; 1992;601-06.
6. Prados M, Scott C, Phillips T, Davis R, Sandler H, Buckner J, et al: Phase III randomized study of radiotherapy plus PCV with or without BUdR for the treatment of anaplastic astrocytoma: RTOG 9404 interim report. *Int J Radia Onco Bio Phys* 1997;39 (2, Supplement):138.
7. Prados M, Seiferheld W, Sandler H, Buckner J, Phillips T, Schultz C, et al: RTOG 99-04: A Phase-3 Randomized Study of Radiotherapy Plus Procarbazine, CCNU, and Vincristine (PCV) With or Without BudR for the Treatment of Anaplastic Astrocytoma. *Int J Radiat Oncol Biol Phys* 2001;51(3S1):207.
8. Brown JM: SR 4233 (tirapazamine): a new anticancer drug exploiting hypoxia in solid tumours. [Review]. *British J of Cancer* 1993;67:1163-70.
9. Brown JM: Therapeutic targets in radiotherapy. *Int J Radiat Oncol Biol Phys* 2001;49(2):319-26.
10. Brown JM, Siim BG: Hypoxia-Specific Cytotoxins in Cancer Therapy. *Semin Radiat Oncol* 1996; 6(1): 22-36.
11. Del Rowe J, Scott C, Werner-Wasik M, Bahary JP, Curran WJ, Urtasun RC, et al: Single arm, open-label phase II study of intravenously administered tirapazamine and radiation therapy for glioblastoma multiforme. *J Clin Oncol* 2000;18(6):1254-59.
12. McKenna WG, Weiss MC, Bakanauskas VJ, Sandler H, Kelsten ML, Biaglow J, et al: The role of the H-ras oncogene in radiation resistance and metastasis. *Int J Radiat Oncol Biol Phys* 1990; 18(4):849-59.
13. Alapetite C, Baroche C, Remvikos Y, Goubin G, Moustacchi E: Studies on the influence of the presence of an activated ras oncogene on the in vitro radiosensitivity of human mammary epithelial cells. *Int J Radiat Biol* 1991;59(2):385-96.
14. James GL, Goldstein JL, Brown MS, Rawson TE, Somers TC, McDowell RS, et al: Benzodiazepine peptidomimetics: potent inhibitors of Ras farnesylation in animal cells. *Science* 1993;260(5116): 1937-42.
15. Kohl NE, Mosser SD, deSolms SJ, Giuliani EA, Pompliano DL, Graham SL, et al: Selective inhibition

of ras-dependent transformation by a farnesyltransferase inhibitor. *Science* 1993;260(5116): 1934-37.
16. Guha A, Feldkamp MM, Lau N, Boss G, Pawson A: Proliferation of human malignant astrocytomas is dependent on Ras activation. *Oncogene* 1997;15:2755-65.
17. Pollack IF, Bredel M, Erff M, Hamilton AD, Sebti SM: Inhibition of Ras and related guanosine triphosphate-dependent proteins as a therapeutic strategy for blocking malignant glioma growth: II— preclinical studies in a nude mouse model. *Neurosurgery* 1999;45:1208-15.
18. von Deimling A, Fimmers R, Schmidt MC, Bender B, Fassbender F, Nagel J, et al: Comprehensive allelotype and genetic anaysis of 466 human nervous system tumors. *J Neuropathol Exp Neurol* 2000;59(6):544-58.
19. von Deimling A, von Ammon K, Schoenfeld D, Wiestler OD, Seizinger BR, Louis DN: Subsets of glioblastoma multiforme defined by molecular genetic analysis. *Brain Pathol* 1993;3(1):19-26.
20. Hale RJ, Buckley CH, Gullick WJ, Fox H, Williams J, Wilcox FL: Prognostic value of epidermal growth factor receptor expression in cervical carcinoma. *J Clin Pathol* 1993;46(2):149-53.
21. Huang SM, Harari PM: Epidermal growth factor receptor inhibition in cancer therapy: biology, rationale and preliminary clinical results. *Invest New Drugs* 1999;17(3):259-69.
22. Bleehen NM, Wiltshire CR, Plowman PN, Watson JV, Gleave JR, Holmes AE, et al: A randomized study of misonidazole and radiotherapy for grade 3 and 4 cerebral astrocytoma. *Br J Cancer* 1981;43(4):436-42.
23. Urtasun R, Feldstein ML, Partington J, Tanasichuk H, Miller JD, Russell DB, et al: Radiation and nitroimidazoles in supratentorial high grade gliomas: a second clinical trial. *Br J Cancer* 1982;46(1):101-08.
24. Sack H, Calcanis A, Godehardt E, Weidtman V, Zulch KJ, Ammon J, et al: [Postoperative radiotherapy of astrocytomas grade 3 and 4 with the radiosensitizer misonidazole. -End results of a multicentric controlled German study]. *Strahlentherapie* 1982;158(8):466-69.
25. Misonidazole in radiotherapy of supratentorial malignant brain gliomas in adult patients: a randomized double-blind study. *Eur J Cancer Clin Oncol* 1983;19(1):39-42.
26. Medical Research Council Working Party of Misonidazole in Gliomas: A study of the effect of misonidazole in conjunction with radiotherapy for the treatment of grade 3 and 4 astrocytomas. *British J Radiology* 1983;56:673-82.
27. Stadler B, Karcher KH, Kogelnik HD, Szepesi T: Misonidazole and irradiation in the treatment of high-grade astrocytomas: further report of the Vienna Study Group. *Int J Radiat Oncol Biol Phys* 1984;10(9):1713-17.
28. Shin KH, Urtasun RC, Fulton D, Geggie PH, Tanasichuk H, Thomas H, et al: Multiple daily fractionated radiation therapy and misonidazole in the management of malignant astrocytoma. A preliminary report. *Cancer* 1985;56(4):758-60.
29. Hatlevoll R, Lindegaard KF, Hagen S, Kristiansen K, Nesbakken R, Torvik A, et al: Combined modality treatment of operated astrocytomas grade 3 and 4. A prospective and randomized study of misonidazole and radiotherapy with two different radiation schedules and subsequent CCNU chemotherapy. Stage II of a prospective multicenter trial of the Scandinavian Glioblastoma Study Group. *Cancer* 1985;56(1):41-47.
30. Nelson DF, Diener-West M, Weinstein AS, et al: A randomized comparison of misonidazole sensitized radiotherapy plus BCNU and radiotherapy plus BCNU for treatment of malignant glioma after surgery: Final report of an RTOG study. *Int J Radiat Oncol BiolPhys* 1986;12:1793-1800.
31. Okkan S, Uzel R, Ober A, Turkan S, Turan N. A: Randomized study of ornizadole as a radiosensitizer in high-grade astrocytomas: preliminary report. *Radiosensitization Newsletter* 1988;7:1-3.
32. Deutsch M, Green SB, Strike TA, Burger PC, Robertson JT, Selker RG, et al: Results of a randomized trial comparing BCNU plus radiotherapy, streptozotocin plus radiotherapy, BCNU plus hyperfractionated radiotherapy, and BCNU following misonidazole plus radiotherapy in the postoperative treatment of malignant glioma. *Int J Radiat Oncol Biol Phys* 1989;16(6):1389-96.

32

Intracranial Metastasis: A Dilemma to Operate or not to Operate

P Singh, AJ Venniyoor

INTRODUCTION

15 to 30% of cancer patients develop brain metastases during the course of their illness.[1] As a result of better care of cancer patients and prolonged survival, incidence of brain metastasis is on the rise. This is reflected by much higher western incidence[2] of brain metastases compared to Indian incidence, which is low.[3] I suppose as cancer care improves in our country this is going to change in coming years.

Intracranial metastases develop when tumour cells originating in the tissue outside the nervous system spread secondarily to the brain. Intracranial metastases may involve the brain parenchyma, cranial nerves, blood vessels (including dural sinuses), meninges and inner table of skull. Of these intraparenchymal metastases being the most common will be the focus of this paper. The most common primary tumours for the brain metastases are lung, breast, melanoma, colon and kidney in adults[1,2,4] and sarcoma, neuroblstoma and germ cell tumours in children.[1] However in 10 to 15% cases primary remains unknown.[2]

Brain metastases though have similar signs and symptoms as of other intracranial tumours/space occupying lesions, they amongst their own group have varied clinical presentations. They generally present features of raised intracranial pressure, progressive neurological dysfunction, seizure and cognitive function impairment. Rarely they can have stroke like presentation usually due to haemorrhage into metastasis such as melanoma, choriocarcinoma, thyroid or renal carcinoma or due to compression/invasion of an artery.

Most of brain metastasis develop after the diagnosis of systemic cancer (metachronous presentation), some may present before the primary is found (precocious presentation) or at the same time (synchronous presentation). In the setting of precocious presentation a thorough clinical examination and detailed investigative work up which includes; an X-ray chest, ultra sound abdomen, CT scan chest; abdomen and pelvis and even bone scan may be required.

Although a contrast enhanced CT scan of brain can detect majority of brain metastases, contrast enhanced MRI is best diagnostic tool as it is more sensitive than contrast enhanced CT scan in detection of lesions in brain metastasis and in differentiating these from other central nervous lesions.[5] Where as 50% lesions are found to have single metastasis with contrast CT scan,[6] only 1/3rd are found to be single lesions with contrast MRI.[5, 6, 7] Since number of lesions has therapeutic implication a contrast MRI is must before choosing a therapeutic modality. Even after a contrast enhanced MRI study 11% false positive cases have been reported, half of these were infections and half were primary tumours.[8] Therefore in doubtful cases biopsy; stereotactic/open is always indicated.

Brain metastasis presents a very peculiar problem to manage, on the one hand we are not absolutely sure how the primary is going to behave due to different histologies from different organs in various stages of progression and on the other hand there is threat to life and its quality from the intracranial disease. Therefore the objective of treating intracranial metastasis remains palliative to relieve neurological symptoms and signs, and at best to avoid or postpone neurological death. To achieve this aim today we have many modalities of treatments to be used singly or in combination:

(a) Steroids

(b) Whole brain radiation therapy

(c) Surgery

(d) Radiosurgery

(e) Chemotherapy

(f) Hormonal therapy

To choose suitable form of therapy to each case may become difficult, and whether to offer surgery or not, is even more difficult particularly in the light recent introduction of radiosurgery for brain metastasis and therefore a dilemma exists whether to operate or not to operate a case with brain metastasis. This dilemma can only be resolved if we have a look on all the treatment options available and then choose a suitable treatment for individual case.

CORTICOSTEROIDS

If a patient with brain metastasis is not treated at all the median survival is for 1-month,[9] with corticosteroids (usually dexamethasone) median survival increases to 2 months.[10] Steroids provide symptomatic palliation by controlling the peritumoural edema and are therefore reasonable choice in patients whose life expectancy due to systemic cancer is 2 or < 2 months. It needs to be made clear that they do not provide any control on intracranial metastatic disease and their prolonged use has well known side effects. They are usually used in daily dosage of 16 mg but dosage of 4 to 8 mg has been found as effective.[11] If 16 mg dosage is found ineffective it can be increased up to 96 mg /day.[12]

WHOLE BRAIN RADIATION THERAPY (WBRT)

WBRT has been the mainstay of treatment for patients with brain metastasis for more than 40 years. It is effective in palliation of neurological symptoms and also significantly decreases the likelihood of death due to neurological causes by achieving local control of intracranial metastatic disease. Median survival increases by 3 to 6 months and overall response rate ranges from 50 to 85%. It is ideally suited in patients with multiple brain metastases **(Fig. 1)**, preferably with radiosensitive histologies and whose expected life expectancy from systemic cancer is more than 2 months. Its role as postoperative adjuvant therapy in single brain metastasis has been well established.[8, 13] Many fractionation/dosage schedules has been tried and response rate and median survival rate do not vary significantly and 30 to 40 Gy over 3 to 4 weeks is usually given.[14] The late toxic effects of WBRT such as leukoencephalopathy, brain atrophy and radionecrosis are well known. Reirradiation can be a reasonable option for a patient with brain metastasis who develops recurrence, have controlled systemic disease, good general condition, unsuitable for radiosurgery or systemic chemotherapy or redo surgery.[15]

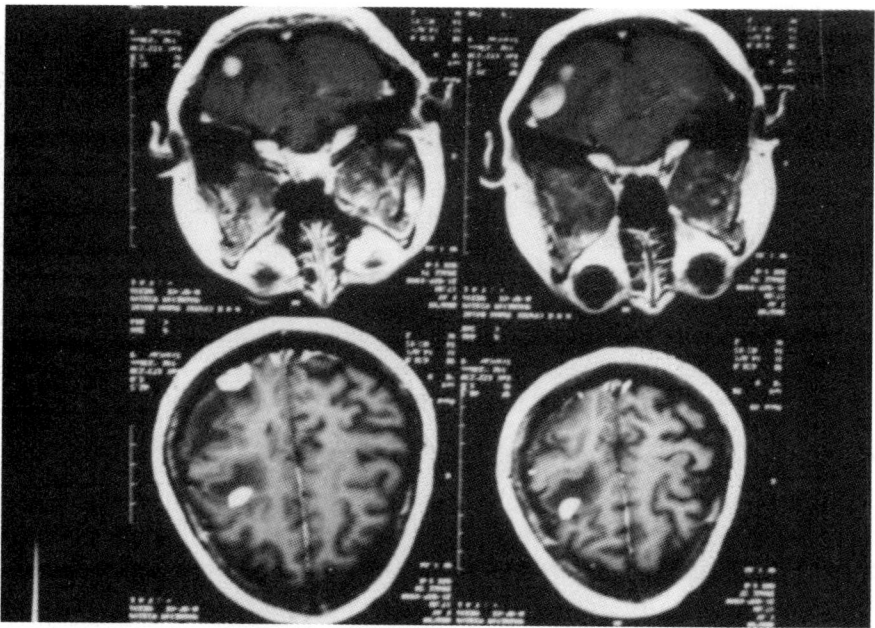

Fig. 1 : Contrast enhanced MRI showing multiple metastases in both cerebral hemispheres and cerebellum in a patient of carcinoma breast. Suitable for WBRT but can also be treated with radiosurgery.

RADIOSURGERY

Radiosurgery is recent addition in our armamentarium for treating brain metastasis. It can be given by linear accelerators, gamma knife and very rarely with charged particles such as protons from cyclotrons.[16] All stereotactic irradiation techniques result in rapid fall off of the dose at the edge of the target volume, resulting in insignificant radiation to untargeted normal tissue. Brain metastases are usually small (<3 cms), radio graphically discrete lesions that are noninvasive, thus rendering them ideal targets for radiosurgery. It is being used more routinely in single and multiple metastases.[17] Many uncontrolled studies have reported good local control (85%) of intracranial metastasis and median survival of 6 to 15 months.[2, 17, 18, 19, 20] Even radioresistant tumour responded equally well.[2, 17] Though there are no randomized studies comparing surgery versus radiosurgery for single metastasis, but Auchter et al [21] identified 122 patients who were treated with WBRT and radiosurgery boost and met the criteria used by Patchell.[8] The overall control rate was 86% with median survival of 56 weeks and median survival Karnofsky performance score (KPS) of >70 of 44 weeks. These results are comparable to the surgery + WBRT arm of treatment studies carried by Patchell et al,[8] Vecht et al[22] and Noordijk et al.[23] However Bindal et al[24] in similar studies reported better results with surgery + WBRT than radiosurgery. Radiosurgery in combination with WBRT has been used in multiple metastases (2 to 4) by Wen and Loeffler[2] and Loeffler et al[17] with better results than WBRT alone. Chen et al[25] compared results of radiosurgery + WBRT with radiosurgery alone in patients with intracranial metastases and found no discernable advantage of adjuvant WBRT. However radiosurgery has its share of failures, side effects and complications. About 10% cases may fail to respond and will progress.[17] 10% cases develop acute side effects such as headaches, nausea, seizures and worsening of neurological deficit. 8 to 16% cases will develop radionecrosis, which may require surgical resection, and 1% can develop cranial nerve palsies.[19] It also needs to be stressed that radiosurgery takes time to control the disease and thus the mass effect and therefore for control of edema corticosteroid may have to be given for 1 to 3 months or even longer.

The advantages with radiosurgery are: (a) it is non-invasive, (b) it can administered to lesions in any location and multiple (3-5) metastasis, (c) and any type of histology. It can also be considered for treating recurrent brain metastasis. Though it has been claimed to be cost effective,[26] at present one sitting of radiosurgery costs about Rest 75,000 at AIIMS Delhi and more than 1 lakh in private set ups. Therefore in our country affording radiosurgery will be difficult to vast majority of our patients with brain metastasis. Radiosurgery has been recommended for young patients with controlled primary, KPS > 70, single lesion < 3 cms having no mass effect and 3-5 multiple lesions.[19] **(Figs. 2 & 3)**

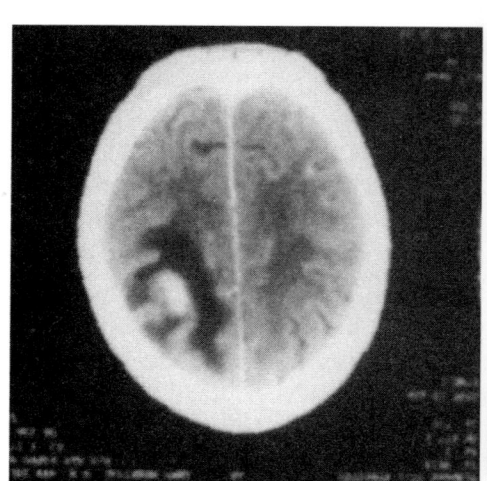
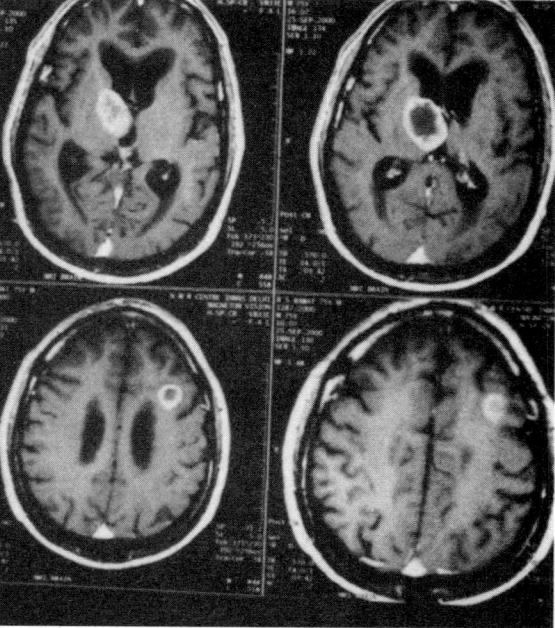

Fig. 2 : Contrast CT scan showing a small left parietal metastasis from non-small cell lung carcinoma. Suitable for radiosurgery.

Fig. 3 : MRI showing multiple metastases (right ganglionic & left frontal lobe). Suitable for radiosurgery.

CHEMOTHERAPY/HORMONAL THERAPY

Some brain metastases such as, those from small cell lung carcinoma (SCLC), breast carcinoma, and germ cell tumours are chemosensitive and selected cases have been treated with chemotherapy.[2, 27] Overall studies suggest that in brain metastases from SCLC, breast carcinoma and choriocarcinoma chemotherapy may a have role for recurrent disease after radiotherapy or possibly as initial treatment in small symptomatic tumours. In some hormone responsive tumours such as breast cancer, tamoxifen and megestrol has been used successful.[28]

SURGERY

Single metastasis

Surgery has been performed in selected patients for last 60 years.[8] Many uncontrolled studies suggested that patients with single metastasis who in addition to WBRT underwent surgical resection had better outcome than those treated with WBRT only.[29-32] These studies had the drawback of selection bias, this was set aside by randomized study by Patchell et al in 1990[8] which showed significant advantage in local control of brain metastasis, increase in median survival and better quality of life measured in terms of Karnofsky performance status in the group treated with surgical resection followed by WBRT than WBRT only. Similar results were reported By Vetch et al[21] and Noordjik et al.[22] However Mintz[33] did not find any difference between surgery + WBRT and WBRT alone. In general there is consensus for the role of surgery for single brain metastasis **(Fig. 4)** provided extra cranial disease is limited or stable, patient is young (age <60 years), has good performance status (KPS>70) surgically accessible lesion and radioresistant histology, long interval between the detection of primary and presentation of brain metastasis. Even in this situation radiosurgery is equally effective and may be preferred provided it is available & affordable. Therefore role of surgery is limited to an accessible respectable supratentorial (superficial & not involving eloquent areas of the brain) single lesion > 3 cms and in resectable infratentorial lesions causing mass effect **(Fig. 5)**. Surgical mortality and morbidity should be definite consideration. These have been reported to range from 0 to 10% and 8% respectively.[8, 34] Today, with the availability of intra operative functional mapping, intra operative ultra sonography, intra operative MRI and image guided surgery the surgical morbidity & mortality has reduced and surgery has become safer. Surgery therefore remains a contender in many situations and preferred

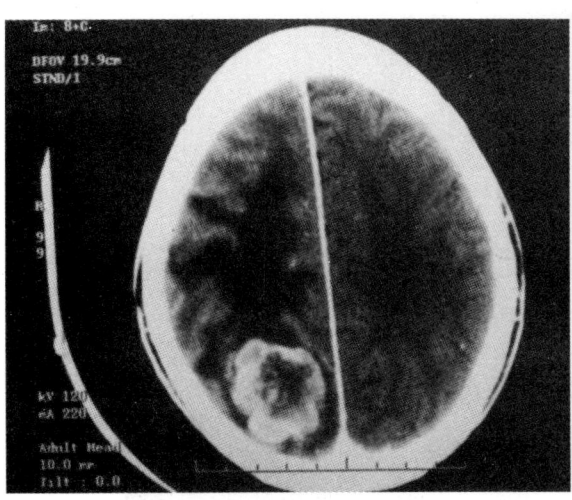 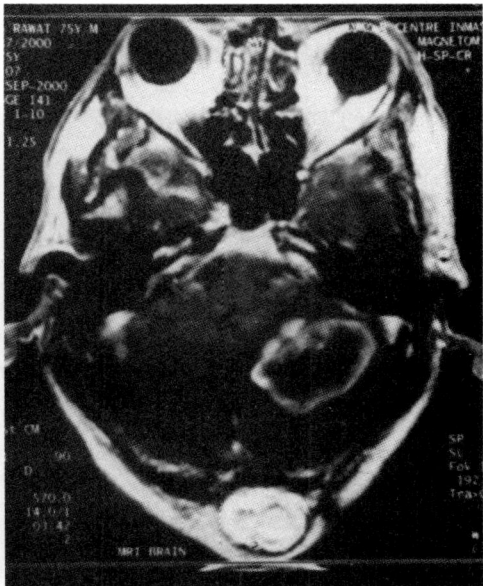

Fig. 4 :   Contrast enhanced CT scan showing single right posterior parietal metastasis. Suitable for surgery but can also be treated with radiosurgery.

Fig. 5 :   Contrast enhanced MRI showing a calvarial and a large size left cerebellar metastases in a case of carcinoma colon. Suitable for surgery.

modality in some.

MULTIPLE METASTASES

Surgical resection has also been used in multiple metastases.[2, 17, 35] Bindal et al[35] proved it to be efficacious, however Hazuka et al[36] did not find any advantage with surgical resection. The role of surgery in multiple metastasis remains controversial, at best remains limited to the resection of a large accessible, symptomatic or life-threatening lesion **(Fig. 6)**. Today radiosurgery or WBRT are better options for managing multiple brain metastases and even chemotherapy in chemosenstive tumours may be considered.

RECURRENT METASTASES

Even recurrent metastatic lesions have been treated with re-operation with good results.[37, 38] Surgical resection can be done in few selected symptomatic recurrent brain metastases; provided patient is young, has good performance status, location is accessible, lesion is single, extra cranial disease is limited, tumour is not radiosensitive and there is long interval between the first operation and the recurrence (>4 months). However other option such as radiosurgery/reirradiation/chemotherapy must be considered on the merit of each case before offering surgery.

HYDROCEPHALUS

Lastly if brain metastasis, single or multiple cause symptomatic hydrocephalus

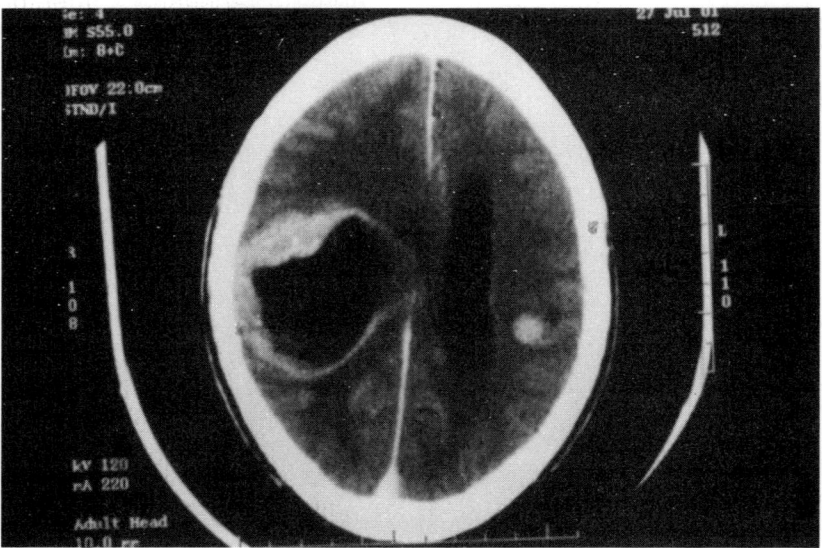

Fig. 6 : Contrast CT scan showing a large cystic right parietal and a small left parietal metastatic tumour in a SCLC. Right parietal lesion requires surgery.

(Fig. 7) which is not likely to be relieved by surgical resection or in those cases in which surgical resection of the metastases is not possible or modalities other than surgical resection are chosen, ventricular shunting will provide good palliation.

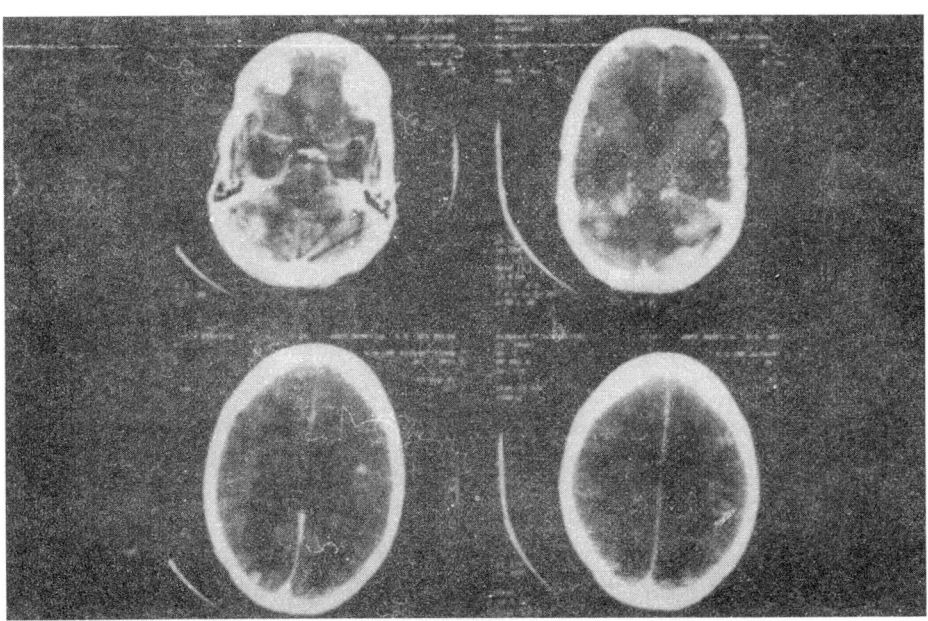

Fig. 7 : Contrast enhanced CT scan showing multiple cerebellar and cerebral metastatic deposits from carcinoma breast, causing hydrocephalus.

CONCLUSIONS

Brain metastasis is a heterogeneous group of varied histologies from different organs which will behave differently, has a very wide spectrum of clinical presentations depending on stage of primary cancer and profile of brain metastasis (single/multiple /different sizes/locations). Many modalities of treatment are available to be used either singly or in combination for providing palliation and achieving local control of the intra cranial disease. Surgical resection is one of them. Prognosis depends on many factors such as, extent of extra cranial, age, general condition, performance status, neurological status of the patient and nature of primary cancer.

Number, size, location, histology of the metastasis and efficacy of various other treatments influence therapeutic decisions. To chart out best therapy for a patient, in addition to contrast enhanced MRI of the brain; requires a detailed work up of the primary cancer to know its status & stage. A team decision by a neurosurgeon, a medical oncologist and a radiation oncologist is ideally desirable and recommended, however it may not be feasible at all places.

TREATMENT FLOW CHART

Single/ few metastatic brain tumours

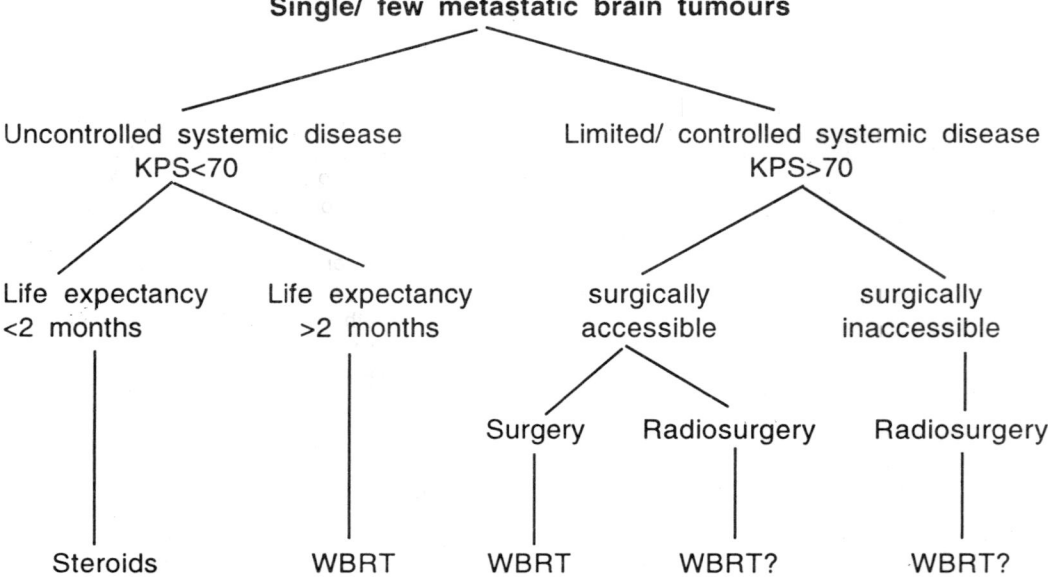

Surgery should be offered to a young patient (<60 years), with absent /limited / controlled systemic cancer, having KPS of >70 having single accessible resectable lesion if >3 cms or even smaller if radiosurgery not available or affordable and primary tumour is not radiosensitive. It should be particularly considered for a lesion in infra tentorial location. It may also be considered for selected cases of multiple metastases in whom 1 or 2 are large & causing mass effect or are life threatening and very rarely for a suitable recurrent metastasis. It has no role in a patient who has progressive/ advanced systemic cancer, poor KPS and elderly patient (age >60 years). It should also be withheld in metastases from radiation sensitive/ chemosensitve tumours such as SCLC, lymphoma and germ cell tumours where these modalities of treatments are more suitable.

REFERENCES:

1. Delattre JY, Krol G, Thaler HT, Posner JB: Distribution of Brain metastases. *Arch Neurol* 1988;45: 741-44.
2. Wen PY, Loeffler JS: Management of brain metastases. *Oncology* 1999;13:941-61.
3. Ramamurthi B: Metastatic deposits. In: Textbook of Neurosurgery, Ramamurthi & Tandon (editors), Second edition, Vol II: Churchill Livingstone, New Delhi 1996;1118-23.
4. Johnson JD, Young B: Demographics of brain metastases. *Neurosurg Clin North Am* 1996;7:337-44.
5. Davis PC, Hudgins PA, Peterman SB, et al: Diagnosis of cerebral metastases: double delayed CT vs contrast enhanced MR imaging. *Am J Neuroradiol* 1991;12:293-300.
6. DeAngelis LM: Management of brain metastases (review). *Cancer Invest* 1994;12:156-65.

7. Sze G, Milano E, Johnson C, et al.: Comparison of contrast enhanced MR with un- enhanced MR and contrast CT scan. *Am J Neuroradiol* 1990;11:785-91.
8. Patchell RA, Tibbs PA, Walsh JW, et al: A randomized trial of surgery in the treatment of single metastases to brain. *N Engl J Med* 1990;322:494-500.
9. Markesbery WR, Brooks WH, Gupta GL young AB: Treatment for patients with brain metastases. *Arch Neurol* 1978;35:754-56.
10. Ruderman NB Hall TC: Use of corticosteroids in the palliative treatment of metastatic brain tumours. *Cancer* 1965; 18:298-306.
11. Vecht CJ, Hoverstadt A, Verbiest HBC, et al: Dose effect relationship of dexamethasone on Karnofsky performance in metastatic brain tumours: a randomized study of doses of 4, 8 and 16 mg per day. *Neurology* 1994;44:675-600.
12. Lieberman A, Brun Y, Glass P, et al: Use of high dose of corticosteroids in patients with inoperable brain tumours. *J Neurol Neurosurg Psychiatry* 1977;40:678-82.
13. Patchell RA, Tibbs PA, Regine WF, et al: Postoperative radiotherapy in the treatment of single metastases to the brain: a randomised trial. *JAMA* 1998; 280:1485-89.
14. Berk L: An over view of radiotherapy trials for the treatment of brain metastases. *Oncology* 1995;9: 1205-11.
15. Cooper JS, Steinfield A, Lerch IA: Cerebral metastases: value of reirradiation in selected patients. *Radiology* 1990; 174:883-85.
16. Phillips MH, Stelzer KJ, Griffin TW, et al: Stereotactic radiosurgery: a review and comparison of methods. *J Clin Oncol* 1994;12:1085-99.
17. Loeffler JS, Barker FG, Chapman PH: Role of radiosurgery in the management of central nervous system metastases. *Cancer Chemother Pharmacol.* 1994; 43(Suppl): S11-14.
18. Flickinger JC, Kondziolka D, Lunsford LD, et al: A multi institutional experience with stereotactic radiosurgery for solitary brain metastases. *Int J Radiat Oncol Bio Phys* 1994;28:797-802.
19. Alexander E, Moriarty TM, Davis RB, et al: Stereotactic radiosurgery for definitive, non invasive treatment of brain metastases. *J natl Cancer Inst* 1995;87:34-49.
20. . Cho KH, Hall WA, Gerbi BJ, et al: Patient selection criteria for the treatment of brain metastases with stereotactic radiosurgery. *J Neurooncol* 1998;40:73-86.
21. Autcher RM, Lamond JP, Alexander E, et al: A multi institutional outcome and prognostic factor analysis of radiosurgery for respectable single brain metastases, *Int J Radiat Oncol Biol Phys* 1996;35:27-35.
22. Vecht CJ, Haaxma-Reiche H, Noordijk EM, et al: Treatment of single brain metastases: radiotherapy alone or combined with neurosurgery? *Ann Neurol* 1993;33:583-90.
23. Noordijk EM, Vecht CJ Haaxma-Reiche H, et al: The choice of treatment of single brain metastases should be based on extra cranial tumour activity and age. *Int J Radiat Oncol Biol Phys* 1994;29: 711-17.
24. Bindal AK, Bindal RK, Hess KR et al: Surgery versus radiosurgery in the treatment of brain metastases. *J Neurosurg* 1996;84:748-54.
25. Chen JCT, Petrovich Z, O'Day S, et al: Stereotactic radiosurgery in the treatment of metastatic disease to the brain. *Neurosurgery* 2000;47:268-81.
26. Mehta M, Noyes W Craig B et al: A cost effective and cost utility analysis of radiosurgery vs resection for single brain metastasis. *Int J Radiat Oncol Biol Phys* 1997;39:445-54.
27. Lesser GJ: Chemotherapy of cerebral metastases from solid tumours. *Neurosurg Clin North Am* 1996;7:527-36.
28. Pors H, Elder von Eyben F, Sorensen OS. et al: Long term remission of multiple brain metastases with tamoxifen. *J Neurooncol* 1991;10:173-77.
29. Henderson FR, Lee MS, Larson M et al: Influence of surgery and radiation therapy on patients with brain metastases. *Int J Radiat Oncol Biol Phys* 1983;9:623-27.

30. Sudaresan N, Galicich JH, Beatre PJ: Surgical treatment of brain metastases from lung cancer. *J Neurosurg* 1983;58:666-71.
31. Patchell RA, Cirricione C, Thaler HT, et al: Single brain metastases: surgery plus radiation or radiation alone. *Neurology* 1986;36:447-53.
32. Mandell I, Hilaris B, Sullivan M, et al: The treatment of single brain metastasis from non oat cell lung carcinoma: surgery and radiation versus radiation alone. *Cancer* 1986;58:641-49.
33. Mintz AP, Kestle J, Rathbone MP, et al: A randomised trial to assess the efficacy of surgery in addition to radiotherapy in patients with single brain metastases. *Cancer* 1996; 78:1470-76.
34. Lang FF: Sawaya Surgical management of cerebral metastases. *Neurosurg Clin N Am* 1996;7:459-84.
35. Bindal RK, Sawaya R, Leaven ME, et al: Surgical treatment of multiple brain metastasis *J Neurosurg* 1993;79:210–16.
36. Hazuka MB, Burloson W, Stroud DN, et al: Multiple metastases are associated with poor survival in patients treated with surgery and radiotherapy. *J Clin Oncol* 1993;11:369-73.
37. Sundaresan N, Suchdev V, DiGiacinto G: Reoperation for brain metastases. *J Clin Oncol* 1988;6:1625-29.
38. Bindal RK, Sawaya R, Leavens ME, et al: Reoperation for recurrent metastatic brain tumours *J Neurosurg* 1995; 83:600-604.

33

Surgical Management of Intracranial Metastases

J Bampoe, M Bernstein

INTRODUCTION

Intracranial metastases from systemic cancer are the most common type of intracranial tumour, with estimates of annual incidence between 100,000 and 400,000 in the U.S.A, that is, approximately one per one thousand of the population.[1,2]

The role of microsurgery in the treatment of brain metastases has come under increased scrutiny in the last decade for several reasons. First, the findings of three randomized clinical trials directly comparing the results in terms of survival and quality of life (functional independence) of patients with single brain metastases were published between 1990 and 1996.[3-5] These patients were treated with either whole brain irradiation (WBI) alone or microsurgical tumour removal in addition to WBI, and the results suggested that microsurgical tumour removal was the optimal treatment for patients in this study group;[3,4] that is, well selected patients with limited systemic disease and a single known brain metastasis on imaging (CT and/or MRI scan).[3,4] Second, further developments in the applications of radiosurgery, specifically its use for the treatment of small discrete malignant brain tumours (the typical appearance of a cerebral metastasis) has resulted in the publication of single arm studies (class III evidence) which have suggested that this modality is at least as efficacious as microsurgery in the management of cerebral metastases.[6-8]

The third factor is an inference from the second; that if radiosurgery is equally efficacious as microsurgery in the management of cerebral metastases, and since radiosurgery is less expensive for the management of the same disease, it is more cost-effective.[2,8-10] Radiosurgery is also more convenient for the patient, requiring in some cases, no hospital admission. It should thus perhaps be adopted as the gold standard for the management of cerebral metastases.[10]

The above inference was made largely on the basis of an "equi-efficacious" presumption; that outcome data from radiosurgery and surgical resection series are equivalent.[2] This assumption could be challenged in the light of more powerful evidence (class I) from randomized studies currently under way which directly compare the two modalities.[11-13] Until the outcome of such studies are known, the best evidence available is class II evidence (from case-control or cohort studies which directly address the issue).[14,15] We undertook to address the topic on the basis of the best evidence currently available.

MATERIALS AND METHODS

A Medline search was performed for studies investigating the management of intracranial, cerebral or brain metastases with surgery or radiosurgery published up until 31st August 2001. The search necessarily included a few studies investigating the role of radiotherapy in the management of this disease. The evaluation of studies and the evidence derived from them was according to previously defined criteria[11,12] with evidence and studies graded from Class I (evidence provided by one or more well-designed, randomized controlled clinical trials) to Class III (evidence provided by expert opinions, nonrandomized historical controls, or case reports.[11] Class II evidence is provided by one or more well-designed clinical studies, such as case-control or cohort studies that directly address the question at hand.[11] According to the strength of the evidence, practice parameters are graded from *standards* that reflect a high degree of treatment certainty to *options* for which the clinical evidence is inconclusive or unclear. *Guidelines* are recommendations that reflect moderate clinical certainty.[11]

RESULTS

Three prospective randomized clinical studies addressed the question of the role of surgery in the management of patients with a single known intracranial or brain metastasis treated with whole brain irradiation (WBI).[3-5] Survival and quality of life (functional independence) data were compared between patient groups treated with either WBI alone, or WBI in addition to surgery. In two studies, (Patchell et al and Vecht et al), the addition of surgical tumour resection significantly improved the benefits in terms of survival and functional independence of patients with a single known brain metastasis treated with WBI[3,4] **(Table 1)**. The benefit is most pronounced in patients with limited (controlled/stable/or absent) systemic disease.[4,5] Surgical resection in patients with a greater systemic tumor load and Karnofsky Performance Status (KPS) score of ≥ 50 (as compared to KPS ≥ 70 in the Patchell study) in the Mintz et al trial resulted in no demonstrable benefit.[3,5] **(Table 1)**.

A single retrospective cohort study (a study which employed well matched cohorts and directly addressed the question at hand, i.e., a class II study)[11] investigated the benefit of surgical resection in the management of patients presenting with multiple brain metastases.[16] The authors concluded that surgical removal of all known lesions (up to three) in selected patients with multiple brain metastases

Table 1: *Whole Brain Irradiation ±Surgery in the Management in the Management of Single Brain Metastases*

Author	Median Survival Surgical Arm	Median Survival Radiosurgical Arm	Functional Independence Surgical Arm	Functional Independence Radiosurgical Arm
Patchell et al 1993	40 weeks P<0.02	15 weeks	38 weeks P<0.005	8 weeks
Vecht et al 1993	10 months P=0.04	6 months P=0.06	7.5 months	3.5 months
Mintz et al 1996	5.6 months P=0.24	6.3 months	0.32% of survival P=0.98	

results in significantly increased survival time and gives a prognosis similar to that of patients undergoing surgery for a single metastasis.[16] The complication rate in patients undergoing multiple craniotomies in this study was not significantly different from those patients who only had a single craniotomy: there were no deaths within 30 days, and a morbidity rate of 7% (compared with 8% morbidity and 0% mortality in patients in group C – who had craniotomy for the removal of a single metastatic lesion during the same time period.[16]

The role of WBI as adjuvant therapy in the management of patients with intracranial metastases treated with a discrete ablative procedure (surgery or radiosurgery) has also been investigated.[17,18] The findings of two randomized clinical trials was that the addition of WBI did not prolong the survival of patients treated with either microsurgical tumour removal[17] or radiosurgery,[18] although it improved the control of brain disease.[17,18]

The relative efficacies of microsurgery and radiosurgery (perhaps the most contentious issue) has been the subject of investigation of two retrospective cohort studies.[14,15] The findings of Bindal et al was that surgery is superior to radiosurgery in the treatment of brain metastasis.[14] Patients who undergo surgical treatment survive longer *P=0.0009* (multivariate analyses). This study matched patients with limited systemic disease and good Karnofsky Performance Status (KPS) scale scores (median preop KPS scores were 80 in both arms), who had surgical removal of brain metastases with patients who had radiosurgical treatment. Patients in the latter group generally had small (< 3 cm in maximum diameter), spherical, well circumscribed lesions.[14] The study by Muacevic et al retrospectively compared patients with single circumscribed brain metastases 3.5 cm or smaller in diameter and stable systemic disease who had microsurgical removal of cerebral metastases with those who had radiosurgery alone.[15] A selection bias was apparent in this study – patients treated with radiosurgery had significantly smaller tumour volumes (p < 0.01).[15] The finding of this study was that in selected patients radiosurgery alone can result in local tumor control rates as good as those for surgery plus WBI. The median survival of patients in the surgical group was 68 weeks and 35 weeks in the radiosurgical group (p = 0.19).[15] Although the difference is not significant, it is to be noted that this closely matches the findings in the Bindal study (median survival 16.4 months [71 weeks] in the surgical arm compared to 7.5 months [33 weeks] in the radiosurgical arm), which was reported as significant.[14] **(Table 2)**

Table 2: *Surgery Versus Radiosurgery in the Treatment of Brain Metastases*

Author	Median Survival Surg. Arm	Median Survival SRS Arm	Local Control Surg. Arm %	Local Control SRS Arm%	Mortality/ Morbidity Surg. Arm%	Mortality/ Morbidity SRS Arm %
Bindal et al 1996	16.4 months P=0.0009	7.5 months	91.9	61.3	0/1.6	3.2/22
Muacevic et al 1999	68 weeks P=0.19	35 weeks	75	83	1.9/7.6	1.8/8.9

Abbreviations: Surg. – Surgery; SRS – Stereotactic Radiosurgery.

DISCUSSION

The appropriate treatment for a patient who presents with limited, controlled, or absent systemic cancer and intracranial metastases is currently a matter of some controversy.[8, 10, 19-22] Of all patients with brain metastases, approximately 10 to 20 percent have true single metastases when investigated with sensitive modern imaging such as high resolution contrast-enhanced magnetic resonance imaging (MRI).[17, 23] The issue of surgical accessibility is relative; improvements in surgical technique and adjuncts such as intraoperative imaging and brain mapping making more patients eligible for surgery.[24-26] The demonstration that the need for multiple craniotomies is not a contraindication to surgical management of brain metastases, that patients with multiple brain lesions (up to three) and limited systemic disease benefit in terms of improved length of survival and quality of life when all lesions are surgically removed[16] would tend to increase the percentage of patients who are eligible for surgery. However, progressive systemic disease usually shortens the life expectancy of most patients with multiple brain metastases.[16, 19, 20] In these patients the appropriate standard treatment is WBI. This is widely available and effective in prolonging survival time from 1 to 2 months with no therapy to 3 to 6 months with WBI.[3-5,19, 20] WBI is also the appropriate treatment for patients with multiple (> 3) intracranial metastases and those with metastases from radiosensitive tumours such as lymphoma, small cell carcinoma of the lung, leukaemia, and germ cell tumours.[19-21]

The importance of patient selection is demonstrated by the results of the three randomized clinical trials which investigated the surgical management of patients with single metastases to the brain.[3-5] The one study out of three which did not show any benefit from surgical resection recruited patients with a greater systemic tumour load, with baseline KPS scores of greater or equal to 50,[5] compared to the study by Patchell et al which recruited patients with KPS scores of greater or equal to 70.[3] Although the study by Vecht et al recruited patients with World Health Organization (WHO) scale scores of greater or equal to 2 (which includes patients who were not functionally independent), the benefits of surgical tumour removal was most pronounced in patients with stable extracranial disease.[4] Mintz and associates also identified extracranial metastases as the most significant variable affecting survival.[5] Another point worth noting is that in the study by Bindal et al which found a useful survival benefit for patients with multiple (less than three) brain metastases treated by surgical removal of all known lesions, patients selected for surgical removal of all known lesions (group B) had a mean KPS score of 76 ± 19.[16]

The recommendation for patients with up to three cerebral metastases is therefore microsurgical removal provided they have controlled, absent or stable systemic disease (that is, that they have an expected median survival of greater than 6 months). This recommendation is valid regardless of the size of the metastases, but is given at the level of a guideline rather than a standard since it reflects the reservation with regards to patient selection.[3-5, 16] **(Fig. 1)**.

Figure 1 : *Treatment Algorythm*

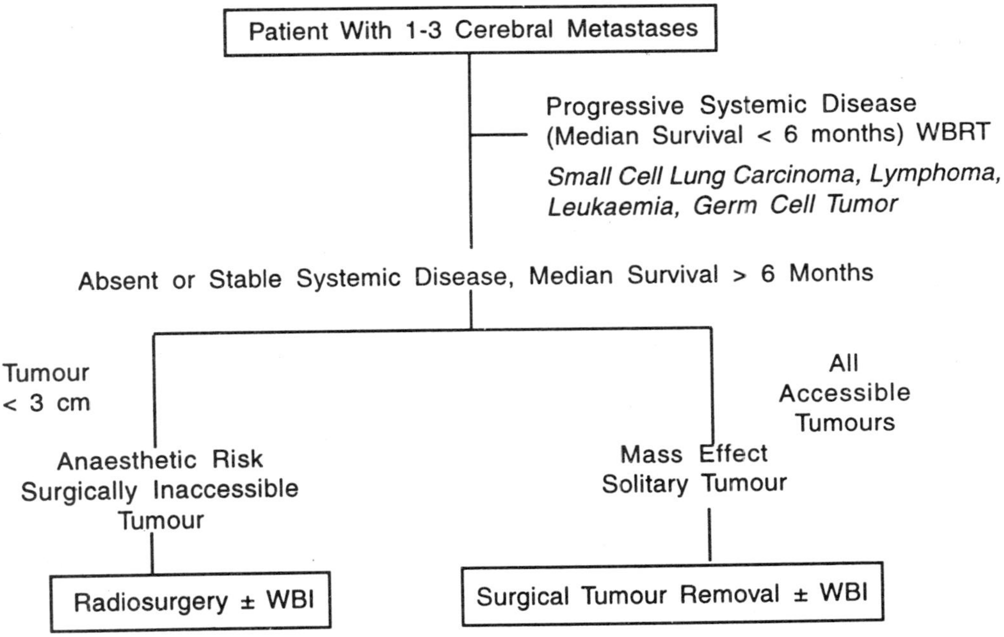

```
            Patient With 1-3 Cerebral Metastases
                          |
                          |── Progressive Systemic Disease
                          |   (Median Survival < 6 months) WBRT
                          |   Small Cell Lung Carcinoma, Lymphoma,
                          |   Leukaemia, Germ Cell Tumor
                          |
        Absent or Stable Systemic Disease, Median Survival > 6 Months
                          |
        ┌─────────────────┴─────────────────┐
    Tumour                              All
    < 3 cm                          Accessible
        |                            Tumours
   Anaesthetic Risk                Mass Effect
   Surgically Inaccessible        Solitary Tumour
   Tumour                              |
        |                               |
   Radiosurgery ± WBI         Surgical Tumour Removal ± WBI
```

There is a further question as to whether patients treated for brain metastases with a discrete ablative procedure (surgery or radiosurgery) should also be subjected to WBI **(Figure 1)**. The rationale for WBI as an adjunct to surgery is its perceived effectiveness in the destruction of residual tumour in the operative bed, as well as the elimination of undetected micrometastases.[17] The risk of indiscriminate use of this modality was highlighted by a study by DeAngelis and associates.[27] In a review of the records of 98 patients who had elective craniotomy between 1978 and 1985 for the resection of single brain metastases, 11% of patients who had received WBI and survived 1 year developed severe radiation-induced dementia.[27] The findings of two recent randomized trials which have addressed this question is that the addition of WBI to either microsurgical or radiosurgical management of discrete brain metastases does not prolong the survival time of patients but improves the control of brain disease (decreases the recurrence rate at the original and other sites and increases the time to recurrence).[17,18] While definitive data on the risk of radiation-induced dementia after WBI are still pending from randomized studies, it would seem prudent to withhold this modality from patients expected to live for more than a year.[20]

The demonstrated efficacy of radiosurgery for the management of discrete small malignant tumours of the brain has given rise to the third major question regarding the role of surgery in the management of intracranial metastases. Several investigators, citing the findings of equivalent or better survival figures for patients with intracranial metastases treated with radiosurgery as compared to those who had microsurgical tumour removal, have questioned whether surgical removal of intracranial metastases is the gold standard for these patients.[8,10] An additional factor apparently in favour of radiosurgical management in this debate is its non-invasive nature. Most patients can be treated as outpatients or discharged home after an overnight hospital stay.[2,8-10] It is therefore an ideal modality for patients more severely affected by their disease, especially for those who are an anaesthetic risk.[14] Radiosurgical management has also been demonstrated to be less expensive than microsurgical, after accounting for the significant costs of obligatory admission for observation after brain surgery.[2,8-10] This, however, is relative, and has been variously estimated at between 1 day by Rutigliano et al,[9] and 6 days by Mehta et al.[2] Of the studies in the literature which directly compare the two treatment modalities, the median hospital stay was determined to be 4 days for the surgical group compared to 1 day for radiosurgery patients ($P < 0.0001$) by Bindal et al,[14] and 10 days for surgery compared to 0 days for radiosurgery by Muacevic et al.[15]

It should be noted that with modern surgical adjuncts, the need for hospitalization can be eliminated in selected patients as has been demonstrated in the series reported by the senior author.[24,26] These included a large proportion of patients with cerebral metastases.[24,26]

The concept of cost-effectiveness in the clinical setting is based on an equi-efficacious presumption, that is, outcome data for radiosurgery and resection are arbitrarily equalized.[2] Until a definitive challenge of this presumption can be made with data from a randomized controlled clinical trial (RCCT) directly comparing the two modalities (class I evidence),[11] the best available evidence is from Class II studies which directly address the question as noted above.[11] The basic necessity of formally demonstrating significant superiority of one modality over the other in terms of either survival or quality of life in the best possible scientific setting (RCCT) has been acknowledged since "It is of course very difficult to assign a monetary value to life and function."[2]

The problem of bias is clearly seen in the single and multi-institutional case series (Class III evidence) on which these cost-effectiveness determinations have been made.[6-8] In their report on a multi-institutional outcome and prognostic factor analysis of radiosurgery for resectable single brain metastases, the median volume of tumours treated was reported by Auchter et al as 2.68 cm^3[6] (median tumour diameter < 1.75 cm). Likewise, the mean tumour diameter reported by Flickinger et al in their multi-institutional (gamma knife) series was 1.87 cm.[7] As noted in *results* above, the apparent equivalence of outcome of radiosurgical and surgical series can be explained by the former treating significantly smaller tumours.[15]

In conclusion, from the available data, the optimum treatment for a patient with a single intracerebral metastastases from systemic cancer with the exception of those which are exquisitely radiosensitive is surgical removal.[3-5,14,15] The benefit of surgery to patients is most pronounced in those with controlled systemic disease and a life expectancy of greater than 6 months.[4, 20] This recommendation is valid regardless of the size of the metastases, and can be extended, with the same reservations, to patients with up to 3 intracranial metastases provided surgical morbidity and mortality from multiple craniotomies can be minimized.[16] The question of tumour accessibility is relative, depending on the experience of the surgeon, but the advantages of radiosurgery for the treatment of small, deep lesions with minimal morbidity should be acknowledged. Likewise, the unique ability of microsurgery to rapidly reverse neurological deficit from larger symptomatic lesions should be recognized.

REFERENCES:

1. Johnson JD, Young B: Demographics of brain metastasis. *Neurosurg Clin N Am* 1996;7:337-44.
2. Mehta M, Noyes W, Craig B, Lamond J, Auchter R, French M, et al: A cost-effectiveness and cost-utility analysis of radiosurgery vs. resection for single-brain metastases. *Int J Radiat Oncol Biol Phys* 1997;39:445-54.
3. Patchell RA, Tibbs PA, Walsh JW, Dempsey RJ, Maruyama Y, Kryscio RJ, et al: A randomized trial of surgery in the treatment of single metastases to the brain. *N Engl J Med* 1990;322:494-500.

4. Vecht CJ, Haaxma-Reiche H, Noordijk EM, Padberg GW, Voormolen JHC, Hoekstra FH, et al: Treatment of single brain metastasis: radiotherapy alone or combined with neurosurgery. *Ann Neurol* 1993;33:593-90.
5. Mintz AH, Kestle J, Rathbone MP, Gaspar L, Hugenholtz H, Fisher B, et al: A randomized trial to assess the efficacy of surgery in addition of radiotherapy in patients with a single cerebral metastasis. *Cancer* 1996;78:1470-76.
6. Auchter RM, Lamond JP, Alexander E, Buatti JM, Chappell R, Friedman WA, et al: A multiinstitutional outcome and prognostic factor analysis of radiosurgery for resectable single brain metastasis. *Int J Radiat Oncol Biol Phys* 1996;35:27-35.
7. Flickinger JC, Kondziolka D, Lunsford LD, Coffey RJ, Goodman ML, Shaw EG, et al: A multi-institutional experience with stereotactic radiosurgery for solitary brain metastasis. *Int J Radiat Oncol Biol Phys* 1994;28:797-802.
8. Alexander E, Moriarty TM, Davis RB, Wen PY, Fine HA, Black PM, et al: Stereotactic radiosurgery for the definitive noninvasive treatment of brain metastases. *J Natl Cancer Inst* 1995;87:34-40.
9. Rutigliano MJ, Lunsford LD, Kondziolka D, Strauss MJ, Khanna V, Green M: The cost effectiveness of stereotactic radiosurgery versus surgical resection in the treatment of solitary metastatic brain tumors. *Neurosurgery* 1995;37:445-55.
10. Sperduto PW, Hall WA: Radiosurgery, cost-effectiveness, gold standards, the scientific method, cavalier cowboys, and the cost of hope. *Int J Radiat Oncol Biol Phys* 1996;36:511-13.
11. Shaw E: Looking through the retrospectoscope in the era of evidence-based medicine. *J Clin Oncol* 1997;15:1289-90.
12. Canadian task force on the Periodic Health Examination: The periodic health examination. *Can Med Assoc J* 1979;121:1193-1254.
13. Boyd TS, Mehta MP: Stereotactic radiosurgery for brain metastases. *Oncology (Huntingt)* 1999;13:1397-1407.
14. Bindal AK, Bindal RK, Hess KR, Shiu A, Hassenbusch SJ, Shi WM, et al: Surgery versus radiosurgery in the treatment of brain metastasis. *J Neurosurg* 1996;84:748-54.
15. Muacevic A, Kreth FW, Horstmann GA, Schmid-Elsaesser R, Wowra B, Steiger HJ, et al: *J Neurosurg* 1999;91:1999.
16. Bindal RK, Sawaya R, Leavens ME, Lee JJ: Surgical treatment of multiple brain metastases. *J Neurosurg* 1993;79:210-16.
17. Patchell RA, Tibbs PA, Regine WF, Dempsey RJ, Mohiuddin M, Kryscio RJ, et al: Postoperative radiotherapy in the treatment of single metastases to the brain. *JAMA* 1998;280:1485-89.
18. Chougule PB, Burton-Willams M, Saris S, Zheng Z, Ponte B, Noren G, et al: Randomized treatment of brain metastasis with gamma knife radiosurgery, whole brain radiotherapy or both. *Int J Radiat Oncol Biol Phys* 2000;48(3 Suppl):114S.
19. Kim M, Bernstein M: Current treatment of cerebral metastasis. *Curr Opin Neurol* 1996;9:414-18.
20. Thornton AF, Harsh GR: Recommendations for treatment of brain metastases. *Neurosurg Clin N Am* 1996;7:559-64.
21. Wen PY, Loeffler JS: Management of brain metastases. *Oncology (Huntingt)* 1999;13:941-54.
22. Chan AW, Loeffler JS: Controversies in the management of patients with brain metastases. *Cancer J* 2001;7:105-07.
23. Davis PC, Hudgins PA, Peterman SB, Hoffman JC: Diagnosis of cerebral metastases: double-dose delayed CT vs contrast-enhanced MR imaging. *AJNR Am J Neuroradiol* 1991;12:293-300.
24. Taylor MD, Bernstein M: Awake craniotomy with brain mapping as the routine surgical approach to treating patients with supratentorial intraaxial tumors: a prospective trial of 200 cases. *J Neurosurg* 1999;90:35-41.
25. Bernstein M, Al-Anazi AR, Kucharczyk W, Manninen P, Bronskill M, Henkelman M: Brain tumor surgery with the Toronto Open Magnetic Resonance Imaging System. Preliminary results for 36

patients and analysis of advantages, disadvantages, and future prospects. *Neurosurgery* 2000;46:900-09.
26. Bernstein M: Outpatient craniotomy for brain tumor: A pilot feasibility study in 46 patients. *Can J Neurol Sci* 2001;28:120-24.
27. DeAngelis LM, Mandell LR, Thaler HT, Kimmel DW, Galicich JH, Fuks Z, et al: The role of postoperative radiotherapy after resection of single brain metastases. *Neurosurgery* 1989;24:798-805.

34
Stereotactic Radiosurgery for Brain Secondaries

K Ganapathy

"The moving finger writes and having writ moves on"
"The only thing that is constant in the universe is change"

INTRODUCTION

Improved systemic care of cancer patients is leading to increased longevity and increased chances of secondaries occurring in the brain.[1] It is estimated that 50 per cent of all cancer patients will develop brain secondaries[2] if they live long enough. Brain metastases will eventually become a problem in India as well. In the United States, with almost 200,000 patients annually suffering from symptomatic brain secondaries it is already significant health-care problem.[3] At the same time brain secondaries need no longer be viewed as untreatable or the harbinger of immediate death. With high-resolution imaging.small, asymptomatic metastases are being increasingly detected Patients with brain secondaries also now expect longer survival and preserved neurological function. Lung, breast, melanoma, renal, and gastrointestinal cancers contribute to the majority of lesions that come to clinical attention. Although median survival may be less than a year, timely therapy can preserve neurological function and minimise neurological complications for the duration of the patient's survival. Death is often due to progression of the primary systemic disease rather than that of the cerebral deposit.

The object of management of brain metastases is **to achieve local control**. Hence surgical resection is often considered. This may be perceived as being needlessly aggressive for patients whose life expectancy is somewhat uncertain.[4] When there is associated brain edema, mass effect and the lesion is in an accessible region open surgery is indicated.[3] Secondary deposits may however be multiple and located in less accessible areas. These factors contribute to some reluctance to recommend surgical resection for all brain metastases. Furthermore brain metastases are physically and biologically ideal lesions to treat with radiosurgery.

Management of brain metastases particularly multiple, will necessarily be influenced by the cultural and philosophical background of the treating neurosurgeon, the patient and the family.[5-7] The socio economic milieu will also play a major role. In the west, secondary deposits form 50% or more of the workload in a radiosurgery unit as contrasted to 2.5% in India. Some radiosurgery units have published data showing reduction in target volume of even 6 to 8 deposits treated with radiosurgery.[8,9] However there is no convincing data that survival is increased. It could always be argued that even a few months of good quality extra life may be meaningful to the individual concerned. In Chennai only 9 cases of secondaries have been treated out of 350 patients who had undergone radiosurgery. This is partly due to the strict selection criteria. Since radiosurgical treatment for secondaries is palliative and not curative it is essential to be very clear in one's mind as to the goals of radiosurgery. Every effort is made to identify the presence of other active systemic deposits. Radiosurgery for brain metastasis will not arrest the general progression of the primary or even the development of new metastasis in the brain, outside the target volume treated. It was generally believed that for every macroscopically visible deposit there could be other microscopic deposits. Hence "sterilisation" with total brain radiation was advocated. Several recent studies have however questioned this concept.[10,2,11,12]

Technical feasibility for stereotactic radiosurgery **should not be the only factor** in the decision-making paradigm[13,6] Karnofsky's status, probable life expectancy, the "virulence" of the primary lesion and the psychological profile of the patient and the family are to be taken into account. Though all are equal it is a fact of life that some are more equal than others. Where the patient has to personally incur the expenses unfortunately economic factors cannot be brushed aside. Previous treatment for a metastatic deposit elsewhere alone is not a contraindication. The number of intra cerebral metastases that should be treated depends on the philosophy of the clinician. Radiosurgery can be used in the initial treatment of relatively radio resistant metastases as in melanoma or renal cell carcinoma. Disappearance or considerable volume reduction of secondary deposits, within 3 to 6 months following treatment have been documented.

Recent investigations support the concept that achieving local control of brain metastases prolongs survival and improves the quality of life. Radiosurgery can be used successfully as salvage therapy in patients who have received prior radiotherapy and in the initial treatment of relatively radio resistant metastases (melanoma or renal cell carcinoma). Disappearance or considerable volume reduction of secondary deposits, within 3 to 6 months following treatment are frequently documented.

Because Radiosurgery can be used for patients who develop new remote tumours (without overlapping prior Radiosurgery volumes), repeat Radiosurgery can be an additional important strategy. In a developing country like India where medical insurance is still limited cost containment however is a major issue. Additional radiation therapy for brain secondaries using steriotactic techniques is, however, a useful addition to the therapeutic armamentorium.[14]

SUITABILITY OF SECONDARIES FOR SRS

1. Metastases are most often conveniently spherical and radiologically distinct.
2. The majority of metastatic lesions are < 3 cm at presentation.
3. Metastases often displace normal brain parenchyma circumferentially outside the potential radiosurgery target volume, reducing the probability of normal brain injury.

Most metastatic lesions are minimally invasive, and the entire extent of disease can be encompassed in the radiosurgery treatment field.

FACTORS INFLUENCING PROGNOSIS IN CEREBRAL SECONDARIES FOLLOWING RADIOSURGERY[2, 4, 10, 15-19]

- Younger age
- High Karnofsky's performance status
- One to three versus more than three lesions
- Longer interval from primary to detection of secondaries
- Presence /absence of neurological deficit (8 m vs 17 m survival)
- Histology of primary e.g. higher rates of local control for deposits from melanoma or renal cell carcinoma).
- Volume of the secondary.
- Larger tumor size, infratentorial metastases
- Extracranial disease control at the time of treatment. (2 m in presence of extracranial disease vs 12 months without)
- Tumors less than 20 mm in diameter
- Tumors treated with a high marginal dose (> or = 18 -23 Gy)
- Collimator size, number of arcs, tumour location, did *not* influence objective response rates

ROLE OF WHOLE BRAIN IRRADIATION

This is perhaps the most contentious subject today in the management of brain metastasis – single and multiple. Some authors believe that after WBRT while local control may be increased, survival is not, and there is an increased risk of late complications. Alopecia and impaired cognitive function also has to be considered.[2, 11, 12] Chronic steroid dependence and increased intracranial edema do not appear to be common problems following WBRT.[2] Whole brain radiotherapy is often not followed by durable control of the disease and carries morbidity[20] WBRT *alone* does not provide lasting and affective care for most patients

Others contend that surgery followed by holocranial radiotherapy is *still* the method of choice for the treatment of metastases.[21, 22] Poor results in terms of local control are in favor of supplementary whole brain irradiation, except for particular cases.[23]

Radiosurgery alone can result in local tumor control rates as good as those for *surgery plus WBRT* in selected patients.[24, 10] Some authors believe that SRS yields better results even in metastatic deposits traditionally considered radioresistant.[25] *No significant survival benefit* could be discerned from *adjuvant whole brain radiotherapy* in some series and *survival* was not statistically different for patients initially presenting with 1-4 metastases at initial treatment.[26, 15]

SRS alone and SRS+WBRT seem better in prolonging life and improving quality of life than WBRT alone for patients with single brain metastasis from lung cancer. SRS+WBRT did not however show significant advantage over SRS alone in improving survival, enhancing local control, and quality of life except for a *more favorable FFNBM (freedom from new brain metastasis)*.[27] In view of the risk of dementia following WBRT it has therefore been suggested that WBRT should be used for salvage treatment as needed and not upfront in the first instance. The omission of WBRT in the initial management of patients treated with RS for up to 4 brain metastases does not appear to compromise survival or intracranial control allowing for salvage therapy as indicated.[28] Still others contend that, in treating single brain metastasis SRS is comparable to microsurgery combined with WBRT. Concerning morbidity and local tumour control, in particular in cases of "radio resistant" primary tumours, SRS is superior. Therefore SRS is advocated except for large tumours (> 3 cm in maximum diameter) and those with mass effect.[29]

Brainstem metastases portend a dismal prognosis. Surgical resection is not part of routine management and radiation therapy has offered little clinical benefit. Although they have slightly lower than the expected survival rates of patients with non-brainstem tumors, patients with brainstem metastases may achieve effective palliation after stereotactic radiosurgery and WBRT.[30] **Skull base metastases** have also been treated by SRS.[31, 32]

Cystic secondary deposits : Stereotactic cyst aspiration and SRS for the solid component leads to palliation of neurologic symptoms and is a low risk treatment for patients with cystic brain metastasis.[33]

Radioprotectors : Omega three fatty acids and bioflavonoids were administered concurrently in patients with metastases in critical brain areas. Survival was prolonged and risks (including radionecrosis) were less. Use of radioprotectors may be yet another option.[34]

Technical factors : 14 Gy delivered at the periphery of a metastasis seems to be a sufficient dose to control most brain metastases, with a minimal toxicity. Better results were obtained for lesions initially treated with radiosurgery, theoretically radio resistant and with a diameter less than 3 cm.[35] Hypofractionated stereotactic radiotherapy with a median dose of 27 Gy, (3 fractions of 9 each) has also been tried out[36] Fractionated SRT has also been used in the treatment of brain secondaries.

Radiosurgery for multiple deposits: How many? How often? Efficacy of the gamma knife in the treatment of patients with 10 or more simultaneous metastatic brain tumors. in 24 patients has been asessed. Acceptable tumor control, low morbidity, and good quality of life during the survival period has been reported.[37, 8] One patient was treated four times and one patient seven times for new lesions[9] 50% of the brain volume received less than 500 cGy for a maximum tumor dose of 40 Gy, and the dose gradient was extremely steep. The Dose Volume Histogram analysis revealed that GKS was a good treatment modality to control local deposits while maintaining normal brain function, even for the large number of brain metastasis treated at different times.[38]

POST SRS EVALUATION

Post SRS *pattern of enhancement* could be a prognostic factor to evaluate FFP of brain metastases treated with SRS, independent of dose and volume. A possible explanation is radio resistance of hypoxic tumor cells associated with necrotic regions, suggesting future investigations with radiosensitizers, hypoxic cell sensitizers, or strategies to improve tumor oxygenation in secondaries[39] Transient enlargement of volume of contrast uptake on MRI after Linear Accelerator (linac) stereotactic Radiosurgery for brain metastases[40] has been reported Changes of regional perfusion in the tumour, peritumoural edematous area and juxta tumoural brain after radiosurgical treatment for metastatic brain tumour were investigated by dynamic SPECT using 1231-IMP. Early improvement of regional cerebral blood flow in the juxtatumour areas after radiosurgery was documented.[41]

RESULTS OF SRS

Are these results due to the therapy alone or can the results be attributed in part to patient selection[42] Radiosurgery provides an effective local control for 90% of treated patients with low morbidity.[43] Several reports discuss the effect of Stereotactic Radiosurgery for specific primaries g non-small cell lung carcinoma[44] Renal Cell Carcinoma[45] and ovarian cancer[46] Patients with "radio resistant" tumours particularly benefit with Radiosurgery. The radiobiological effect of Radiosurgery is vastly different from conventional radiation therapy, because tumour histolo-

CONCLUSION

For many patients, surgery continues to be a treatment option but results are still not encouraging, because even patients with the best prognostic indicators often die within 18 to 24 months. The judicious use of available techniques includiong radiosurgery for treatment of patients with limited systemic disease and brain secondaries provides the best opportunities for palliation and extended survival. For patients with a single lesion, local control and survival rates of radiosurgery compare well with those produced with surgical resection. There is compelling evidence to suggest that aggressive local therapy (surgery or radiosurgery) fo. patients with a single brain metastasis produces superior survival and quality of life compared with treatment with whole brain radiotherapy alone. However, surgery should be restricted to the minority of patients for whom brain metastases represents an immediate threat to life. For an asymptomatic or mildly symptomatic patient with a lesion smaller than 3 cm in diameter, radiosurgery is an excellent alternative to surgery. Although radiosurgery is a noninvasive procedure, the same selection criteria should be considered as for those patients undergoing surgical resection.[3]

ACKNOWLEDGMENTS

I wish to thank Mrs. Vijayalakshmi Ganapathy and Mr. G. Murali Krishnan for secretarial assistance.

REFERENCES:

1. Kondziolka D, Patel A, Lunsford LD, Kassam A, Flickinger JC: Stereotactic Radiosurgery plus whole brain radiotherapy versus radiotherapy alone for patients with multiple brain metastases. *Int J Radiat Oncol Biol Phys* 1999;45(2):427-34.
2. Boyd TS, Mehta MP: Radiosurgery for brain metastases. *Neurosurg Clin N Am* 1999;10:337-50.
3. Alexander E 3rd, Loeffler JS: The case for radiosurgery. *Clin Neurosurg* 1999;45:32-40.
4. Kondziolka D, Dade Lunsford L, John C Flickinger: Brain metastases in renal cell carcinoma: management with gamma knife radiosurgery. *Cancer J* 2000;6(6):372-76.
5. Ganapathy K: Radiosurgery. Advances in Clinical. *Neurosciences* 1995;5:369-86.
6. Ganapathy K: Decision making in radiosurgery. *Neurosciences Today* 1997;1:72-77C.
7. Ganapathy K: Radiosurgery in Text Book of Neurosurgery 2nd edition (eds) Ramamurthi B, Tandon PN. Orient Longmans Delhi 1995.
8. Suzuki S, Omagari J, Nishio S, Nishiye E, Fukui M: Gamma knife radiosurgery for simultaneous multiple metastatic brain tumors. *J Neurosurg* 2000;93 Suppl 3:30-1.
9. Amendola BE, Wolf AL, Coy SR, Amendola M, Bloch L: Brain metastases in renal cell carcinoma: management with gamma knife radiosurgery. *Cancer J* 2000;6(6):372-6.
10. Muacevic A, Kreth FW, Horstmann GA, Schmid-Elsaesser R, Wowra B, Steiger HJ, Reulen HJ: Surgery and radiotherapy compared with gamma knife radiosurgery in the treatment of solitary cerebral metastases of small diameter. *J Neurosurg* 1999;91(1):35-43.

11. Nakagawa K, Tago M, Terahara A, Aoki Y, Sasaki T, Kurita H, Shin M, Kawamoto S, Kirino T, Otomo K: A single Institutional outcome analysis of Gamma Knife Radiosurgery for single or multiple brain metastases. *Clin Neurol Neurosurg* 2000;102(4):227-32.

12. Helfre S, Pierga J: Cerebral metastases: radiotherapy and chemotherapy. *Neurochirurgie* 1999; 45(5):382-92.

13. Ganapathy K: Stereotactic Radiosurgery for malignant gliomas. *Indian Clinical Neurosurgery* Vol. 2 p.. Editors Anil K Singh, et al. CBS Publishers & Distributors

14. Tokuuye K, Akine Y, Sumi M, Kagami Y, Murayama S, Nakayama H, Ikeda H, Tanaka M, Shibui S, Nomura K: Fractionated sterotactic radiotherapy of small intracranial malignancies. *Int J Radiat Oncol Biol Phys* 1998;42(5):989-94.

15. Chen JC, O'Day S, Morton D, Essner R, Cohen-Gadol A, MacPherson D, Giannotta SL, Petrovich Z, Yu C, Apuzzo ML: Stereotactic Radiosurgery in the treatment of metastatic disease to the brain. *Stereotact Funct Neurosurg* 1999;73(1-4):60-63.

16. Joseph J, Adler JR, Cox RS, et al: Linear accelerator based sterotaxic Radiosurgery for brain metastases. The influence of number of lesions on survival. *J Clin Oncol* 1996;14:1085-92.

17. Cho KH, Hall WA, Gerbi BJ, Higgins PD, Bohen M, Clark HB: Patient selection criteria for the treatment of brain metastases with stereotactic radiosurgery. *J Neurooncol* 1998;40(1):73-86.

18. Jeremic B, Becker G, Plasswilm L, Bamberg M: Activity of extracranial metastases as a prognostic factor influencing survival after radiosurgery of brain metastases. *J Cancer Res Clin* Oncol 2000: 126(8):475.

19. Maor MH, Dubey P, Tucker SL, Shiu AS, Mathur BN, Sawaya R, Lang FF, Hassenbusch SJ: Stereotactic Radiosurgery for brain metastases: results and prognostic factors. *Int J Cancer* 2000;90(3):157-62.

20. Marcou Y, Lindquist C, Adams C, Retsas S, Plowman PN: What is the optimal therapy of brain metastases? *Clinical Oncology* 2001;13:105-11.

21. Bartumeus F, Clavel P: Surgical treatment of brain metastases. *Rev Neurol* 2000;31(12):1247-49.

22. Fuller BG, Kaplan ID, Adler J, et al: Sterotaxic Radiosurgery for brain metastases: the importance of adjuvant whole brain irradiation. *Int J Radiant Oncol Biol Phys* 1992;23:413-18.

23. Vendrely V, Prie L, Benyoucef A, Chemin A, Kantor G: Radiosurgery of single brain metastasis without combined total cerebral irradiation. Results of a consecutive series of 12 cases. *Cancer Radiother* 1998;2(4):375-80.

24. Ueki K, Matsutani M, Nakamura O, Tanaka Y: Comparison of whole brain radiation therapy and locally limited radiation therapy in the treatment of solitary brain metastases from non-small cell long cancer. *Neurol Med Chir (Tokyo)* 1996;36(6):364-69.

25. Feuvret L, Germain I, Cornu P, Boisserie G, Noel G, Hardiman C, Tep B, Hasboun D, Faillot T, Duffau H, Valery C, Delattre JY, Poisson M, Marsault C, Philippon J, Fohanno D, Baillet F, Mazeron JJ: First treatment for brain metastases by stereotactic radiosurgery. *Bull Cancer* 1999; 86(7-8):666-72.

26. Chen JC, Petrovich Z, O'Day S, Morton D, Essner R, Giannotta SL, Yu C, Apuzzo ML: Stereotactic Radiosurgery in the treatment of metastatic disease to the brain. *Neurosurgery* 2000;47(2):268-279; discussion 279-81.

27. Noel G, Bleichner O, Mazeron JJ: Stereotactic radiosurgery plus whole brain radiotherapy versus radiotherapy alone for patients with multiple brain metastases. *Cancer Radiother* 2000;4(1):90-91.

28. Sneed PK, Lamborn KR, Forstner JM, McDermott MW, Chang S, Park E, Gutin PH, Phillips TL, Wara WM, Larson DA: Radiosurgery for brain metastases: is whole brain radiotherapy necessary? *Int J Radiat Oncol Biol Phys* 1999;43(3):549-58.

29. Schoggl A, Kitz K, Reddy M, Wolfsberger S, Schneider B, Dieckmann K, Ungersbock K: Defining the role of stereotactic radiosurgery versus microsurgery in the treatment of single brain metastases. *Acta Neurochir (Wien)* 2000;142(6):621-26.

30. Huang C, Kondziolka D, Filckinger J, Lunsford D: Stereotactic Radiosurgery for brain stem metastases. *Journal Neurosurgery* 1999;91(4):563-68.
31. Chang SD, Tate DJ, Goffinet DR, Martin DP, Adler JR Jr.: Treatment of nasopharyngeal carcinoma: stereotactic radiosurgical boost following fractionated radiotherapy. *Stereotact Funct Neurosurg* 1999;73(1-4):64-7.
32. Iwai Y, Yamanaka K: Gamma Knife radiosurgery for skull base metastasis and invasion. *Stereotact Funct Neurosurg* 1999;72(Suppl 1):81-7.
33. Uchino M, Nagao T, Seiki Y, Shibata I, Terao H, Kaneko I.: Radiosurgery for cystic metastatic brain tumor No Shinkei Geka 2000;28(5):417-21.
34. Gramaglia A, Loi GF, Mongioj V, Baronzio GF: Increased survival in brain metastatic patients treated with stereotactic radiotherapy, omega three fatty acids and bioflavonoids. *Anticancer Res* 1999; 19(6C):5583-86.
35. Feuvret L, Germain I, Cornu P, Boisserie G, Dormont D, Hardiman C, Tep B, Faillot T, Duffau H, Simon JM, Dendale R, Delattre JY, Poisson M, Marsault C, Philippon J, Fohanno D, Baillet F, Mazeron JJ: Importance of radiotherapy in stereotactic conditions (radiosurgery) in brain metastases: experience and results of the Hospital Pitie-Salpetriere Group Cancer Radiother 1998; 2(3):272-81.2.
36. Manning MA, Cardinale RM, Benedict SH, Kavanagh BD, Zwicker RD, Amir C, Broaddus WC : Hypofractionated stereotactic Radiosurgery as an alternative to Radiosurgery for the treatment of patients with brain metastases. *Int J Radiat Oncol Biol Phys* 2000;47(3):603-8.
37. Chang SD, Tate DJ, Goffinet DR, Martin DP, Adler JR Jr.: Treatment of nasopharyngeal carcinoma: stereotactic radiosurgical boost following fractionated radiotherapy. *Stereotact Funct Neurosurg* 1999;73(1-4):64-7.
38. Yang CC, Ting J, Wu X, Markoe A: Dose volume histogram analysis of the gamma knife radiosurgery treating twenty-five metastatic intracranial tumors. *Stereotact Funct Neurosurg* 1998;70(Suppl 1):41-9.
39. Goodman KA, Sneed PK, McDermott MW, Shiau CY, Lamborn KR, Chang S, Park E, Wara WM, Larson DA: Relationship between pattern of enhancement and local control of brain metastases after radiosurgery. *Int J Radiat Oncol Biol* Phys 2001;50(1):139-46.
40. Huber PE, Hawighorst H, Fuss M, Van Kaick G, Wannemacher MF, Debus J: Transient enlargement of contrast uptake on MRI after Linear Accelerator (linac) stereotactic Radiosurgery for brain metastases. *Int J Radiat Oncol Biol Phys* 2001;49:1339-49.
41. Suguo N, Shibata I, Nemoto A, Nemoto M, Kushida T, Mitou T, Ohishi H, Kuroki T, Terao H, Takkano M, Takahashi H: Changes of regional perfusion in metastatic brain tumour and peritumoural area after Radiosurgery: a study by 123I-IMP dynamic SPECT. *Kaku Igaku* 1996;33(2):123-30.
42. Gasper L, Scott C, Rotman M, Asbell S, Phillips T, Wasserman T, McKenna WG, Byhardt R: Recursive partitioning analysis (RPA) of prognostic factors in three Radiation Therapy Oncology Group (RTOG) brain metastases trials. *Int J Radiat Oncol Biol Phys* 1997;37(4):745-51.
43. Simonova G, Liscak R, Novotny J Jr., Novotny J.: Solitary brain metastases treated with the Leksell gamma knife: prognostic factors for patients. *Radiother Oncol* 2000;57(2):207-13.
44. Kim YS, Kondziolka D, Flickinger JC, et al: Stereotactic Radiosurgery for patients with non-small cell lung carcinoma metastatic to the brain. *Cancer* 1997;80:2075-83.
45. Mori Y, Kondziolka D, Flickinger JC, Logan T, Lunsford LD: Stereotactic Radiosurgery for Brain Metastases from Renal Cell Carcinoma. *Cancer* 1998;83:344-53.
46. Corn BW, Mehta MP, Buatti JM, Wolfson AH, Greven KM, Kim RY, Dunton CJ, Loeffler JS: Stereotactic Irradiation: potential new treatment method for brain metastases resulting from ovarian cancer. *Am J Clin Oncol* 1999;22(2):143-46.

35
Surgical Management of Medulloblastoma

BA Chandramouli

Bailey and Cushing coined the term medullobalstoma in 1925.[1] It accounts for 3.74% of all Intracranial tumours. Choux et al[2] in their literature survey found that the medulloblastoma constitutes 18.22% of all paediatric brain tumours and 28.69% of all the posterior fossa tumours in children. Under the WHO classification the medulloblastoma is included under primitive Neuroectodermal tumours (PNET). Medulloblastoma constitutes 90-95% of PNETS. Due to its distinct clinical behavior, and response to different modalities of treatment it is considered as a distinct entity.[3]

Rapid strides have taken place in the management of medulloblastoma over the decades leading to prolonged survival of patients. Still it remains a challenge as the treatment modalities are not optimized, which has led to the serious disabilities among the long-term survivors.[4] Medulloblastoma is sensitive to both radiotherapy and chemotherapy,[4] still surgery remains critical part of the treatment though it is not curative. Residual tumour after surgery has been found to be a prognostic factor and predicts the long-term survival. In patients with localized disease with post operative residual tumour of less than I.5 Cm^2 and more than I.5 Cm^2 the 5 Yrs. Progression free survival is 78% and 54% respectively.[5-7] Similarly Bourne[8] have reported 100% survival among patients who had no contrast enhancement in postoperative CT scan compared 41% among those patients who had residual contrast enhancement on CT scan. Thus radical excision of medulloblastoma has distinct advantage over the partial excision.

Radical excision of the tumour should be the goal of surgery, however this may not be possible in all cases as significant number 33 to 38%[9-11] of patients have brainstem infiltration by the tumour. In this group of patients radical or total excision of tumour is not feasible and any attempt towards that would lead to serious post operative neurological morbidity. Gajjar et al[6] have reported that in such patients there is no survival advantage by attempting radical excision. It is reported that radical excision can be achieved in 80-90% of cases.[2,12] Thus in patients with localized disease and without brainstem infiltration the aim of surgery should be radical excision of tumour to give patient best chance of long-term survival with adjuvant therapy.

MANAGEMENT OF HYDROCEPHALUS

Hydrocephalus at he time of diagnosis is most common finding varying from 70% to 80%.[2] At the present time the literature is in favour of not doing permanent CSF diversion procedure. This has been advocated to avoid complications of shunt surgery and in patients of medulloblastoma there are reports of systemic metastasis following shunt surgery. It is advocated to have external ventricular drainage intra operatively and in post operative period. However varying number (5-47%) patients require permanent shunt placement inspite of radical tumour excision due to persistence of hydrocephalus in post operative period.[2, 3, 13, 14] Though it would be a sound principle to avoid and shunt procedure in our situation it may not be feasible to undertake surgical excision of tumour in 24 to 48 hrs. after diagnosis due to logistical reasons. Hence still majority of our patients undergo ventriculoperitoneal shunt preoperatively. However one should make conscious effort to avoid shunt surgery not only to reduce the chances of spread of tumour but also to avoid complications of shunt surgery.

SURGICAL MANAGEMENT

Advances in microsurgical techniques, anesthesia, post operative intensive care and use of surgical microscope have made surgery for medulloblastoma a safe procedure. Radical resection of tumor is possible in majority of cases. The goal of surgery is to establish diagnosis, radical excision of tumor whenever possible and to establish CSF pathway.

The surgery is done with patient either in prone position or sitting position. Both have their merits and demerits. The major problem in sitting position is the air embolism this can be minimized by carefully following surgical technique and can be effectively managed by prompt detection and intervention by anesthesiologist. This requires additional monitoring equipments and swan-ganz catheter or central venous catheter to treat the air embolism when it occurs. The advantage being no pooling of blood and CSF hence operative field remains dry and in view of dependent position the tumor will get progressively displaced downward with internal decompression. The advantage of prone position is it doesn't require additional monitoring, or insertion of central lines, it is comfortable for the surgeon. The disadvantage is the pooling of blood and CSF at the operative site. The head is fixed in may filed head clamp however in children below the age of 2 yrs. head is fixed on horse shoe head rest. The medulloblastoma is located in midline in 65.5% of cases hence midline approach is most often preferred.[2] The skin incision is from the external occipital protuberance to the level of spinous process of axis vertebra. The dissection is carried in midline, it is crucial to identify midline and dissect in the same plane this would avoid blood loss, which is be very crucial in young children. The muscle is dissected sub-periosteally to expose the occipital bone, the foramen magnum and posterior arch of atlas. The exposure of suboccipital dura is done by either craniectomy or by craniotomy.

Craniotomy is possible with the help of craniotome. The dural exposure should be from transverse sinus to the level of posterior arch of atlas. The posterior rim of foramen magnum should be removed as far lateral as possible. This would facilitate approach to the tumor from lower pole. There is no need to remove the posterior arch of atlas in every case. It may be necessary if there is gross tonsillar herniation below the level of posterior arch of atlas. The spinous process and lamina of Axis vertebra should not be removed this might lead to instability of spine.[15, 16]

The dural opening is done close to lateral limit of bone removal. The dural opening extends from just below the transverse sinus on either side, parallel to the bone margins and at the level of foramen magnum cutting the dura as far laterally as possible and curve the incision below the level of foramen magnum will reduce the chances of bleeding from the junction of occipital sinus with the marginal sinus. Bleeding from the occipital sinus is controlled by ligating the sinus, the coagulation of dura should be avoided to the maximum, to prevent shrinking of the dura, this would facilitate dural closure. The cervical dura is incised in midline up to C1 arch from the lower end of the cranial dural opening. The dural opening sometimes becomes very difficult due to large venous plexus within the dura. This is encountered in young children, in such a situation one may have to open the dura to the best possible extent and do decompression of tumor. After dural opening the rest of the surgiacal procedure is done under microscope. Observation has to be made regarding the seedling over the arachnoid. The arachnoid is opened in midline exposing the vermis, tonsils and Paramedian parts of the cerebellum. In large tumours, the tumour may be protruding through the foramen of magendie into the cisternamagna between the tonsils. The tumour can be approached by three ways depending on the disposition of the tumour. If the tumour is not protruding through the foramen megendie the tumour can be approached by transvermian approach or by cerebellomedullary fissure (CMF) approach. If the tumour has grown into the cisterna magna the tumour can be excised through the widened foramen megendie.

TRANSVERMIAN APPROACH

This is the well recognized and widely practiced approach. Most of the time due to the presence of the tumor the vermis will be expanded. A vertical incision is made over the lower part of the vermis, incision deepened to reach the tumour and internal decompression of tumour is done.

CEREBELLOMEDULLARY FISSURE (CMF) APPROACH

Anatomy: The cerebellomedullary fissure[17, 18, 19] is between the anterior surface of tonsil and the medulla, the fissure extends cranially between the tonsil and the lower part of the vermis. The ventral surface of the tonsil cover the telachoroidea, the inferior medullarry velum and the nodule[17, 18] by opening this fissure one would be widening the foramen of megendie. The Posterior inferior cerebellar artery (PICA) is in close relationship with the tonsil at the inferior pole and ascends on the medial side of tonsil in the CMF, care should be taken to protect the same. The CMF is opened by lifting the tonsil off the medulla oblongata, flimsy arachnoid strands would be crossing the fissure, the dissection in further carried on the medial aspect between the tonsil and the vermis. This exposes the telachoroidea, the Choroid plexus can be seen easily and anterior to it tumor will be visible. The telachoroidea is cut after coagulation. The CMF dissection is done on both sides this gives wide exposure of the tumor.

Transforamen Megendie approach: It is reported that in 68% of cases the tumor will be visible on opening the dura.[2] Larger tumours, grown through the foramen of megendie and occpy the space between the tonsils and the cisternamagna. The tumor by its growth widenes the foramen sufficiently large to do surgical resection after the internal decompression.

Once the tumor is exposed the margins of the tumour should be defined in the exposed part and tumour should be dissected from the surrounding structures. This would facilitate dissection of the tumour in later part after the internal decompression. The brain stem should be separated from the tumour and a cotton patty should be placed between the tumour and brainstem to protect the brainstem and also it acts as a reference during further course of surgery. when once the tumour is well defined, internal decompression is done. Medulloblastoma is usually soft, friable and vascular and can easily be debulked by using suction, however ultrasonic surgical aspirator will be of use in situations where the tumor may be firm and fibrous. It is very vital for the surgeon to have a three dimensional picture of the tumour and its relation to the surrounding structures. This will facilitate proper internal decompression and help in avoiding injury to normal neural structures surrounding the tumour. Following the tumour debulking the tumor is dissected from the surrounding structures. The dissection is done inferiorly, laterally and superiorly where the tumour is exposed. The blind area of the surgery is the dorsal aspect and the lateral aspects particularly in prone position. Where the angle of vision is down and towards the vertex. Effort should be made to deliver the tumor from the undersurface of vermis.

The part of the tumour covering brainstem should be removed last, because of large space created by tumour excision this enhances the visibility. This part of surgery is very vital as nearly in one third of patients there will be brainstem infiltration.[9,11] It is important to recognize this and leave portion of the tumour infiltrating brainstem. Any aggressive attempt to excise this part would lead to postoperative morbidity. With the radical excision of tumor the CSF pathway gets established. However in situation where part of the tumour is left behind effort should be made to establish CSF pathway by excising the part of the tumor blocking the aqueduct.

During the surgery conscious effort should be made to use retraction of cerebellum to the minimum and where necessary only. Haemostasis is achieved with usual techniques and it would be preferable to confirm the Haemostasis with valsalva maneuver.

Dural closure should be done meticulously if required pericranial graft can be used to close the dura. The closure of the wound with good approximation of muscle layers is vital in preventing post operative CSF leak.

Most of the time patients are extubated following surgery. However if there has been brainstem infiltration or there has been cardiovascular changes such has bradycardia, hypertension, indicative of insult to brainstem these patients would benefit from elective ventilation. The other management would include administration of steroids, fluid and electrolyte management and antibiotics according to the policy adopted. Patients have to be specifically evaluated for fresh neurological deficits particularly if there has been brainstem invasion.

COMPLICATIONS

Complications following excision of medulloblastoma is well recognized. In addition to general complications like operative site haematoma, infection certain complications are observed more frequently in midline posterior fossa tumors particularly medulloblastoma.

CHEMICAL MENINGITIS

This manifests with fever, headache, meningeal signs, usually noticed as the steroid is being tapered in the post operative period. The symptoms are due to presence of blood products and release of cytokines which elicit meningeal reaction.[3] The diagnosis is made by excluding pyogenic meningitis after lumbar CSF examination. The management include administration of high dose corticosteriods and if hydrocephalus is present CSF drainage procedure has to be done using appropriate method depending on the extent of tumor excision.

NEUROLOGICAL MORBIDITY

Neurological morbidity in the form of ocurrence of fresh neurological deficits have been reported 26% to 44%.[10,15] The neurological morbidity could be fresh cranial nerve palsy particularly the abducent and facial nerve, worsening of cerebellar dysfunction or major neurological deficit in the form of bulbar palsy, spastic quadriparsis or hemiparesis[15] due to injury to a the brainstem. These are also labeled as the "Floor of IV ventricle syndrome"[3] 14% of patients may recover the neurological deficits to pre operative level[15] in majority of patients defecits remain permanent.

CEREBELLAR MUTISM

Acquired mutism is defined as condition of complete absence of speech that is not associated with other aphasic symptomatology or alteration of consciousness. It may occur in various neurological conditions with different aetiologies. Neurosurgical intervention such as callosotomy, resection of supplementary motor cortex of dominant hemisphere and cerebellar mass lesions are associated with post operative mutism.[20] Mutism following posterior fossa surgery was initially described by Hirsch[21] in 1979. However Rekate[22] 1985 drew attention to this specific syndrome, and attributed to excessive involvement of vermis and the cerebellar hemispheres and the deep nuclei of cerebellum. Cerebellar mutism occurs between 5 and 30%.[9, 23-25] Classically the syndrome occurs 1-2 days following radical resection of large midline posterior fossa mass usually seen in children. In literature review of Ersahin[26] only 9% of patient were adults rest were all children and occurred in patients following either total or subtotal resection of tumour. In usual situation child would be conversing normally in the immediate post operative period and after 1 or 2 days child becomes irritable and mute. The symptoms are variable from decreased verbal output to severe behavioral disturbances global cerebellar dysfunction and weakness of limbs. Some of them have cranial nuropathies.[3] This is self limiting and takes 4 days to 4 months to recover.[3, 26] In majority the recovery is not complete and are left with speech disturbances cerebellar dysfunction and cognitive defects.[27, 28] children are managed with supportive nursing care, and tube feeding when necessary.

The anatomical substrate for this syndrome appear to be the damage to the dentate nucleus and the dentate-thalamo-cortical connection. This is substantiated by a the fact that this occurs in majority of times, following radical excision of large, vermian mass and which necessitates incision of vermis and lateral retraction of the dentate nucleus. SPECT studies of these patients in acute phase has revealed hypo-perfusion in the thalamus, medial frontal lobe and the fronto-temporal and fronto-parietal regions with the clinical recovery the perfusion of these areas improves.[29, 30] Doxey[23] in their review of their cases of posterior fossa syndrome found in addition to mutism all children had additional neurological deficits, In their series all patients had brain stem involvement by tumor and patient who had undergone incomplete resection also developed mutism, suggesting that extent of excision need not be the cause for mutism, probably the involvement of brainstem by the tumor may play major role in this syndrome, 44% of their patients with brainstem involvement developed mutism and they conclude patients with brainstem involvement are at high risk of developing mutism.

POST OPERATIVE SEIZURE

Seizure can develop following posterior fossa surgery in 1.8 to 5.9% of patients. Its incidence among pts of medulloblastoma varies from 7.2% to 8.5%.[31, 32]

MORTALITY

The post operative mortality among patients of medulloblastoma has reduced significantly over the 3 decades. In 1970's the mortality was between 20% and 30%. This has reduced to around 10%. This has been possible with improvement in surgical techniques anesthesia and intensive care. This further came down to 1-2% in different series.[9,33] Presently the post operative mortality varies from 0-10%.[2,9,33,34]

REFERENCES:

1. Bailey P, Cushing H: Medulloblastoma cerebelli: A common form of midcerebellar glioma of childhood. *Arch Neurol Psychiatr* 1925;14:192-224.
2. Choux M, Lena G, Gentet JC, et al: Medullobalstoma in:Pediatric Neurosurgery: Surgery of the developing nervous system, Mc Lone DG (Ed): W.B. Saunders Company, Philadelphia, 2001; 804-21.
3. Siffert J, Allen JC: Medulloblastoma in: Handbook of Clinical Neurology, Vinken PJ, Bruyn GW(ED): Elsevier Science. B.V. Amsterdam, 1997;Vol. 68: 181-209.
4. Chintagumpala M, Berg S, Blaney SM: Treatment controversies in medulloblastoma. *Curr Opin Oncol* 2001;13:154-59.
5. Zeltzer PM, Boyett JM, Finlay JL, et al: Metastasis stage,adjuvant treatment, and residual tumor are prognostic factors for medulloblastoma in children: Conclusions from the Children's Cancer Group 921 randomized phase III study. *J Clin Oncol* 1999;17:832-45.
6. Gajjar A, Sanford RA, Bhargava R, et al: Medulloblastoma with brainstem involvement: the impact of gross total resection on outcome. *Pediatr Neurosurg* 1996;25:182-87.
7. Zeltzer P, Boyett J, Finlay J: Prognostic factors for survival in high risk Primitive neuroectodermal tumors (PNET) in children: report fom Children's Cancer Study Group CCG-921. *Proc Am Soc Clin Oncol* 1993;12:415.
8. Bourne JP, Geyer R, Berger M, et al: The prognostic significance of operative residual contrast enhancement on CT scan in pediatric patients with medulloblastoma. *J Neuro oncol.* 1992;14: 263-70.
9. Mainprize TG, Taylor MD, Rukta JT: Pediatric brain tumours: a contemporary prospectus. *Clin Neurosurg* 2000;47:259-302.
10. Albright Al, Wisoff JH, Zeltzer PM, et al: Current neurosurgical treatment of medulloblastomas in children. A report from the Children's Cancer Study Group. *Pediatr Neurosci* 1989;15:276-82.
11. Choux M, Lena G: Medulloblastoma Neurochirurgie 1982; 28(Suppl):1-229,
12. Albright Al, Wisoff JH, Zeltzer PM, et al: Effects of medulloblastoma resections on outcome in children: a report from the Children's Cancer Study Group. *Neurosurgery* 1996;38:265-71.
13. Modha A, Vassilyadi M, George A, et al: Medulloblastoma in children-the Ottawa experience. *Childs Nerv Syst* 2000;16:341-50.
14. Dias MS, Albright Al: Management of hydrcephalus complicating childhood posterior fossa tumors. *Pediatr Neurosci* 1989;15:283-89.
15. Cochrane DD, Gustavsson B, Poskitt Kp, et al: The surgical and natural morbidity of aggressive resection for posterior fossa tumors in childhood. *Pediatr Neurosurg* 1994;2019-29.
16. Renier D, Daussange J, Rigaqult P, et al: Children's cervical spine instability after posterior fossa surgery. *Childs Brain* 1981;8:77-78.
17. Matsushima T, Inoue T, Inamura T, et al: Transcerebellomedullary fissure approach with special reference to methods of dissecting the fissure. *J Neurosurg* 2001;94:257-64.
18. Rhoton AL Jr.: Cerebellum and fourth ventricle. *Neurosurgery* 2000;47(Suppl):S7-27.
19. Kellogg JX, Piatt JH Jr.: Resection of fourth ventricle tumors without splitting the vermis: the cerebellomedullary fissure approach. *Pediatr Neurosurg* 1997;27:28-33.

20. Crutchfield JS, Sawaya R, Meyers CA, et al: Postoperative mutism in Neurosurgery. Report of two cases. *J Neurosurgery* 1994;81:115-21.
21. Hirsch JF, Renier D, Czernichow P, et al: Medullobalstoma in childhood. Survival and functional results. *Acta Neurochir (Wien)* 1979;48(1-2):1-15.
22. Rekate HL, Grubb RL, Aram DM, et al: Muteness of cerebellar origin. *Arch Neurol* 1985;42:697-98.
23. Doxey D, Bruce D, Skalr F, et al: Posterior fossa syndrome: identifiable risk factors and irreversible complications. *Pediatr Neurosurg* 1969;31:131-36.
24. Pollack IF, Poliko P, Albright AL, et al: Mutism and pseudobulbar symptoms after resection of posterior fossa tumors in children:Incidence and pathophysiology. *Neurosurgery* 1995;37:885-93.
25. Siffert J, Allen J, Epstein F: The posterior fossa syndrome following tumour resection:incidence. clinical features and long term outcome. *Ann Neurol* 1995;38:187(Abst.)553.
26. Ersahin Y, Mutluer S, Cagli S, et al: Cerebellar mutism: report of seven cases and review of literature. *Neurosurgery* 1996;38:60-65.
27. Riva D, Giorgi C: The cerebellum contributes to higher functions during development: evidence from series of children treated posterior fossa tumors. *Brain* 2000;123:1051-61.
28. Vadeinse D, Hornyak JE: Linguistic and cognitive defecits associated with cerebellar mutism. *Pediatr Rehabil* 1997;1:41-44.
29. Sagiuchi T, Ishii K, Aoki Y, et al: Postoperative mutism in Neurosurgery. Report of two cases. *J Neurosurg* 1994;81:115-21.
30. German A, Baldair S, Caruso G, et al: Reversible crebral perfusion alterations in children with transient mutism after posterior fossa surgery. *Childs Nerv Syst* 1998;14:114-19.
31. Suri A, Mahapatra AK, Bithal P: Seizures following posterior fossa surgery. *Br J Neurosurg* 1998;12:41-44.
32. Lee ST, Lui TN, Chang CN, et al: Early postoperative seizures after posterior fossa surgery. *J Neurosurg* 1990;73:541-44.
33. Tomita T, McLone DG: Medulloblastoma in childhood:results of radical resection and low-dose neuraxis radiation therapy. *J Neurosurg* 1986;64:238-42.
34. Helseth E, Due-Tonnessen B, Wesenberg F, et al: Posterior fossa medulloblastoma in children and young adults (0-19 years):Survival and performance. *Childs Nerv Syst* 1999;15:451-55.

36

Posterior Fossa Ependymoma

M Husain, DK Vatsal

Ependymomas are tumors of the central nervous system that derive from the ependymal cells lining the cerebral ventricles and the central canal of the spinal cord and from ependymal rests in cortical white matter. Intracranial ependymomas comprise about 1.2 to 9% of all intracranial tumors.[1-6] They occur most commonly in young children, constituting 6 to 10 per cent of primary brain tumors in the pediatric population.

Most, previously reported retrospective clinical series do not distinguish tumor location in their evaluations of treatment modalities.[7-9] Infact there are wide variations between series in the percentage of high grade tumors and in percentage of patients.[8,9] This is a problem because it has been reported that biological activity is related to intracranial location. Supratentorial tumors are more mitotically active than those in the posterior fossa.[10,11]

HISTORICAL PERSPECTIVES

The first description of tumors arising from ependymal cells was made by Virchow in 1863.[12] The later observation that the cells of both the normal ependymal lining and ependymomas contain blepharoplasts was made by Mallory in 1902 and confirmed the common histogenetic origin of these tissues.[13] In a series of publications from 1924 to 1932, Bailey and Cushing described in detail the histological nature and classification of the ependymal tumors as distinct from other gliomas.[14,15]

INCIDENCE

Ependymomas constitute between 1.2 to 9 per cent of the intracranial tumors in most series.[1-5] Although the peak age of diagnosis is between 5 and 15 years, new cases are identified at a slowly declining incidence throughout adulthood, and documented cases have been diagnosed in patients in their seventh and eighth decades of life.[16,17] In most large series of tumors in both adult and pediatric patients, the mean age ranges from 8 to 25 years. Ependymomas are relatively more common in the pediatric population than in adults, and they represent approximately 13 per cent of all reported intracranial tumors in children.[17]

Analyses of reported series show that ependymoma has a slight predominance among males. Intracranial tumors can be classified by their location as well. The majority of ependymomas occur beneath the tentorium. One third occurs supratentorially and two third occur infratentorially. Almost all infratentorial ependymoma occur in midline, often involve the floor of fourth ventricle.[18,19]

PATHOLOGY

Histological classification is very controversial. World Health Organization Scheme and Russel and Rubinstein's scheme divide the tumors in to nonanaplastic and anaplastic ependymoma. Kernohan proposed a four-tier grading system.

Histological Classification Schemes for Eprndymoma

World Health Organization[20]	Kernohan[21]	Russell and Rubinstein[22]
Ependymoma	Grade	*Ependymoma*
Cellular islands Uniform, dense cells Rosettes Pseudorosettes Mineralization Necrosis (occasionally)	I. Typical architecture, uniform cells II. Less distinct but uniform; slight pleomorphism, hyperchromatism	Essentially similar to World Health Organization Scheme
Anaplastic ependymoma Epithelial arrangement of cells Loss of pattern Disordered stroma High mitotic activity and cellularity Nuclear atypia Ependymoblastoma	III. Fragmentary but discernible architecture; half of cells may be pleomorphic; one mitotic figure per two high-power fields IV. Only remnants of architecture, rare ependymal-appearing cells; marked pleomorphism; 4-5 mitotic figures per high-power field	*Anaplastic ependymoma* At least focally recognizable architecture Increased cellularity Cellular pleomorphism Nuclear hyperchromatism Numerous mitotic figures May have extensive areas of undifferentiation Ependymoblastoma

On gross examination, most ependymomas appear reddish, lobulated, and friable. Specimens collected at the time of surgical resection are often well vascularized. Supratentorial tumors may appear more grayish in colour and according to Svein and associates, are more commonly cystic (39 per cent of supratentorial tumors versus 10.7 per cent of infratentorial tumors).[23]

The microscopic appearance of low-grade ependymomas is somewhat variable; however, frequently the overall architecture, which can be described as "leopard skin" when viewed at low magnification, is suggestive of the diagnosis.[24] The perivascular areas are nuclear free zones composed of long, filamentous, eosinophilic cellular processes that extend from the region of the nuclei to the central blood vessels. These are the "pseudorosettes" commonly seen in abundance in at least the focal areas of most tumors. True "ependymal rosettes," in contrast, are less frequently observed but are diagnostic when present. In general, ependymomas are moderately cellular and have a low mitotic index.[25]

The nuclei of ependymomas typically are rather large, round or oval, and uniform, with abundant coarse chromatin on staining. The cytoplasm usually is not prominent. Mineralization or necrosis is relatively often identified.

When ependymomas occur in the fourth ventricle they tend to grow down through the foramen magnum and out of the foramen of Lushka into the cerebellopontine angle. These tumors usually are solid and well demarcated, and limited infiltration of adjacent brain tissue usually is observed.

Anaplastic ependymomas are characterized principally by increased cellularity and loss of the overall architecture of the nonanaplastic tumor. These tumors have variable nuclear atypia, marked mitotic activity, and often prominent vascular proliferation.[25] These tumors may seed the cerebrospinal fluid pathways.

World Health Organization classification lists ependymoblastomas as a distinct entity.[25] In any case, these are rare tumors that clinically behave like medulloblastomas and have a poor prognosis.

METASTATIC EPENDYMOMA

Some intracranial ependymomas metastatize within the subarachnoid space of the craniospinal axis or to extraneural sites. The potential for metastasis becomes important in the staging and treatment of primary intracranial ependymomas.

Overall reported rates determined from a compilation of published reports are approximately 11 to 12% of all cases of intracranial ependymoma. Important determinants of metastatic propensity include tumor grade and location and the success of local tumor control.[26] Highest risk of spinal metastases is observed in those patients with high-grade lesions located in the posterior fossa, especially if treatment at the primary site of tumor origin is not successful.[27]

Metastasis of intracranial primary central nervous system neoplasms to sites outside the central nervous system is even more rare, occurring in 0.4 to 2.0% of such lesions.[28] Ependymoma is one of the least common tumor types that is observed to produce such extraneural metastases and accounts for less than 10% of the total. More than 90% of intracranial ependymomas associated with extraneural metastases are from the supratentorial compartment.[29] Common sites of spread include the lungs (63%), scalp, skull or dura (57%), lymph nodes, especially cervical (50%); and rarely heart, skeletal bones, kidneys or skin. Although longer survival times and multiple operative interventions are postulated but unproven risk factors for such metastases, ventriculoperitoneal shunts have also been implicated as a slight risk factor for the spread of intracranial ependymomas to peritoneal locations.[29, 30]

ETIOLOGY

Etiology of ependymomas is unknown. Cytogenetic analysis of chromosomal changes in ependymomas has characterized the loss of chromosome 22 as the most frequently identified genetic abnormality. Loss of chromosomal material identified in association with a tumorous state is consistent with the possibility that some of the lost genetic information is responsible for suppressing the altered phenotype. These missing regulatory sequences have been termed "tumor-suppressor genes". As more than 50% of ependymomas have lost or altered chromosome 22 sequences, the possibility exists that a tumor-suppressor gene important in ependymomas is located on this chromosome. However, to date none has been identified.[31]

PRESENTATION

Patients with infratentorial ependymomas often have long clinical history and insidous onset. Initial symptoms associated with infratentorial ependymomas include intermittent nausea and vomiting and headache. These symptoms are usually followed by gait disturbance, vertigo and dysphagia.[31]

Signs of infratentorial lesions include ceerbellar dysfunction (approximately 70%), papilledema (upto 72%), cranial neuropathy (20 to 36%), and abnormal deep tendon reflexes (upto 23%) and cranial deep tendon reflexes (upto 23%).[31]

Cranial nerve dysfunction may be due to compression or invasion of the floor of the fourth ventricle (typically affecting cranial nerves VI and VII), to tumor spread into the cerebellopontine angle (affecting cranial nerves V, VII, or VIII) or more rarely to tumor spreading into the region of the lateral foramen magnum (potentially affecting cranial nerves IX, X or XI). Truncal ataxia (52%) is more common than limb ataxia (32%); this reflects the predominant midline fourth ventricular origin of most infratentorial ependymomas. The most common sign leading to the diagnosis of a fourth ventricular ependeymoma is gait abnormality.

INVESTIGATION

1. **Plain X-ray Skull:** may show signs of raised intracranial pressure. Calcification in fourth ventricular region favour the diagnosis of ependymoma.

2. **CT Scan:** Plain CT scan shows an isodense to mildly hyperdense mass. Contrast study demonstrates irregular enhancement **(Fig. 1)**. Areas of hypodensity with in the tumor may be due to cyst or necrosis. Calcification may be visualized.

3. **MRI.:** MR reveals a hypointense mass in T1 weighted images and hyperintense on T2 weighted images. Marked heterogenicity may be seen due to necrosis, cystic degeneration, acute, subacute or chronic hemorrhage. Gadolinium administration produces heterogenous and irregular contrast enhancement **(Figs. 2 & 3)**.

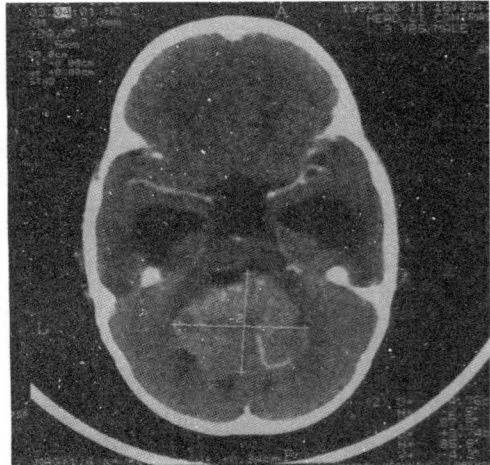

Fig. 1 : Hyperdense, irregularly enhancing mass in fourth ventricular region with dilated ventricles

TREATMENT

(a) **Surgery:** Total removal of the tumor is the goal of surgery but the bulk of the tumor, location of tumor and infiltration or invasion of normal brain structure preclude total removal in most cases. Therefore, the goal of surgery are –

(i) to reduce the bulk of tumor and when possible excise it;

(ii) to establish the tissue diagnosis;

(iii) to open up the C.S.F. pathways.

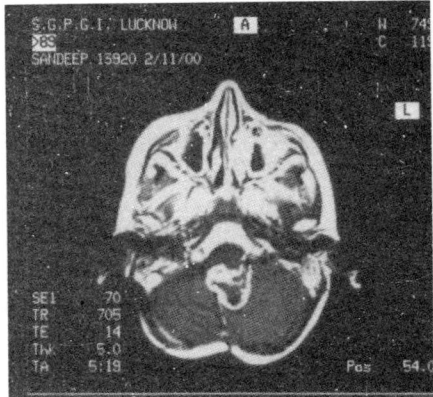

Fig. 2 : Irregular with heterogenous Gd enhancement mass lesion in fourth ventricular region T1 weighted MR image - axial cut

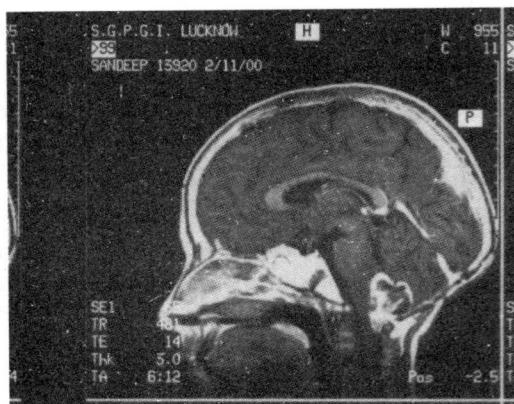

Fig. 3 : Irregular with heterogenous enhancement mass lesion in fourth ventricular region T1 weighted MR image - sagittal section

(b) **Postoperative Therapy:**

Radiation Therapy : There is no doubt that the postoperative radiation therapy has the great impact on the survival of ependymoma patients. Though most controversial aspects of postoperative care has been the form and extent of such therapy.

Radiation doses from 4,500 to 6,000 rad to the primary tumor site over 5 to 6 weeks are necessary for the postoperative treatment of patients with subtotally resected nonanaplastic ependymomas, and most radiation therapists recommend a dose between 5,000 and 5,500 rad.[32, 33] Higher doses (5,500 to 6,000 rad) over 6 to 7 weeks are neeed for those with more aggressive ependymomas and ependymoblastomas.[34] Tumor doses above 4,500 rad have been associated with better local control and survival than lower radiation doses have been.[20, 35] No studies have shown a dose-reprophylactic craniospinal irradiation does not appear necessary for nonanaplastic intracranial ependymomas, as most relapses occur within the primary radiation therapy field.[36] Thus, at our institution, craniospinal axis irradiation is reserved for patients with high grade ependymomas (malignant ependymomas and ependymoblastomas) or with evidence of leptomeningeal spread.

Despite the use of doses of 5,000 rad or greater to the tumor bed, recurrence is seen in patients with subtotally resected ependymomas and in approximately 50 per cent of those with high grade ependymomas.[31] With radiation doses of less than 4,500 rad to the tumor site, patients are significantly more likely to have a relapse.[38] Because the most important prognostic factors in ependymoma treatment are the degree of resection, radiation therapy and the histological grade of the tumor. The role of radiation therapy in the treatment of anaplastic ependymoma and ependymoblastoma is less well defined.[35]

Chemotherapy: Indication –

- Recurrence after surgery and radiotherapy.
- In infant after incomplete removal.

Published case reports have shown responses in patients with single-agent chemotherapy with carmustine, lomustine, or etoposide.[39, 40] High-dose cyclophosphamide and oral dibromodulcitol have also been shown to increase progression-free survival in some patients. Cis-platinum has been most carefully studied in regard to ependymoma, with several reports showing complete and partial responses of from 3 months to 3 years or more.[41] Combination chemotherapy regimens are also advised by Michael N. Needle et al.[42] The chemotherapeutic agents in this regimen are Carboplastin, Vincristin, Ifofsamide and Etoposide.

Flow Chart of Management

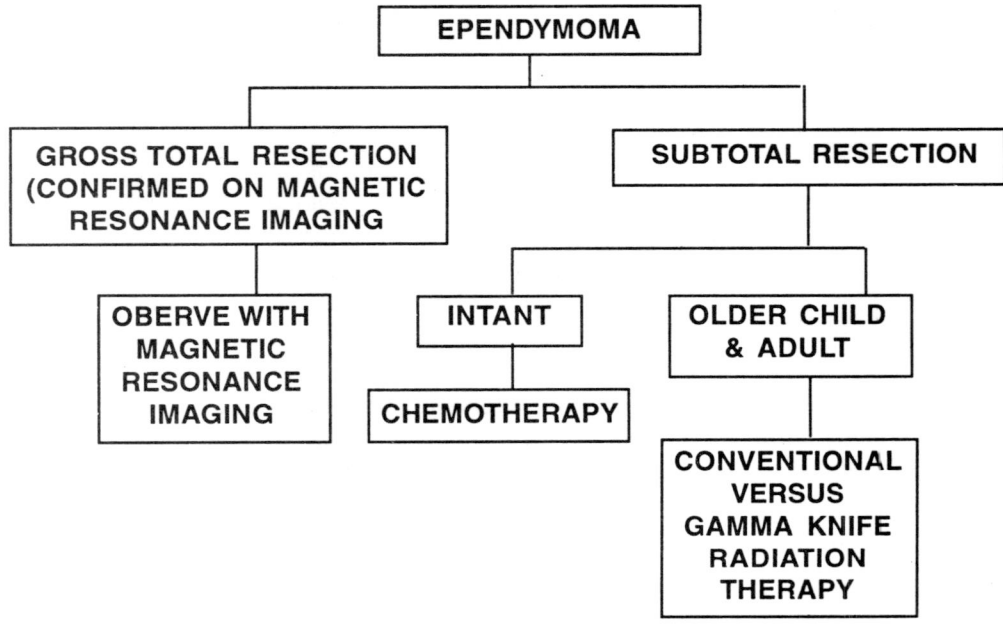

PROGNOSIS

Surgery combined with radiation therapy gives five year survival rate 49-83% for adult patients.[43-44] The important prognostic factors are:[45, 46]

(i) Age of patient : Younger the age worse the prognosis.

(ii) Duration of symptom : Better with longer duration.

(iii) Grade of tumor : More the grade poorer is outcome.

(iv) Tumor location : Near brain stem or infiltrating into it associated with poor prognosis.

(v) Radiation therapy : improve outcome.[47]

(vi) Other factors : that may influence prognosis in patients with ependymoma are gender and race, although definitive studies of either of these factors have not been performed.

REFERENCES:

1. Bailey P, Cushing H: A classification of the tumors of the Glioma Group on a Histogenetic Basis with a Correlated Study of Prognosis. Philadelphia JB Lippincott 1926.
2. Coulon RA, Till K: Intracranial ependymomas in children – A review of 43 cases. *Childs Brain* 1977;3:154-68.

3. Mork SJ, Loken AC: Ependymoma: A follow up study of 101 cases. *Cancer* 1977;40:907-15.
4. Packer RJ, Siegel KR, Shunt L, Bruce DA, Sutton LN, Litmann P: Central nervous system spread of childhood brain tumors at diagnosis or at initial disease recurrence. *Concepts Pediatr Neurosurg* 1985;6:16-24.
5. Pierre-Kahn A, Hirsch JF, Roux FX, Renier D, Sainte-Rose C: Intracranial ependymomas in childhood: Survival and functional results of 47 cases. *Childs Brain* 1983;10:145-56.
6. Read G: The treatment of ependymomas of the brain or spinal canal by radiotherapy: A report of 79 cases. *Clin Radiol* 1984;35:163-66.
7. Bloom HJG: Intracranial tumors. Response and resistance to therapeutic endeavours. 1970-1980. *Int J Radiat Oncol Biol Phys*, 1982;8:1083-1113.
8. Chin HW, Maruyama Y, Markesbery W, Young AB: Intracranial ependymoma: Results of radiotherapy at the University of Kentucky, *Cancer* 1982;49:2276-80.
9. DiMarco A, Campostrini F, Pradella R, Reggio M, Palazzi M, Grandinetti A, Garusi GF: Postoperative irradiation of brain ependymomas: Analysis of 33 cases. *Acta Oncol* 1988;27:261-67,
10. Zulch KJ: Brain tumors: Their Biology and Pathology. New York, Springer-Verlag, 1986, ed. 3,258-76.
11. Mark K Lyons, Patrick JK: Posterior fossa ependymomas: Report of 30 cases and review of the literature. *Neurosurgery* 1991;28(5):659-65.
12. Virchow RLK: Cellular pathology as based upon Physiology and Pathological Histology translated from the second edition of the Original by Frank Chance. Philadelphia, JB Lippincott, 1863.
13. Mallory, FB: Three gliomata of ependymal origin: Two in the fourth ventricle, one subcutaneous over the coccyx. *J Med Res* 1902;8:1-10.
14. Bailey P: A study of tumors arising from ependymal cells. *Arch Neurol Psychiatry* 1927;11:1.
15. Bailey P, and Cushing H: A classification of the tumors of the glioma group on a Histogenic basis with a correlated study of prognosis. Philadelphia, JB Lippincott, 1926.
16. Birgisson S, Blondal H, Bjornsson J, et al: Tumors in Iceland: 15. Ependymoma: A clinicopathological and immunohistological study. *APMIS,* 1992;100:294-300.
17. Helseth A, and Mork SJ: Neoplasms of the central nervous system in Norway: III Epidemiological characteristics of intracranial gliomas according to histology. *APMIS* 97:547-55.
18. Shuangshoti S, Panyathanya R: Ependymomas: a study of 45 cases. *Dis Nerv Syst* 1973;34:307-14.
19. Svien HJ, Mabon RF, Kernohan JW, Craig WM: Ependymoma of the brain: pathologic aspects. *Neurology* 1953;3:1-15.
20. Kim EE, Domstad PA, Choy YC, et al: Differential accumulation of Tc-99m DTPA and Tc-99m pyrophosphate within cerebral and cranial lesions: Concise communication. *J Nucl Med* 1980; 21:838-40.
21. Kernohan JW, Mahon RF, Svien JH, et al: A simplified classification of gliomas. *Proc Staff Meet Mayo Clin* 1949;24:71-75.
22. Russell D and Rubinstein L: Tumors of central neuroepithelial origin. In Russell, DS and Rubinstein LJ eds.: Pathology of Tumours of the Central Nervous System. Baltimore, Williams and Wilkins 1989;192-206.
23. Svien H, Mabon R, Kernohan J, et al: Ependymoma of the brain: Pathological aspects. *Neurology* 1953;3:1-15.
24. White L, Johnston H, Jones R, et al: Postoperative chemotherapy without radiation in young children with malignant non-astrocytic brain tumours: A report from the Australia and New Zealand Childhood Cancer Study Group (ANZCCSG). *Cancer Chemother Pharmacol* 1993;32:403-06.
25. Kleihues P, Burger PC, Scheithauer BW eds.: Histological typing of tumours of the Central Nervous System. 2nd ed. Kleuheus, et al. are eds. In: International Histological Classification of Tumours. Berlin/Heidelberg, Springer-Verlag, 1993;112.
26. Pierre-Kahn A, Hirsch J, Roux F, et al: Intracranial ependymomas in childhood: Survival and functional results of 47 cases. *Child's Brain* 1983;10:145-56.

27. Vanuytsel L and Brada M: The role of prophylactic spinal irradiation in localized intracranial ependymoma. *Int J Radiat Oncol Biol Phys* 1991;21:825-30.
28. Pasquier B, Pasquier D, N'Golet A, et al: Extraneural metastases of astrocytoma and glioblastomas: Clinicopathological study of two cases and review of literature. *Cancer* 1980;45:112-25.
29. Itoh J, Usui K, Itoh M, et al: Extracranial metastases of malignant ependymoma: Case report. *Neurol Med Chir (Tokyo),* 1990;30:339-45.
30. Hoffman HJ, Duffner PK: Extraneural metastases of central nervous system tumors. *Cancer* 1985;56:1778-82.
31. Lyons MK, Kelly PJ: Posterior fossa ependymomass: Report of 30 cases and review of the literature. *Neurosurgery* 1991;28:655-59.
32. Kun LE, Kovnar EH, Sanford RA: Ependymomas in children. *Pediatr Neurosci* 1988;14:57-63.
33. Leibel SA and Sheline GE: Radiation therapy for neoplasms of the brain. *J Neurosurg* 1987;66:1-22.
34. Marsh WR, Laws E Jr: Intracranial ependymomas. *Prog Exp Tumor Res* 1987;30:175-80.
35. Marks J, Adler S: A comparative study of ependymoma by site of origin. *Int J Radiat Oncol Biol Phys* 1982;8:37-43.
36. Healey EA, Barnes PD, Kupsky WJ, et al: The prognostic significance of postoperative residual tumor in ependymoma. *Neurosurgery* 1991;28:666-72.
37. Lesser GJ, Grossman SA: The chemotherapy of adult primary brain tumors. *Cancer Treat Rev* 1993;19:261-81.
38. Goldwein JW, Leahy JM, Packer RJ, et al: Intracranial ependymomas in children. *Int J Radiat Oncol Biol Phys* 1990;19:1497-1502.
39. Chiu JK, Woo SY, Ater J, et al: Intracranial ependymoma in children: Analysis of prognostic factors. *J Neurooncol* 1992;13:283-90.
40. Levin VA: Chemotherapy of primary brain tumors. *Neurol Clin* 1985;3:855-66.
41. Corden BJ, Strauss LC, Killmond T, et al: Cisplatin, ara-C and etoposide (PAE) in the treatment of recurrent childhood brain tumors. *J Neurooncol* 1991;11:57-63.
42. Michael N Needle, Joel W Goldein, Jeffrey Grass, et al: Adjuvant chemotherapy for the treatment of intracranial ependymoma of childhood. *Cancer* 1997;80(2): 341-47.
43. Bloom HJ: Intracranial tumours: Response and resistance to therapeutic endeavours, 1970-1980. *Int J Radiat Oncol Biol Phys* 1982;8:1083.
44. Garrett PG, Simpson WJ: Ependymomas results of radiation treatment. *Int J Radiat Oncol Biol Phys* 1983;9:1121.
45. Dohrmann GJ, Farwell JR, Flannery JT: Ependymomas and ependymoblastomas in children. *J Neurosurg* 1976;45:273.
46. Mork SJ, Loken AC: Ependymomas: A follow up study of 101 cases. *Cancer* 40:907, 1977.
47. Salazar OM, Rubin P, Bassno D, et al: Improved survival of patients with intracranial ependymomas by irradiation. *Cancer* 1975;35:1563.

37
Controversies in Management of Malignant Posterior Fossa Tumour

AK Mahapatra

INTRODUCTION

Posterior fossa tumours are not uncommon lesions, however, in children these lesions may constitute 2/3rd of all intracranial tumours.[1-3] In adults a significant proportion of brain tumours arise from the posterior fossa structure, considering a smaller volume of posterior fossa as compared to supratentorial volume. Malignant tumours of posterior fossa from a major portion of intracranial tumour in children. Among them Medulloblastoma, malignant astrocytoma and malignant brainstem tumours form the major group.[4-7] Over the years the outlook in management of malignant tumours of posterior fossa has changed significantly, be it, Medulloblastoma or brainstem tumours, which are no more consider neurosurgical challenge with overall poor prognosis. Inspite of changed outlook and better outcome, there remains a considerable controversies in the management of the Malignant tumours of the posterior fossa. This write up bring home some of these controversies and tried to rationalise the appropriate treatment.

The malignant lesions in the posterior fossa can arise from the neural structure such as brainstem, cerebellum or from the clivus, petrous or occipital bone. Among the intraaxial malignant lesion, Medulloblastoma tops the list followed by brainstem gliomas and malignant gliomas of the cerebellum **(Table 1)**.

MEDULLOBLASTOMA

Medulloblastoma is the most common malignant tumour of the posterior fossa. It constitute 15-25% of all intracranial tumour in children and 33% of all tumours of the posterior fossa. Large number of controversies still dominate the managementof Medulloblastomas. Some of them are like, (a) role of preoperative shunt or external ventricular drainage, (b) Role of preoperative steroids and anticonvulsant, (c) operating in sitting or prone position, (d) postoperative radiotherapy in children and screening of asymptomatic children for recurrence or metastasis.

Controversies stills dominate regarding the genetic proponderance of medulloblastoma.

Table 1 : *Malignant tumour of posterior fossa*

A.	Axial	(a)	Cerebellum
		-	Medulloblastoma
		-	Ependymomas
		-	Malignant astrocytoma
		-	Glioblastoma
		-	Metastatic cerebellar tumour
		(b)	**Brainstem tumours**
		-	Malignant astrocytoma
		-	Glioblastoma
		-	Ependymoma
B.	Extraaxial	-	**Malignant Meningioma**
		-	C.P. angle
		-	Clivus
		-	Petrous
		-	Meningeal metastasis

In some situation there is a history of familial involvement, and can be associated with Ataxia talangetasia, Rubeinstein – Taybi syndrome and Gorlin syndrome. Recently, some authors have reported different gene expression in Medulloblastomas.[8,9,10] Some authors have demonstrated Isochromosome 17q in 30% patients with Medulloblastomas. However, tumour suppression gene P53 is isolated in short arm of chromosone 17 in mutation of P53 which lead to malignancy.[8,10] Loss of heterozygoity observed in chromosome 17 in Medulloblastomas reported by Srinivasan et al and Batra et al.[8,10] Batra et al[10] had tried to correlate the long term survival with chromosome 17P detection in medulloblastomas. However, more study and long term follow up is necessary to prove the genetic aspect and outcome.

Preoperative investigation in patients of medulloblastoma remains unclear. No doubt, large number of patients with medulloblastoma have spinal metastasis through C.S.F. Many centres routinely study lumbar CSF for malignant cell and others routinely perform spinal MRI. Based on CSF positivity medulloblastoma is staged as Mo, M1, M2 and M3[11] **(Table 2)**. However, postoperative MRI may not be even possible in all the patients routinely in poor countries like India, where most of the patients are operated only on CT Scan basis. Thus, routine spinal MRI preop remains a matter of controversy. Similarly, for postoperative evaluation routine spinal MRI in frequent intervals may be ideal but may not be feasible. Preoperative or postoperative screening of asymptomatic cases remains a big question.[12-15] However, there is no doubt routine screening in asymptomatic children helps in early diagnosis and proper management which is associated

Table 2 : *Staging of medulloblastoma*

(A)	Tumour size	(B)	Metastasis
T_1	- Less than 3 cm diameter	M_0	- No CSF or blood metastasis
T_2	- More than 3 cm diameter invasion to adjacent fissure and partially filling up IV ventricle	M_1	- CSF microscopic +ve for malignant cells
$T_3 A$	- IV ventricle completely filled up with gross hydrocephalus	M_2	- gross nodule in cerebellum, IIIrd or lateral ventricle
$T_3 B$	- Tumour arising from floor of IV Ventricle and filling of IV Ventricle	M_3	- Spinal nodules
T_4	- Tumour spreading through Aqueduct to 3rd ventricle	M_4	- Extradural metastasis- Chest, Liver, bone etc.

with significantly long survival.[13,15] Thus, early diagnosis of recurrence provides a critical window for new therapy.

A subject for continuous controversy is the ventriculoperitoneal shunt in medullobalstoma cases.[16-18] No doubt, there is a small risk of peritoneal metastasis via V.P. Shunt. However, shunt surgery helps in reduction of raised ICP and enhance the chance of good outcome. However, off late pediatric neurosurgeons feel the necessity of shunt less frequently after the radical surgery, when the tumour can be cleared and CSF circulation can be reestablished. Some authors also have suggested temporary external ventriculostomy to avoid shunt related complication.[17,18] Va queso et al[17] reported intratumoural bleed in patients of posterior fossa tumour, following the ventricular drainage.

Position during the surgery for posterior fossa lesions was controvertial till recently. For a long time patients were operated in sitting position, inspite of risks of hypotension, air embolism and pneumocephalus. Now, there is a great deal of concensus regarding the surgery in prone position. Carrying out craniectomy or craniotomy in posterior fossa still is a matter of personal choice, rather than on scientific ground.

Overall, long term survival in patients with medulloblastoma has improved because of radical surgery, and postoperative radio and chemotherapy. However, radiotherapy in small children is not out of risk. Large number of authors have reported growth disturbance and other endocrine problems with high dose radiation for Medulloblastoma.[19-22] Long term problem of learning disability to severe Neurocognitive disorders are reported in 2/3rd of the cases, those who received radiation.[21,22] Quality of life and intellectual outcome also depends on age of the patient, amount and quality of radiation.[20] Endocrinal problems are reported in

50% children, in the form of hypopituitarysm, mainly impairment of growth, due to direct damage to vertebral cartilage.[23] Other problems like hypogonadism, hypothyroidism and hypocorticism are also reported.[24,25] Hence, radiotherapy is best avoided in children under 10 years of age.

Recently, to reduce the neurotoxicity and endocrine problem, low dose radiation was tried by some authors.[26,27] However, result and long term survival was inconclusive, whereas hyperfactionated craniospinal radiation[28,29] have been helpful in reducing neurotoxicity. Allen et al used 1 Gy twice daily dose as hyperfactionated irradiation.

Chemotherapy in medulloblastoma is a well established adjunt therapy. There is no controversy today regading the avoidance of radiation to children under 3 years of age. Hence, chemotherapy is the only mode of adjunct available for this subgroup.[30]

BRAINSTEM TUMOURS

Braintem tumours were considered inoperable in the past.[31-33] Only for over last two decades, brainstem tumours have been considered operable by many authors. After the availability of MRI scans clear delineation of brainstem pathology became possible, and surgery has become rewarding.[34-39] However, like any other conditions in neurosurgery there are controversies, which includes, biopsy for diffuse pontine glioma, extent of surgery, role of chemotherapy etc. In 1969 Matson[40] stated "regardless of specific histology, they must all be classified malignant, since their location in itself render them inoperable". Since that time the things have changed significantly except few type which is still consider inoperable **(Table 3)**. Currently only type I **(Fig. 1)** which is diffuse pontine glioma is considered inoperable, while other type not only can be operated, but also considered to have a good

Table 3: *Classification of Brainstem Glioma on CT/MRI basis*

	Types	Nature in MRI
1.	Type I	Intrinsic tumour, diffuse hypodense on CT. Low intensity in T1 weighted MR image. No contrast enhancement
2.	Type II	Intrinsic focal tumour — cystic or solid
3.	Type III	Exophytic — Dorsally / Laterally
4.	Type IV	Cervicomedullary

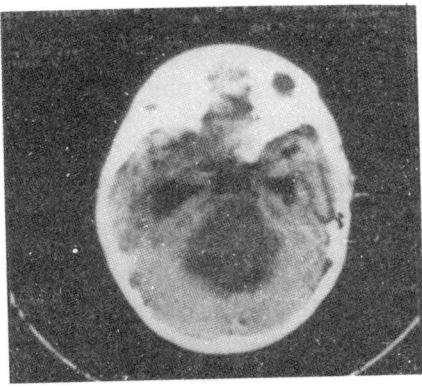

Fig. 1 : Contrast enhanced CT scan axial cut diffuse hypodensity suggestive of Type I Brainstem glioma

prognosis. Generally, in a benign tumour history is long, while in a malignant tumour history is shorter than 3 months. Over 2/3rd of the Type II **(Figs. 2a & b)**, **(Fig. 3)** Type III **(Fig. 4)**, Type IV tumours are benign. The acturial long term survival of Type IV tumour is around 90%.

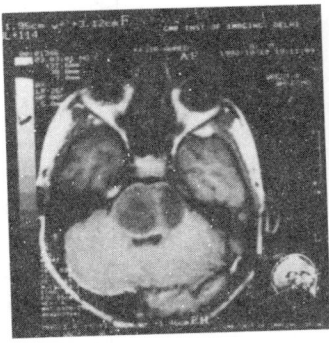 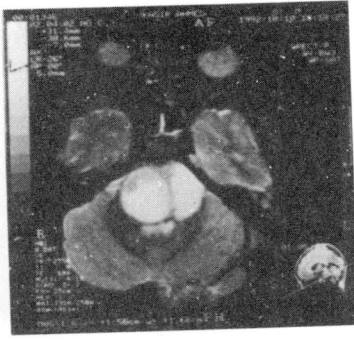

Figs. 2 : (a-b) MRI axial cuts both T1 and T2 weighted image show pontine glioma

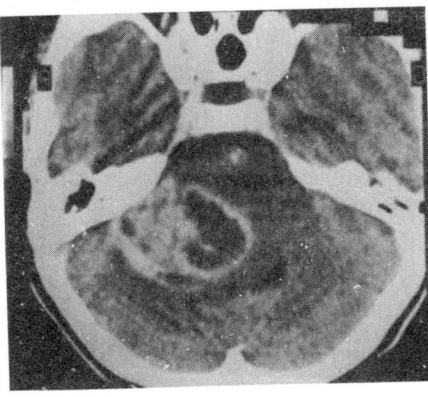 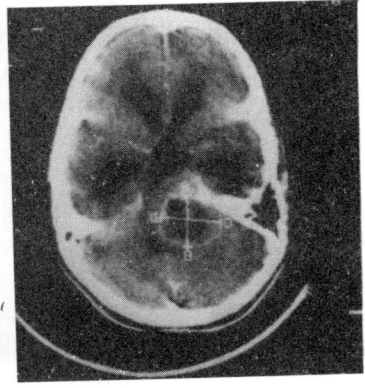

Fig. 3 : Contrast CT axial cut show focal cystic glioma of pons (Type II)
Fig. 4 : Contrast enhanced axial CT magnified view show cystic cum solid laterally exophytic mass in pons going to C.P. angle (Type III)

It was generally believed that the diffuse pontine gliomas or brainstem tumours as a whole do not produce raised ICP or hydrocephalus. Currently this has been proved wrong, and hydrocephalus is recorded in 20-55% cases.[36,38-41] In the series reported by Choux et al raised ICP was recommended in 48% cases. In our study 25% patients with brainstem glioma had hydrocephalus.[36] In patients with tectal tumour incidence of hydrocephalus is higher (82-100%).

Over the years, neurophysiological studies have proved to be helpul in assessing the brainstem function, both preoperatively and intraoperatively. Brainstem auditory evoked potentials (BAEP) is one of the most useful technique to assess brainstem function.[42-46] Some authors have used monitoring of motor nuclei in floor of the fourth ventricle during surgery.[45-46] BAEPs are resistant to pharmacologic suppression, hence are particularly helpul in intraoperative monitoring during the surgery of brainstem tumours. Inspite of intraoperative monitoring of BAEPs and normal intraoperative recording, clinical worsening can still occur.[47] Some authors have monitored SSEP during the surgery of cervicomedullary lesion, and SSEP may not be of much help in pontine lesion.[48] Brainstem trigeminal evoked potentials are also monitored during surgery.[43,49] However, the usefulness is not very reliable.

Biopsy from the Type I brainstem glioma is still a matter of debate. These lesions uniformly expand the brainstem, have a characterstic CT and MRI morphology, and is generally believed to have a poor prognosis. No doubt, these lesions are treated on an emperical ground rather than scientific basis. Hence, many authors have performed stereotactic biopsy to establish the histological diagnosis.[50-52] Albright in 1996[55] laid down the criteria for the stereotactic biopsy of diffuse brainstem tumour. In developing countries the possibility of these diffuse lesion as inflammatory pathology does exist[53] justifying biopsy to provide appropriate treatment. The incidence of non neoplastic masses in these lesion is as high as 10%, which include encephalitis, abscess, neuroeithelial cyst, radionecrosis, neurocysticercosis and tuberculosis. In the series reported by Rajsekhar et al[53] 3 out of 52 masses were non neoplastic in nature. According to Epstein,[53] biopsy is only helpful if the biopsy reveals malignancy. Overall, it seems wise to rely on clinical features and high quality MRI. Thus, biopsy appeared to have no role in case of typical diffuse pontine glioma. Biopsy can be restricted those whose clinical and radiological feature are not classical of pontine glioma. Albright in[55] 1996 had outlined indications for biopsy. The need for routine tracheostomy in postoperative period, in patients with brainstem glioma is not clear. Patients having preoperative medullary dysfunction in whom a total resection has been performed or attempted may need elective tracheostomy, while others follow wait and watch policy. Choux et al performed tracheostomy as and when necessary rather than indiscriminate tracheostomy.

Need for total excision vs partial excision in brainstem glioma depends upon type of tumour, experience of the surgeon, intraoperative monitoring facility and type of postoperative ICU care. Aim of the surgery, however, is to remove as much tumour as possible to relieve the compression on normal brainstem structures and make the adjunct chemotherapy and radiotherapy more effective.[56-58] No doubt, radical or total excision is the choice, in many situations, it may not be possible and carries more postoperative morbidity. Surprisingly, many of these intrinsic and localized gliomas can be totally excised without much deterioration. Abbott et al[56] in 1996 reported total or near total excision in 18 out of 24 patients with medullary gliomas.

Overall, there is a need for adjunct therapy in braintem glioma.[59-61] Some authors have advocated hyperfractionated radiotherapy.[60] Basically hyperfractionated radiation reduce the neurotoxicity. There is also a debate not to give radiation to children under 3 years of age, in this group chemotherapy is preferred to radiation.

MALIGNANT ASTROCYTOMAS AND GLIOBLASTOMA OF CEREBELLUM

Cerebellar astrocytomas are most common posterior fossa tumour in childhood. Rarely, a malignant astrocytoma and GBM is reported in the cerebellum. Very very rarely malignant transformation is described in the literature following radiation therapy.[61, 62, 63] Butka in 1975[63] reported a case in whom malignant evolution occurred 28 years after the radiation for the cerebellar astrocytoma. This only strengthen the concept that radiation in astrocytomas of cerebellum is not only unnecessary but also is dangerous as malignant transformation can occur even after long time following radiation.[63] Thus, radiation in recurrent astrocytomas of cerebellum remains debatable and surgery is always preferred.

PROBLEMS WITH EPENDYMOMAS OF POSTERIOR FOSSA

Ependymomas are relatively rarer tumours of the posterior fossa as compared to medulloblastomas and astrocytoma. On the other hand two thirds of epenedymoma occur in posterior fossa and median age in around 5 years. The incidence of malignant ependymoma is very varied and reported 7-89% in various series.[64-66] There is a considerable disagreement in grading ependymoma.[64] Frequently, posterior fossa ependymomas exhibit a biphasic patterns. Histologically there may be a overall benign picture with small areas of malignancy. The prognosis depends on the brain infiltration rather than island of malignancy. Infiltration of floor of 4th ventricle, make it difficult to be excised totally. Less than 10% ependymoma do have metastases.

It is generally agreed that the ependymoma is a surgical disease and need to be totally excised[64-66] and the other modalities like radiotherapy and chemotherapy are less effective than medulloblastoma. Hence, general concept is aggressive surgical excision, which is difficult and is only possible less than 50% cases.[65] Overall the long term survival depends upon the extent of surgical removal. Thus, surgical resection is the most important prognostic factor.[64, 65]

Radiotherapy is an useful adjuvant treatment and local radiation is consider sufficient. However, optimal dosage of radiation is still not clear. Generally, chemotherapy is believed to have no or very little role in malignant or anaplastic ependymoma. There is a general agreement to have a relook surgery for residual tumour or repeat surgery for recurrent local tumour if not contra indicated on medical grounds. Thus, there is a overall good agreement in management of ependymoma.

MALIGNANT TUMOURS OF CLIVUS OR PETROUS BONE

Malignant tumours of the clivus are rare and considered as surgical challenges.[67,68] In a series reported by Spetzler et al,[67] 4 out of 46 patients had malignant tumour. In study conducted by Mahapatra and Gupta[68] 2 out of the 20 cases had malignant lesions. Almost more than 60% are benign lesion such as meningioma or chordoma, however, malignant bony lesions **(Fig. 5)** are not uncommon.[69-72] They include chondrosarcoma, malignant chordoma, giant cell tumour **(Fig. 6)** and metastatic tumours.[73-74] Rarely, lesions like lymphoma, osteoblastoma **(Fig. 7)** malignant mesenchymal tumour and multiple myeloma. These lesions have un-

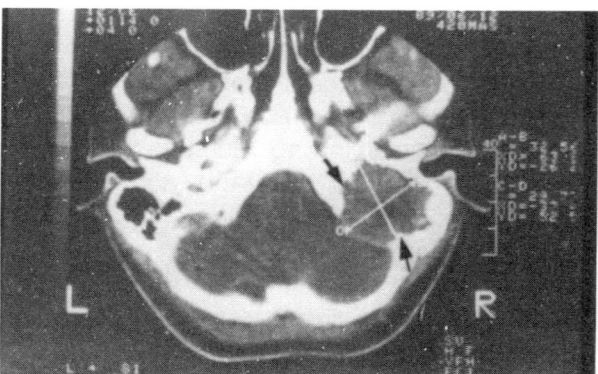

Fig. 5 : CT scan posterior fossa magnified cut show petrous tumour which was a malignant tumour on histology.

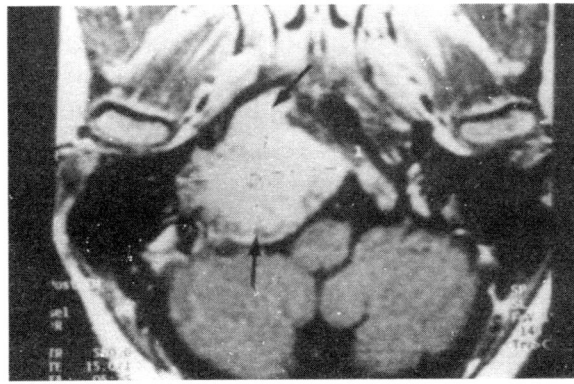

Fig. 6 : Contrast MRI, axial cut posterior fossa shows a large petroclival lesion, which was on histology was a giant cell tumour.

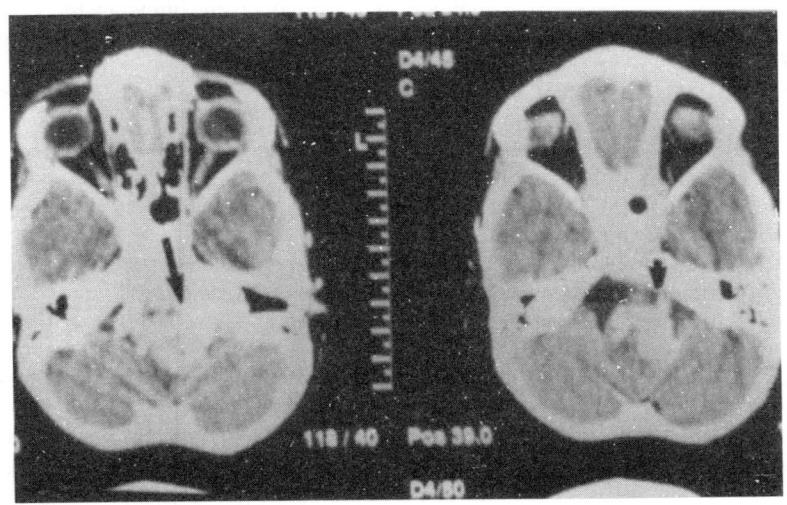

Fig. 7 : Contrast enhanced CT cuts showed hyperdense posterior fossa tumour, which was malignant mesenchymal tumour.

predictable course and require radical surgery followed by radiation theapy and chemotherapy.

CONCLUSIONS

Malignnant tumours of the posterior fossa are common lesion and constitute a significant proportion of the intracranial tumours in childhood. Medulloblastoma, Malignant astrocytoma of cerebellum and brainstem, and ependymoma form the majority proportion. However, rarely metastatic lesion of cerebellum and malignant lesion of petrous and clivus contribute to the list. Over the years the management of the common lesion like medulloblastoma and ependymoma has undergone radical change. However, malignant lesions of petrous and clivus are still challenging situation, with unpredictable course.

REFERENCES:

1. Zimmerman RA, Bilanuik LT, Bruno L, et al: Computed tomography of cerebellar astrocytoma. *Ame J Roentgenol* 1978;130:929-933.
2. Mishra BK, Tandon PN, Banerji AK, Bhatia R: Intracranial tumours in infancy, childhood and adolescence. *Ind J Cancer* 1984;21:63-66.
3. Farewell JR, Dohrmann FJ, Fammery JT: Central nervous system tumour in children. *Cancer* 1977;40:3123-30.
4. Throne RN, Pearson AD, Nicolla JA, et al: Decreasing incidence of Medulloblastoma in children. *Cancer* 1994;74:3240-44.
5. Albright A: Tumours of the pons. *Neurosurg Clin North Ame* 1973;43:529-33.
6. Hoffman HJ, Becker L, Craven MA: A clinical and pathologically distinct group of benign braintem gliomas. *Neurosurgery* 1980;7:243-48.

7. Konovalov A, Atich J: The surgical treatment of primary brainstem tumours. In: Schmideck HH and Sweet WH (eds). Operative neurosurgical technique (2nd ed) New York. Grune and Stratta 1988; 709-39.
8. Srinivasan J, Berger MS, Silber JR: P53 gene expression in four human medulloblastoma cell lines. *Child's Nerve Syst.* 1996;12:76-80.
9. Biegel J, Rooke L, Packer R, et al: Iso chromosome 17q in primitive neuroectodermal tumours of CNS. *Cancer* 1989;1:139-41.
10. Batra AK, McLendon RE, Koo JS: Prognostic implication of chromosone 17 P detection in human medulloblastoma. *Neuro Oncol* 1995;24:29-35.
11. Chang CH, Housepian EM, Herbert C, et al: An operating staging system on megavoltage radiotherapy technique for cerebellar medulloblastoma. *Radiology* 1969;93:1351-59.
12. Torres CF, Rebsamen S, Silber JH, et al: Survillence scanning of children with medulloblastoma. *N Eng J Med* 1994;330:892-95.
13. Shaw DWW, Geyer JR, Berger J, et al: Asymptomatic recurrence detection with survillence scanning in children with medulloblastoma. *J Clin Concol* 1997;15:1811- 13.
14. Bouffer E, Genter JC, Doz F, et al: Metastatic medulloblastomas: the experience of the French cooperative M. Group. *Eur J Cancer* 1994;30A:1478-83.
15. Mendel E, Levy ML, Raffel C, et al: Survillance imaging in children with primitive neuroectodermal tumours. *Neurosurgery* 1996;38:692-95.
16. Hoffman AJ, Henderick EB, Humphrey RP: Management of Medulloblastoma in childhood. *Clin Neurosurg* 1982;30:236-45.
17. Vaquero J, Cabezu do J, Desoda R: Intratumoural haemorrhage in posterior fossa tumour after ventricular drainage. Report of two cases. *J Neurosurg* 1981;54:406-08.
18. McLaurin R: On the use of precraniotomy shunting in management of posterior fossa tumour in children. A cooperative study. In: Champ man P (ed) Concepts in *Pediatric Neurosurg* Vol. 6 S. Karger Basel 1985; 1-5.
19. Grabenbaaer GG, Beck JD, Eahardt J, et al: Postoperative radiotherapy of medulloblastoma. Impact of radiation quality on treatment outcome. *Ame J Clin Oncol* 1996;19:73-77.
20. Hoope-Hirsch E, Brunet L, Laroussiniz F: Intellectual outcome in children with malignant tumours of the posterior fossa. Influence of the field of radiation and quality of surgery. *Child Nerv Syst* 1995;11:340-45.
21. Mulhern RK, Hancok J, Fairdough D, et al: Neuropsychological status of children treated for brain tumour-a critical review and integrative analysis. *Med Pediatr Oncol* 1992;20:181-90.
22. Chintagumptla M, Berg S, Blaney SM: Treatment controversies in Medulloblastoma. *Current opinion in Oncology* 2001;13:154-59.
23. Shalegt SM, Beardwell CG, Aarous BM, et al: Growth impairment in children treated for brain tumours. *Arch Dis Child* 1978;53:491-94.
24. Brauner R, Rappaport R, Prevot C, et al: A prospective study of the development of growth harmone deficiency in children given statural growth. *J Clin Endocrinol Metab* 1986;68:346-51.
25. Rappaport R, Brauner R, Czernichow P, et al: Effect of hypothalamic and pituitary irradiation on pubertal development in children with cranial tumours. *J Clin Endcrinol Metab* 1982;54:1164-68.
26. Tomita T, McLone DG: Medulloblastoma in childhood – results of radical surgery and low dose radiation therapy. *J Neurosurg* 1986;64:238-42.
27. Brand WN, Schneider PA, Tokars RP: Long term results of a pilot study of low dose cranial spinal irradiation for cerebellar medulloblastoma. *Int J Radiol Oncol Biol Physio* 12 (Suppl) 1986;144.
28. Allen JC, Donahe B, DaRosso R, et al: Hyperfractionated craniospinal radiotherapy and adjunct chemotherapy for children with newly diagnosed medulloblastoma and other premitive neuroectodermal tumours. *Int J Radiol Oncol Biol Physio* 1996;36:1156-61.
29. Duffner P, Horwitz M, Krischer J, et al: Postoperative chemotherapy and delayed radiation in

children less than 3 years of age with malignant brain tumour. *New Eng J Med* 1993;328:1725-31.

30. Ater J, Woo S, Van Eyes J: Update on MOPP chemotherapy as primary therapy for infant brain tumour. *Pediatr. Neurosci* 1988;14:153-55.
31. Bailey P, Buchanan DN, Bucy PC: Intracranial tumour of infancy versus childhood. Chicago University Press, 1939;188-241.
32. Lassman LP, Arjona VE: Pontine gliomas of childhood. *Lancet* 1967;1:913-15.
33. Koos WT, Miller MH: Intracranial tumour of infant and children. Stuttgart; Thieme 1971;346-50.
34. Hoffman HJ, Becker L, Craven MA: A clinically and pathologically distinct group of benign brainstem gliomas. *Neurosurgery* 1980;7:243-48.
35. Epstein FJ, McCleary EL: Intrinsic brainstem tumours of childhood: Surgical indications. *J Neurosurg* 1986;64:11-15.
36. Kansal S, Jindal A, Mahapatra AK: Brainstem glioma – A study of III case. *Ind J Cancer* 1999;36:99-108.
37. Epstein FJ, Wisoff J: Intraaxial tumours of the cervico-medullary junction. *J Neurosurg* 1987;67:483-87.
38. Pierre-Kahn A, Hirsch JF, Vinchon M, et al: Surgical management of brainstem tumours in children, results and statistical analysis of 75 cases. *J Neurosurg* 1993;79:845-52.
39. Bricolo A, Turazzi S, Cristofori L, et al: Direct surgery for brainstem tumours. *Acta Neuroschir (Suppl)* 1991;53:148-58.
40. Matson DD: Neurosurgery of Infancy and childhood. 2nd (ed). Springfield, Illinois: Thomas 1969;469-77.
41. Tokuiski Y, Handa H, Yamashita J, et al: Brainstem glioma: an analysis of 85 cases. *Acta Neurochir (Wien)* 1986;79:67-73.
42. Sclabassi RJ, Kalia KK, Sekhar L, et al: Assessing brainstem function. *Neurosurg Clin North Am* 4:255-60.
43. Mahapatra AK: Electrophysiology in Neurootology.
44. Soustiel JF, Hafner H, Chistyakov AV, et al: Monitoring of brainstem evoked potentials. Clinical applications in posterior fossa surgery. Electroencephalogr. *Clin Neurophysiol* 1993;88:225-60.
45. Morota N, Deletis V, Epstein F, et al: Brainstem mapping: Neurophysiological localization of motor nuclei on the floor of the IV ventricle. *Neurosurg* 1995;37:922-30.
46. Strauss E, Romstock J, Nimsky C, et al: Intraoperative identification of motor areas of rhomboid fossa using direct stimulation. *J Neurosurg* 1993;79:393-399.
47. Guy G, Jan M, Guegan Y: Les Leions chirurgicales dutronc cerebral. *Neurochirugie* 35 (Suppl 1) 1989;1-133.
48. Ogata N, Wieser HG, Yonekawa Y: Approaches to dorsal pontine lesions via the IV ventricle with SSEP monitoring: a report of two cases. *J Clin Neurosciencs* 1996;4:373-78.
49. Soustiel JF, Feinsol M, Hafner H: Short latency trigeminal evoked potentials: normative data and clinical correlations. *Electroencephalogr Clin Neurophysiol* 1991;80:119-25.
50. Coffey RJ, Lunsford LD: Stereotactic surgery for mass lesion of midbrain and pons. *Neurosurgery* 1985;17:12-18.
51. Ismat F, Acebes JJ, Conesa G: The significance of image guided stereotactic surgery in the management of braintem tumours. *Crit Rev Neurosurg* 1997;7:123-28.
52. Steek J, Friedman WA: Stereotactic biopsy of braintem mass lesions. *Surg Neurol* 1993;43:563-68.
53. Rajsekhar V, Chandy MJ: Computerized tomography guided stereotactic surgery for brainstem masses: a risk and benefit analysis in 71 patients. *J Neurosurg* 1995;82:976-81.
54. Epstein FJ: Comments. *Neurosurg* 1982;12:301-02.
55. Abbott R, Shiminski-Maher T, Epstein FJ: Intrinsic tumours of the medull: predicting outcome after surgery. *Pediatr. Neurosurg* 1996;25:41-44.
56. Hoffman HJ: Dorsally expophytic brainstem tumour and midbrain tumours. *Pediatr Neurosurgery* 1996;24:256-62.

57. Pollack I, Hoffman HJ, Humphireys RP, Becker L: The longterm outcome after surgical treatment of dorsally exophytic brainstem glioma. *J Neurosurg* 1993;78:859- 63.
58. Kalifa C, Hartmann O, Demeoca F, et al: High dose busulfan and thiotepa with autologus bone marrow transplant in childhood malignant tumours: a phase II study: Bone Marrow Transplant 1982;9:227-33.
59. Packer RJ, Boyett JM, Zimmerman R: Outcome of children with brainstem gliomas after treatment with 7800 C Gy of hyperfractionated radiotherapy. A children's Cancer Group study Phase I/II trial. *Cancer* 1994;74:1827-34.
60. Freeman CR, Bourgouin PM, Sanford RA, et al: Longterm survivors of childhood brainstem gliomas treated with hyperfractionated radiotherapy. *Cancer* 1996;77:555-62.
61. Wisoff HS, Lena JF: Glioblastoma multiforme of the cerebellum five decade after irradiation of a cerebellar tumour. *J Neurooncology* 1989;7:339-44.
62. Schwartz AM, Ghatak NR: Malignant transformation of benign cerebellar astrocytoma. *Cancer*, 190;15:333-36.
63. Butka H: Partially resected and irradiated cerebellar astrocytoma of childhood malignant evolution after 28 years. *Acta Neurochir (wien)* 1975;32:139-46.
64. Nazar GB, Hoffman HJ, Becker LE: Infratentorial ependymoma in childhood. Prognostic factors and treatment. *J Neurosurg* 1990;72:408-17.
65. Rousseau P, Habrand J, Sarrazin D: Treatment of intracranial ependymoma of children: review of 15 years experience. *Int J Radiation Oncology Biol Physi* 1993;28:381-86.
66. Ross GW, Rubinstein LJ: Lack of histopathological correlation of malignant ependymomas with postoperative survival. *J Neurosurg* 1989;70:31-35.
67. Spetzler RF, Daspit P, Pappas CTE: Combined supra and infratentorial approach for lesions of the petrous and clival region: experience with 46 cases. *J Neurosurg* 1992;76:588-99.
68. Mahapatra AK, Gupta PK: Petroclival tumour surgery. A personal experience of 20 cases. (In press).
69. Jaiswal A, Tripathi M, Saratchandra P, Sharma MC, Mahapatra AK: An unusual case or primary skull base lymphoma extending from C.P.angle to cavernous sinus and orbit. *J Neurosurg Science* 2000;44:145-49.
70. Nishimur T, Uchida Y, Fukuoka M, et al: Cerebellopontine angle lymphoma: A case report and review of the literature. *Surg Neurol* 1998;58:480-86.
71. Kanemura N, Tsurumi H, Hara T, et al: Multiple myeloma complicated by bilateral abducens nerve palsy due to a tumour in the clivus. Rinshoketsueki 2001;42:218-20.
72. Cheng IH, Yen KL, Jian JJ, et al: Examining prognostic factors and patterns of failure in nasopharyngeal carcinoma following concomitant radiotherapy and chemotherapy mpact on future clinical trials. *Int J Radiol Oncol Biol Physio* 2001;1:717-26.
73. Mahapatra AK, Gupta PK, Lad SD: Duct carcinoma of breast with cerebellar metastasis. *Pan Arab Neurosurg Journal* (2001 in press)
74. Shirane R, Kumabe J, Yoshida Y, et al: Surgical treatment of posterior fossa tumours via the occipital transtentorial approach: evaluation of operative safety and results of 14 patients with anterior superior-cerebellar tumours. *J Neurosurg* 2001;94:927-35.

38

Anaesthetic considerations in Paediatric Posterior Fossa Surgery

M Abraham, MS Tandon, P Ganjoo

INTRODUCTION

Brain tumours are the most common solid tumours occurring during childhood and rank second only to leukaemia as the most common malignancy in children. Although most central nervous system (CNS) tumours in adults are supratentorial, in children, they are mainly located in the posterior fossa and are predominantly intraaxial in origin. Primary neoplasms of the posterior fossa neuroaxis can be further subdivided into tumours affecting the cerebellum, those affecting the fourth ventricle and tumours affecting the brainstem. Cerebellar astrocytomas and medulloblastomas are the most frequent lesions and account for almost half of all paediatric CNS tumours. Malignancies affecting the brain stem (astrocytoma and ganglioneuroma) and fourth ventricle (ependymoma) occur less frequently.

Anaesthesia in children for posterior fossa lesions can be particularly intimidating not only because of the mere size of the child and the equipment necessary for anaesthetizing infants and children, but because posterior fossa surgery itself is fraught with hazards for a number of reasons. Firstly, surgical impingement in the region in and around the brain stem can cause temporary or permanent disturbances in cardiovascular, respiratory and autonomic function, and it can also lead to potentially serious postoperative airway difficulties. The second most important reason why posterior fossa surgery in children is particularly challenging is the fact that these operations are performed in unusual positions each of which imposes its own particular problems regarding anaesthesia and monitoring.

PREOPERATIVE EVALUATION.

A thorough preoperative evaluation of the paediatric patient is of utmost importance to the anaesthesiologist and goes a long way in reducing the risks of anaesthetizing these children.

Preoperative considerations for posterior fossa in children include the following:

- Mental status changes and symptoms of increased intracranial pressure, i.e nausea, vomiting and headache should be especially enquired in the preanaesthetic check up. Posterior fossa pathology and raised intracranial pressure increases the gastric emptying time in children and makes the child prone to regurgitation at induction. Increased intracranial pressure is liable to occur early because of rapid exhaustion of the compensatory mechanisms due to the small size of the posterior fossa. Blockade of the narrow aqueduct results in obstructive hydrocephalus and ventricular enlargement requiring a cerebrospinal fluid shunt. Because induction agents and techniques may be altered if ICP is elevated, this should be well considered in the preoperative evaluation. The MRI should be scanned for upward engagement as this sign will help in deciding on the administration of diuretic agents. Furthermore, the distance between the pons and clivus should be > 2 mm; if not, it indicates an increased ICP in the infratentorial space.

- Posterior fossa tumours in children have a higher incidence of brain stem involvement and brain stem compression may cause upper airway dysfunction with inspiratory stridor. The incidence of lower cranial nerve involvement, also, is higher in children and this could present as vocal cord palsy or impairment of gag and swallowing and can predispose to perioperative airway problems.

- A history of recurrent pneumonitis, dysphagia, aspiration and difficulties in coughing and clearing secretions should be specifically enquired. These symptoms are often present in patients with brain stem pathology affecting the lower cranial nerves. Central sleep apnoea may be present and may persist postoperatively.

- The presence of an intracardiac defect or a patent foramen ovale constitute enough evidence to contraindicate sitting position for surgery because of the risk of paradoxical air embolism. A probe patent foramen ovale was found to be present in 20% to 35% of the normal population at autopsy[1] and this can open up if pressure in the right atrium exceeds the left atrium during anaesthesia. In yet another study, a patent foramen ovale was demonstrable during a Valsalva maneuver using contrast echocardiography in 18% of young healthy subjects.[2] However, the role of routine preoperative echocardiographic examination in children for posterior fossa surgery in the sitting position is controversial.[3,4] The presence of a functioning ventriculo-atrial shunt is a contraindication for posterior fossa surgery in the sitting position because of the risk of paradoxical venous air embolism (VAE).

- Preoperative attempt to reduce the ICP may generate electrolyte imbalance and intravascular volume contraction. Prolonged vomiting and poor appetite can also contribute to inadequate hydration.

- Significant blood loss may occur during craniotomy particular with vascular tumours. Therefore, laboratory data should include a haematocrit and blood should be typed and cross matched and one adult sized unit should generally be adequate.

- Anticonvulsant medications may increase nondepolarizing muscle relaxant requirements and this should be kept in mind.

PRE-OPERATIVE PREPARATION

The child's condition should be optimized preoperatively by the following measures:

- Patients with increased intracranial pressure due to tissue swelling should be treated with decongestive measures including corticosteroids. Those with marked hydrocephalus may benefit from a cerebrospinal diversionary procedure such as a ventriculoperitoneal shunt which reduces intracranial pressure, stabilizes the vital signs and affords a smoother intraoperative and postoperative course.
- In children with chest infection due to aspiration, the airway must be protected from further aspiration by an endotracheal tube if necessary along with prophylactic antibiotics and chest physiotherapy to minimize perioperative morbidity and mortality.
- If signs of dehydration are present, the intravascular volume should be adequately replenished.

CONDUCT OF ANAESTHESIA

The anaesthesiologist must develop an anaesthetic plan that meets the requirements of the neurosurgeon, the neurophysiologist at the same time keeping patient safety uppermost in mind. It should also tailored to alleviate the fears and concerns of the child and parents.

PREMEDICATION

Premedication makes parental separation more tolerable & also facilitates anaesthetic induction. It should be administered in the preoperative holding area so that the patient can be constantly monitored for respiratory depression or other adverse events. Premedication should be avoided in obtunded patients and patients with raised intracranial pressure because of the risk of respiratory depression, hypercarbia and worsening ICP. Children with functioning ventriculo-peritoneal (VP) shunts and those without evidence of elevated ICP can be safely premedicated. Midazolam (0.5-0.7 mg/kg, orally or intranasally) is an excellent anxiolytic and amnesic and is usually given approximately 20 minutes prior to induction.

INDUCTION

Anesthetic induction is aimed at preserving cerebral perfusion pressure, avoiding an increase in the ICP and providing an appropriate depth of anesthesia. The type of induction is influenced by the neurological status and the age of the patient.

In smaller children, with functioning V.P shunts or without evidence of raised intracranial pressure, an inhalational induction with sevoflurane or halothane is rapid and pleasant. Respiration is gradually assisted and controlled as soon as the patient is adequately anesthetized. An IV access is then obtained.

Intravenous induction is preferable in older children, children with malfunctioning shunts and in children with evidence of increased ICP. Thiopentone (5-7 mg/kg) is the most commonly used i.v induction agent. It decreases the CBF, $CmrO_2$, ICP and preserves autoregulation and CO_2 reactivity. Propofol (3-4 mg/kg) can also be used. Tracheal intubation is facilitated with a nondepolarising muscle relaxant (atracurium, vecuronium or pancuronium), and an opiate (fentanyl, pethidine or morphine. Mild hyperventilation lowers the ICP and supplemental thiopentone blunts the hemodynamic responses to laryngoscopy and intubation.

In patients with risk of aspiration, e.g those with lower cranial nerve palsies, a rapid sequence intubation with succinylcholine and cricoid pressure is preferable.

After endotracheal intubation the patient is mechanically ventilated to maintain a $PaCO_2$ in the range of 30-35 torr.

A gastric tube is usually inserted to suction the gastric contents and to drain passively during the surgery. Eyes are lubricated with an ophthalmic ointment and taped shut and care should be taken to ensure no undue pressure on the orbit.

MAINTENANCE OF ANESTHESIA

The selection of the anesthetic depends on how severely the intracranial compliance is compromised and the requirement of the surgical procedure. The aim is to provide a slack brain which will reduce the amount of retractor pressure and allow adequate cerebral perfusion in a hemodynamically stable patient.

Most pediatric anesthesiologists choose a balanced anesthetic technique, so that early awakening and extubation even after lengthy neurosurgical procedures is possible which will facilitate an early neurological assessment This technique combines nitrous oxide, oxygen, an opiate (usually fentanyl), a nondepolarising muscle relaxant (atracurium, vecuronium or pancuronium), supplemented with low dose inhalational agents (isoflurane, halothane).

All halogenated inhalational agents decrease CMR but increase CBF (and ICP) in a dose dependent manner. These deleterious effects on ICP can be ameliorated by concomitant hyperventilation to induce hypocapnia ($PaCO_2$-30-35 torr). Isoflurane is preferred to halothane because it causes less increase in CBF.

If the intracranial compliance is severely compromised, it is better to avoid inhalational agents at least until the dura is opened and instead use an intravenous agent e.g a propofol infusion. Propofol causes a dose dependant decrease in CBF which parallels a less marked decrement in $CMRO_2$. It causes acceptable brain relaxation and provides good quality of recovery even following long operations.

After skin preparation, infiltration of the scalp with bupivacaine or xylocaine with adrenaline1:200,000, minimizes scalp blood loss and provides supplemental analgesia. Mannitol (0.5 mg/kg) preceeded by furosemide (0.3 mg/kg) may be given to decrease the ICP if the patient is being operated in the horizontal position.

POSITIONING

Position for surgery is usually decided by the surgeon, but it is the joint responsibility of the surgeon and anaesthesiologist to ensure both patient comfort and safety.

Most pediatric neurosurgeons prefer the prone position for posterior fossa surgery particularly for children less than 2 years.[5] **(Fig. 1)** Because of the high incidence of hydrocephalus, the risk of ventricular collapse and subdural collection is lower in this position. The sitting position is still occasionally used by some, usually for children more than 3 years of age with midline lesions. Sitting position should be avoided in patients with congenital heart disease, patent foramen

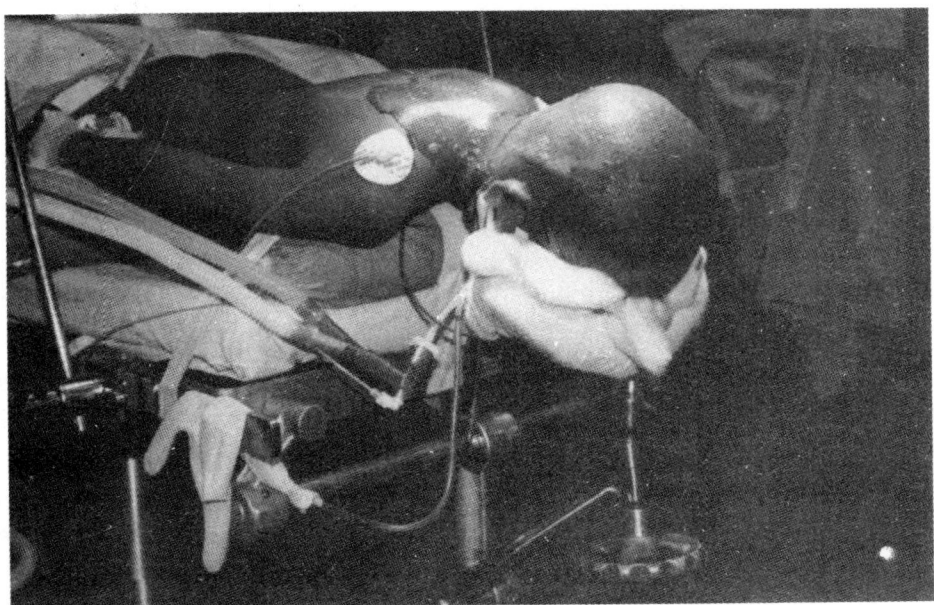

Fig. 1 : Child positioned for posterior fossa surgery in the prone position

ovale and in patients with hemodynamic instability. Lateral, park bench and supine may also be used in some cases.

The anesthesiologists responsibility is to apply special care during positioning with regard to position of the endotracheal tube, ventilation, pressure points and stretch injuries.

It should be ensured that flexion of the head does not occlude jugular venous drainage nor displace the tube to an endobronchial position. Careful auscultation of the breath sounds is mandatory and adequate distance between the chin and the sternum should be ensured. **(Fig. 2)** Care should be taken that the endotracheal tube is well secured to the face as it can become dislodged during positioning or later as oral secretions loosen tape attachments. All monitoring lines should be secure and accessible and all pressure points should be padded.

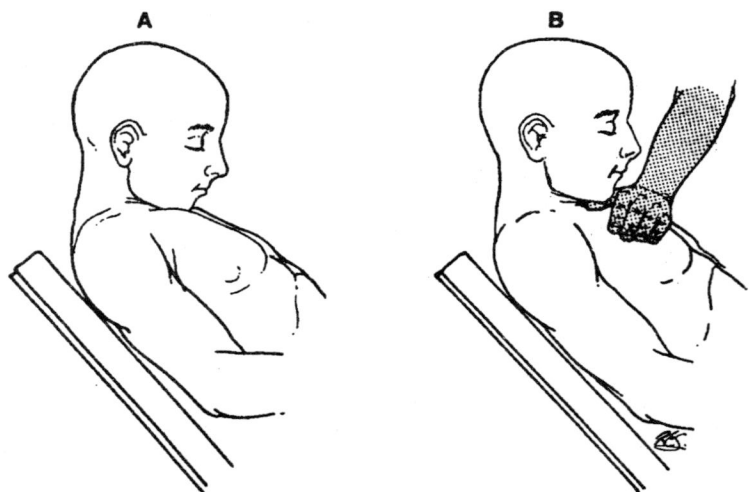

Fig. 2 : Child position with excessive neck flexion (A). Correct position of head (B).

In the prone position, chest and pelvis rolls are placed so that the thorax and abdomen is not compressed and ventilation and venous return are not impeded. In the prone and lateral decubitus position, the head should be elevated to improve venous drainage.

Preloading with crystalloids, elastic bandaging and elevation of the legs to improve venous return with gradual elevation to the sitting position are recommended to prevent VAE and haemodynamic instability.

The method of head fixation depends on the age of the patient (thickness of the skull) and also to the surgeons needs. In the prone position, horseshoe rests are appropriate but the face of the patient must be padded carefully and the eyes are carefully checked to ensure that there is no undue pressure on the orbit. In children more than 3 years of age, a multipin head holder is often used. Prior infiltration of the pin sites with local anesthetic will reduce the nociceptive and hemodynamic responses to pin fixation which is a very noxious stimulus.

Provisions for maintaining the body temperature should be made, including the use of blanket warming devices when necessary.

MONITORING

Standard monitoring includes ECG, temperature, inspired oxygen concentration (FiO_2), oxygen saturation (SaO_2), end tidal CO_2 ($etCO_2$) and noninvasive blood pressure monitoring. The utility of continuous auscultation with an oesophageal stethoscope cannot be overemphasized, as connections can get dislodged and small endotracheal tubes can get either occluded by secretions or displaced during lengthy neurosurgical operations.

Most of the neurosurgical procedures cause significant blood loss. Besides, haemodynamic disturbances may occur during surgery near the brainstem. So apart from standard monitoring, invasive monitoring is mandatory. An arterial line placement for hemodynamic monitoring and blood chemistry sampling is highly recommended. A 22 G cannula is adequate and can be placed in the radial, dorsalis pedis or posterior tibial artery. Patents in whom blood loss could be significant (e.g medulloblastoma), those with expected hemodynamic instability or those at increased risk for venous air embolism should have a central venous catheter. Cannulation of the brachial vein is a better route in older children and the femoral vein can be cannulated in children of all ages. The catheter tip should be placed in the superior vena cava - right atrial junction for maximal retrieval of air in the event of VAE.

A peripheral nerve stimulator should be used to assess the depth of neuromuscular blockade to reduce the risk of patient movement during surgery and the risk of excessively deep blockade at the conclusion of surgery. A urinary catheter is mandatory because of the duration of the surgery and the use of diuretic drugs.

Precordial Doppler should preferably be used for all posterior fossa surgery for detection of VAE irrespective of the position of the child as VAE can occur in both upright and horizontal positions.

Monitoring of brain stem function is essential particularly if the surgery is the vicinity of the brain stem.

BRAIN STEM CONSIDERATIONS

The pontomedullary area of the brain stem has important cardiorespiratory control centers and is the site of lower cranial nerve nuclei. Stimulation of these centers during dissection of the tumour can cause profound haemodynamic and respiratory changes. **(Fig. 3)**

Stimulation of the sensory portion of the V cranial nerve results in significant hypertension and reflex bradycardia while traction on the X cranial nerve can cause bradycardia and hypotension. Cardiac arrhythmias (severe bradycardia, ventricular extrasystoles, ventricular tachycardia and idioventricular rhythm) and hypertension or hypotension can also occur with stimulation of cardiovascular control centers in the pons and medulla. Cerebellar traction can also cause hypertension. Pontomedullary brain stem stimulation may elicit the gasp response and irregular respiration in a spontaneously breathing patient. In lightly paralysed patient, stimulation of the motor components of the V, VII and XII cranial nerves may yield jaw, facial and shoulder movements. **(Fig. 4)**. Management of these brain stem occurrences include immediately informing the surgeon as cessation of the surgical manipulation usually resolves these events. These signs should be taken as warning signals and preferably not treated pharmacologically unless they are severe and life threatening.

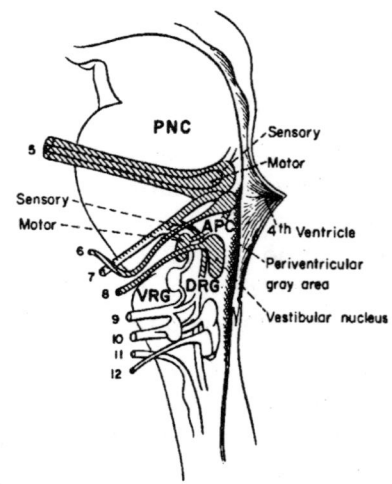

Fig. 3 : Location of the cardiovascular centers (hatched areas), respiratory centers (bold-lettered areas) and cranial nerve nuclei (slant-lined areas). Respiratory centers are designated by the following abbreviations: PNC-pneumotaxic center; APC-apneustic center; DRG-dorsal respiratory group; VRG-ventral respiratory group.

Cranial Nerve Stimulation during Posterior Fossa Surgery

Cranial Nerve or Area	Classic Activity Description-Physical Signs	Electornic Device
V	Motor—jaw jerk	EMG
	Sensory—hypertension	Arterial line
	bradycardia	ECG
VII	Facial twitch	EMG
X	Hypotension	Arterial line
	Bradycardia	ECG
XI	Shoulder jerk	EMG
Pons	Brainstem compression	BAER, SEP
	Ectopic cardiac foci	ECG
	Hypertension/hypotension	Arterial line
	Tachycardia/bradycardia	ECG
	Gasp. irregular respirations	Respirator trigger

Abbreviation: BAER, brainstem aditory evoked potential, SEP, sensory evoked potential

Fig. 4 : Effect of Cranial Nerve Stimulation during Posterior Fossa Surgery

Besides cardiovascular parameters, the other methods of detecting intraoperative brain stem dysfunction is use of spontaneous ventilation and electrophysiological monitoring. The disadvantage of spontaneous ventilation in children is that it imposes an increased work of breathing due to anaesthetic circuits and it can also cause respiratory fatigue and failure in small children. Electrophysiological monitoring includes somatosensory evoked potential, brain stem auditory evoked potential and either spontaneous or evoked electromyogram of the V, VII, XI and XII cranial nerves. These modalities are now widely used for detection of surgical encroachment on the brain stem, particularly for brain stem tumours.

BLOOD LOSS AND FLUID MANAGEMENT

Blood loss is difficult to assess during a craniotomy because of insidious bleeding from the scalp edges, use of copious amounts of irrigating solution, and the inability to collect lost blood or easily visualize the entire surgical field. Urine output in the presence of aggressive diuresis may be a misleading indicator of adequate volume replacement. Central venous pressure monitoring and serial hematocrits are a good indicator of blood loss and adequate volume replacement.

Fluid administration depends on the type of brain pathology. A patient with increased intracranial pressure requires a fluid regimen that must balance adequate intravascular volume against any effort to dehydrate the patient.

The goal of fluid management is to maintain an isovolemic, isoosmolar intravascular volume and to maintain the cerebral perfusion pressure. Since normal saline is slightly hypertonic to blood (osmolality of 308 mOsm/kg), it may be a better fluid replacement to Ringers lactate (272 mOsm/kg). Dextrose containing solutions are associated with a poorer neurological outcome and are best avoided unless hypoglycemia has been confirmed (stressed neonates with depleted glycogen stores).

Serum sodium & potassium levels must be checked intraoperatively because furosemide and mannitol can cause hypokalemia and hypernatremia.

EMERGENCE AND POSTOPERATIVE CARE

Surgical closure for suboccipital craniectomy is faster than supratentorial craniotomy as the bone flap does not have to be replaced. Hence tapering of the anaesthetic drugs has to be done accordingly and inhalational agents discontinued at scalp closure so as to have a fully awake patient at the end of the surgery. Elevation of blood pressure is treated pharmacologically and ventilation and temperature are adjusted to achieve normocarbia and normothermia at the time of reversal. In patients without any major intraoperative complication, extubation and restoration of spontaneous ventilation is usually anticipated at the end of the surgery except in certain situations which are elaborated in the section on complications.

REVERSAL OF ANAESTHESIA

Reversal of neuromuscular block should be accomplished only after removal of skull holder and application of dressing. Unlike adults, most children do not develop hypertension after a craniotomy if adequate doses of narcotics are administered. Neostigmine and glycopyrrolate/atropine are used to reverse the effect of neuromusular blockade. The patient should be extubated only when he/she is fully alert, breathing spontaneously and gag reflex is present.

POSTOPERATIVE PAIN MANAGEMENT

This can be dealt with postoperatively with various narcotic and non-narcotic drugs taking care not to overtly affect the neurological status, respiratory drive and pupil size of the patient with suitable monitoring in a neurointensive care unit.

Guidelines for management of postoperative pain:[6]

- Intravenous fentanyl/morphine perioperatively

- Loading dose of paracetamol PR in the operating theatre (40 mg/kg) and regular dose of paracetamol (15 mg/kg) orally or PR.

- Continue administration of morphine infusion in the ICU

    Diclofenac/ibuprofen where NO BLEEDING problem is present

- ALL PAIN RELIEF IS TO BE REVIEWED DAILY

COMPLICATIONS OF POSTERIOR FOSSA SURGERY IN CHILDREN

Most of the complications seen in children are similar to those seen in adults and are usually related to the surgical and anaesthetic management of the disease. Besides these, there are some problems which are specific to the paediatric age group.

VENOUS AIR EMBOLISM (VAE)

VAE is a long recognized complication of posterior fossa surgery which may occur because of entrainment of air from an open venous sinus or blood vessel at the surgical field due to a vertical gradient between the open vessel and the level of the heart. The attachment of the venous sinuses to the dura prevents them from collapsing, encouraging continuous flow of air to the heart. Small volumes of air, if detected in time, are cleared without major sequelae. However, larger air volumes may increase pulmonary artery pressure and right ventricular afterload leading to impaired cardiac performance and low cardiac output and even circulatory arrest. Late consequences of VAE may result in persistent pulmonary perfusion deficits and acute respiratory distress syndrome.

The sitting position adopted for posterior fossa surgery is particularly blamed for the occurrence of VAE. However, VAE can also occur in various horizontal positions, namely the prone, lateral and supine position, though with a lower incidence.

Infants and children are not only at increased risk for developing VAE but also its consequences have been found to be more severe in the child. It has been found that although the frequency of Doppler detected VAE in the sitting position in children was not found to be significantly different from that in adults (33% versus 45%) the incidence of haemodynamically significant VAE was found to be much more in children (69% versus 36%).[7]

Why are children at greater risk for developing VAE and its adverse effects?

a) Firstly, the head of infants are relatively large as compared to the rest of the body and so there is a higher negative pressure gradient between the operative site and heart even in the horizontal position. This causes a greater entrapment of air due to a higher negative pressure gradient.

b) Also, in infants the occipital bone is highly vascularized and venous sinuses are poorly developed which enhances the risk of air embolism.[5]

c) Prolonged preoperative fasting and hypovolaemia due to excess blood loss during craniotomy can decrease the venous pressure producing a higher negative pressure gradient.

d) The volume of entrained air is larger relative to the cardiac volume causing more hypotension.[7]

e) Moreover, the chances of successful aspiration of air from the right atrial catheter is less in children as compared to adults (38% versus 68%) as correct placement of the catheter in the right atrium is difficult in the child.

Young infants may, in addition, be at greater risk for paradoxical air embolism also (PAE). This is because the incidence of recently closed congenital heart defects are more in small children and air can traverse through these defects to the left side of the heart and cause systemic embolization. This may be manifested in the postoperative period as unexplained neurological deficits. Also, since the pulmonary vascular resistance is high in small children, it rapidly responds to stimuli such as hypoxia and acidosis.

MONITORING FOR VAE

Early detection of VAE is of utmost importance so as to avoid major sequelae.

The various monitoring devices used for this purpose vary in their sensitivity for detection of air **(Fig. 5)**. In order of decreasing sensitivity these are the following:

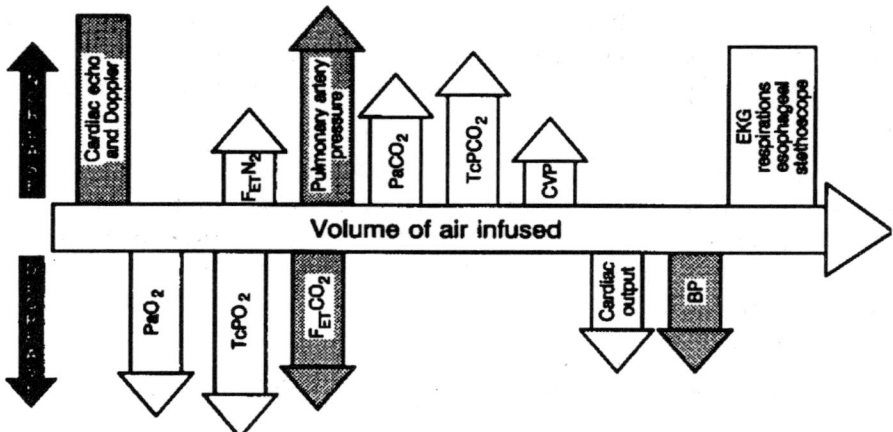

Fig. 5 : Changes in detection parameters for VAE with increasing volume.

- *Transoesophageal echocardiography (TEE)* is the most sensitive and can detect very small volumes of air. It is also the only monitoring modality that can pick up PAE. However, its prohibitive cost restricts its routine use and small transducers needed for children are not readily available.

- *Precordial Doppler* is the preferred monitor for detection of VAE in children. It is usually placed in the 3rd to the 6th intercostal space along the right or left parasternal border. Its correct placement is confirmed by rapid injection of agitated saline into a central or peripheral venous catheter which produces a noise simulating intravascular air. But the drawback with precordial Doppler is its inaccessibility and chances for dislodgement in the prone position. Soriano et al found that placement of the Doppler posteriorly between the two scapulae was quite feasible in children less than 10 kg.[8]

- *End tidal carbon dioxide($EtCO_2$)* A decrease in the $EtCO_2$ is a commonly observed phenomenon with the onset of VAE, but it can also occur with hypotension and hypovolaemia and this should kept in mind.

A combination of capnography and precordial Doppler has been found to be very effective in the detection of VAE.

- *End tidal nitrogen(EtN_2)* An increase in EtN_2 is very specific to VAE.

- *Transcutaneous oxygen (TcO_2)* This is as sensitive as $EtCO_2$ and is particularly useful in infants.

- *Right atrial catheter (RAC)* Whenever the head is elevated to more than 20% or the operative site is more than 15 cm above the heart, a RAC should be placed and positioned at the superior vena cava-right atrial junction for optimum air aspiration.

- Oesophageal stethoscope. It is useful in detecting the mill wheel murmur, which is usually a late occurrence, but a non-specific murmur may occur much earlier.

MANAGEMENT OF VAE

If VAE is suspected the following steps should be promptly taken:

- Discontinue N_2O, as it is known to expand venous air bubbles. Ensure amnesia in the absence of N_2O by other means.

- Notify the surgeon immediately so that the source of air can be identified and necessary measures like flooding the field with saline and applying bone wax to open bony surfaces can be undertaken.

- Gentle compression of the jugular veins also helps in identifying an occult bleeding site from where air may have entered. Bilateral jugular venous compression was found to be effective in elevating the cerebral venous pressures in children and thus decreasing the degree of air entrainment. (ref) However, the surgeon should be warned of the possibility of brain bulging during jugular compression.

- Prompt aspiration of air through the right atrial catheter should be done which, however, may not be successful.

- Lowering the head end and turning the patient to the left lateral decubitus position so as to trap air in the apex of the right ventricle is recommended if air entrainment continues unabated. Surgery should then be continued only in the prone or lateral position.

- The role of PEEP in preventing and treating VAE is controversial as it can increase the risk of PAE by elevating the right heart pressures. Moreover, it can cause hypotension and it was not found to be effective in raising the dural sinus pressures.[7] But Meyer et al found that a combination of lower body positive pressure using antishock trousers and PEEP was a satisfactory method of preventing VAE in children.[9]

- Support of the circulation and resuscitation may be necessary in severe episodes.

Some of the safety precautions for preventing VAE and its disastrous effects include proper screening for intracardiac defects preoperatively and increased awareness by both the surgeon and the anaesthesiologist so that early detection and prompt management is possible. Other measures include maintenance of high venous pressures by fluid loading, crepe bandaging of the lower limbs or lower body positive pressure and avoiding hyperventilation in the sitting position which can decrease the cerebral blood flow and local blood pressure.

CARDIOVASCULAR INSTABILITY

This is usually seen in hypovolaemic patients when they are placed in the sitting position. Prevention is by keeping the CVP around 4 torr and wrapping the legs in elastic bandages. It may also occur secondary to VAE due to decreases in cardiac output associated with larger volumes of intravascular air. This necessitates aggressive management of VAE as already discussed.

BRAIN STEM DYSFUNCTION AND ITS SEQUELAE.

Intraoperative damage to the respiratory control center and one or more cranial nerve nuclei can cause pathological breathing patterns postoperatively and vocal cord paralysis.[10] This could manifest as stridor and partial or complete airway obstruction. The diagnosis can be confirmed by indirect or fibreoptic laryngoscopy. If airway obstruction is severe, anaesthesia should be reinduced and the airway secured by conventional means. For patients not meeting the clinical criteria for extubation, adequate level of sedatives and opiates should be given and the patient transferred to the intensive care unit for postoperative mechanical ventilation.

The indications for postoperative ventilation following posterior fossa surgery are:

- Intraoperative profound vital sign changes.
- Extensive tumour dissection.
- Prolonged surgery.
- Preoperative obtunded patient with lower cranial nerve involvement.

In patients with surgery for intrabulbar ependymoma, although the respiratory efforts appear in the first hour postoperatively, they rapidly decrease later due to pathological edema of the brain. So it is recommended that after a neurological assessment, the patients should be ventilated for the next 12 to 24 hours.

PNEUMOCEPHALUS

Hyperventilation, cerebrospinal fluid drainage and use of diuretics decreases brain size and in the head up position, air gets trapped in the frontal areas. After dural closure, continued administration of N_2O, re-expansion of the brain and rehydration and rewarming during postoperative period can cause the gas to expand and lead to tension pneumocephalus. The incidence of tension pneumocephalus in one large series was 3% and all patients responded to conservative treatment.[11] Development of this syndrome is suspected if there is delayed return of consciousness, seizures and neurological deterioration in the postoperative period and early CT scan or Xray skull should be done to rule out this complication. Immediate twist drill aspiration of air through burr holes on either side of the vertex is indicated if a tension pneumocephalus is diagnosed.

Although this complication is commonly associated with the sitting position, its occurrence is not limited to the upright posture and it can also occur in the prone and parkbench position also.

One can prevent its occurrence by avoiding excessive hyperventilation and diuresis so as to avoid a very slack brain especially during surgery in the sitting position. It is also recommended that a functioning CSF shunt should be externalized before surgery in the sitting position. The role of N_2O in its causation is controversial and even its discontinuation prior to dural closure was not found to be effective in its prevention. Other methods of prevention include flushing the subdural space with saline to displace gas and maintenance of normothermia intraoperatively.

MACROGLOSSIA & FACIAL SWELLING

Facial and tongue swelling can particularly occur in the sitting and prone positions and may be so severe as to preclude extubation. **(Fig. 6)** This may be related to excessive neck flexion which compromises venous and lymphatic drainage of the face and tongue as well as use of oropharyngeal airway and long duration of surgery. Infants may be at particular risk because of the high anterior larynx, small tracheal diameter and relatively large tongue. If undetected and the patient is extubated there may be airway obstruction which may be particularly dangerous since endotracheal intubation may be impossible in the face of tongue swelling.

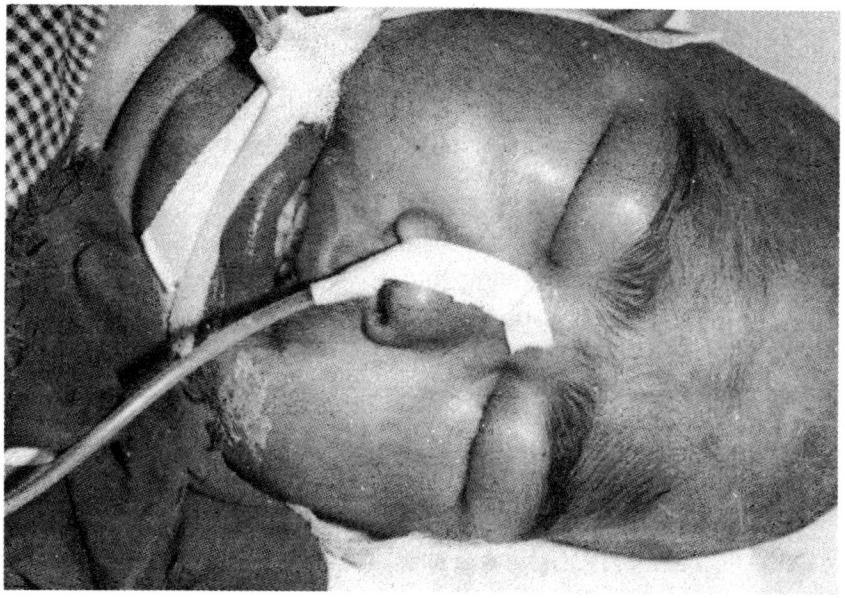

Fig. 6 : Facial swelling in a child who underwent posterior fossa surgery in the prone positon.

EYE INJURY

Eye compression in the prone and lateral position may occur and blindness has even been reported following retinal artery compression. This is more likely with the use of horseshoe head holders in the prone position with inadvertent compression of the eye and can be prevented with the use of pin fixation.

ENDOTRACHEAL TUBE MIGRATION

The incidence of endotracheal tube migration into the bronchus or pharynx is more in children as compared to adults and stability of the endotracheal tube is of utmost importance particularly in the prone position. Perspiration, salivation, blood and irrigating fluids may all cause the tape securing the endotracheal tube to come loose resulting in accidental extubation. Excessive neck flexion may cause the tube to move endobronchially with resultant unequal ventilation. Accessibility to the endotracheal tube during surgery is limited and hence dislodgement of the tube can lead to disastrous consequences. Some anaesthesiologists prefer the use of a nasal endotracheal tube particularly in the prone position as there are less chances of kinking and accidental extubation.[12]

SUBDURAL HAEMATOMA

There is an increased incidence of subdural haematoma in the sitting position due to rupture of the bridging veins and one should avoid the use of excessive hyperventilation and osmotic diuretics particularly in the sitting position.

INCREASED BLOOD LOSS

There is a potential for abrupt blood loss during posterior fossa surgery. As estimation of blood loss in children is difficult, severe hypovolaemia may occur. Continuous monitoring of the arterial blood pressure and CVP with prompt replacement with warm blood is essential to prevent haemodynamic instability due to blood loss.

HYPOGLYCAEMIA

Due to prolonged preoperative fasting and long duration of surgery, children are particularly prone for hypoglycaemia. Intraoperative blood glucose monitoring and glucose supplementation if necessary should be done.

HYPOTHERMIA

Children are prone for intraoperative hypothermia due to their large head surface area (18% of BSA), large body surface area to weight ratio and less subcutaneous fat.

ORTHOPAEDIC, DERMATOLOGICAL AND NEUROLOGICAL INJURIES

The sitting and prone positions for posterior fossa surgery can cause bone and skin injuries. Compression peripheral neuropathy is also known to occur. Extreme stretching of the neck during flexion in the sitting position can predispose to quadriplegia, though it is a rare occurrence.

POSTOPERATIVE BRAIN SWELLING

This can cause depression of the consciousness and airway maintenance may be difficult. Upward tentorial herniation may also be associated with brain swelling.

REFERENCES:

1. Hagen PT, Scholz DG, Edwards WD: Incidence and size of patent foramen ovale during the first 10 decades of life: An autopsy study of 965 normal hearts. *Mayo Clin Proc* 1984; 59:17. (Mayberg 37)
2. Lynch JJ, Schuchard GH, Gross CM, et al: Prevalence of right-to-left shunting in a healthy population: Detection by Valsalva maneuver contrast echocardiography. *Am J Cardiol* 1984;53:1478. (Mayberg 54)
3. Black S, Muzzi DA, Nishimura RA, et al: Preoperative and intraoperative echocardiography to detect right-to-left shunt in patient undergoing neurosurgical procedures in the sitting position. *Anesthesiology* 1990;72:436.(Mayberg10)
4. Fuchs G, Schwarz G, Stein J, et al: Doppler color-flow imaging: screening of a patent foramen ovale in children scheduled for neurosurgery in the sitting position. *J Neurosurg Anesthesiol* 1998;10: 5-9. (Vienna 51)
5. Humphreys RP, Creighton RE, Hendrick EB: Advantagees of the prone position for neurosurgical procedures on the upper cervical spine and posterior cranial fossa in children. *Child's Brain* 1975;1:325-26. ((bone highly vascularized)
6. Pollock JSS: Pain relief in children after neurosurgery. *Anaesthesia* 2000;55:410.
7. Cucchiara RF, Bowers B: Air embolism in children undergoing suboccipital craniotomy. *Anesthesiology* 1982;57:338-39.
8. Soriano SG, McManus ML, Sullivan LJ, et al: Doppler sensor placement during neurosurgical procedures for children in the prone position. *J Neurosurg Anesthesiol* 1994;6:153-55.
9. Meyer P-G, Cuttaree H, Charron B, Jarreau M-M, Perie A-C, et al: Prevention of venous air embolism in paediatric neurosurgical procedures performed in the sitting position by combined use of MAST suit and PEEP. *Br J Anaesth* 1994;73:795-800.
10. Artru AA, Cucchiara RF, Messick JM: Cardiorespiratory and cranial nerve sequelae of surgical procedures involving the posterior fossa. *Anesthesiology* 1980;52:83-86.
11. Standefer M, Bay JW, Trusso R: The sitting position in neurosurgery: A retrospective analysis of 488 cases. *Neurosurgery* 1984;14:649-58.
12. McLeod ME, Creighton RE, Humphreys RP, et al: Anaesthetic management of cerebral arteriovenous malformations in children. *Can Anaesth Soc J* 1982;29:299-306.

39

Malignant Tumors of the Posterior Fossa – Radiotherapy and Chemotherapy

S Hukku, S Halder

Anatomically the posterior fossa consists of the cerebellum and the brain stem and the malignant tumors arising from these structures are more common in children and young adults. They require a special and careful treatment planning in view of the long term effects of radiation and cytotoxic drugs, both of which are used extensively in their management.

These tumors differ in their natural history from other brain neoplasms. Unlike other intracranial neoplasms, medulloblastoma metastasizes extra cranially to bone, lung, and lymph nodes and also has a high propensity for cerebro spinal fluid spread. All these factors have to be considered in their management. In addition, the tumors of the brain stem produce significant morbidity due to their location. Also, these tumors are not amenable to biopsy without significant morbidity and so most of the times the treatment has to be initiated without a histological proof and this is based on clinical findings and imaging.

The common histological types encountered in the pediatric age group include astrocytoma, medulloblastoma (PNET), ependymoma, and brain stem glioma, and those in the adults include astrocytoma, glioblastoma multiforme, ependymoma, brain stem glioma, and metastatic tumors.

Poor prognostic factors include male gender and age less than 5 years, locally extensive tumors, CSF spread, sub total excision or biopsy, post operative residual more than 1.5 cc, and high grade histology of PNET other than medulloblastoma.

Headache, nausea and vomiting, ataxia, nystagmus are the presenting features of the cerebellar tumors and cranial nerve palsies, unilateral limb ataxia, ataxic gait, gaze disorders, and hemisensory syndrome are more common in brain stem tumors. Amongst the various imaging modalities, MRI is the best suited due to the lack of artifacts and availability of high quality sagittal and coronal sections and its effectiveness in evaluating the whole spine for deposits. In addition the information obtained is very helpful in radiation treatment planning.

The central nervous system because of its slowly replicating cells is sensitive to both radiation and chemotherapy. In order to keep the morbidity low, strict adherence to the radiation tolerance parameters is a must. The radiation tolerance of the brain is dependent on the volume irradiated. One third of the brain tissue and the whole brain tolerate a dose of 60 Gy, and 40 Gy respectively. Whereas, one third of the brain stem and the whole brain stem will tolerate 60 Gy and 50 Gy respectively. Infarction and necrosis are the result of radiation induced morbidity and should be avoided. Factors associated with a higher morbidity include higher total dose and dose per fraction, (more than 2 Gy daily), co morbid conditions viz. hypertension and diabetes, high LET radiation viz. protons and neutrons, and concomitant use of central nervous system toxic drugs viz. methotrexate.

CEREBELLAR ASTROCYTOMA

This is a separate entity as the prognosis is better than those arising in the cerebrum and the brain stem. Histologically, most of them are low grade tumors arising from the vermis. They are common in the first two decades of life.

Surgery is the primary treatment and provides histology and decompression. This is followed by radiation therapy except in a small percentage of completely resected low grade tumors. The treatment is initiated after two weeks of surgery.

In low grade tumors, the radiation field includes the CT defined tumor with 2 cm margin and a dose of 54 Gy is delivered. In high grade tumors the radiation therapy is divided in two phases. The first phase a 2 cm margin is taken around the tumor and the edema and a dose of 40-46 Gy is delivered. Second phase uses a margin of 2 cm around the tumor and an additional dose of 14-20 Gy is delivered, thus delivering a total dose of 60 Gy in 30-33 fractions.

Chemotherapy is used in high grade tumors. Although the response rates range from 44%-83% in anaplastic astrocytomas and 21%-73% in glioblastoma, contribution of chemotherapy towards a better survival is uncertain.

BRAIN STEM GLIOMA

As a group these tumors carry a poor prognosis. Exophytic tumors have better results than diffuse and infiltrative varieties.

Because of the high morbidity associated with surgical intervention, radiotherapy plays a significant role in their management. Radiation therapy improves survival, and stabilizes and/or reverses neurological dysfunction in most patients.

In focal lesions radiotherapy field incorporates a margin of 1-2 cm around the tumor but in diffuse lesions it is increased to 2-3 cm. A dose of 50-60 Gy in 27-33 fractions is used. Hyperfractionated regimes, using 2-3 fractions per day, have been used delivering higher radiation dose (upto 78 Gy) with no benefit in survival over conventional fractions.[1]

Chemotherapy has a questionable usefulness in these tumors. CCG compared radiation alone and radiation with CCNU, PCV, and prednisolone with no benefit. High dose chemotherapy with bone marrow rescue has also not shown any benefit. Neoadjuvant chemotherapy shows clinical and radiological response but no increase in survival. Most results are available for children as these tumors are rare in adults. Median survival recorded in various studies varies from 34-40 months. Median time to progression is 19 months. There is no significant difference in the results between patients with biopsy proven gliomas and those on whom diagnosis was made radiologically.[2]

MEDULLOBLASTOMA

This tumor is similar to primitive neuroectodermal tumors than glioma. It originates from the germinal neuroepithelial cells in the roof of the fourth ventricle and is highly proliferative in nature. Most tumors occur in the first decade of life but there are two peaks (5-9 years and 25-30 years).

The tumor is typically situated in the cerebellum mostly in the midline and vermis. The propensity to spread to CSF is 30% and to extracranial sites is 5%.

Postoperative craniospinal radiation is a must in all cases except very young children where chemotherapy is used. Phase one radiation includes the whole brain using a german helmet technique which adequately covers the cribriform plate and the posterior fossa which are the commonest sites of failure. It also includes radiation to the whole spine upto the level of S2-S3 junction. The radiation dose to the whole brain is 36 Gy in 20 fractions and to the spine is 36 Gy in 24 fractions. The primary tumor with 2 cm margin is boosted in the second phase delivering an additional dose of 18 Gy in 10 fractions, taking the total dose to the primary tumor to 54 Gy.[3] Doses are reduced by 10 Gy for children younger than 2-3 years. Because of the morbidity associated with craniospinal radiation, a close monitoring is essential which includes twice weekly blood counts.

Given the neuroendocrine and cognitive sequele of moderate doses of craniospinal radiation, attempts have been made to omit or reduce the spinal radiation. Single institution trials reported good results with a lower dose to the spine (24-25 Gy) in selected patients. Randomized trials (CCSG, POG) have however shown an excess of early CSF relapses (30% versus 15%) in patients receiving low dose radiation. The combination of chemotherapy and low dose radiation is currently being tested.[4] Trials substituting chemotherapy for cranial or craniospinal radiation have been unsuccessful. Several randomized trials have examined the use of chemotherapy in addition to radiation. Subset analysis suggests a benefit in T3/T4 disease, CSF dissemination, and in patients with sub total resection. Currently trials are examining neoadjuvant chemotherapy followed by radiotherapy in high risk patients. A variety of agents have been used in treatment which include vincristine, nitrosoureas, cyclophosphamide, and platinum compounds. The response rate with the combination chemotherapy ranges from 25%-100%.

The treatment for medulloblastoma in adults has generally paralleled that for children. Surgery followed by postoperative radiotherapy is recommended. Role of chemotherapy is less well defined.

Surgery followed by craniospinal radiation results in a disease free survival of 50%-70%. The disease free survival for good risk patients is 60%-90% and for poor risk patients is 20%-40%. Adjuvant chemotherapy increases the overall survival in high risk patients to 50%-60%.[5,6]

EPENDYMOMA

It arises from the ependymal lineage. Of the infratentorial sites, the fourth ventricle is the commonest. Subarachnoid invasion is seen in 50% patients. Encasement of the medulla and cord can occur. Histologically they are classified as well differentiated, anaplastic, and ependymoblastoma. The last two categories have a higher propensity to disseminate to the CSF.

Post operative radiotherapy is used in all cases. Low grade tumors receive a limited field radiation with a dose of 54 Gy in 27-30 fractions. Craniospinal radiation is not indicated unless CSF involvement is present or there are subarachnoid metastases. It is also indicated in high grade tumors and ependymoblastoma. Chemotherapy has little role in the management of these tumors. 5-year survival of 50%-80% is achieved with postoperative radiation.[7] Spinal radiation does not increase survival significantly.[8]

METASTATIC CANCER

30% patients with systemic cancer have brain metastases. Lung, prostate, breast, kidney, and gastrointestinal cancers frequently metastasize to the brain.

Treatment is essentially palliative and radiation doses of 30 Gy in 10 fractions produce a response rate of 70%-90%. Recent studies indicate the use of temozolamide along with radiation but long term results are not available.

FUTURE DIRECTIONS

Newer radiotherapy techniques include stereotactic fractionated radiotherapy and radiosurgery, 3D conformal radiotherapy, and Intensity modulated radiation therapy (IMRT). All these techniques are aimed at reducing the dose to the normal brain tissue and thereby increasing the tumor dose in order to achieve better results.

REFERENCES:

1. Shrieve DC, Warn WM, Edwards MS, et al: Hyper fractionated radiation therapy for gliomas of the brain stem in children and in adults. *Int J Radiat Oncol Biol Phys* 1992;24:599-610.
2. Guieny MJ, Smith JG, Hughes P, et al: Contemporary management of adults and pediatric brain stem gliomas. *Int J Radiat Oncol Biol Phys* 1993;25:235-41.

3. Cun LE, Constina LS: Medulloblastoma: caution regarding new treatment approaches. *Int J Radiat Oncol Biol Phys* 1991;20:897.
4. Deutsch M, Thomas PRM, Feischer J, et al: Results of a prospective randomized trial comparing standard dose neuraxis radiation with reduced neuraxis radiation in patients with low stage medulloblastoma. *Pediatr Neurosurg* 1996;24:167-77.
5. Jenkin D: The radiation treatment of medulloblastoma. *J Neurooncol* 1996;29:45-54.
6. Cohen BH, Paiter RJ: Chemotherapy for medulloblastoma and primitive neuroectodermal tumors. *J Neurooncol* 1996:29:55-68.
7. Phillips TL, Sheline GE, Bodrey E: Therapeutic considerations in tumors affecting the central nervous system: Ependymomas. *Radiology* 1964:83:98-105.
8. Vanuystel J, Ashley SE, Bloom JG, et al: Intracranial ependymomas: long term results of a policy of surgery and radiotherapy. *Int J Radiat Oncol Biol Phys* 1992;23:313-19.

40

How I do it — Pineal Tumor

Occipital Transtentorial Approach = Poppen - Jamieson approach

VK Khosla

INTRODUCTION

Descartes has described pineal gland as the seat of soul. Dandy termed it as the seat of consciousness. This area encompasses pineal tumors, pineal region tumors, and posterior 3rd ventricular tumors. The challenge for neurosurgeons in this region is due to the great depth of the pineal irrespective of the route of approach. Due to the hazardous surgical results in the earlier days various alternatives were sought. Direct Radiotherapy, Chemotherapy and surgery were used with variable results. The advent of operating microscope, micro-instruments, newer modalities of imaging has tilted the balance in favor of surgery. However, no final verdict has been reached yet regarding the ideal method of management for lesions in this region. Tumors of pineal region account for 1% to 8% of all brain tumors in children and adults respectively. Tumors of pineal region account for 1% to 8% of brain tumors in children and adults respectively. Germinomas account for 1% to 4% of all intracranial tumors in the western and Japanese population respectively.[1] Tumors in this region are classified by Rubinstein as:[2]

a) Germ cell tumors (65%-72%)

   Benign

           i) Epidermoid
           ii) Dermoid
           iii) Teratoma

   Malignant

           i) Germinoma
           ii) Embryonal carcinoma
           iii) Choriocarcinoma

b) Pineal cell tumors (20%-30%)
   - i) Pineocytomas — Benign
     - Malignant
   - ii) Pineoblastomas
c) Supporting cell tumors
   - i) Astrocytomas
   - ii) Meningiomas
   - iii) Ependymomas
   - iv) Oligodendroglioma
   - v) Hemangiopericytoma
   - vi) Choroid plexus papilloma
d) Metastatic tumors
e) Non neoplastic lesions
   - i) Pineal cyst
   - ii) Arachnoid Cyst
   - iii) Cysticercus cyst
   - iv) Vein of Galen Malformation

IMAGING

MRI scan is superior to CT scan due to better resolution for anatomical and pathological details. Both studies should have contrast. Spinal MRI or myelogram to look for silent tumor spread is suggested by several authors.[3]

The exact knowledge of the type of pathology greatly helps the neuro-oncologists in using the appropriate treatment modality, from shunt, stereotaxy, radiotherapy, microsurgery, and chemotherapy either singly or in combination.

Tumor Markers

Tumor markers have not been systematically studied in this area.[4] However, CSF and serum alpha-fetoprotein (AFP) and human chorionic ganadotrophin B sub-unit (BHCG) should be. Evaluated as elevated AFP rules out a pure germinoma and may have yolk sac component. AFP has an upper reference limit of approximately 15ug/L (~ 10kU/L) after the first year of life. Mild to moderate elevation of HCG (< 2000 mIU/mL) may be seen in germinoma with trophoblastic component. Significantly high HCG is seen in non-germinomatus tumors like choriocarcinoma.[3] The sensitivity of BHCG is 0% to 43%.[4,5] The mean blood to CSF ratio in cranial germinomas is 1:10 which is very different to 286:1 reported for systemic GCT.[5] However, ventricular CSF appears to be disconnected from lumbar CSF, the latter being preferred site for sample.[5]

As negative tumor marker does not preclude a diagnosis of a tumor the most important use of it is in the follow-up of positive patients after therapy. A safe level of HCG is probably below 50 mIU/mL.

STEREOTAXY

In the past, patients with pineal tumors have been diagnosed with a trial of focal radiotherapy without biopsy. If the tumor was radiosensitive, it was presumed to be a germinoma. With the advent of microsurgical techniques and reduction of morbidity and mortality from 30%-70%[6] to around 1%,[7] most clinicians recommend a tissue diagnosis.

Open biopsy or resection is favored over staxy because of their anatomic location and because microscopic morphology may be heterogenous missing mixed GCT. The expertise to diagnose from a small piece is also not easily available. In addition, tumor removal is the preferred mode of therapy in non-germinomatus tumors, which is not possible by staxy.[3]

ROLE OF SHUNT

There has been concerns that shunt may disseminate germinomas. However analysis at present shows that shunt is not a contraindication.[8]

ROLE OF RADICAL RESECTION

In germinomas there is no significant difference in outcome related to the extent of surgery. However as mentioned earlier the chance of missing a mixed component is higher in small specimens and radical excision is the method of choice for tumors, other than germinomas which may be 30% to 40% of all pineal region tumors.

TYPES OF SURGERY

There are mainly two types of approaches utilized at present. The Poppen/Jamieson occipital transtentorial approach and the Infratentorial supracerebellar Krause/Stein approach.

The advantages of Poppen's approach are

a) Good wide exposure

b) Greater mobilization of tumor (space)

c) Excellent anatomical visualization

d) Veins are hardly sacrificed while in Stein's approach the superior cerebellar veins may get sacrificed.

The disadvantage is a possible visual field deficit due to retraction of the occipital lobe.

The advantages of the Steins approach are:

a) Minimal retraction due to sitting position.

b) Safer approach for tumors arising below the Galenic system.

However it appears that as any other approach the familiarity to one method makes the surgeon biased towards a particular approach.

POPPEN'S APPROACH[9,10]

Preparation

Discussion with the patient about the surgery and possible complications like visual field deficit is essential. Preoperative field charting is desirable. Dexamethasone at the dose of 4 mg every 6 hours is started. Decision of shunt or perioperative external ventricular drain is decided as per patients' neuroradiological status.

Anesthesia

General anesthesia with central venous line along with ETCO2 / Trans esophageal Doppler to detect air embolism is used.

Position

The sitting, semiprone or parkbench position may be used.

The sitting position has the advantage of normal anatomical orientation, which gets reversed, in semiprone position. Air embolism and visual field deficit is higher in sitting position. Children tolerate sitting position poorly. However due to gravity the field remains cleaner in sitting position.

Park bench position has the advantage that the anatomy is not fully reversed. The occipital lobe tends to fall away spontaneously avoiding retraction and chances of air embolism and chest compression are low. I prefer to use this position with the right side dependent.

Instrumentation

The single most important instrument is the operating microscope. It is our practice to bring it in just after the dural opening. Long, fine, bayonet shaped instruments, bipolar cautery and ultrasonic aspirator are essential.

Self-retaining retractors with appropriate blade is indispensable. We use the Sugita head frame and retractor system though a Mayfield head frame may also be used.

Incision

The right occipital side is kept downside. With a 7-8 cm-incision starting across to left of midline beginning at the level of the external occipital protuberance. The incision is turned down again about 6-7 cm laterally to the level of the mastoid groove. The skin is reflected along with the periosteum. **(Fig. 1)**

Burr Holes

Usually 6 burr holes are made. Two each on left and right of midline to expose superior sagittal sinus and just above the transverse sinus. Two holes 6-7 cm lateral to midline are made and removed by Gigli saw or craniotome. **(Fig. 2)**

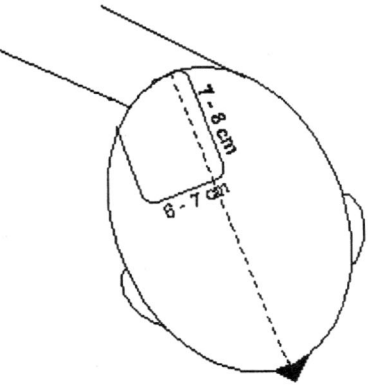

Fig. 1 : Position of the patient with ipsilateral side dependent and marking of the flap.

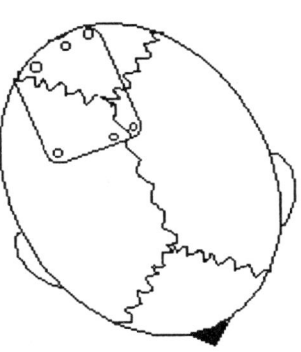

Fig. 2 : Bone flap with six burr holes

The removal of bone over the transverse sinus is done using rongeurs. Sometimes in older patients the burr holes to expose SSS are made to the right of midline and the bone over SSS is rongeured off as the dura may be densely adherent to the bone, thus injuring the SSS. The sinuses are covered with gelatin sponge and/or oxidized cellulose and bone flap wrapped in an abdominal sponge. In principle the SSS is exposed off the bone to help retracting the midline dural structures. This reduces brain retraction.

Dura

The dura should always be lax and the ventricles may be tapped to do so. The dura may be opened in a U shape manner based on SSS in which the fold tends to hide the midline. It can be cruciate but I prefer an L shape incision along SSS and transverse sinus. **(Fig. 3)** Traction sutures are used to maximize exposure.

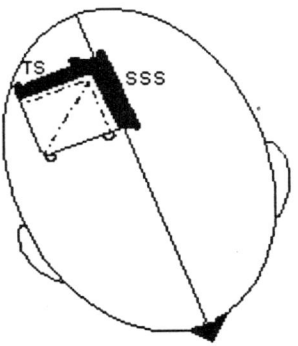

Fig. 3 : Dural incision along superior sagittal sinus (SSS) and Transverse sinus (TS)

Retraction

The occipital lobe in park bench position tends to fall away requiring minimal retraction more towards the pole rather than the posterior interhemispheric fissure to avoid calcarine cortex. Brain should be absolutely lax otherwise chances of retraction injury will be more. The brain is covered with soft patties for retraction.

Tent

The tent is cut parallel to the straight sinus from torcula to incisura. Use of bipolar avoids bleeding from dural lakes and causes retraction due to shrinkage. Cutting of the tent wide open is critical and retraction sutures on these are used to maximize exposure.

Display

The arachnoid over the cerebellum protecting the superior cerebellar veins should be protected and arachnoid over the tumor and vein of the Galen should be "hugged" throughout to remain within the tumor the arachnoid all around the tumor is dissected to lay the tumor "naked" off the arachnoid. Frozen section is sent at this stage, as if it comes out to be a germinoma no heroic attempt should be made for radical excision. Ultrasonic aspirator is used throughout for decompression and further arachnoid dissection.

After total excision the following structures may be identified

- ✓ Vein of Galen
- ✓ Vein of Rosenthal
- ✓ Internal cerebral vein
- ✓ Precentral veins
- ✓ Superior Vermis
- ✓ Quadrigeminal plate
- ✓ Splenium
- ✓ Pineal gland in non-tumorus pathology

Cutting into the splenuim should be avoided and as much tumor removal as possible should be done with the aim of possible total excision.

When to stop

In germ cell tumor as much tumor as easily possible should be removed. In meningiomas or gliomas as much decompression as possible should be done. Meningiomas should be totally excised. In a teratoma a gross total removal is usually safe. In younger patients it is safer to be more aggressive and opening the posterior III ventricle should be tried as a ventriculostomy. However, the degree of aggression should be tempered with ones own limitations, experience and facilities.

Closure

After hemostasis, dura should be closed water tight with graft if necessary. I do not attempt to suture the tent. I use subcuticular sutures to avoid suture removal at a later date and I usually avoid a suction drainage.

Complications

The most disastrous possible complication is damage to the Galenic system with possible mutism and death. Hemianopia may occur if calcarine cortex is damaged. Disconnection syndrome is possible if splenuim is breached. Infections, bleed, Osteomyelitis are complications, which are inherent to craniotomy.

Chemotherapy

Chemotherapy is being used as both primary therapy and to treat recurrence. Chemo alone or neoadjuvant followed by radiotherapy is being attempted in multi centric trials.[3] The possibility of eliminating radiation is compelling especially in treating children and young adults because of the possibility of neuro hormonal and cognitive deficits associated with radiotherapy. The agents being used are carboplatin, etoposide, and bleomycin, cyclophosphamide Vinblastine, Iosfamide in various combinations and cycles.

SUMMARY

The Poppen's approach to the pineal region has proven its place in the last 3 decades and has reduced the morbidity and mortality to acceptable levels compared to older approaches. Familiarity of the regional anatomy, proper instruments can make it effective, safe and relatively simple. However the management of these tumors is still evolving.

REFERENCES:

1. Jennings MT, Gelman R, Hochberg F: Intracranial germ cell tumor. Natural history and pathogenesis. *J Neurosurg* 1985;63:155-67.
2. Rubinstein LF: Cytogenesis and differentiation of Pineal neoplasms. *Hum pathol* 1981;12:441-48.
3. Paulino AC, Wen BC and Mohideen MN: Controversies in the Management of Intracranial Germinomas. *Oncology* 1999;13:513-21.
4. Sawamura Y, Ikeda J, Shirato H, et al: Germ cell tumors of the CNS. Treatment considerations based on 111 cases and their long term clinical outcomes. *Eur J cancer* 1998;34:104-10.
5. Rogers PB, Plowman N: Blood to cerebrospinal fluid human chorionic gonadotropin beta ratios in intracranial germ cell tumors. *Neurosurg focus* 1998;5 article 4.
6. Baumgartner JE, Edwards MSB: Pineal tumors. *Neurosurg Clin North Am* 1992;3:853-62.
7. Regis J, Bouillot P, Rouby- volot F, et al: Pineal region tumors and the role of stereotactic biopsy. Review of the mortality, morbidity and diagnostic rates in 370 cases. *Neurosurgery* 1996;39:907-12.

8. Berger MS, Baumeister B, Geyer R, et al.: The risks of metastasis from shunting in children with primary central nervous system tumors. *J Neurosurg* 1991;74:872-77.
9. Poppen JL, Marino R Jr: Pinealomas and tumor of the posterior portion of the third ventricle. *J Neurosurg* 1968;28:357-64.
10. Clark KW: Occipital transtentorial approach. In surgery of the third ventricle. Apuzzo MLJ ed. Williams and Wilkins Baltimore 591-610.

41
Surgery for Pineal Region Tumours in Children

A Sharma

Pineal region tumours constitute 3% to 8% of intracranial tumours in children with major racial and geographical differences in the incidences of different histological types.[1-2] For a long time, surgical management of pineal region tumours was influenced by earlier unfavourable results with high mortality and morbidity from direct surgical attempts. Early attempts at surgery in this region had poor outcome with operative mortality approaching 90%.[3] This resulted in a change in attitude with a preference for a cerebrospinal fluid diversion by a shunt followed by an empirical radiation therapy.[4-5] If the patient did not show any response to radiation therapy, a direct surgical procedure to remove the tumour was undertaken. This strategy resulted in unnecessary radiation to benign lesions and significant morbidity, particularly in children.[6] Further, radiation induced scarring, arachnoidal thickening and subsequent adhesion of tumour to the surrounding structures prevented radical removal of an otherwise, removable benign lesion.[7] With the advent of microsurgical techniques, safer paediatric neuro-anaesthesia, stereotactic and endoscopic procedures and biochemical tests for tumour markers, direct surgery for pineal region lesions is undertaken with less than 5% mortality and the practice of empirical radiation therapy with out a tissue diagnosis has been abandoned.

CLINICAL PRESENTATION

The most common clinical presentations of pineal tumours in children, result from a raised intracranial pressure caused by an obstruction of cerebrospinal fluid flow by the tumour. A Child may present with Headache, vomiting, diminished vision or double vision. In our country, significant number of children with pineal tumour demonstrate papilledema and sixth nerve palsy and altered consciousness at the time of presentation. Compression of surrounding structures results in ophthalmic symptoms. These include impaired conjugate upgaze, pupillomotor paralysis, retractory nystagmus on convergence described as Perinaud Syndrome. Large invasive tumours may extend in midbrain tegmentum resulting in third nerve palsy and hemiparesis or may invade cerebellar peduncle to cause ataxia and nystagmus. Children with germ cell tumours may present with endocrine disturbances in form of diabetes insipidus or precocious puberty resulting from concurrent suprasellar tumour or hydrocephalus.

RADIOLOGIAL INVESTGATIONS

For children suspected to have a pineal region tumour, Computed tomography (CT) and Magnetic Resonance Imaging (MRI) are the current radiological investigations of choice. Pineal calcification in plain skiagram skull of a child younger than 10 yrs of age, is indicative of a pineal tumour. In older children calcification of more than 10 mm or shift of a calcified pineal is also suggestive of a pineal tumour. Cerebral angiography, which may demonstrate tumour blush or displacement of normal vasculature is undertaken before a stereotactic procedure. CT and MRI have revolutionized the diagnosis of pineal region tumours with their exquisite sensitivity in detecting the tumours[8] **(Fig. 1)**. High-resolution MRI with gadolinium is the investigation of choice **(Fig. 2)**. Tumour characteristics such as size, vascularity, its anatomical relation with surrounding structures, particularly in sagittal plane are provided by MRI. CT is useful in detecting a calcification in the tumour as it is at least eight times more sensitive than skull skiagram in detecting pineal calcification.[9]

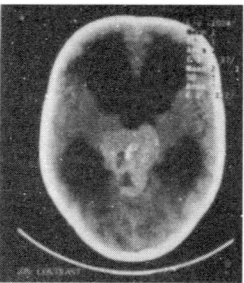 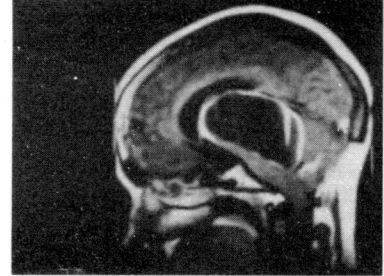

Fig. 1 : CT Scan (non Contrast) showing a hyperdense lesion with intratumoral Calcification in a pineocytoma.

Fig. 2 : Sagittal MRI showing peripheral Contrast enhancement in a Germinoma

Although, on MRI, there are some radiographic features associated with specific tumours, however, there is too much variability and overlap to be confident of the histology on the basis of the radiologic appearance **(Fig. 3)**. *Pineocytoma and pineoblastomas* are hypo/isointense on T1 and show increased signal on T2-weighted images with homogenous enhancement. *Germinomas* are isointense on T1, slight hyperintense on T2 weighted images and show homogenous enhancement. Cysts may also be visible. *Astrocytomas* are hypo/isointense on T1, with increased signal intensity on T2 weighted images and show variable enhancement. *Teratomas* are heterogenous on T1 and multilocular on T2 weighted images and show irregular enhancement. Calcification surrounds the pineal gland

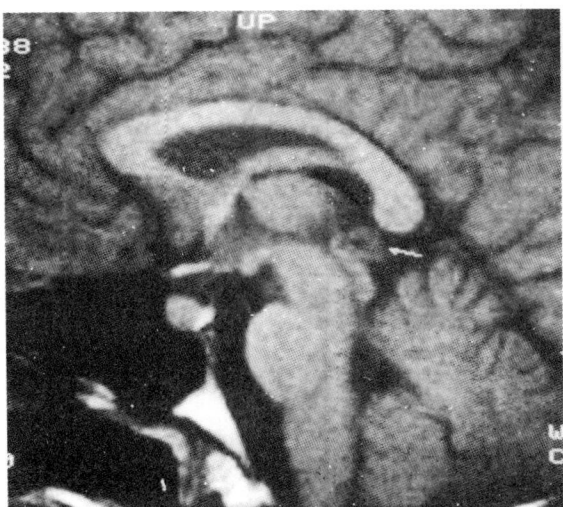 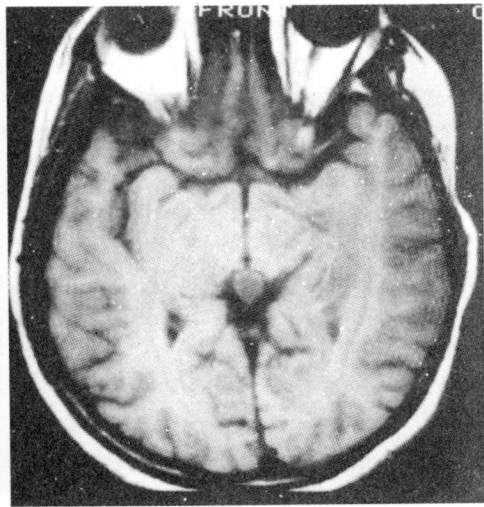

Fig. 3 : (A & B) MRI Sagittal and axial cut demonstrating a small tumour (Astrocytoma)

as the germinoma grows, as opposed to the intratumoral calcification within a pineocytoma.[10]

LABORATORY DIAGNOSIS: TUMOUR MARKERS.

Measurements of Cerebrospinal fluid (CSF) and serum tumour markers are an important component of preoperative evaluation and have been most helpful in children with germ cell tumours.[11] Expression of embryonal proteins such as α-fetoprotein (AFP) and β-hCG are indicative of malignant germ cell elements. Serum and CSF measurements can be used for diagnosis or for monitoring the response to therapy.[12]

Other biological markers for germ cell tumours include Lactic dehydrogenase isoenzymes and placental Alkaline Phosphatase (PLAP).[13-14] Pineal parenchymatous cell tumour markers include melatonin and the S antigen and the use of melatonin levels in the follow-up of patients with pineocytoma after surgical treatment has been reported.[15]

MANAGEMENT

Management of children with pineal tumour is aimed at treating symptomatic hydrocephalus and establishing a diagnosis of pathology of the tumour. At the time of presentation, most of the children with a pineal region tumour have hydrocephalus, resulting from aqueductal compression. Hence, first step is the management of hydrocephalus. Further, at the same time CSF may be collected for cytology and tumour marker estimations.

Management of Hydrocephalus

The hydrocephalus can be managed by a implanting a ventriculo-peritoneal shunt. This provides time for ventricular system to compensate and intracranial pressure to stabilized. However, in addition to its usual complications, it carries a potential risk of tumour dessimination.[16]

Hydrocephalus can also be managed effectively by temporary placement of external ventricular drain, third ventriculostomy and aqueductoplasty prior to surgical approach to the pineal tumour. Improved endoscopic techniques have made endoscopic third ventriculostomy an easy and reliable method, which can be done stereotactically or free hand.[17] Finally, a successful resection of the tumour may remove the obstruction to CSF flow, obviating the need of any kind of CSF diversion.

Management of the tumour :

The decision to perform open procedure versus a biopsy (stereotactic) for the pineal region tumour has been a matter of debate in the literature, however the choice of the procedure is based on surgeon's personal bias and his experience. Stereotactic biopsy has been described as a procedure of choice in patients with disseminated disease, invasive malignant tumour and with multiple medical problems. Early experiences with stereotactic biopsies were unsatisfactory, however, recent studies have shown it to be safe and efficient means of obtaining a tissue diagnosis.[18] In addition to the inherent risks of this procedure with vascular tumours, decompression to reduce the bulk of the tumour mass is not achieved and a subsequent operative procedure becomes necessary in cases of benign lesions.

Open resection of the pineal tumour is relatively safe and removal of most, if not all, of the tumour is possible **(Fig. 4)**. For patients with benign lesions, surgical resection is curative. Even with most of the malignant tumours, there exists a potential for evacuating a cyst and satisfactory decompression of tumour, which may improve response to postoperative adjuvant therapy.[19]

Surgical Management

Improvement in microsurgical techniques and neuroanaesthesia, has markedly lowered the morbidity and mortality associated with surgery for pineal region tumours. The surgical approaches to the pineal region can be divided into supratentorial, infratentorial and a combined supratentorial/ infratentorial approach. Supratentorial approaches include parietal interhemispheric approach described by Dandy and Occipital transtentorial approach first described by Horrax and later modified by Poppen and others.[20-21] Infratentorial Approach, an infratentorial supracerebellar approach was originally described by Krause and later popularised by Stein.[22-23] Of these approaches occipital transtentorial and infratentorial supracerebellar techniques are presently in vogue.

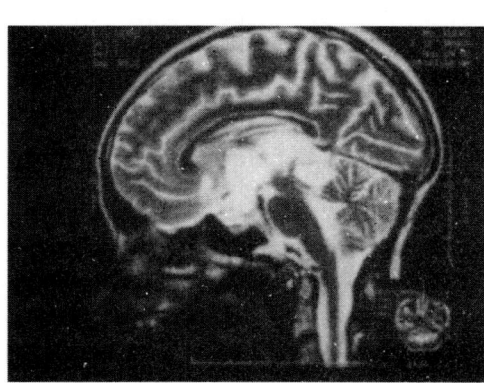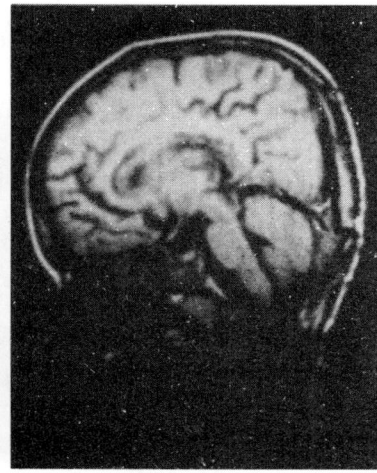

Fig. 4 : (A & B) Follow-up Sagittal M.R.I. Showing absence of the lesion.

Supratentorial Approach is best suited for tumours extending supratentorially or laterally into the trigone of the lateral ventricle. It provides a wide exposure however it is difficult to remove the tumour that lies below the convergence of the deep venous system.[22-23]

Infratentorial Approach is a direct midline approach and provides an easily recognized orientation to the surgeon. It is suitable for tumours that are midline, and grow into both posterior 3rd ventricle and posterior fossa and displace anterior lobe of cerebellum and quadrigeminal plate. *Advantages are* Midline trajectory for approach of the tumour and deep venous system caps the dorsal aspects of the tumour and remains away from the area of approach. The technical difficulties include limited access to the tumour that extends above deep venous system and for lateral extension of tumour, tumour growth in trigone and large tumours. It requires long instruments because of depth of the lesion.

Complications following pineal region surgery, irrespective of the approach include, extraocular movement dysfunction, ataxia and altered mental status. Seizures and hemianopsia may result from supratentorial approach.[24]

ADJUVANT THERAPY

Current treatment protocol for children older than 3 years with malignant pineal tumours include radiation therapy however, chemotherapy is considered as first line treatment in very young children. Germinomas are among the most radiosensitive tumours. A long term tumour-free survival of greater than 90% is reported in most published series. Nongerminomatous malignant germ cell tumours are less responsive to radiation with 5 years survival of 40%.[25] The use of prophylactic spinal irradiation for malignant pineal tumours is controversial. Spinal irradiation should be given only for documented seeding. Standard radiotherapy protocols for children with malignant pineal cell tumours use 4000 cGy of whole brain irradiation followed by 1500 cGy to the pineal region,[26] with a daily fraction of 180 cGy. For Children with malignant germ cell tumours standard therapy is focal radiotherapy followed by irradiation of the ventricle field. With platinum-based chemotherapeutic regimens a response rate of 80%-100% was reported in germinomas and nongerminomatous germ cell tumours.[27]

PROGNOSIS

Children with pineal region tumour require a life-long follow-up as these tumour may recur as late as 5 years after the diagnosis. Further, the children who have received irradiation may present with new tumour formations later in life.[28] Serial M.R.I. and tumour marker estimation surveillance must be employed. The prognosis depends on the histological tumour type. Children with Germinomas have an excellent prognosis because of the radiosensitivity of these tumours. Children with pineal cell tumours and non-germinomatous germ cell tumours have a significantly worse prognosis. There is no standard protocol for the management of recurrent pineal tumour. Recurrent germ cell tumours have been shown to respond to chemotherapy.[29] Radiosurgery may be considered for recurrence less than 3 cms in diameter.

REFERENCES:

1. Hoffman HJ, Yoshida M, Becker LE, et al: Pineal region tumours in childhood. Experience at the hospital for sick children, in Humphreys RP (ed). Concepts in pediatric Neurosurgery 4, *Basel S Karger* 1983;360-86.
2. Oi S: Recent advances and racial differences in therapeutic strategy to the pineal region tumour: an editorial. *Child's Nerv Syst* 1998;14:33-35.
3. Borit A: History of tumours of the pineal region. *Am J surg Pathol* 1981;5:613-20.
4. Camins MB, Schlesinger EB: Treatment of tumours of the posterior part of third ventricle and pineal region: a long term follow-up. *Acta Neurochir* 1978; 40:131-43.
5. Torkildsen A: Should extirpation be attempted in case of neoplasm in or near the third ventricle of the brain. Experiences with a palliative method. *J Neurosurg* 1948;4:249-75.
6. Edwards MS, Hudgins RJ, Wilson CB, et al: Pineal region tumours in children. *J Neurosurg* 1988;68: 689-97.
7. Stein BM: Surgical treatment of pineal tumours. In Carmel PW (ed) Clinical Neurosurgery 26, Baltimore, William & Wilkins, 1979;496-510.
8. Zimmerman R: Pineal region masses: imaging in Wilkins R, Rangachary S. (eds) Neurosurgery. New York, McGraw-Hill. 1996;1003-18.
9. Norman D, Diamond C, Boyd D: Relative detectability of intracranial calcification on computed tomography and skull radiography. *J Comput Assist Tomograph* 1978;2:61-64.
10. Chang T, Tang Mm, Guo Wy: CT of pineal tumours and intracranial germ cell tumours. *Am J Roentgenol* 1989;153:1269-74.
11. Bruce JN, Stein BM: Surgical management of pineal region tumours. *Acta Neurochir* 1995;134: 130-35.

12. Arita N, Ushio Y, Hayakawa T, et al: Serum levels of alpha fetoprotein, human chorionic gonadotrophin and carcinoembryonic antigen in patients with primary intracranial germ cell tumours. *Oncodev Biol Med* 1980;1:235-40.
13. Fetell MR, Stein BM: Neuroendocrine aspects of pineal tumours. *Neurol Clin North Am* 1986;4:877-905.
14. Talerman A: Germ cell tumours. *Ann Pathol* 1985;5:145-57.
15. Erlich SS, Apuzzo ML: Pineal gland: anatomy, physiology, and clinical significance. *J Neurosurg* 1985;63:321-41.
16. Devkota J, Brooks BS, el Gammel T : Ventriculoperitoneal shunt metastasis of a pineal germinoma. *Comput Radiol* 1984;8:141-45.
17. Ellanbogen RG, Moores LE: Endoscopic management of a pineal and suprasellar germinoma with associated hydrocephalus. *Minim Invasive Neurosurg* 1997;40:13-15.
18. Popovic E, Kelly P: Stereotactic procedures for lesions in pineal region. *Mayo Clin Proc* 1993;68:965-60.
19. Rout D, Sharma A, Radhakrishnan VV, Rao VRK: Exploration of pineal region : Observations and Results. *Surg Neurol* 1984;21:135-40.
20. Dandy W: Operative experience in cases of pineal tumor. *Arch Surg* 1936;33:19-46.
21. Poppen J: The right occipital approach to a pinealoma. *J Neurosurg* 1966;3:1-8.
22. Stein BM: Supracerebellar- infratentorial approach to pineal tumour. *Surg Neurol* 1979;11:331-37.
23. Kobayashi S, Sugita K, Tanaka Y, Kyoshima K: Infratentorial approach to the pineal region in the prone position: Concorde position. *J Neurosurg* 1983;58:141-43.
24. Bruce J: Management of pineal region tumours. *Neurosurg Q* 1993;103-19.
25. Fuller BG, Kapp DS, Cox R: Radiation therapy of pineal region tumours: 25 new cases and a review of 208 previously reported cases. *Int J Radiat Oncol Biol Phys* 1994:28:229-45.
26. Kang JK, Jeun SS, Hong YK, et al: Experience with pineal region tumours. *Child's Nerv Syst* 1998;14:63-68.
27. Jennings MT, Gelman R, Hochberg F: Intracranial germ cell tumours : natural history and pathogenesis. *J Neurosurg* 1985;63:155-67.
28. Starshank RJ: Radiation induced meningioma in children: report of two cases and review of the literature. *Pediatr Radiol* 1996;26:537-41.
29. Choi JU, Kim DS, Chung SS, Kim TS: Treatment of germ cell tumours in the pineal region. *Child's Nerv Syst* 1998;14:41-48.

42

Medical Management of Gliomas and Terminal Care

D Chowdhury

INTRODUCTION

Gliomas are primary tumors arising from the neural tube, particularly the glial cells. Epidemiological studies indicate that the annual incidence of the primary CNS tumors, including gliomas is increasing in the West.[1,2] Further, the prognosis of these tumors remain poor, with a median survival of only 5 years. Imaging of tumors by static and functional imaging like CT scan, MRI, MR-Spectroscopy and PET scans is increasing our knowledge of tumor biology and the extent of the disease. Improvement in the operative procedures, radiotherapy planning, tumor targeting and repositioning for the treatment have all improved the initial tumor management and reduced the mortality and the morbidity. Chemotherapy for advanced and recurrent disease although improves quality of life has less desirable effects in prolongation of survival. With new discoveries and increasing knowledge of the physiology and molecular biology of these tumors the potential for targeting therapy at a genetic level is becoming increasingly promising. Given the above background, the term "**Medical Management**" needs qualification. In a restrictive sense, it refers to only drug therapy employed to manage the symptomatology of the tumor effects. In a broader sense however, the term encompasses all the management strategies which are "**non-surgical**". Obviously, it includes the symptomatic drug therapy, anaesthesia and the related therapeutics, radiotherapy, chemotherapy and management of complications related to the various therapies. Since there are separate articles on the management of these aspects of gliomas in this book, I would restrict myself in this article on some specific issues concerning medical management in the first section.

Overall poor prognosis in managing gliomas obviously means that at some point in the disease course, most of the modalities of treatment will get exhausted and the patient will go in the terminal stage for the inevitable. In the last decade and half, much attention has been paid to the best possible management strategies which can be adopted for this stage. Section 2 in this article will discuss these issues.

SECTION 1

I. MANAGEMENT PLAN AND MANAGEMENT TEAM

Diagnosis

For successful management of any disease, a proper plan is essential. The first thing of-course is to diagnose and confirm the diagnosis. Certain points are worth emphasizing:[3]

1. The image diagnosis although is fairly suggestive many a times, should never be considered as the accurate predictor of the tumor type. Histological confirmation of the diagnosis must be actively considered and is usually undertaken. Confirmation of the diagnosis is necessary to provide a prognosis for the patients and their family, to decide on appropriate treatment and to allow the valid comparisons of the outcome. Sometimes in exceptional situations like in the frail elderly patients with severe neurological deficits, such benefits of clear diagnosis need to be balanced against risk of biopsy which with modern techniques is appreciably low.

2. The time interval between imaging and the biopsy of that lesion should be as short as possible.

3. Where biopsy is considered in preference to more extensive resection, image directed biopsy techniques should be used rather than freehand needle biopsy.

4. Histological results should be reported by a pathologist with appropriate training. Histological or cytological reporting should be available at the time of the operation. One of the accepted grading systems should be used.

Breaking the news of the diagnosis to the patient and the family

The manner in which bad news about a serious illness is broken affects the adjustment and coping that the patient can subsequently make. It is the humanity in the doctor which allows the patient or relative to suffer most easily the bad news. Multiple team care approach in glioma management can result in shortfalls in communication between the professionals and the patients. Following points need special consideration:[3]

1. Critical points in the diagnostic pathway at which the patients and the relatives need to be told new information, for example before and after imaging, before and immediately after biopsy, and later as the formal histopathological reports become available need to be identified beforehand.

2. News of the diagnosis and information about the prognosis should be imparted in an appropriate environment.

3. After each consultation, a brief summary of what has been disclosed should be written in the case notes.

4. The patients and their relatives should be given sufficient information to allow them to have a clear enough idea of the prognosis in order to be able to weigh up for themselves the relative advantages and disadvantages of treatment.

Deciding on appropriate treatment plan

Depending on the clinical details, radiology and the histopathology, a treatment plan which is optimum for the patient needs to be devised. This may include surgical resection, radiotherapy and chemotherapy in various combinations. Apart from these primary interventions, a good care plan will need to consider support, advice on rehabilitation for disability and when appropriate, palliative care. This will involve health professionals other than doctors alone.

Management team

Glioma management is a team effort. It often starts from the general practitioner who refers the patient to the specialist (physician, paediatrician, general surgeon, neurologist or neurosurgeon). Later, radiologist, pathologist, oncologist, anaestheologist, psychiatrist and psychologist join in. Other health professionals like physiotherapist, nurses (especially trained nurses), social workers are also indispensable members of the team. It is also be emphasized that the liaisoning with patient's general practitioner should be maintained through and through as ultimately during terminal care, his active participation will be of great help for the patient and the relatives. Although often forgotten, the relatives of the patient are also important members of the management team and should adequately be communicated with and taken into confidence.

II. SPECIFIC ISSUES IN SYMPTOMATIC MEDICAL MANAGEMENT

1. Management of Peritumoral edema

Peritumoral edema is defined as an area surrounding a tumor that contains an increase in plasma infiltrate in the extracellular space. Typically, the edema is present immediately surrounding the tumor, but at times it can also be seen distant to the tumor, seeming to follow the white matter tracts. Peritumoral edema is classified as vasogenic as the basic abnormality is an abnormal blood brain barrier accounting for the acumulatation of the water and protein in the interstitial space. However, necrosis from ischaemia which frequently occurs in malignant gliomas may also add a cytotoxic component to it.

The functional breakdown of blood brain barrier, possibly results from the structural changes in the tumor miocrovasculature. Important changes have been documented in the endothelium and includes increased pinocytosis, trapping of fluid by long microvillous processes, and attenuation and fenestration of endothelial cells. Changes in the tight junctions do not play a role. Various mechanisms have been proposed which bring about these changes. Del Maestro et al, have summarized it into four dynamic processes:

1. Permeability secondary to tumor angiogenesis.
2. Permeability due to factors secreted by tumors.
3. Immunologic mechanisms.
4. Inflammatory processes.

With the background of above information, the following treatment options have been found to be useful in the management of peritumoral edema:

A. Corticosteroids

Mechanisms of action

(1) Alters vascular permeability (2) Has a possible cytotoxic effect (3) Decreases rate of tumor formation (4) Decreases CSF formation. Interestingly, one study showed a mean reduction of edema volume by 30% and tumor volume by 15% when using Methyl-prednisolone in a variety of tumor types.[4]

Type of corticosteroids, Dose and Response

Usually parenteral preparations are used. Dexamethasone is the preferred drug. A typical starting dose is 16 mg/day in divided doses. However, smaller doses have also been found to be useful. Typically the response occurs within 24-48 hours following which the drug should be to slowly tapered to the minimum possible level. If no significant response occurs, efficacy can be attained by doubling the dose and redoubling it every 48 hours upto a dose of 100 mg/day.

Prophylactic role of H-2 antagonists with steroids

Although H-2 antagonists are often prescribed prophylactically along with steroids to prevent gastrointestinal irritation and haemorrhage, scientific benefit for such a treatment is lacking.

B. Mannitol

Mechanisms of action

(1) Osmotic effect (2) Rheologic effects (3) Diuretic effect. The osmotic effect pulls free water out of the extracellular space and into the vasculature. The rheologic effects allow for the increased blood cell deformity and better oxygen delivery to tissue as well as for dilution of hemoglobin and fibrinogen, which lowers the viscosity of blood. This allows for less total cerebral volume, yet still meets the nutritional needs of the brain. Less blood volume lowers the intracranial pressure. Mannitol is also a diuretic that will cause an eventual negative fluid balance and dehydration of the brain if urine fluid loss is not replenished.

Dose

A typical dose of mannitol is 0.25 to 1.0 g/kg every 6-12 hours. The osmolality should be 310-320 mOsm and $2/3^{rd}$ of the diuresed volume should be replaced.

Rebound effect of Mannitol

In brain tumors the blood brain barrier is not intact in certain regions. Over time mannitol can accumulate in extracellular space causing an osmotic gradient that draws water into the extracellular space. A second possible cause of rebound may be related to dehydration and lowering of systemic blood pressure which in turn may critically lower the cerebral perfusion pressure. This leads to vasodilatation which increases the blood volume in the intracranial cavity and increases intracranial pressure.

Indication

In geneal, mannitol should not be considered as a form of definitive treatment of peritumoral brain edema but, it is readily effective in patients with critically elevated intracranial pressure as a short term management.

C. Furesemide

It is mainly used as an adjunct and has an additive effect with dexamethasone. Its main mechanism of action is secondary to its dehydration effects although a possible role in decreased CSF production has also been suggested. A typical dose is 10 to 20 mg every 6-12 hours.

D. Other agents:

Carbonic anhydrase inhibitors like **Acetazolamide**, hyperosmotic agent like **Glycerol** have also been tried with variable success. The effects are modest, slow to develop, show tolerence (Acetazolzmide) and have side effects like gastric irritation, acidosis. Nonsteroidal anti-inflammatory agents (NSAIDs) and Lipid peroxidation inhibitors have not been found to be useful.

Management of Coexisting clinical conditions which have bearing on Peritumoral edema

1. Fluid and Electrolyte imbalance

Overhydration and hyponatremia can lead to increase in peritumoal edema and cause clinical deterioration. These need to be identified and corrected.

2. Hypertension

It can exacerbate peritumoral edema as the hydrostatic pressure in the capillary bed is increased and hence the rate of formation of edema. On the other hand lowering blood pressure precipitously can decrease the cerebral perfusion pressure leading to clinical deterioration. Hence a proper balance must be maintained.

3. Seizures

There occurs increase metabolic need following seizures which increases the blood flow and hence an increase in peritumoral edema. Therefore, seizures need to be properly controlled.

2. MANAGEMENT OF SEIZURES

Seizures in the context of gliomas can be viewed from different angles. In general, seizures as presenting symptom occur in approximately 1/3rd of gliomas. They are present in 50-70% of cases at some stage of the disease. Of these, about 50% of these cases have focal seizures and the rest have generalized seizures. Conversely, 10-20% of adult patients with new onset seizures have brain tumors. Also, among the patients who undergo resective surgery to treat intractable seizures, tumors are found in 10-20% of adults and in 25-46% of the children. Slow growing low grade gliomas are particularly likely to cause seizures whereas glioblastomas have a lower frequency of seizures, possibly because the decreased life span allows less time for development of epileptic focus. Contrarily, patients with malignant gliomas who present with seizure tend to have a better prognosis; possibly these tumors are diagnosed at an earlier stage than tumors presenting with a mass effect or focal deficits. Location of the tumors also affects the frequency of seizures. Supratentorial tumors are far more likely to cause seizures than infratentorial ones.

Use of Anticonvulsants

Management of patients with gliomas who present with seizures involves use of standard anticonvulsants in the appropriate loading followed by maintenance dose. All the four standard antiepileptics, namely Phenytoin, Phenobarbital, Carbamazepine and Sodium Valproate have all been found to be useful in this setting. However side effects may limit their utility. It has been reported that glioma patients receiving phenytoin and cranial irradiation, a morbilliform rash develop in 15-20% cases out of which a small number may develop Stevens-Johnson syndrome. Similarly, patients on Carbamzzepine may develop shoulder-hand syndrome. Futher, drug interactions may occur like phenytoin may decrease the level of corticosteroids (hepatic induction), chemotherapeutic agent carmustine may decrease the phenytoin level leading to breakthrough seizures. Newer agents which have been tried to obviate some of these side effects include Lamotrigine, Gabapentine, Clobazam and Topiramate.

Role of prophylactic medication?

This is a controversial issue. Seizure risk in patients with infratentorial gliomas is small enough to warrant prophylactic antiepileptics. However, in supratentorial gliomas, many surgeons give prophylactic antiepileptics as they are perceived to be at a higher risk from seizures. Scientific evidence to support this view however is lacking. It has also been suggested that chances of seizures increase after craniotomy and so prophylactic treatment should be given. There is also no evidence in favor of such an approach.

III. ASSESSMENT OF FUNCTIONAL STATUS

Before progressing any further, a word or two about the functional status is important. As mentioned previously, the management plan for a patient of glioma depends on various factors. The major ones are age, tumor histology and grade, tumor location and functional status. Functional status essentially reflects the effects of the disease on the activities of daily living. A patient with good score on functional status will not only have a better prognosis but also may be a potential candidate for more aggressive therapy as compared to a patient with low functional score. Functional status assessment thus helps in prognostication, plan the treatment, follow the progression of the disease and test the efficacy of a particular treatment. Although many scales for functional status assessment are available, the most commonly used one is Karnofsky's Performance Status (KPS). It is a 0 to 100% scale denoting death and normality at two ends respectively.

IV. PREOPERATIVE ASSESSMENT AND MANAGEMENT OF THE PATIENT

Surgical resection is one of the commonly employed method in the management of malignant gliomas. Before the actual operation, preoperative assessment and preparation of the patient is very important. Briefly, these are as follows:[5]

1. **General Medical status assessment to exclude complicating medical conditions**
    i) Cardiovascular Disorders (CAD, HT).
    ii) Respiratory Disorders (COPD, Asthma)
    iii) Renal Disorders (Chronic renal failure)
    iv) Hepatic Disorders (Hepatitis and Cirrrhosis)
    v) Endocrine Disorders (Diabetes , Hypothyroidism)
    vi) Neurologic Disorders (TIA, Stroke)
    vii) Haematologic Disorders (Coagulopathies)
    viii) Genetic/ Metabolic Disorders (Previous anaesthetic reactions, Pseudo-cholinesterase deficiency)

2. **Evaluation of patient's medications**
    i) Aspirin
    ii) Warfarin
    iii) Antiepileptics

3. **Psychological and Emotional support**

V. IMMEDIATE POST-OPERATIVE MANAGEMENT

Immediate post-operative period is very critical and needs intensive monitoring. Few issues in the context of glioma surgery merit slight elaboration. These are:

1. Intracranial pressure monitoring

Patients with large amounts of preoperative brain edema surrounding the supratentorial tumors should be considered particularly at risk for developing intraoperative intracranial hypertension. In fact these cases should be preoperatively worked up thoroughly by the management strategies already outlined. Intraoperative management should be aimed at maintaining the patient's CPP above 60 mm Hg. An unexpected or sudden bulging of brain through the craniotomy should alert the surgeon to the possibility of intracranial haemorrhage. Pharmacologic and ventilatory therapy are the mainstay in managing raised intracranial pressure intraoperatively.

In the post operative period, intracranial pressure monitoring has been found to be useful and advantageous in two studies in 80's.[6,7] Further. These were also found to be almost risk free. The case therefore may exist for routine use of ICP monitoring following glioma surgery; current literature however is insufficient to provide any rigid guideline.

2. Fluid and Electrolytes

The normal adult fluid requirement is approximately 35 ml/kg/day. In the immediate postoperative period, consideration of fluid replacement should include intraoperative fluid losses. Given the normal renal functions, the adult patients require about 75 mEq of sodium, 40 mEq of potassium and 150 gm of glucose per day. The most common electrolyte problem in post craniotomy patients is overhydration during surgery.

3. Diabetes Insipidus

This condition manifests as polyuria with secondary polydipsia. It is commonly seen after neurohypophyseal injury but can also occur with hypothalamic tumors. Treatment with fluid replacement or by vasopressin therapy is usually effective.

4. Syndrome of Inapproriate Secretion of Antidiuretic hormone

The syndrome consists of excessive water retention and hyponatremia in the face of continuing high renal sodium excretion. Treatment is by fluid restriction.

5. Use of Antibiotics

A number of well controlled, randomized clinical trials have demonstrated that antibiotic prophylaxis is effective in reducing the risk of infection following elective neurosurgical procedures.[8,9] The relative risk of infection is 5.6 times greater without antibiotic prophylaxis than in the antibiotic treated group.[8,9] No single antibiotic regimen has been demonstrated to be superior to another. The choice of antibiotic must be based on knowledge of the common infecting organisms at a given institution.

6. Venous Thromboembolism

Glioma patients are particularly prone to venous thromboembolism. The incidence of deep venous thrombosis (DVT) and pulmonary embolism (PE) are striking when subclinical events are considered. It is 40% or DVT and 18% for PE.[10] The reasons for such predisposition are unknown. In particular, supratentorial surgery, malignant gliomas and paraperesis and hemiparesis have been found to be significant risk factors for thronboembolic events. Therefore, thromboembolism prophylaxis should be considered in all glioma patients. Low dose heparin or currently available low molecular weight heparins can be used safely in this setting.

7. Post operative seizures

A number of supratentorial glioma patients may have de-novo seizures in the immediate postoperative period. Hence, prophylactic use of antiepileptics are advocated by many. A recent study however, found a high correlation between the incidence of seizures and CT findings of intracerebral haemorrhage, cerebral edema, and cerebral infarction, suggesting that major operative complications, not the tumor had a role in triggering postoperative seizures.[11]

VI. TREATMENT INDUCED COMPLICATIONS

Whatever the modality of treatment that is adopted for a glioma patient, each has it's own share of complications. Detailed discussion on these complications is beyond the scope of the present article. These have been specifically referred to by individual authors in the respective chapters on surgery, radiotherapy and chemotherapy. Complications related to the use of anticonvulsants, corticosteroids and decongestants have already been mentioned. The complication which has received major attention in the last decade is the cognitive deficits seen in relatively younger glioma patients who are long time survivors. Possible role of combined insults resulting from surgery, radiotherapy and/or chemotherapy have been highlighted. Among these patients, it is difficult to define the costs in terms of quality of life for achieving increased duration of survival.

VII. FOLLOW UP STRATEGIES AND MANAGEMENT ISSUES

The model of continuing care depends largely on local factors and the prevalent practices. Some of the important guidelines which have been proposed are:[3]

1. Palliative cancer care model

The mainstay of the management of patients with malignant glioma is a palliative model of cancer care. Fragmented care with separate visits to neurologists, neurosurgeons and oncologist is best avoided since such traditional outpatient follow up for cancer patients has little proven value. Both treatment and care should instead be planned in a coordinated fashion. Further, rehabilitation, support and counseling to the patients and their relatives must be offered early in the course of the disease and the treatment. The provision of hospital and community services should be formulated and monitored through multidisciplinary meetings.

2. The management of Transitions from hospital to community setting

Involvement of the patient's general practitioner, the role of primary health care team for continuity of care and the needs of relatives or carers are the focus of attention at this stage.

3. Issues related to the Terminal care

This is discussed in the next section.

SECTION 2

I. TERMINAL CARE

1. Concept

The terminal phase can be defined as the last few days prior to death in a cancer patient with progressive incurable disease for whom no further specific anticancer treatment is possible or appropriate.[12] It is not always easy to know when a patient has entered the terminal phase. Some patient with advanced disease may appear to be dying, but may infact have an easily treatable, reversible cause for their deterioration, for example, hypercalcemia or uremia. Conversely, it is sometimes clear to patients, their families and to other healthcare professionals that an individual is dying while medical staff are unable to recognize or to admit this fact and alter the goals of therapy.

2. Importance

Most of the terminally ill glioma patients die in hospital. The care of terminally ill patients often falls to junior doctors who have had little specific training in the care of the dying. If the management of terminal phase is poor, it can be a source of great distress for patients (who may suffer unnecessary discomfort), for their families (who may have more difficulty in coming terms with their bereavement) and healthcare professionals (who may feel that they have failed in their duty of care). Further, it is to be noted that expertise in caring for dying patients has developed in hospices and specialist palliative care teams. Although, trials in such settings are difficult to conduct to find out about the effectiveness of such endeavors, a recent systematic review concluded that specialist palliative care teams can improve the outcome of cancer patients.

3. Duration and prediction of survival time

To plan for terminal care, it becomes important to predict the survival time and define the terminal period which may vary from country to country. Clinical prediction of survival for patients in the advanced stages of cancer has therapeutic and psychological implications for the patient and the family, and it also influences the use of health and social resources. Besides clinical prognosis including KPS score, asthenia and many Quality of Life Indices have been found to be independent predictors of survival time.

4. Altering the goals of therapy

The recognition that an individual is dying should lead to a change of emphasis in their care. The futility of any further anticancer therapy should be acknowledged and the comfort of the patient should become paramount. Investigations and interventions should be kept to a minimum. High quality nursing care is essential. Routine observations, such as temparature, pulse and blood pressure, should be discontinued, but regular patient contact is still required in order to monitor any pain or discomfort and to keep relatives and carers informed of developments. Treatments should be reviewed and many can be discontinued. The question of continuing parenteral administration of fluids to dying patients unable to drink is a controversial issue. Some advocate its continuation while others prefer to withhold it. Infact, they argue that a degree of dehydration may well be beneficial in reducing edema, pulmonary congestion, gut secretions and the need for toileting.

5. Alternative routes of drug administration

As it becomes increasingly difficult for patients to take essential medications by mouth, alternative routes of administration must be considered. The intravenous route is generally avoided. Rectal, transdermal and subcutaneous routes are increasingly employed.

6. Communication

Good communication and interdisciplinary team working are an essential part of ensuring that the terminal phase is managed well. If a patient is still conscious as death approaches, it may be appropriate to involve him/her in the decision making. More often than not, the patient will be unable to participate in discussions about their prognosis or management and regular liaison with the family/carers must be established or maintained.

II. MANAGEMENT OF COMMON SYMPTOMS DURING TERMINAL CARE

1. Pain

It is the most common and most feared symptom. Opiods can be used liberally. Diamorphine can be given as sub-cutaneous infusions and is quite effective. Additional doses for Breakthrough analgesia can also be given. NSAIDs are generally avoided. In rarer cases sedation can also be used.

2. Terminal restlessness

It encompasses the agitation, anxiety, fear, mental distress, general discontent and unease. Sedation is often used for symptomatic relief. Midazolam, levomepromazine, haloperidol are commonly used for this purpose.

3. Dyspnoea

It is a difficult symptom to treat and becomes increasingly severe during the terminal stages. Opioids are effective in the palliation of dyspnoea.

4. Noisy secretions

As death approaches, it is not uncommon for patients to lose the ability to clear secretions from their upper airways. This can lead to noisy and labored respirations that are often referred to as the "death rattles". Antimuscarinic agents to dry the secretions are often used like hyiscine butylbromide, glycopyrronium.

5. Nausea/Vomiting

Haloperidol, levomepromazine and cyclizine are the agents commonly employed for controlling nausea and vomiting.

6. Convulsions

These are rarely a problem in the terminal phase. Phenytoin and sodium valproate have relatively long half lives such that drug levels may remain therapeutic for some time after the cessation of oral drugs. Thereafter, seizures can be controlled by rectal diazepam.

III. ETHICAL ISSUES IN TERMINAL CARE

There are many ethical issues involving terminal care of cancer patients. Two issues, however, stand out. First is the desire of hastened death which has been separately dealt with in this book. The second is issue of determinants of place of death.[13] Approximately 2/3rd of cancer patients, when asked about the preferred place of death, wish to die in their own homes. However, majority die in the hospitals. To maintain dignity, terminal cancer patients must be able to do things in their own way, to make their own decisions and preside over their own dying. By making the decision about their own preferred place of death, terminal cancer patient can remain much more in control of their own death. This can alleviate concerns about the unknown process of dying.

CONCLUSIONS

The following concluding points can be generated from the above discussion:

1. Glioma management is a team approach. Contribution of each member of the team is essential for optimum outcome.
2. Goals of management should be set first before any action is undertaken.
3. Goals should be realistic.
4. Evaluation of performance status is very important for the management plan and setting a realistic goal.
5. There should be a strong emphasis on palliative care model.
6. Terminal care is not a terminal concept. It should be planned early and patient given opportunity to choose from various options.
7. Honesty, understanding, compassion and above all the humanity of doctor are the key to the successful management of this difficult condition.

REFERENCES:

1. Greig NH, Ries LG, Yancik R, Rapoport SI: Increasing annual incidence of primary malignant brain tumors in the elderly [see coments]. *J Natl Cancer Inst* 1990;82:1621-4.
2. OPCS, Cancer Statistics- Registration (1998). Series MB1 1993; No. 21 (HMSO).
3. Davies E, Hopkins A: Good practice in the management of adults with malignant cerebral glioma: clinical guidelines. *Br J Neurosurgery* 1997;11:318-30.
4. Leiguarda R, Sierra J, Pardal C, et al: Effect of large doses of methyl-prednisolone on supratentorial intracranial tumors: A clinical and CAT scan evaluation. *Eur Neurol* 1985;24:23-32.
5. Wen PY, Black PM: Clinical presentation, evaluation, and preoperative preparation of the patient. In: Mitchel S. Berger and Charles B. Wilson Eds, *The Gliomas*, Philadelphia;W.B. Saunders 1999; 328-43.
6. Constantini S, Cotev S, Rappaport ZH, et al: Intracranial pressure monitoring after elective intracranial surgery: A retrospective study of 514 consecutive patients. *J Neurosurg* 1988;69:540-44.
7. Pappada G, Formaggio G, Regalia F, et al: Course of intracranial pressure after extirpation of posterior fossa tumors. *Acta Neurochir* 1984;70:11-19.
8. Mollman HD, Haines SJ: Risk factors for postoperative neurosurgical wound infections: A case control study. *J Neurosurg* 1986;64:902-6.
9. Haines SJ: Antibiotic prophylaxis in neurosurgery: The controlled trials. *Neurosurg Clin North Am* 1992;3:355-58.
10. Muchmore JH, Dunlap JN, Culicchia F, et al: Deep venous thrombophlebitis and pulmonary embolism in patients with malignant gliomas. *South Med J* 1989;82:1352-6.
11. Fukamachi A, Koizumi H, Nukui H: Immediate postoperative seizures: Incidence and computed tomographic findings. *Surg Neurol* 1985;24:671-76.
12. Stone P, Rees E, Hardy JR: End of life care in patients with malignant disease. *European Journal of Cancer* 2001;37:1070-75.
13. Tang ST, McCorkle R: Determinants of place of death for terminal cancer patients. *Cancer Investigation* 2001;19:165-80.

43

Neuropsychiatric Aspects of Brain Tumors: A Focus on Glioma

N Ahuja, K Arora, RC Jiloha

INTRODUCTION

Psychiatric symptoms are the initial and occasionally the only symptoms of intracranial tumors (Davidoff 1930; Andersson 1970). Such symptoms may precede the more obvious motor or sensory symptoms by days or even months. Despite this general consensus, patients with gliomas (especially low grade) may still mistakenly carry the diagnosis of a primary psychiatric disorder months or years before the discovery of the tumor (Henry 1932).

This has led to the controversy regarding whether or not to use neuroimaging in all or most of patients presenting with primarily psychiatric manifestations. Most clinicians by now advocate CT or MRI scanning in older patients with a new occurrence of neurobehavioral symptoms or signs and patients of any age who have these features accompanied by headache, nausea and vomiting, papilledema, seizures or focal deficits.

Keeping this in view, it becomes important to know the possible clinical indicators of an underlying glioma. Compounding this is the fact that the behavioral changes occurring are due to the fluctuating interplay of several forces that include psychiatric symptoms directly caused by the glioma, psychiatric symptoms occurring as an epileptic disorder, and psychiatric symptoms as a response to the resultant cerebral loss.

Besides occurring as initial or ongoing manifestations of gliomas, psychiatric symptoms may also be encountered post diagnosis, as a reaction to the revelation of having the tumor (adjustment disorders), due to the raised intracranial pressure as the tumor progresses, and even as a side-effect of chemotherapy, radiotherapy or surgical intervention. These factors have made the scrutiny and assessment of the neuropsychiatric aspects of gliomas increasingly relevant and even essential.

This chapter aims to highlight the salient features that might help a psychiatrist to assess when a neurosurgical referral may be relevant and a neurosurgeon to decide when to send a patient of a glioma to a psychiatrist.

COMMON CLINICAL PRESENTATIONS OF GLIOMA

A. HEADACHE

At least a third of all glioma patients will present with headache as the initial manifestation. The number increases to more than two thirds at a later stage of the tumor. There are certain characteristics of the headache that may alert the clinician to the possibility of an underlying tumor. It is generally a deep, non-pulsatile nature, with a typical worsening in the morning and a history of comparative relief as the day progresses. Headache of such a variety present immediately on awakening should be taken as an important clinical indicator. There might be history of worsening of the headache with exertion or coughing and bending over and kneeling. In the later stages, projectile vomiting may be associated with headache, along with pain in the neck and blurring of vision.

B. SEAIZURES

Epileptic seizures constitute a well-recognized early symptom of gliomas, where seizures may precede other symptoms by several years. Seizures are known to occur in 50% of patients, with seizures occurring as the presenting symptom in about 33%.

Seizures in adults, with onset after 30 years of age, serve as an important indicator of brain tumor (Mulder & Daly 1952). The possible offered explanations for seizures include mass effects, tumor localization, tumor infiltration and raised ICT. The glia are known to be involved in the uptake of glutamate and GABA. Thus interference of the tumor with the normal mechanisms of GABA/glutamate uptake and metabolism in the surrounding cortex has been suggested as one possible cause, as gliomas associated with epilepsy have higher concentrations of glutamine (Bateman et al 1988).

For the treatment of the seizures, whether they are initial manifestations or consequent to surgery, anticonvulsants are routinely prescribed. In the presence of mood or cognitive symptoms, anticonvulsants like carbamazepine or valproate may be better choices.

C. DEPRESSION

The depressive disorders are quite common in patients with gliomas and can occur due to various reasons. They may be due to a direct effect of the tumor (right sided tumors more often than the left hemispheric tumors), on a less focal basis in obstructive hydrocephalous due to the tumor, as part of an adjustment disorder after the patient learns of his condition, as a side effect of the treatment method (chemotherapeutic, radiation or surgical procedure). A vague anxiety and unease has been reported for long as the most common initial presentation of a tumor. Prominent anxiety features, including panic attacks, may occur as part of the depressive syndrome or as stand-alone features.

The existent myth that someone with a brain tumor would "naturally" be depressed and would remain so if the tumor remains is just that...a myth. Depressive disorder due to any of the above outlined causes is eminently treatable and in fact, markedly improves the QOL of the patient.

While evaluating a patient with glioma, it should be kept in mind that an initial phase of helplessness, amotivation, and sadness may occur after revelation of the presence of tumor but it need not be a "disordered" state. Teaching stress management and early detection of syndromal depression may help immensely toward a better outlook on treatment and also better compliance.

An important issue in the evaluation would be the differentiation of depression from the apathetic state of frontal lobe involvement. The quality of sad mood and "endogenous" depressive features (biological alterations of function, diurnal variation in mood, etc) are the discerning factors.

The treatment issues include prescribing drugs that have least interactions with ongoing primary medication(s) and have the least likelihood of inducing delirium and nausea. Cognitive behavior therapy (CBT) has been found immensely useful in such patients.

D. DELIRIUM

An acute or chronic confusional state with impaired arousal, attention and concentration is an event that may occur at any stage of the glioma. It may occur as an initial manifestation, especially with a fast growing cerebral hemisphere glioma or as a result of the raised ICT. The mass effect with neuronal loss and post surgical damage of neuronal mass may also cause such a deficit. Delirium can also co-occur with ongoing radiotherapy and/or chemotherapy and as a secondary metabolic complication of a chronic illness. It appears to be the final common pathway of varied pathophysiology that may or may not be directly attributable to the glioma. Whatever may be the cause, acute delirium needs immediate treatment of the primary cause, with simultaneous symptomatic treatment by administering low dose antipsychotics and by environmental manipulation.

E. COGNITIVE DYSFUNCTION

It is a well known fact that brain function can be altered before anatomic evidence of change occurs. The available data suggest that cognitive dysfunction is found in the majority of, if not all, patients with brain tumors (Scheibel et al 1996). Impairments due to the tumor itself will vary from patient to patient, varying with the site of the lesion. In contrast, the impairments due to treatment tend to be related to frontal-subcortical white matter dysfunction including deficits of information processing speed, frontal lobe executive functions, memory, sustained attention and bilateral motor coordination.

Clinically a patient would present with slight bewilderment, slowness in comprehension, loss of capacity to sustain continuous mental activity, persistent lack of application to daily tasks, slowing of thought processes, slowed reaction time, faulty insight, reduced range of mental activity and an indifference to common social practices.

The most commonly used tests for global cognitive decline (e.g. the MMSE) are insensitive tools when required to measure these cognitive deficits but researchers are coming up with more sensitive tools to detect initial symptomatology as also the residual deficits relating to the tumors. In fact recent studies suggest that cognitive function might be a unique prognostic factor in predicting survival in patients with recurrent malignant glioma (Meyers et al 2000). Cognitive dysfunction seems also to have a direct correlation with the QOL and the ability to carry out activities of daily living. This is important when decision-making regarding the type of treatment needs to be made. For instance, cognitive deficits tend to worsen with radiotherapy in older patients, in those who have received more than 30 months of radiation, have received whole brain rather than local radiation, and have received a high dose of radiation (versus a lower dose). Similarly, there seems to be a correlation between the lesion location, its histopathology and the neurosurgery undertaken and the resultant cognitive dysfunction.

Though data in this area is as yet tentative, there is awareness that cognitive dysfunction is an important dimension of gliomas and needs sensitive and appropriate tools for measurement, as it could provide one of the several end points in treatments of gliomas (Archibald et al 1994).

F. OTHER CLINICAL PRESENTATIONS

The important clinical presentations of gliomas include adjustment Disorders with anxiety and/or depression, accentuation of existing personality traits, a "shrinking ability to adapt to the environment", anticipatory grief, catastrophic reaction (when provoked) (Goldstein 1959), dementia (especially with corpus callosum or bilateral involvement (Selecki 1964)), mania (with involvement of orbito-frontal and basotemporal lobes, or diencephalon), panic and anxiety symptoms (with involvement of insula, claustrum, and parahippocampal gyrus) (Lilja & Salford 1997), organic catatonia (with tumors around the 4th ventricle and frontal lobe tumors (Neuman et al 1996; Ahuja 2000)), and schizophreniform psychosis (with involvement of hippocampus, uncus, amygdala, cingulate gyrus, frontal lobe and temporal lobe (Malamud 1967)).

SYMPTOMS RELATED TO LOCATION

A. HEMISPHERIC MANIFESTATIONS (LATERALIZATION)

Patients with tumors in the left hemisphere may have language deficits, difficulty with verbal learning and memory, problems with verbal reasoning and impaired right-sided motor dexterity. A higher incidence of depressive disorders (along with dementia) and psychotic symptoms is also known.

On the other hand, right hemispheric high-grade gliomas may produce visual perceptual difficulties, difficulty in facial recognition and defects of left sided motor dexterity. These tend to produce more depression (without dementia) and mania.

B. LOBAR MANIFESTATIONS (LOCALIZATION)

I. Frontal Lobe

Frontal lobe pathology often results in profound changes in personality and behavior, but at times large lesions can be accommodated with little clinical disturbance. This discrepancy has on occasion led to the delayed diagnosis of treatable pathology.

Psychiatric symptoms are the first to appear in over a third of those with anterior tumors. The other clinical symptoms besides psychiatric features are headaches and seizures. In those presenting with psychiatric features, there is usually a combination of features but broadly based on the localized function disruption, there are three categories or behavioral complexes.

1. *Orbito-frontal involvement* is associated with some degree of impulsivity, lack of inhibition, tendency to make puerile jokes with silly laughter (referred to as "*witzelsucht*"), and a lack of concern for their illness and planning for the future. Despite the marked behavioral alterations, the patient may have few or no neuropsychological deficits, with intact memory, language and visuo-spatial skills.

2. *Medial frontal involvement* is more likely to manifest as an inflexible attitude, apathy, abulia and impaired motivation. Ranging from specific attentional impairment, loss of initiative and limited gesturing to a state of akinetic mutism in larger lesions may be present. The syndrome of organic catatonia has been recognized with tumors affecting this part (Ahuja 2000).

3 *Frontal convexity tumors* may reveal motor programming deficits with abnormalities of sequential behavior including perseveration, impersistence, impaired executive function and abstraction. Impaired speech initiative and production may also be present. Depression frequently accompanies the frontal convexity syndrome.

Symptoms resembling these might be present in non-frontal lobe tumors as well because of the disruption of fronto-temporo-parietal circuits; therefore localization of the tumor according to theses clinical features is not reliable.

II. Temporal Lobe

Seizures are the most common symptom of a tumor in this location. Partial seizures, with olfactory and gustatory hallucinations, déjà vu, feeling of fear and pleasure may be the initial presentation of a temporal lobe glioma. The ability to recognize sounds or the source of sounds may be affected. Impaired visual recognition of objects and aphasias can also be associated features. Amnesia, emotional disturbances and disturbances of sense of self can also occur.

Inferior temporal or temporo-limbic involvement leads to prominent psychiatric symptoms. Both mania and depression can occur, more often with non-dominant lesions as compared with dominant lesions. Personality changes, explosive affect, impulsivity, delusions, auditory and visual hallucinations, and schizophreniform psychosis can occur. The psychotic features are more common with dominant as compared with non-dominant lesions.

Panic attacks and anxiety can occur with gliomas involving the insula, claustrum, and para-hippocampal gyrus. It is important to note that these symptoms responded to treatment in 7 out of 8 cases (Filley et al 1995).

III. Parietal Lobe

Sensory disturbances and partial seizures (sensory) are prominent with parietal lobe involvement. Pure psychiatric symptoms are however uncommon.

IV. Occipital Lobe

Gliomas affecting the occipital lobe are rather uncommon. Visual disturbances, including blindness (if tumor is located medially) may occur. Psychiatric symptoms are uncommon but may be present as visual hallucinations and visual agnosia (parieto-occipital junction), simultagnosia or color agnosia.

V. Brainstem

Brainstem gliomas are more common in children as compared to adults. These mostly arise in pons. In > 90% cases, cranial nerve palsy (VI or VII) occurs. If medulla is involved, uncontrollable hiccups and disturbances of cardiac and respiratory rate can occur. However, pure psychiatric symptoms are uncommon.

Organic catatonia has been described with tumors near the 4th ventricle (Neuman et al 1996; Ahuja 2000). Even small tumors here can cause death.

VI. Cerebellar

Cerebellum is a common site in children, e.g. medulloblastoma. Not only cerebellar signs but also cognitive dysfunction characterize the clinical picture of cerebellar gliomas (Dennis et al 1996). Other psychiatric symptoms are relatively uncommon and occur late. These are predominantly related to an increase in the ICT.

SYMPTOMS RELATED TO TREATMENT

A. PHARMACOLOGICAL TREATMENT

Most of the chemotherapeutic agents used for treatment of gliomas have side effects of psychosis, delirium and cognitive deficits. Steroids, administered for the edema and inflammation, are well known to cause hypomania, mania, depression, psychosis and delirium. The anticonvulsants also may cause psychosis and aggression, in addition to cognitive side effects.

B. RADIOTHERAPY

Till a few years ago, it was an accepted maxim that radiation therapy in glioma patients produced significant cognitive disturbances. Current researchers are instead trying to find the variables within the radiation therapy that may be responsible for the apparent increase in cognitive dysfunction post-irradiation.

The putative variables that may adversely affect the occurrence of dysfunction are:

1. Whole brain (versus focal) therapy,
2. High total dosage administered,
3. Longer duration of radiation exposure, and
4. Older age of patients.

One study concluded that in low-grade gliomas, focal radiotherapy had no impact on the neurological, functional, cognitive and affective status of their sample of patients (Taphoorn et al 1994).

C. NEUROSURGICAL TREATMENT

As the surgery of gliomas involves removal of brain tissue, along with resection of the tumor, it results in focal deficits mostly depending upon the location. One of the commonly seen manifestations is delirium in the post-operative period, which may be due to the trauma caused to the local tissue or to the raised intracranial pressure due to edema and inflammation.

OUTCOME MEASURES

The final objective of medical intervention of any variety is not just survival but restoration of health related *quality of life* (QOL). With this realization, the end points of clinical trials of cancer treatment are now being reconsidered. The traditional primary end points of therapy, namely survival, tumor response, time to progression and progress or disease free intervals are giving way to the concept of quality of life (QOL).

QUALITY OF LIFE (QOL)

QOL is a multifaceted concept that encompasses physical, occupational, psychosocial and spiritual aspects of a person's life and the influence of disease & treatment-related symptoms. As against the conventional view of rating the level of a disease or intensity of symptomatology, this measures the amount of normalcy attained or the level of well being.

QOL may be qualified as a "state of well being" having two components:

1. The physical, psychological and social components that enable performance of everyday activities, and

2. The patient satisfaction with the levels of functioning and control of disease, and treatment of its symptoms.

The concept of QOL affords several advantages when incorporated into clinical trials of glioma therapy. For instance, when comparing treatment modalities, most conventional measures give paramount importance to efficacy of treatments. QOL measures could provide clear evidence of differential toxicity that could determine the overall superiority of one of the regimen. Furthermore, glioma therapy being a still expensive option, QOL measures might go far in forming the cost-benefit analyses of competitive and new therapies. Lastly, certain QOL instruments and techniques (e.g. Q-TWIST, QALY) offer the opportunity to integrate the overall therapeutic experience by integrating QOL and efficacy into a single parameter. Consequently, many patients of gliomas consider QOL to be a paramount consideration when selecting treatment options (Gilbert et al 2000).

QOL TOOLS IN PATIENTS WITH BRAIN TUMORS

There are several tools available for assessing the general QOL as well as those that specifically target aspects related to the morbidity and treatment toxicities that are common in patients with malignant gliomas. An optimum approach would require the use of both of the above.

A. GENERAL QOL TOOLS

1. *WHO-QOL* and *QOL-Bref*: WHO Quality of Life Instruments.

2. *Karnofsky's Performance Score (KPS)*: This is a 100-point index of a person's physical performance capability. The values range from 0 (dead) to 100 (normal activity; no impairment). Although it has been widely used in the clinical trials of

patients with cancer, it does not address all the issues of quality of life, apart from the physical function. This is particularly seen in high-grade glioma patients, where typically the actual life quality decreases but the KPS scores are relatively maintained for a long period of time before the eventual death.

3. *Functional Living Index – Cancer (FLIC):* This is a subjective measurement instrument that has global, role, social, emotional, pain and nausea scales. It lacks the physical and cognitive components, though it incorporates a unique component of "hardship".

4. *Activities of Daily Living (ADL)*

5. *European Organization for Research and Treatment of Cancer Quality of Life Questionnaire-30 (EORTC QLC 30):* This is a 30 item, self assessment, questionnaire that incorporates five functional scales (physical, role, cognitive, emotional and social), three symptom scales (fatigue, pain and nausea/vomiting) and six single items (dyspnea, insomnia, anorexia, constipation, diarrhea and financial impact). The scores can be interpreted in terms of small, moderate, or large changes in the QOL. It has been used in a number of anticancer therapy trials. However, it does not address the issues specific to malignant gliomas.

B. QOL TOOLS SPECIFIC TO BRAIN TUMORS

1. *Brain Cancer Module:* This is an instrument designed to be used with the EORTC QLC-30 or other general questionnaires. It contains 4 multi-item scales (future uncertainty, visual disorder, motor dysfunction and communication deficit) and seven single items (headache, seizures, drowsiness, hair loss, aching, weakness of legs and difficulties with bladder control). It can be used quite effectively when a modified version is integrated with the QLC-30.

2. *Functional Assessment of Cancer Therapy-Brain (FACT-BR):* This is a subjective instrument designed to evaluate the QOL in patients with brain tumors. It is an effective tool when used in combination with general QOL measures.

3. *Preston Profile*: This brain tumor specific QOL instrument addresses the following different domains: physical, emotional and social functioning; relationship with family; tomorrow and the future; ongoing needs and limited neurological deficits. This does not evaluate cognitive impairments and has not undergone validity or reliability assessments.

4. *Q-TwiST:* Besides quantifying the cross sectional effects, approaches that integrate and quantify in time, the overall therapeutic experience, are now gaining respect. One such approach or method is the *Q-TWiST* (Quality Time without Symptoms or Toxicity), which examines the quality adjusted survival by dividing it into 3 components:

a. Time spent without symptoms or toxicity
   b. Time spent after relapse, and
   c. Time spent without toxicity.

Quantification as Quality adjusted life years (*QALY*) or by the *Q-TWiST* methodology can be used to compare treatments and to determine whether a new treatment will improve quality adjusted survival.

C. LIMITATIONS IN QOL INTERPRETATION

It is seen especially in brain tumors that the impact of the cancer and its treatments is related to several variables. For example, the reported decreases in QOL in a survivor of a glioma therapy trial may be related to the effect of the tumor itself, prior neurosurgical intervention, chemotherapy, radiation treatment, effects of simultaneous seizure medications, or even inability of cognitively impaired patients to discriminate among the choices in the QOL questionnaires. Some patients with high-grade gliomas may even be unable to respond to the questionnaires, verbally or in written.

FAMILY BURDEN

There has been virtually no research work in the important area of family burden or caregiver burden, unlike in the case of other disorders with fatal or serious outcome, e.g. Alzheimer's disease. Since outcome may be death in high-grade gliomas, it is of utmost importance to give professional psychological and psychiatric support and care to the family, and to prepare them to deal with the disease and the individual in a supportive environment.

Keeping in mind the natural course of malignant gliomas and their social conceptualization as "brain tumors", it is essential that the family and caregivers be given adequate information and participation in the management process. The caregivers need as much support and understanding as the patient during the process of diagnosing, investigating and managing the glioma and its myriad manifestations. It is also necessary to explore with them multiple options currently available regarding pain relieving drugs, chemotherapy, radiotherapy, possibilities of yet experimental treatments, alternative therapies and surgery, along with their possible side effects, consequences and financial implications for the family.

PSYCHIATRIC INTERVENTIONS

The possible management that can be provided by a psychiatrist depends on the manifestations of the glioma, the appearance of symptoms, the stage of illness and the cost benefit ratio. The various options available are:

A. SOMATIC TREATMENTS

I. Psychopharmacological Treatment

In the event of manifest psychiatric disorders or specific behavioral symptoms, the appropriate medications (antipsychotics, antidepressants, anti-anxiety drugs, mood stabilizers, cognition enhancers) need to be administered in coordination with the ongoing primary therapeutic measures. Caution about possible drug-drug interactions and masking of tumor is essential and close collaboration between the oncologist, neurosurgeon and psychiatrist is preferable.

II. Electroconvulsive Therapy (ECT)

The presence of brain tumor and increased intracranial pressure has long been considered an absolute contraindication to the electroconvulsive therapy. Recently, however, the American Psychiatric Association (APA) Task Force Report (1990) questioned the absolute nature of this contraindication and it appears that though further evidence needs to be accumulated, ECT may be considered a viable treatment option with appropriate modification, even in the presence of clinical evidence of increased intracranial pressure (Patkar et al 2000).

B. PSYCHOSOCIAL TREATMENTS

I. Counseling and Psychotherapy

The first step in the management of any potentially terminal or perceived fatal illness is free provision of information. Most patients are told less than they would like to know. Though it usually falls on the primary clinician to disseminate this information, reinforcement of available knowledge at various stages and discussing the fears and emotions of the patients and caregivers with a therapist may prove beneficial in decreasing stress due to the tumor.

Furthermore, it may help the patient to communicate his distress in a situation where he already knows the probable outcome but everyone shies away so he cannot discuss his fears (of pain, of indignity, or that his family will not cope). There may be many affairs for the patient to be put in order. Moreover, education about the illness and treatment(s) would enable him to judge whether the unpleasant therapy is worthwhile.

Another issue surfacing often in recent literature is that of the *survivor guilt*. While the survivor guilt is not experienced by everyone, and may vary a great deal in intensity, it appears to be a common experience. Caregivers can experience their own brand of *survivor guilt* for not being the one with the brain tumor. Patients suggest that it is easier to cope being the one with the tumor than the one who stands by, feeling helpless.

Pre-operative counseling goes far in balancing expectations from the procedure for both the patient and the family. Reduction of preoperative anxiety levels is known to decrease the chances of post-operative delirium and psychosis.

II. Stress Management and Enhancement of Coping Skills

This encompasses many aspects of an ongoing process:

1. It could help with managing the normal feelings of anxiety, sadness, anger, worry, or depression that can be associated with diagnosis and treatment.
2. *Addressing Communication Problems*: It would help in increasing constructive interaction and sharing of emotional burden within the family. For example, the patient may not know what to tell people about his illness, or how to explain cancer to his children and family. Therapy could provide help in dealing with feelings and changes in family life to accommodate a chronic illness.
3. *Crisis Intervention*: Techniques of problem solving; realizing the options available; and weighing, judging and then acting, are some methods that could go far in increasing the family's QOL. The various situations may warrant such a strategy (e.g. in response to acute symptoms, or a side effect of a therapeutic regimen, or emergent aggression or behavioral disturbances).
4. Teaching the patient specific techniques, such as relaxation techniques that may relieve the physical (nausea) or emotional (anxiety) side-effects of some of the treatment, can be very useful.

III. Group Therapy

The knowledge that one is not alone in suffering is probably the biggest stress reducer of all. The patients report that group therapy sessions, with interaction and reliving of experiences with others so afflicted, is an opportunity to gain inspiration and on the other hand feel worthwhile despite being diseased. So far, there is no scientific proof that a person's attitude will guarantee or influence survival. Certainly a positive attitude will affect the quality of the patient's life (QOL), since people who feel positive and hopeful are happier than those who feel hopeless.

IV. Supportive Psychotherapy

Probably the best thing that we can ever offer the caregivers and the patients is the assurance of the treating team's availability and accessibility, and a realistically hopeful outlook so that the patient is encouraged to live as near a normal life as possible. Numerous voluntary organizations provide support to specific type(s) of cancer patients and their families. Being in touch with such groups and being aware of the helplines, apart from the hospital-based teams, help decrease dependence and increase self-styled solutions.

V. Cognitive Retraining and Remediation

Available data suggests that though aggressive multi-modality treatments for high-grade gliomas have resulted in a larger number of long-term survivors, most long-term survivors have significant cognitive dysfunction. For such patients with cognitive dysfunction, either as an initial manifestation or as a consequence of treatment, cognitive remediation has been tried.

Cognitive remediation is based on the hypothesis that direct training of cognitions will generalize to broader measures of social performance and clinical improvement. The computer has been used variously as a rehabilitative instrument for direct remediation of memory, vigilance, reaction time, discrimination or concept formation. Repeated rehearsals and trials on these tasks, accompanied by positive reinforcement, do seem to show significant improvements in the cognitive tasks but the application of this to broader social, interpersonal and emotional areas that are affected in glioma patients is as yet tentative.

The other strategy is to directly apply cognitive therapy principles to treat emergent symptoms of gliomas like depression, psychosis, aggression and apathy.

CONCLUSIONS

Neuropsychiatric symptoms are a very common manifestation of the brain tumors, particularly gliomas. Fortunately, these neuropsychiatric symptoms are amenable to treatment and their successful treatment can lessen the burden on the care-givers and considerable improve the quality of life of the patients with gliomas.

In addition, the data from patients with brain tumors, in addition to the virtually parallel data from stroke literature, is an important source of information for the study of brain-behavior relationship. A better understanding of the brain-behavior relationship can in turn help us in improved treatment of the neuropsychiatric symptoms of the gliomas.

REFERENCES:

1. Alpers BJ: A note on the mental syndrome of the corpus callosum tumors. *J Nerv Ment Dis* 1936;84:621-27.
2. Andersson PG: Intracranial tumors in a psychiatric autopsy material. *Acta Psychiatr Scand* 1970;46:213-24.
3. Henry GW: Mental phenomena observed in cases of brain tumor. *Am J Psychiatry* 1932;12:415-63.
4. Mulder DW, Daly D: Psychiatric symptoms associated with lesions of temporal lobe. *JAMA* 1952;150:173-76.
5. Bateman DE, Hardy J, Mcdermott JR: Amino acid neurotransmitter levels in gliomas and their relationship to the incidence of epilepsy. *Neurol Res* 1988;10:112-14.
6. Scheibel RS, Meyers CA, Levin VA: Cognitive dysfunction following surgery for intra-cerebral glioma: Influence of histopathology, lesion location, and treatment. *J Neurooncol* 1996;30:61-69.
7. Meyers CA, Hess KR, Yung WK, Levin VA: Cognitive function as a predictor of survival in patients with recurrent malignant glioma. *J Clin Oncol* 2000;18:646-50.

8. Archibald YM, Lunn D, Ruttan LA, Macdonald DR: Cognitive functioning in long-term survivors of high-grade glioma. *J Neurosurg* 1994;80:247-53.
9. Goldstein K: Functional disturbances in brain damage. In: Arieti S. Ed. American Handbook of Psychiatry. Vol. I. New York: Basic Books, 1959.
10. Selecki BR: Cerebral mid-line tumors involving the corpus callosum among mental hospital patients. *Med J Aust* 1964;2:954-60.
11. Lilja A, Salford LG: Early Mental changes in patients with astrocytomas with special reference to anxiety and epilepsy. *Psychopathology* 1997;30:316-23.
12. Neuman E, Rancurel G, Lecrubier Y, Fohanno D, Boller F: Schizophreniform catatonia on 6 cases secondary to hydrocephalus with subthalamic mesencephalic tumor associated with hypodopaminergia. *Neuropsychobiology* 1996;34:76-81.
13. Ahuja N: Organic Catatonia: A Review. *Indian J Psychiatry* 2000;42:327-46.
14. Malamud N: Psychiatric disorder with intracranial tumors of limbic system. *Arch Neurol (Chic)* 1967;17:113-23.
15. Filley CM, Kleinschmidt-DeMasters BK: Neurobehavioral presentations of brain neoplasms. *West J Med* 1995;163:19-25.
16. Dennis M, Spiegler BJ, Hetherington CR, Greenberg ML: Neuropsychological sequele of the treatment of children with medulloblastoma. *J Neurooncol* 1996;29:91-101.
17. Taphoorn MJ, Schiphorst AK, Snoek FJ: Cognitive functions and quality of life in patients with low-grade gliomas: The impact of radiotherapy. *Ann Neurol* 1994;36:48-54.
18. Gilbert M, Armstrong T, Meyers C: Issues in assessing and interpreting quality of life in patients with malignant glioma. *Semin Oncol* 2000; 27 (Suppl 6): 20-26.
19. Weitzner MA, Meyers CA: Cognitive functioning and quality of life in malignant glioma patients: A review of the literature. *Psycho-Oncology* 1998;7:141-45.
20. Lyons GJ: The 'PRESTON Profile': The first disease-specific tool for assessing quality of life in patients with malignant glioma. *Disabil Rehabil* 1996;18:460-68.
21. American Psychiatric Association. The practice of ECT: Recommendations for treatment, training, and privileging. A Task Force Report of the APA. Washington DC: APPI, 1990.
22. Patkar AA, Hill KP, Weinstein SP, Schwartz SL: ECT in the presence of brain tumor and increased intracranial pressure: Evaluation and reduction of risk. *J ECT* 2000;16:189-97.

44

Tumor Markers in Neuro-Oncology– Diagnostic and Prognostic Significance in Recurrent Gliomas

V Santosh, SK Shankar, S Dubey, BA Chandramouli

INTRODUCTION

Despite recent advances in the diagnosis and surgical technique, radio or chemotherapy, the treatment and prognosis of many neuroepithelial tumors, particularly malignant gliomas remain dismal. Irrespective of several therapeutic modalitis employed for the treatment of many primary neuroepithelial tumors, recurrences are inevitable. In fact, the problem of recurrence is an area which is of greater dilemma to the neurosurgeon than the management of the initial tumor. This is often compounded by absence of broad universal guidelines for therapy.[1,2]

The duration of survival of patients with malignant neuroepithelial tumors, particularly glioblastomas and anaplastic astrocytomas who are treated by surgery, radiation and chemotherapy depends on the extent of tumor resection at the initial surgery. Patients who undergo gross total or subtotal removal of the tumor respond better to adjuvant therapy than those who have had limited resection or biopsy only. By extensive resection, the goal achieved is internal decompression, which slows the progression of the neurological deficit by removing the tumor bulk and necrotic debris from the lymphatic deficient brain. Maximal cytoreduction enhances the effect of subsequent treatment by reducing the tumor to a small number of cells that are vulnerable to radiation and chemotherapy. However, this does not appear to be true in several patients with gliomas in whom inspite of adequate debulking of the tumor, there is invariably recurrence. The chance of a long, recurrence free survival is also closely associated with the biology of the neoplasm. Significant indicators of anaplasia in gliomas include nuclear atypia, mitotic activity, increased cellularity, vascular proliferation and necrosis. For practical purposes, these characteristics are condensed into a grading scheme. Currently the malignancy scale of WHO classification and ST. Anne Mayo grading systems have proved to be highly reproducible and predictive of patient survival.[3,4] Patient survival also depends on a variety of clinical parameters such as patient's age, and condition as reflected by Karnofsky performance score, tumor location, extent of surgical resection, post-operative radio or chemotherapy.[5,6]

Tumor markers: By definition, tumor markers are biochemical indicators of the presence of a tumor and it's metabolic, proliferative and molecular genetic nature. They include cell surface antigens, cytoplasmic proteins, enzymes, hormones and can be detected in tissue sections, plasma and other body fluids. Diagnostic neuro oncology has benefited in the last two decades from the incorporation of newer techniques such as immunohistochemistry using various antibodies to tumor markers. The continual development of new, defined reagents to identify epitopes associated with stages of cell lineage, cell cycle, oncogene and suppressor gene products or cell activation, is helping to clarify the nature of cellular maturation, tumor differentiaton and predict the prognosis for the patient.[7,8] The development of these marker defining reagents has resulted not only in improving diagnostic capability but also in clarifying developmental pathways within and between cell lineages and thus has resulted in the revision of historical classifications.[8]

TUMOR MARKERS USED IN NEURO-ONCOLOGY

These include

1. Markers used to define cell lineages and cell differentiation such as intermediate filament proteins, neuroendocrine related markers, oncofetal antigens etc.

2. Markers used to define cell cycle status - which include nuclear proteins which are involved in cell proliferation.

3. Markers associated with the expression or action of oncogene or tumor suppressor genes such as growth factor receptors, p53 protein etc.

4. Markers used to identify cell senescence or cell death-detection of apoptosis.

In this review, the various tumors markers will be discussed with respect to their expressions in recurrent neuroepithelial tumors.

MARKERS OF CELL DIFFERENTIATION

Among the diverse proteins expressed by developing and mature cells in the nervous system, the family of Intermediate filament protein (IFP) subunits are of particular interest for tumor diagnosis. The IFPs are intracellular filaments measuring 10 nm in diameters and form a major part of the cytoskeleton.[9,10]

These IFPs are divided into several classes that include Nestin, Vimentin, GFAP, Neurofilament proteins, S-100 protein α-internexin, peripherin etc.

GLIAL DIFFERENTIATION AND GLIAL FIBRILLARY ACIDIC PROTEIN (GFAP) EXPRESSION IN RECURRENT GLIOMAS:

When patients with low grade, diffuse astrocytoma develop a recurrent lesion, the second surgical biopsy often shows histopathological evidence of increased nuclear atypia, hyperchromasia, mitotic activity and eventually vascular proliferation with or without necrosis, i.e. features of glioblastoma. The acquisition of anaplastic features, though an inherent property of diffuse astrocytomas, is largely unpredictable clinically and histopathologically, in particular with regards to the time over which these changes take place. While some astrocytomas show no change in histological grade over more than 10 years following the first operation, others show a rapid transition to malignancy within 1-2 years.[11]

In the recurrent tumor, anaplastic changes are usually diffuse, although it is often possible to distinguish within the surgical biopsy, areas with low and high grade features. Occasionally, the transition in the grade of histology is abrupt. A sharply delineated, often round focus of anaplastic tumor cells appears, with high mitotic activity and lack of GFAP expression. It is assumed that these foci represent new clones of neoplastic astrocytes with an additional genetic alteration.[11,12] GFAP expression by tumor cells characterises glial differentiation. This is a class III intermediate filament protein of 55k Da mol. weight, and a major cytoskeletal component of mature astrocytes, and hence a useful marker for astroglial cells,[8] although it can co-express with other intermediate filaments, particulary Vimentin and S-100 protein in several gliomas such as pilocytic astrocytomas.[13] Intensity of GFAP staining is inversely related to anaplastic changes in astrocytic neoplasms. The increased expression of GFAP is indicative of cellular differentiation. When there is progression of astrocytoma from low to high grade malignancy, as depicted in recurrent tumors, the cells tend to become more dedifferentiated and hence lose their GFAP expression.

CELL PROLIFERATION MARKERS IN RELATION TO RECURRENT GLIOMAS

A high cell proliferation rate is associated with aggressive clinical behaviour and therefore, determination of mitotic activity has been conventionally used to detect cycling cells in primary CNS tumors. Various labeling methods that determine indices related to tumor cell proliferation, like the immunohistochemical detection of the incorporation of the thymidine analogs, Bromodeoxyuridine (BrdU) and Iododeoxyuridine (IdU) or the expression of proliferation associated nuclear proteins such as Ki-67, PCNA/Cyclin or DNA Polymerase a have been investigated as adjuncts to classical histological grading of CNS neoplasms.[14]

Immunostaining for PCNA (proliferating cell nuclear antigen) and Ki-67 molecules, expressed in cycling cells are readily applicable tests for assessing proliferative potential of tumor cells. MIB-1 antibody which detects the Ki-67 antigen, expressed in all the phases of the cell cycle except the G_0 and the transition between G_0 and G_1 phase, is prognostically significant and the most informative proliferation parameter tested.[15] Ki-67 labeling index (L1) is conveniently expressed as the percentage of positive cells in defined tumor fields. The best method is to count in areas where high density of labelled tumor cell nuclei are found, which follows the principle of considering the most malignant part of the tumor for diagnostic and prognostic purposes. PCNA and MIB-1 L1 have been shown to increase with the grade of the tumor and reflect the clinical progression. However, several authors are of the opinion that a single kinetic determinant such as proliferation index alone cannot serve as the sole predictor of the complex cellular behaviours involved in neoplastic progression and therapeutic sensitivity.[16]

DETECTION OF TUMOR SUPPRESSOR GENE, ONCOGENE RELATED PROTEIN PRODUCTS IN CNS TUMOR

Some of the molecular genetic alterations underlying the malignant progression of gliomas have been analysed.[17] This is particularly with regard to the p53 suppressor gene, the epidermal growth factor receptor (EGFR), vascular endothelial growth factor (VEGF) expression, in relation to tumor cell proliferation and apoptosis.[18] Association of p53 gene alterations and mutant protein expression has been consistently observed with progression of gliomas while the EGFR gene amplification has been demonstrated mainly in glioblastomas. Some authors have shown that in a relatively benign tumor the extent of angiogenesis can predict the future malignant transformation of the tumor.[19]

The overexpressed and/or altered expression of proteins resulting from altered transcription of either oncogenes or tumor suppressor genes can be detected in the tumor tissues by immunohistochemistry.

ALTERED SUPPRESSOR GENE P53 PROTEIN LOCALISATION

The p53 protein is a 53-kDa nuclear phosphoprotein that controls the elements of cell cycle. It is encoded by a tumor suppressor gene localised to human chromosomes 17p, which is frequently mutated in glioblastomas and thought to be one of the earliest alterations in human astrocytoma progression.[20-23] The wild type p53 protein is localised in the nucleus in homo-dimeric or homo-tetrameric complexes,[23] which have a short half life, rapid turn over and negligible accumulation usually undetectable by immunohistochemistry.[24] On the other hand, mutated gene products generally have a slower turnover rate, resulting in nuclear and occasionally cytoplasmic accumulation, which is sufficiently stable to allow immunohistochemical localisation.[23,25] The assumption of a correlation between detectable p53 protein and TP53 gene mutation led to the development of several monoclonal antibodies (Mabs) and anti-p53 polyclonal antisera for ready immunohistochemical localisation of cells presumably expressing mutant p53 protein. The Mabs DO-1, DO-7,[26] and CM-1 are found to be the most reliable reagents for analysis of archival (routinely formalin-fixed and paraffin embedded) material.

Immunoreactivity for p53 protein is primarily associated with astrocytic cells. Nearly two-thirds of adult astrocytomas express p53 in the tumor cells and there is a general trend for increasing frequency of p53 protein immunoreactivity with increasing grade of malignancy. Well differentiated astrocytomas have been reported to have a lower incidence (18-46%), as compared with anaplastic astrocytomas (29-57%) and glioblastomas (49-70%).[27,28] This higher incidence of Tp53 gene mutation and protein accumulation, noted in high grade gliomas in contrast to the low grade gliomas suggests an additional role of p53 alterations in glioma progression.

Sideransky et al have noted that in recurrent gliomas, an increasing number of p53 immunoreactive cells in the recurred tumor as compared to the initial tumor (irrespective of the grade of astrocytomas).[20] They suggested that this could indicate the clonal expansion of p53 immunopositive and mutant tumor cells during the glioma progression. Hayashi et al[29] have reported in their series respectively that in a small number of cases, p53 protein expression was noted only in the recurrent tumor and not in the respective primary tumor. However, this observation was not replicated by other authors. Jain et al[30] have shown that p53 alterations are not critical for recurrence of gliomas and probably there are other underlying factors associated with tumor progression. Iuzzolino et al[31] have reported that in a long term follow up study of 52 patients with low grade astrocytomas, the mean survival times for p53 positive versus p53 negative cases were not significantly different, although a more aggressive course was seen in p53 immunoreactive tumors, 3-4 years after surgery. Therefore p53 protein expression appears to be of some value in the prognostic assessment of gliomas. Given the relative ease of the methodology versus standard molecular techniques, p53 immunostaining will continue to have a significant role in understanding the biology of gliomas.

ONCOGENES AND RELATED PRODUCTS OF IMPORTANCE IN ASTROCYTOMAS PROGRESSION AND RECURRENCE

The oncogenes are not found in normal cells, but are generated by the activation of their corresponding proto-oncogenes during tumor development. Proto oncogenes are normal cellular genes and their products are important components of intracellular signaling pathways involved in the growth cycle. Proto oncogenes can mutate and become oncogenes and keep the pathways continuously active when they should be otherwise still. The consequences may be-

(a) Overproduction of growth factors

(b) Flooding of cells with replication signals

(c) uncontrolled stimulation in the intermediary pathway

(d) Unrestricted cell growth driven by elevated levels of transcription factors.

ROLE OF GROWTH FACTORS IN TUMORIGENESIS AND GLIOMA PROGRESSION

Growth factors are polypeptides that are active at nanomolar concentration and mediate their effects via binding to cell surface receptors with a high degree of specificity. Most growth factors are Protein Tyrosine Kinases (PTK) involved in the process along with their Transmembrane Receptors (TMR). The TMR conveys the growth factor signal into cytoplasm and the enzyme activity of the intrinsic kinase domain is absolutely essential for signal transduction.

Growth factors are classified into four groups:

(1) Those that promote cell proliferation – such as Epidermal Growth Factor Receptor (EGFR) and its ligand EGF and Platelet Derived Growth Factor (PDGF).

(2) Those involved in glioma cell motility such as EGFR and Transforming Growth Factor α (TFG α)

(3) Those with inhibiting functions - such as TGF-β

(4) Those involved in tumor angiogenesis - such as Vascular endothelial Growth Factor (VEGF).

Amplification of EGFR and overexpression has been observed in 40-50% of high grade astrocytomas but not in low grade astrocytomas.[32] This is noted mainly in primary denovo glioblastoma. In the series of Reinfenberger,[17] none of the recurrent tumors revealed EGFR gene amplification, though others have shown this in glioblastomas which subsequently recurred.

TUMOR ANGIOGENESIS

Malignant glioma is one of the most vascularised human tumors and has been used as a model system for angiogenic factors. The growth of a solid tumor in the CNS requires that tumor cells be able to induce neovascularisation of the tumor. The vascular state of the tumor is the end result of interplay between a number of microvessel growth regulators. Among the angiogenic proteins, VEGF (Vascular endothelial growth factor) and βFGF (Fibroblast growth factor) are most commonly found in gliomas. VEGF brings about endothelial cell proliferation resulting in progression of the malignant grade of the tumor gliomas, particularly high grade gliomas produce VEGF. This growth factor exists in one of the four molecular species based on the number of aminoacids (VEGF-121, 165, 189, 206). VEGF-165 is the most predominant isoform secreted by variety of normal and transformed cells. This can be detected in the tumor tissue by immunohistochemical techniques. VEGF production is stimulated by low oxygen concentrations in the cultured cells and this is the presumed mechanism of its increased production in the presence of ischemia/necrosis, as in glioblastomas.[33]

Microvascular density has been found to be an important indicator of malignant behaviour of many different cancers. Microvessel density has been associated with tumor recurrence and mortality in a heterogenous group of brain tumors. There is a correlation between microvessel density, VEGF expression and malignant progression of gliomas and further survival of patients. Low-grade fibrillary astrocytomas with increased VEGF expression have been shown to recur and progress to higher grade tumors.[19]

ASTROCYTOMA INVASION

When glioma cells penetrate neural parenchyma both normal glial and neuronal cells are gradually displaced and compressed by tumor cells. Diffuse invasion, particularly into grey matter and leptomeninges is an inherent property of gliomas which turn malignant. The biological basis of astrocytoma invasion is a complex process, involving cell to cell and cell to extracellular matrix (ECM) interactions, tumor cell motility and remodelling of extracellular space by protease mediated degradation of ECM constituents.

The tumor invasion now is perceived as three staged process. The initial step requires receptor mediated adhesion of tumor cells to the matrix proteins; the second is the degradation of the matrix by tumor-secreted metalloproteinases. This process creates an intercellular space into which invading cells can move by an active mechanism that requires membrane synthesis, receptor turnover and rearrangement of cytoskeletal elements as the third step.[34] Astrocytoma cells produce several classes of matrix degrading proteases, such as serene proteinases, cysteine proteinases, aspartic proteinase and endoglycosidases that have been implicated to play a role in astrocytoma invasion. Identification of these specific metalloproteinases could be an early indicator for tumor progression, invasion and recurrence.

CELL CYCLE PROTEINS

The cell cycle is regulated by an orderly activation of complexes called cyclins and Cyclin Dependent Kinases (CDKs) accompanied by the simultaneous action of inhibitors (CDK Is) such as p16, p21 and pRb (retinoblastoma protein). These normally regulate DNA synthesis, spindle assembly, mitosis etc. Abrogation of cell cycle checkpoints, either by aberrant expression of positive regulators (cyclins, CDKs) or loss of negative regulators, (p16, p21 or pRb) can lead to tumorigenesis and tumor progression since the cell is no longer able to respond to important internal and external cues that check its growth. Overexpression of some of the cyclins, CDKs or presence of mutated suppressor gene proteins such as pRb can be detected by immunohistochemical techniques. These have a bearing on astrocytoma progression and recurrence.

DETECTION OF CELL DEATH

Necrosis and apoptosis represent the two major forms of cell death.[35] Necrosis is considered to be the end product of irreversible cell injury and thought to be a passive rather than an active pathological process. In contrast, apoptosis is an active form of cell death and is initiated as a result of transcription and/or translation of mediators of cell death process and plays an important role in many physiological and pathological processes. It has been implicated as a critical factor in controlling tumor progression[36] and has often been observed in human medulloblastomas.[37] Loss of spontaneous apoptosis and uncontrolled cell proliferation probably plays an important role in glioma progression leading on to recurrence of the tumor.

Recent advances in molecular biology have led to the development of enzyme mediated methods for 'in situ' labelling of the 'nicked' ends of fragmented DNA.[38] The in situ end labelling (ISEL) family of assays depends upon the use of enzymes such DNA polymerase or terminal deoxynucleotidyl transferase (tdt: TUNEL) to insert labelled nucleotides in the 3'-hydroxy termini of endonuclease induced DNA breaks. Labelling is then revealed by calorimetric method.

INDIAN LITERATURE

Indian literature on recurrent neuroepithelial tumors in limited. There are very few published series dealing with clinical and pathological aspects of recurrent neuroepithelial tumors. Tandon et al,[39] studied the clinicopathological aspects of patients with recurrent supratentorial gliomas in two hundred patients in whom follow-up was available for one to sixteen years. All patients were treated initially with radical excision of the tumor followed by radiotherapy. Even though recurrence occurred at a later stage in the relatively benign groups of tumors as compared to the more malignant gliomas, the period of recurrence could not be correlated to the histological grade or presence or absence of residual tumor in a post operative CT scan. Few other studies on p53 alterations in recurrent gliomas have been by Jain et al[30] and Ghosh et al.[40]

In order to evaluate some biological factors that govern recurrences, a preliminary study was conducted at NIMHANS, on a group of patients with recurrent neuroepithelial tumors. Various markers of cell differentiation (GFAP, synaptophysin), cell proliferation (MIB-1), oncogene and tumor suppressor gene products (EGFR, VEGF, p53 protein, Bcl-2, pRb protein), apoptosis (TUNEL method) were used. The findings in the initial and recurrent tumors were compared and an attempt was made to correlate the same with the duration of recurrence and patients survival characteristics. Further, the findings were compared to those found in a group of patients without any tumor recurrence for 10 years or more (work under publication).

In keeping with the Western literature, it was noted that all neuroepithelial tumors become more undifferentiated on recurrence and in more than 60% of cases, the grade of the tumor also became higher. MIB-1 labelling index (proliferative potential of tumor cells) seemed to be a strong predictor of outcome. It also helped to predict survival in patients with the same grade tumor and thus can be a strong indicator suggesting the need for close follow up or adjuvant therapy.

P53 protein expression was noted in nearly 90% of astrocytomas of all grades that recurred and a clonal expansion of mutant tumor cells expressing p53 protein was observed in the recurrent tumor. In contrast, less than 40% of the nonrecurrent astrocytomas revealed p53 protein expression. P53 protein expression in all grades of astrocytomas and EGFR overexpression in glioblastomas predicted early recurrences and worse prognosis. A combination of p53, pRb mutations and EGFR overexpression indicated the worst survival. In addition, VEGF expression in gliomas of all grades was also an important factor for predicting early recurrences.

Our study showed that p53 alterations along with other genetic changes, particulary EGFR overexpression, influence malignant progression and recurrence, probably by (1) Promoting tumor associated angiogenesis (2) Preventing tumor cells from entering the apoptotic pathway and (3) Allowing further mutations to occur.

The above review has focused mainly on diagnostic and prognostic significance of various tumor markers used in neurooncology and the utility of marker studies to understand some of the multistep mechanisms involved in the formation and progression of gliomas leading to recurrences. Such studies help to identify early prognostic markers that can serve as a tool to choose appropriate post-operative adjuvant therapy more aggressively in certain groups of tumors.

REFERENCES:

1. Harsh GR IV, Levin VA, Gutin PH, Seager M, Silver P, Wilson CB: Reoperation for recurrent glioblastoma and anaplastic astrocytoma. *Neurosurgery* 1987;21:615-21.
2. Muller W, Afra D, Schroder R: Supratentorial recurrences in gliomas. Morphological studies in relation to time intervals with astrocytomas. *Acta Neurochir* 1977;37:75-91.
3. Daumas Duport C, Scheithaur BW, U'Fallon J, Kelly P: Grading of astrocytomas. A simple and reproducible method. *Cancer* 1988;62:2152-65.
4. Kim TS, Halliday AL, Hedley White ET, Convery K: Correlates of survival and Daumas Duport grading system for astrocytomas. *J Neurosurg* 1991;74:27-37.
5. Sneed PK, Prados MD, McDermott MW, Larson DA, Malee MK, Lamborn KR, et al: Large effect of age on survival of patients with glioblastoma treated with radiotherapy and brachytherapy boost. *Neurosurgery* 1995;36:898-904.

6. Ammirati M, Vick N, Liao YL, Ciric I, Mikhael M: Effect of the extent of surgical resection on survival and quality of life in patients with supratentorial glioblastomas and anaplastic astrocytomas. *Neurosurgery* 1987;21:201-06.
7. Bonnin JM, Rubinstein LJ: Immunohistochemistry of central nervous system tumors. Its contribution to neurosurgical diagnosis. *J Neurosurg* 1984;60:1121-33.
8. Keihues P, Kiessling M and Janzer RC: Morphological markers in neurooncology. Current topics in Pathology. 1987;77:307-38.
9. Lazaridus E: Intermediate filaments: a chemically heterogenous developmentally regulated class of proteins. *Annal Review of Biochemistry* 1982;51:219-50.
10. Steiner PM, Roop DR: Molecular and cellular biology of intermediate filaments. *Annal Review of Biochemistry* 1988;57:593-625.
11. Cavenne WK, Bigner DD, NewComb EW, Paulus W, Kleihues P: Diffuse astrocytomas. In: Pathology and Genetics. Tumors of the Nervous System (Ed) Kleihues P and Cavenee WK. International Agency for Research on Cancer, Lyon 1997; 2-9.
12. Schmitt: Rapid anaplastic transformation in gliomas of adulthood. "Selection" in neurooncogenesis. *Pathol Res Pract* 1983;176:313-23.
13. Cutarelli PE, Roessmann VR, Miller RH, et al: Immunohistochemical properties of human optic nerve glioma. Evidence of type I astrocyte origin. *Investigative ophthalmology and visual science* 1991;32:2521-24.
14. Onda K, Davis RL, Wilson CB, Hoshino T: Regional differences in bromodeoxyuridine uptake, expression of Ki-67 protein and nucleolar organiser region counts in glioblastoma multiforme. *Acta Neuropathologica* 1994;87:586-93.
15. Sallinen PK, Haapasalo MK, Visakorpi T, et al: Prognostication of astrocytoma patient survival by Ki-67 (MIB-1) PCNA and S-phase fraction using archival paraffin embedded samples. *Journal of Pathology* 1994;174:275-82.
16. Morimura T, Kitz K, Budka H: In situ analysis of cell kinetics in human brain tumors. A comparative immunocytochemical study of S phase cells by a new in vitro bromodeoxyuridine -labelling technique and of proliferative pool cells by monoclonal antibody Ki-67. *Acta Neuropathologica* 1989;77:276-82.
17. Reinfenberger J, Gudrum UR, Gies U, et al: Analysis of p53 mutation and epidermal growth factor receptor amplification in recurrent gliomas with malignant progression. *J Neuropathol Exp Neurol* 1996;55:822-33.
18. Jaros E, Perry RH, Adam L, Kelly PJ, et al: Prognostic implication of p53 protein, epidermal growth factor receptor and Ki-67 labelling in brain tumors. *Br J Cancer* 1992;66:373-85.
19. Abdulrauf SI, Edavardsen K, Ho KL, et al: Vascular endothelial growth factor expression and vascular density as prognostic markers of survival in patients with low grade astrocytomas. *J Neurosurg* 1998;88:513-20.
20. Sideransky D, Mikkelson T, Schwechheimer K, et al: Clonal expansion of p53 mutated cells is associated with brain tumor progression. *Nature* 1992;355:846-47.
21. Newcomb EW, Bhalla SK, Parrish CL, et al. Bcl-2 protein expression in relation to patient survival and p53 gene status. *Acta Neuropathol* 1997;94:369-75.
22. Van Meyel DJ, Ramsay DA, Casson AG, et al: P53 mutation expression, and DNA ploidy in evolving gliomas;evidence of two pathways of progression. *J Natl Cancer Inst* 1992;84:845-55.
23. Louis DN, Von Deimling A, Chung RY, et al. Comparative study of p53 gene and protein alterations in human astrocytic tumors. *J Neuropathol Exp Neurol* 1993;52:31-38.
24. Louis DN: The p53 gene and protein in human brain tumors. *J Neuropathol Exp Neurol* 1994;53:11-21.
25. Iggo R, Gatter K, Bartek J, et al: Increased expression of mutant forms of p53 oncogene in primary lung cancer. *Lancet* 1990;335:675-79.
26. Vojtesek B, Bartek J, Midgley CA, Lane DP: An immunochemical analysis of the human nuclear

phosphoprotein p53, New monoclonal antibodies and epitope mapping using recombinant p53. *Journal of Immunological methods* 1992;151:237-44.
27. Bruner JM, Saya H, Moser RP: Immunocytochemical detection of p53 in human gliomas. *Modern Pathology* 1991;4:671-74.
28. Ellison DW, Gatter KC, Steart PV, lane DP, Weller RO: Expression of p53 protein in a spectrum of astrocytic tumors. *Journal of Pathology* 1992;168:383-86.
29. Hayashi Y, Yamashita J, Yamaguchi K: Timing and the role of p53 gene mutation in the recurrence of gliomas. *Biochem Biophys Res Common* 1991;180:1145-50.
30. Jain KC, Chattopadhyay P, Sarkar C, Sinha C, Mahapatra AK: A pilot study of recurrence of human glial tumors in light of p53 heterozygosity status. *J Biosci* 1999;24:477-81.
31. Iuzzolino P, Ghimenton C, Nicolato A, et al: P53 protein in low grade astrocytomas. A study with long term follow up. *Br J Cancer* 1994;69:586-91.
32. Collins VP: Gene amplification in human gliomas. *Glia* 1995;15:189-296.
33. Shweiki D, Itin A, Soffer D: Vascular endothelial growth factor induced by hypoxia may mediate hypoxia initiated angiogenesis. *Nature* 1992;359:843-45.
34. Giese A, Westphal M: Glioma invasion in the Central Nervous System. *Neurosurg.* 1996;39:235-52.
35. Farber E: Programmed cell death necrosis versus apoptosis. *Modern pathology* 1994;7:605-9.
36. Arends MJ, McGreger AH, Wyllie AH: Apoptosis is inversely related to necrosis and determines net growth in tumors bearing constitutively expressed myc, ras and HPV oncogenes. *Am J Pathol* 1994;144:1045-57.
37. Schiffer D, Cavalla P, Chio A, et al: Tumor cell proliferation and apoptosis in medulloblastoma. *Acta Neuropathol.* 1994;87:362-70.
38. Migheli A, Cavalla P, Marino S, Schiffer D: A study of apoptosis in normal and pathologic nervous tissue after in situ and labeling of DNA strand breaks. *J Neuropathol Exp Neurol* 1994;53:606-16.
39. Tandon PN, Mahapatra AK, Sarkar C, Khosla A: Recurrence of supratentorial gliomas. *Neurol India* 1997;45:141-49.
40. Ghosh M, Dinda A, Chattopadhyay P, Sarkar C, et al: Rearrangement of p53 gene with loss of normal allele in a low grade recurrent mixed glioma. *Cancer Genet Cytogenet* 1994;78:68-71.

45

Recurrence versus Radionecrosis: An Imaging Challenge

I Sen, P Pankaj, H Mahajan

One of the most challenging dilemmas for the neurosurgeon or interventional neuro-radiologist is to manage the patient of brain tumor who has undergone surgery and /or high dose radiotherapy and returns with a recurrence of symptoms. Physical changes which happen within the treated volume in response to high dose radiotherapy and chemotherapy render interpretation of conventional CT or MRI difficult. Differentiation between residual or recurrent tumor and radio-necrosis then requires an appropriate combination of different imaging modalities to maximize information which can be obtained non invasively. Following is a brief resume of the different imaging techniques available and the relative merits and constraints of different methods

CONTRAST ENHANCED CT AND MRI

The introduction of paramagnetic agents for MRI has considerably expanded the contrast range of this technique. Small and subtle distortions of the anatomy caused by local pathology can be demonstrated. An abnormal enhancement may however be seen both in residual/recurrent tumour, as well as in post-radiotherapy or post chemotherapy changes. This enhancement basically represents as breach in blood-brain-barrier and this may be seen in both the above conditions. Hence, in a given case it may not be possible to differentiate the two.

PERFUSION-WEIGHTED MRI

This is performed using dedicated hardware and software on a high field strength scanners. This is significantly better than routine CE MRI and viable tumor tissue shows increased perfusion in the early stage after contrast injection (Gadolinium-DTPA 0.01 mmol/kg body weight).This is now done on standard MRI equipment using EPI (echoplanar imaging) sequences and compares quite favorably with the results of FDG PET scans.

MR SPECTROSCOPY

This helps in differentiating between viable tumor and radionecrosis by showing high levels of choline in viable tumor. However, there is a degree of overlap which also depends on voxel placement. With multi-voxed MR spectroscopy the entire area of abnormality can be sampled and this gives more definite answers. MR spectroscopy at the present time is not an absolute method but along with CE MRI increases the confidence level in predicting about viable tumor.

SPECT

SPECT scans have been used since the early sixties to evaluate tumour viability. However these studies lacked the spatial resolution for exact clinical interpretation. In 1984 Kaplan et al administered radioactive Thallous chloride (TL-201) in a series of patients who had been treated for malignant gloms and demonstrated that TL-201 showed accumulation in persistent or recurrent tumour in sufficient concentration to effect differential diagnosis between viable tumor and radionecrosis. The mechanism of uptake was the ATPase mediated sodium potassium pump within cell membranes, which transported the thallium, a potassium analogue, into intact cells in high concentration relative to extracellular space. For thallium accumulation, adequate blood flow need be present for uptake to occur. Technical improvements in hardware and computer interfaces and evolution of the multi-detector Gamma Camera systems allowed workers to improve the specificity of Thallium uptake by quantification of the uptake and establishing ratios established between the region of interest and its mirror image on the opposite side of the brain. A T/N ratio of $\geq 3.5:1$ was found in high grade lesions & moderate to low uptake of TL-201 with ratios < 3.5 indicated low likelihood of tumor persistence or recurrence. Tl-201 may occasionally produce false positive studies in the presence of necrosis, possibly due to increased permeability of the blood brain barrier. Performing a 99m Tc HMPAO rCBF SPECT in addition to TL-201 further enhanced the sensitivity of detecting persistent on recurrent tumors as low uptake of Tc HMPAO in addition to low TL uptake values less than 50% of cerebellar activity indicated very low likelihood of tumor presence. A moderate uptake Tl-201 uptake and high 99m Tc – HMPAO rCBF values are not predictive.

With the development of newer radiotracers like 99m Tc labelled Sesta MIBI the avenue opened for images with even higher resolution than before. Uptake of SestaMIBI, a lipophilic cation is strongly influenced by the negative charges on the mitochondria and the resultant transmembrane potential. **(Figs. 1, 2 & 3)** The uptake is dependent on the density of mitochondria which explains the avid uptake in actively replicating tumor cells. Like thallium the uptake is also dependent on adequate blood supply. However unlike Thallium the high photon flux of technetium allows better quality images and higher resolution studies.

PET

PET or Positron Emission Tomography assesses physiology and biochemistry rather than the anatomy seen by conventional imaging. Its high cost and the limited access to radiopharmaceutical has limited its diffusion through the clinical community, even though it has been available as a research tool since the early 1970s. PET is excellent for detecting recurrent brain tumor from postoperative scarring and radiation necrosis based on the hypermetabolic uptake of glucose. These studies are based on glucose utilization by the glucose analog FDG (Fluorodeoxyglucose in which Fluorine-18 is the radionuclide). Since viable tumors utilize more glucose than does scar tissue, PET can differentiate between the recurrent tumor and radionecrosis. Studies have demonstrated a sensitivity as high as 92% with an 83% specificity for recurrent tumor following treatment. Correlation has been shown between the uptake of FDG and the histopathological grade of gliomas as well as between the uptake of FDG & Prognosis.

A ratio òf FDG uptake greater than 1.4:1 in comparison to contra lateral white matter indicates omnivores prognosis.

IMAGE FUSION

The large number of medical imaging technologies has provided information on very different attributes of body tissues, including physical parameters and measures of biologic functions. The information is provided over wide ranges of spatial resolution or localisation accuracy (approximately 1mm for MR and CT, 5-10 mm for PET and SPECT) Features present in one modality may be absent in another. The same imaging device may provide very different information according to the study performed. Accurate registration and combined display provides additional useful clinical information in several areas of medical imaging. For example accurate three dimensional representation of the patient's anatomy is obtained by combining the complementary information contained in MR, CT, PET/SPECT to provide the surgeon or the interventional radiologist and radiotherapist with an aid to planning and increasingly intraoperative guidance. Another application is the use of high resolution, accurately registered anatomic maps derived from MR or CT to aid in interpretation and quantitation of lower resolution functional images, usually from PET or SPECT.

CONCLUSION

Frequently, a consensus emerges for a particular combination of imaging studies for a specific disease status but technical advances appear which question the validity of the recently achieved consensus. This change in the imaging studies recommended not only reflects progress and a dynamic evolution of brain imaging technology, but also poses real and significant challenges when decisions have to be made about patient management and cost control. These conflicts are more apparent today than in the past. It is only through experience and an extensive database that a logical and justifiable consensus can be achieved regarding the optimum investigations in a patient.

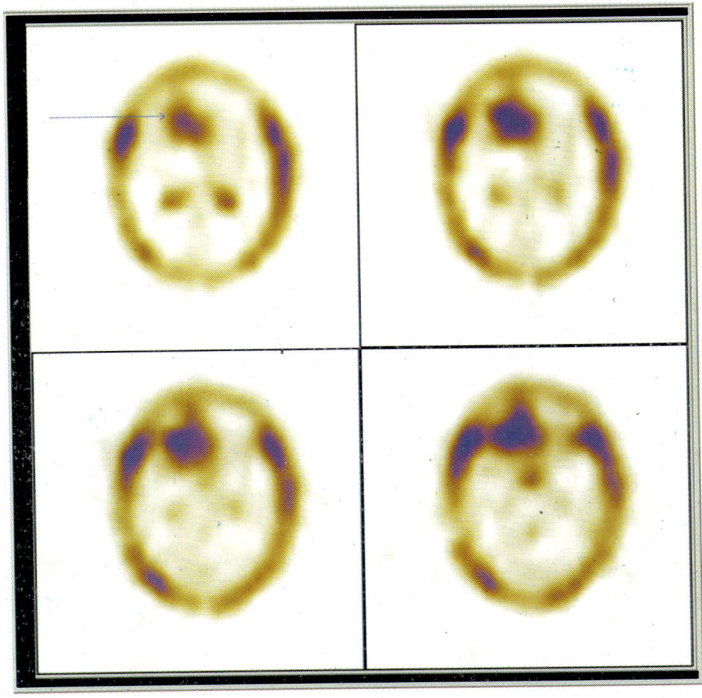
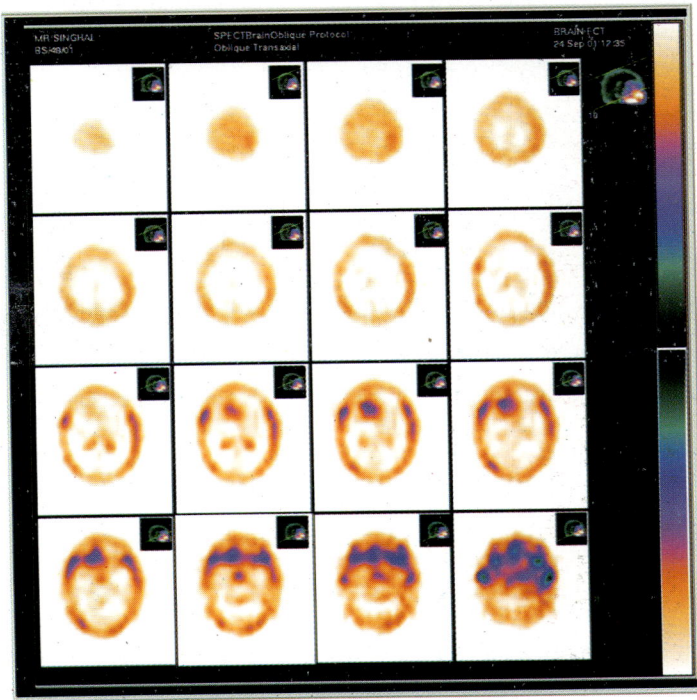

Figs.1 & 2: Intensely increased uptake of Sesta MIBI seen in the right frontal region of a patient of operated astrocytoma suggesting tumour recurrence

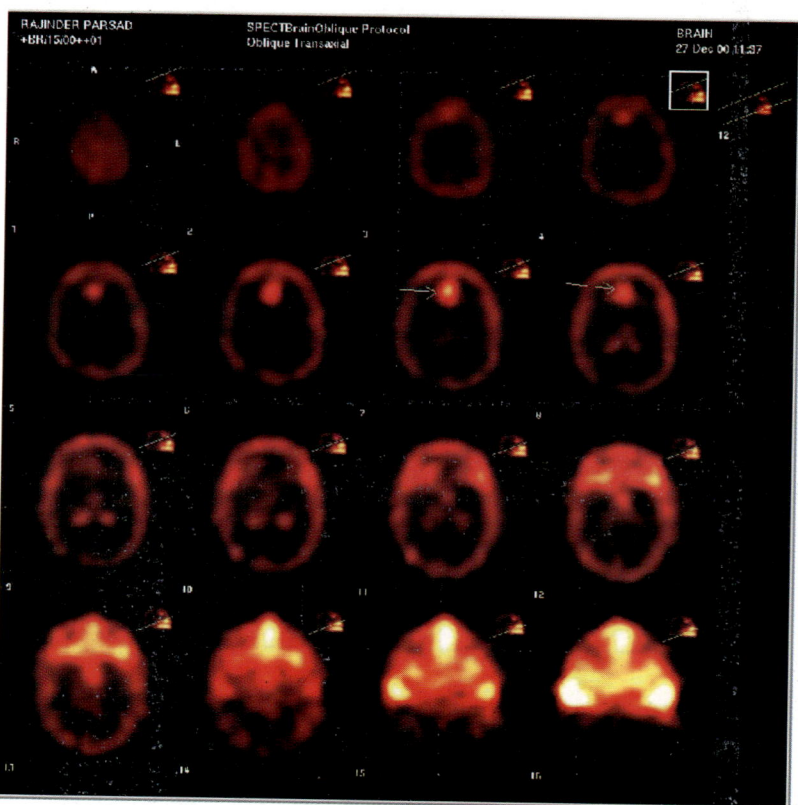

Fig. 3 : Post operative and post radiotherapy patient of frontal Glioma with recurrence of clinical symptoms. Contrast CT/MRI unable to differentiate between radionecrosis & tumour recurrence. 99m Tc Sesta MIBI scan shows viable tumour tissue in the frontal region suggesting tumour recurrence

REFERENCES:

1. Peters TM, Clark JA, Olivier A, et al: Integrated stereotaxic imaging with CT, MR imaging and digital subtraction angiography, *Radiol* 1986;161:821-26.
2. Pelizzari CA, Chen GTY, Spelbring DR, Weichselaum RR, Chen C: Accurate three dimensional registration of CT, PET and/ or MR images of the brain. *J Comput Assist Tomogr* 1989;13:20-26.
3. Black KL, Hawkins RA, Kim KT, Becker DP, Lerner C, Mariano D: Use of thallium-201 SPECT to quantitate malignancy grade of gliomas. *J Neurosurg* 1989;71:342-46.
4. Mecapinlac H, Finaly J, Caluser C, et al: Comparison of TL-201 SPECT and F18FDG PET imaging with MRI (Gd-DTPA) in the evaluation of recurrent supratentorial and infratentorial brain tumors (Abstract). *J Nucl Med* 1992;33:867-68.
5. Baillet G, Alburquerque L. Chen CR, Poisson M, Delaftne JY: Evaluation of SPECT imaging of supratentorial brain gliomas with TC 99m sesta MIBI. *Eur J Nucl Med* 1994;21:1061-66.

46

Reoperation for Recurrent Gliomas—Rationale and Techniques

AK Singh, V Gupta, S Sinha

The malignant behavior of the gliomas is characterized by the tendency to recur, despite often very aggressive anti tumoral therapy, including combination of radical surgical resection, radiotherapy and chemotherapy. Despite recent advance in the diagnostic and therapeutic modalities the treatment of gliomas remains a frustrating chapter. For all practical purposes recurrence appears to be inevitable irrespective of the therapeutic modality employed for treatment of the primary tumor. Recurrence is the most common cause of death in a case of a supratentorial glioma. More than 90% of malignant gliomas recur within 2cm of their original site of presentation after adequate therapy.[1,2,3] Five to 10% of patients may develop recurrence at some distance from the original tumor either in ipsilateral hemisphere or opposite side. The problem of recurrence is beset with greater dilemma for the surgeon than the management of initial tumor.[4] There is relative paucity of data concerning the efficacy of reoperation in the treatment of recurrent glioma, probably for this very reason there is considerable controversy regarding the proper role of surgery in the context of primary treatment failure. In general, the frequency of reoperation for recurrent glioblastoma has been low and undertaken only in a sporadic fashion.[5-13] Historically speaking published reoperation rates have been low in the range of 0 to 10 percent, and the surgical mortality for such procedures has been high in the range of 10 to 20 percent. Additionally most surgeons have been highly selective in their choice of patients for reoperation, so the available data is biased by an inordinate number of low grade astrocytoma and oligodendroglioma.

Re operations are usually carried out in patients who are relatively young and in good neurological condition, in those who have tumors that are favorably situated and in patients in whom recurrence has occurred long after operation. Even fewer patients with glioblastoma have been subjected to re operation, on the possible erroneous assumption that surgery has nothing to offer the patient with a malignant glioma. This almost universal and nilhistic approach to recurrent glioblastoma is probably justified because of poor prognosis of all modalities of management. But there are several valid grounds for questioning this assumption.

1. The rough correlation between the length of post operative survival and extent of surgical resection in patients with malignant astrocytoma and medulloblastoma.
2. Importance of early and radical surgery in the treatment of solid cancers elsewhere in the body.
3. The existence of cell compartments within malignant gliomas that are inherently resistant to all other treatment modalities.
4. Possible potentiation of other treatment by mechanical cytoreduction.

Therefore the rational for primary surgery may apply equally well to reoperation, especially if sampling of the tissue is adequate and a fresh evaluation of both the tumor and the effects of previous therapy can be made.[5]

Renewed growth of a lesion at the site of previous surgery calls for considering following issues.[6]

1. Is the mass a recurrence of the original tumor?
2. Why did the tumor grow?
3. Where threat to the patient's neurological function and survival does the regrowth pose?
4. What additional therapy is adequate?

DIAGNOSIS OF RECURRENCE

When recurrence of a glioma is suspected clinically or radio logically, a plain and contrast computed tomography scan of the head, a plain and contrast MRI of the head should be obtained and compared with the pre operative films, immediate post operative films and post radio therapy films.

DIFFERENTIAL DIAGNOSIS

An enlarging lesion at the site of previously treated glioma most likely represents renewed growth in a residual tumor rather than the development of a new pathological entity.

However there are few exceptions as follows:

- A tumor of related histology may supplant the original tumor (e.g. the astrocytic component may replace the oligodendrocytic component as the predominant subtype of a mixed glioma or a gliosarcoma may arise from a previously treated glioblastoma)
- A distinctly new tumor may arise near the site of an eradicated tumor.
- Non-neoplastic lesion induced by treatment of the original tumor may mimic tumor growth (e.g. an abscess at the site of tumor resection or radiation necrosis following focal high dose radiation.)

Recurrent gliomas have imaging features similar to those of original lesion. On computed tomography it is likely to show central low intensity, rim enhancement with surrounding hypo density on contrast enhancement. On magnetic resonance imaging they are hypo intense on T1 and hyper intense on T2 with rim enhancement.

However some subtle difference may point to different diagnosis:
- a more spherical, sharply demarcated contrast enhancing rim may suggest abscess rather than recurrent glioma.
- A more diffuse, irregularly marginated pattern of surrounding edema may indicate radio necrosis rather than recurrent tumor.

However there is a practical difficulty in the diagnosis of recurrent tumor and to differentiate it from radio necrosis

RADIOLOGICAL FEATURES OF RECURRENCE

When there is a recurrence in a case of low-grade glioma after surgery, approximately half remain non-anaplastic tumors, but other 50% progress to a more malignant form.[7-11] Enlarging low-grade tumors are likely to resemble the original tumor on imaging studies. When progression in grade has occurred the new tumor may also resemble the old one, especially if the original tumor enhanced with contrast. Enhancement is highly predictive of recurrence, low-grade enhancing tumors are 6.8 times more likely to recur as compared to non-enhancing ones.[10] Most commonly a new malignant growth in a previously non-enhancing lesion enhances, and is thus readily identified.

In one study, only 30% (16/42) of low-grade tumor enhanced initially but 92% (22/24) enhanced at recurrence.[10] However, occasionally an enlarging mass which may though be malignant, may not enhance. In such a case it should be seen as a region of hyper metabolism on 2-deoxyglucose-positron emission tomogram study (PET) or a functional MRI and recently on diffusion MRI. But histological confirmation shall always be needed to verify malignant transformation.

RADIATION EFFECTS

Radiation necrosis generally causes diagnostic difficulty in case of recurrence of a glioma. CT & MR are often not very helpful in distinguishing between the two.

Radiation can cause tumor enlargement by one of the following ways.[8]
- *Early reaction:* which is likely to be edema occurring during or shortly after irradiation.
- *Early delayed reaction:* that involves edema and demyelination arising a few weeks to a few months after radiation.

- *Lake delayed reaction:* that occurs 6-24 months after reduction and reflects radiations induced necrosis.[8]

Only occasionally large, very malignant tumors grow sufficiently fast to show significant enlargement during or within three months of completing a course of radiation. In most cases tumor enlargement from early or early delayed effects represents edema is transient and responds to a short course of corticosteroids. In contrast radiation induced necrosis appears at about the time malignant tumors might be expected to recur. It is thus more likely to be mistaken for recurrence.[6]

The risk of radiation necrosis increases with the volume treated, dose delivered and fraction size.[12] Regional tele-radiotherapy to a dose of 60 Gy is the current standard radiation treatment for most gliomas.[13] It has a low risk of inducing radiation necrosis.[14] Radiation induced changes following regional teletherapy are relatively diffuse and are distinguishable from the more focal appearance of a recurrent tumor. In contrast, radiation necrosis following focal radiation treatments, such as brachytherapy and radio surgery is more difficult to distinguish from recurrent tumor. A variety of functional neuro diagnostic imaging techniques are currently being studied for their ability to distinguish between these two possibilities. These are PET Scan, Thallium Studies and cerebral blood volume mapping. Regions of high activity are thought to distinguish recurrent tumor from relatively metabolically inactive and hyper vascular radiation necrosis.[15-18]

However in spite of these modern studies the diagnosis may be revealed either by clinical course or by analysis of a pathology specimen. At reoperation for presumed radiation necrosis following focal radiation treatment of a malignant glioma, necrosis was found in 59% of cases, tumor above in 20% and mixture of radiation necrosis and tumor in 66%.[19] In almost all cases, the tumor that was seen was of reduced viability.[20-21] However Muller et al reported that recurrent tumor will show the appearance of more malignant biological activity than the initial tumor.[10]

CAUSES OF RECURRENCE

Recurrence of glioma indicates failure of therapy for control of tumor. Failure is due to number of factors that limit the efficacy of each modality.

RECURRENCE AFTER SURGERY

The total extirpation of the tumor at primary operation is the goal however it may not be achieved due to anatomical considerations such as:

- Involvement of vital neurovascular structures such as MCA, ACA, cranial Nerves.
- Involvement of diencephalons, internal capsule, brain stem or invasion of the eloquent areas of the brain.

An error of judgment on the surgeon part to underestimate the amount of tumor that can be safely removed may lead to residual tumor and hence an nidus for regrowth.

RECURRENCE AFTER RADIATION

Radiation therapy may fail because of inadequate targeting, under utilization of tolerable dose, or radiation resistance of the tumor cells.

RECURRENCE AFTER CHEMOTHERAPY

Chemotherapy fails as a result of inadequate drug delivery, toxic effects or cell resistance. The Blood Brain Barrier is deficient in contrast enhancing region of the tumor, but surrounding brain usually has an intact blood brain barrier, lipid soluble drug thus have limited access to tumor cell infiltrating peripheral regions.

RATIONALE FOR REOPERATION

Resection of recurrent malignant gliomas was practiced by Cushing[22] and others in the earlier part of the century but Bronson Ray[23] is credited with emphasizing the beneficial role of reoperation in the event of a recurrence in a malignant glioma. Pool[24] in 1968 highlighted the technical difficulties associated with re-operation. However most of literature on this topic has appeared in the last 15 years. [2-8, 10, 24-25, 27] This may indicate the increasing feasibility of undertaking resection of a recurrence as well as earlier and more precise diagnosis. Historically speaking published reoperation rates have been low in the range of 0 to 10% and the surgical and the surgical mortality has been high in the region of 10-20%. The improved ability of Neurosurgeons to perform such resections, combined with the belief that they are beneficial has led to reoperation rates as high a 58%.[25] A national survey of malignant gliomas treatment in USA revealed a reoperation rate of 14%.[26]

No randomized trials on resection of recurrence malignant gliomas have been reported.[26] To best of our knowledge, no retrospective study has demonstrated a statistically significant survival benefit for patients who undergo resection for a recurrent glioma, when survival is measured from the date of tumor recurrence.

Strombald et al[27] compared survival between 58 patients with malignant gliomas who underwent reoperation and 85 patients with malignant gliomas who did not undergo reoperation, who were treated from the time of diagnosis using a prospective protocol. They reported no significant difference in survival. Rostomily et al[28] determined no difference in survival time between patients who did not undergo resection of recurrent malignant gliomas immediately before entry into a Phase II Chemotherapy trial, although patients who underwent resections before chemotherapy had longer time to tumor progress.

In 1981 Young and Colleagues[29] published a series of 24 patients examined for statistical significance for identifying prognostic factors that effect outcome following reoperation over a period of 8 years. Pre-operative Karnofsky performance status more than 60 and time interval between first surgery and reoperation more than or less than 6 months were significant predictors of increase in median survival (22 weeks versus 9 weeks). Age, sex & tumor location were not significant variables.

Salcman et al[30,31] studied prospectively a consecutive series of 74 patients entered into a multimodality therapy program for malignant astrocytoma. 40 patients underwent reoperation for recurrence. They found mean survival after reoperation of that of 37 weeks and single most critical factor for determining outcome was age under 40 years. In 1987 Ammirati et al[32] published data of 55 patients undergoing reoperation at Memorial sloan Kettering and found survival of 29 weeks for GBM and 61 weeks for anaplastic astroctoma. Preoperative Karnofsky score, pathology & extent of tumor resection were statistically significant factors. Harsh & Colleagues[33] reviewed retrospectively a series of 70 patients who underwent reoperation over 9 year period and reported median survival of 36 & 88 weeks respectively for glioblastoma & anaplastic astrocytoma. Preoperative Karnofsky score did influence the outcome significantly in terms of quality of life but not survival after reoperation. Younger patients with anaplastic astrocytoma also had better quality of life but not survival after reoperation. Operative morbidity was 5.7% and was acceptable. This analysis supported reoperation as modus operandi for sustaining better and meaningful quality of existence. Moser et al[34] and Vick et al[14] published survival results after reoperation of improved clinical status of patients thus supporting that reoperation may impact significantly on the quality of life, thus allowing time for further therapy to ensue. Barker et al[35] studied the selection factors for and results of second resections for recurrent glioblastomas. 46 patients were prospectively studied and younger age and more extensive initial resection but not Karnofsky Performance Scale (KPS) score predicted a higher chance of selection for reoperation. The median survival time for 43 patients who underwent reoperation was 42 weeks (19 weeks more than who did not underwent reoperation). They observed the reoperation were more likely to benefit the KPS scores of those patients who had symptomatic recurrence than in those who were asymptomatic.

Now the crucial decision to be made by a clinician is who should be offered reoperation and how many times. Patrick J Kelley[36] has succinctly summed up the surgical philosophy being practiced today "there is no question that selected patients harbouring grade IV astrocytomas derive significant benefit from aggressive tumor resection prior to external brain radiotherapy". Patients who have mass lesion in which tumor tissue mass (defined by contrast enhancing volume) represents a large portion of their global tumor volume (contrast enhancement plus edema) will benefit from resection. However many apparently large tumors with little or no mass effect (because tumor tissue has replaced brain parenchyma) debulking may not help in such cases. He advocates individualization of these cases in terms of expected benefit. Young GBM patients with high KPS scores who have done well (survived more that 1 year) following resection, radiation / chemotherapy and now present with a large recurrent contrast enhancing mass lesion, will gain useful time following a second resection provided that some effective adjuvant therapy remains an option post operatively. Older patients whose tumor is now recurring 4 months after radical resection will not benefit from a second procedure any more than they did from the first. Overall the critical factors associated with survival following reoperation were age, duration of symptoms before I[st] operation and time to tumor progression before reoperation.

The policy being followed by the author's center is by and large on similar lines. All gliomas patients who have recurrence after a reasonable long time after initial resection, have good KPS score and have tumor in a favorable location are again subjected to aggressive cyto-reduction.

Results of reoperation for recurrent gliomas in literature

Series	Year	No	Histology	Age	Survival (wks) From Dx	Survival (wks) From Re-Sx	Mortality CSF leak	Complications Morbidity	Complications Infection
Young	1981	24	GBM+AA	50.5	62	14	17%(4) 4.2%(1)	54%(13)	20.8%(5)
Salcman	1982	40	32GBM+8AA	43.4	91	37	0 1.6%(1)	8.3%(3)	6.7%(4)
Sadfari	1986	27	GBM+AA	59		139	-	-	-
Ammirati	1987	55	35GBM+20AA	48	76	29	1.8%(1) 1.8%(1)	18.2%(10)	7.3%(4)
Harsh	1987	70	39GBM 31AA	45.5 31	82 264	36 88	4.3%(3) -	5.7%(4) -	-
Fadul	1988	62	GBM+AA	52	-	-	3.3%	26%(16)	-
Vick	1989	15	10GBM 5AA	43	77 148	22 52	0 0	0	0
Berger	1992	70	56GBM+14AA						
Kaye	1992	50	50GBM				0%	6%	
Scharfen	1992	45	18GBM 27AA			52 153			
Rostomily	1994	31	22GBM 9AA	42		30 202	0 -	0 -	
Ryken	1994	22	12GBM 10AA	44	63.5	28	- -	-	-
Barker	1998	46	GBM	54	42	0	-	-	

Technique for Re-operation

Corticosteroids are started before surgery and should be continued. The procedure should be planned in advance by specifying the location of tumor by its relationship to the margins of craniotomy on CT scan or the cortical pattern of gyri and sulci on as MRI Scan. It is to ensure adequate exposure to expose the recurrent mass. Usually the old incisions need enlargement due to the increased extent of the tumor.

The skin incision from the previous operation is usually used. The skin opening can be increased by introducing additional incisions they should be external to previous incisions, avoid its base and vascular pedicles. The margins of prior craniotomy flap should be defined. Only rarely previous kurf need to be recut. Dissection is done in the epidural space using curet and penfield dissectors. Craniotomy flap is raised after stripping dura from its inner surface. After raising the craniotomy flap, if needed an additional segment of bone can be removed using a craniotome after stripping the dura from the craniotomy margins.

The durotomy should be planned so as to minimize traversing the cortical adhesions and should be done external to the cortical adhesions. Flapping the dura then puts traction on the adhesions so that they can be coagulated and dissected from the cortex. The prior incision of durotomy should be traversed perpendicularly as it is often the site of densest adhesions.

The cortical surface in then examined for abnormal color, consistency, vascularity and for surface presentation of the tumor. Localization of the sub cortical extent of the tumor is then undertaken. In it preoperative imaging studies, stereotectic technique[37] and Transcortical ultra sonography[38] can be of immense value. Electrocorticography can be useful in mapping eloquent areas and may reduce chance of inflicting a neurologic deficit and may help in more extensive resection. This technique is often more difficult at the time of re-operation because of cortical disruption by the tumor and prior surgery.

The tumor mass can be internally debulked using CUSA or laser as initial step. It can lead to significant blood loss sometimes. Enucleation of mass can also be done by dissecting in the pseudo plane about the rim of the solid tumor. Arteries supplying and veins draining the tumor can be coagulated and cut. Often the tumor can be removed without putting much retraction of surrounding normal brain. If needed retraction of tumor is preferable to retraction of normal brain.

Once the tumor has been removed the margins of resection should be inspected to verify total excision of tumor. Biopsies of surrounding brain should be sent for frozen section to ensure absence of tumor.

After the tumor resection, complete hemostasis should be done by coagulating all the small blood vessels appearing as small thin walled strands rather than tamponading with hemostatic packing which may lead to deeper dissection of hematoma. Hemostasis should be confirmed by filling tumor cavity with saline and during Valsalva maneuver. The patient's blood pressure should be allowed to rise at least to the pre-op level. The cavity is then lined with single layer of surgicel and filled with saline. Hyperventilation is then reversed to allow brain expansion.

Dural closure is needed to be watertight. Normally a primary closure is possible. If dura is deficient it can be supplement by pericranial graft or artificial dural substitute. Hitch sutures to dura are applied. Bone fragments are wired together and then to craniotomy plate. The wound is then closed in layers of muscle, galea and skin.

Post operatively patient should be monitored for signs of raised ICT from hematoma or edema which may necessitate need for urgent CT scan and reopening if needed. Dehydration and corticosteroids should continue for 72 hrs. The patient should be mobilized as early as possible

REFERENCES:

1. Hochberg FH, Pruitt A: Assumptions in the radiotherapy of glioblastomas. *Neurology* 1980;30:907-11.
2. Choucair AK, Levin VA, Gutin PH, Davis RL, Silver P, Edwards MS, Wilson CB: Development of multiple lesions during radiation therapy and chemotherapy in patients with gliomas. *J Neurosurgery* 1986;65: 654-58.
3. Liang BC, Thornton AF Jr., Sandler HM, GreenBerg HS: Malignant astrocytomas: Focal tumor recurrence after focal external beam radiation therapy. *J Neurosurgery* 1991;75:559-63.
4. Tandon PN, Mahapatra AK, Sarkar C, Khosla A: Recurrence of supratentorial glioma. *Neurology India* 1997;45:141-49.
5. Salcman M: Supra tentorial gliomas: Clinical features and surgical therapy. In Neurosurgery Vol. I Wilkins RH, Rengachary SS (eds) 2nd Edition Mc graw Hill 1996;777-88.
6. Harsh GR IV: Management of Recurrence gliomas. In The Gliomas. Berger MS and Wilson CB (eds). WB Saunders Company, Philadelphia 1999;649-59.
7. Kaye AFI: Malignant Brain tumors. In RothenBerg RE (ed) Reoperative Surgery, New York Mc Graw Hill 1992;51-76.
8. Laws ER jr, Taylor WF, Clifton MB, Okazaki H: Neurosurgical management of low grade astrocytoma of the cerebral hemispheres. *J Neurosurgery* 1984;61:665-73.
9. Mc Cormack BM, Miller DC, Budzilovich GN, Voorhees GJ, Ransohoff J: Treatment and survival of low grade astrocytoma in adults 1977-1988 *Neurosurgery* 1992;31:636-42
10. Muller W, Aftra D, Schroder R: Supratentorial recurrences of gliomas: Morphological studies in relation to time intervals with astrocytomas. *Acta Neurochir (Wein)* 1997;37:75-91
11. Wilson CB: Reoperation for primary tumors seminar Oncology 1975;2:19-20
12. Marks JE, Baglan RJ, Prassad SC, Blank WF: Cerebral radio necrosis: Incidence and risk in relation to dose, time, fractionation and volume. *Int J Radiat Oncol Biol Physic* 1981;7:243-52.
13. Walker MD, Alexandar E Jr., Hunt WE, MacCarthy CS, Mahaley MS Jr., Mealey J Jr., Norrell HA, Owens G, Ransohoff J, Wilson CB, Gehan EA, Strike TA: Evaluation of BCNU and radio therapy in the treatment of anaplastic gliomas A cooperative clinical trial. *J Neurosurg* 1978;49:333-43.
14. Vick NA, Ciric IS, Eller TW, Cozzens JW, Walsh A: Reoperation for malignant astrocytoma *Neurology*;1989:39:430-32.

15. Alavi JB, Alavi A, Chawluk J, Kushner M, Powe J, Hickey W, Reivick M: Positron emission tomography in patients with gliomas; A prediction of prognosis. *Cancer* 1988;62:1074-78.
16. Di Chiro G, Brooks R, Baira man D: Diagnostic and prognostic value of positron emission tomography using {18F} fluorodeoxy glucose in brain tumors. In Reivich M, Alavi A (ed) positron Emission Tomography, New York, *Alan R Liss*, 1985:291-309.
17. Le Bihen D, Douek M, Argyropoulou M, et al: Diffusion and perfusion magnetic resonance imaging in brain tumors. *Magn reson Imaging* 1993;5:25-31.
18. Valk PE, Budinger TF, Levin VA, Silver P, Gutin PH, Doyle WK: PET of malignant cerebral tumors after interstitial brachytherapy: Demonstration of metabolic activity and correlation with clinical outcome. *J Neurosurg* 1988;69:830-38.
19. Scharfen CO, Sneed PK, Wara WM, Larson DA, Phillips TL, Prados MD, Weaver KA, Malec M, Acord P, Lamborn KR: High activity iodine-125 interstitial implant for gliomas. *Int J Radiat Biol Phys* 1992;24:583-91.
20. Daumas-Duport C, Blond S, Vedrenee C, et al: Radio lesion versus recurrence: Bioptic data in 30 gliomas after interstitial implant or combined interstitial and external radiation treatment. *Acta NeuroChir* 1983;30 (Suppl);291-99.
21. Rosenblum ML, Chur Lui H, Dewis RL, et al: Radiation necrosis versus tumor recurrence following interstitial brachytherapy: Utility of tissue culture studies. *Proc Am Asso Neurol Surg* 1985;53:264.
22. Cushing H: Intracranial tumors. Springfields, Charles C Thomas 1932.
23. Ray BS: Surgery of recurrent intracranial tumors. *Clin Neurosurg* 1964;10:1-30.
24. Pool JL: The management of recurrent Gliomas. *Clin Neurosurg* 1968;15:265-87.
25. Salcman M, Scholtz H, Kaplan RS, Kulik S: Long-term survival in patients with malignant astrocytomas. *Neurosurgery* 1994;34:213-20.
26. Mahaley MS Jr., Mettlin C, Natrajan N, Laws ER Jr., Peace BB: National Survey of patterns of care for brain-tumor patients. *J Neurosurg* 1989;71: 826-36.
27. Stromblad LG, Anderson H, Malmstrom P, Salford LG, :Reoperation for malignant astrocytomas: Personal experience and a review of literature. *Br J Neurosurg* 1993;7:623-33.
28. Rostomily RC, Spence AM, Duong D, McCormick K, Bland M, Berger MS: Multimodality management of recurrent adult malignant gliomas: result of phase II multi agent chemotherapy study and analysis of cyto-reductive surgery. *Neurosurgery* 1994;35:378-88.
29. Young B, Old field EH, Markesbery WR, Haack D, Tibbs PA, McCombs P, Chin HW, Maruyama Y, Meacham WF: Reoperation for glioblastoma. *J Neurosurg* 1981;55:917-21.
30. Salcman M, Kaplan RS, Ducker TB, Abdo H, Montgomery E: Effect of age and reoperation on survival in the combined modality treatment of malignant astrocytomas. *Neurosurgery* 1982; 10:454-63.
31. Salcman M: Resection and reoperation in neuro-oncology Rationale and approach. *Neurol Clin* 1985;3:831-42.
32. Ammirati M, Galicich JH, Arbit E, Liao Y: Reoperation in the treatment of recurrent intracranial malignant gliomas. *Neurosurgery* 1987;21:607-14.
33. Harsh GR 4th, Levin VA, Gutin PH, Seager M, Silver P, Wilson CB: Reoperation for recurrent glioblastomas & anaplastic astrocytoma. *Neurosurgery* 1987;21:615-21.
34. Moser RP: Surgery for Glioma relapse factors that influence a favorable outcome. *Cancer* 1988; 62:381-90.
35. Barker FG, Chang SM, Gutin PH, Malec MK, Mc dermott MW, Prados MD, Wilson CB: Survival and functional status after resection of recurrent glioblastoma multiforme. 1998;42:709-20.
36. Kelly PJ, Rappa port ZH, Bhagwati SN, Ushio Y, Vapalahti M, Tribolet N: de. Reoperation for recurrent malignant gliomas. What are your indications. *Surg Neurol* 1997;47:39-42.
37. Kelly PJ: Stereotactic biopsy and resection in thalamic astrocytomas. *Neurosurgery* 1989; 25:185-95.
38. Le Roux, PD Berger, MS, Ojemann GA, Wang K, Mack LA: Correlation of intraoperative ultrasound tumor volumes and margins with preoperative computerized tomography scans. *J Neurosurgery* 1989;71:691-98.

47

Stereotactic Irradiation for Recurrent Gliomas

K Ganapathy

INTRODUCTION

The multi faceted approach necessary to try to reduce the relentless progression of malignant gliomas and the possible use of Stereotactic Radiosurgery has been described.[1] While technological advances in surgery and radiation techniques have been impressive, these have not translated into equal improvements in survival for malignant gliomas, particularly the recurrent ones. Unequivocal evidence of significant benefits following use of chemotherapy, different types of radiotherapy, or repeat surgery is not yet forthcoming. Re-treatment of recurrent malignant gliomas is essentially palliative. A risk – benefit analysis must be done, with special reference to quality of life (QOL) before embarking on any treatment for recurrent gliomas. Based on previous treatment in each specific case and the likely prognosis, individually tailored re-treatment strategies are possible. Performance status at the time of re treatment is the single most important predictive factor.[2]

LIMITATIONS AND COMPARISON WITH INTERSTITIAL BRACHYTHERAPY

The observation that most malignant gliomas recur within a few centimeters of the original site prompted the use of focal therapies. In selected cases stereotactic radiosurgery appears to prolong survival for recurrent malignant gliomas. Post radiation symptomatic necrosis following non-invasive SRS is less than that following invasive brachytherapy. Being non invasive, and with a lower reoperative rate, SRS appears to be a better treatment option.[3] Interstitial brachytherapy requires a second surgical procedure, a stay of 5-6 days in the hospital, and a necessity for reoperation in up to 50% of patients due to radiation necrosis. Review of the literature suggests that at present the role of Stereotactic irradiation for recurrent gliomas – probably already operated, irradiated, and exposed to chemotherapy and immunotherapy, is very limited.

Defining Target Volume: The precise demarcation of actively dividing malignant cells after high-dose radiation therapy is frequently limited by the lack of metabolic discrimination available with conventional imaging methods. Conventionally the contrast enhancing region on MRI scans with a safety margin of 2 to 5 mm is chosen.[4] Improved tumor contrast and delineation of the target volume can also be obtained with contrast media-supported FLAIR imaging.[5] Current image software matching algorithms allow the surgeon to superimpose data from magnetic resonance spectroscopy (MRS), position emission tomography (PET), and single photon emission computed tomography on MRI studies.

ROLE OF MAGNETIC RESONANCE SPECTROSCOPY IN SRS FOR RECURRENT GLIOMAS

Proton magnetic resonance spectroscopic imaging may help in prognostication following SRS for recurrent gliomas. More extensive initial metabolic abnormalities are suggestive of reduced survival time.[6] Characterisation of tissue response to radiation is critical. Evaluation of metabolic changes with proton MR spectroscopy, and structural changes with MR imaging, improve tissue discrimination. In one study, Graves et al were even able to provide correlation with histologic findings.[7] Response within the radiosurgery target was observed as a reduction of Cho levels and an increase in lactate / lipid levels, typically within 6 months of treatment. Increases in Cho correlated with poor radiologic response and suggested tumor recurrence. Thallium-[201] SPECT scan is sensitive and specific in detecting tumour recurrence and radiation necrosis (94% and 63% respectively).[8]

TARGET SELECTION

Differentiation of radiation necrosis from tumour on follow-up MRI in patients who have received prior external beam radiation therapy is a major problem. Stereotactic biopsy, besides removing the uncertainty, helps in prognostication. There is a significant difference in the survival of patients with recurrent tumour versus those with a mixture of tumour and necrosis, the latter group living longer.

Guidelines for Stereotactic Irradiation:

- Prior to treatment, informed consent is obtained. This should include a discussion of possible acute, early, and late-delayed radiation side effects that may follow Radiosurgery. After treatment planning is complete, I.V. dexamethasone (10 mg) is give routinely. Follow up contrast MRI is obtained at 2-month intervals.
- Designing a highly conformal treatment plan, limiting the high dose to the region of solid tumour tissue, and thus limiting complications should be attempted. There is a higher risk of acute toxicity when the **target volume is > 8.2 ml**; and a maximum dose: **prescription dose ratio of ≥ 2**, which is a measure of dose inhomogeneity.
- Brain adjacent to the solid tumour tissue contains viable, infiltrating tumour cells that are presumed responsible for marginal failures after focal treatments. MRS is a method used in an attempt to discriminate between solid tumour, necrosis,

and normal brain tissue. Patients with abnormal spectra outside the contrast-enhancing volume require additional treatment.

- The objective of stereotactic irradiation is to increase overall survival; reduce the frequency and severity of toxicity; and improve neurological outcome and quality of life.

RESULTS:

- Failure at the site of the original treatment occurs in about 76% of cases.
- Reoperation for increasing mass effect following SRS is necessary in about 19%.
- Reoperation did not influence survival.
- Only age had a significant influence on out-come; not histology, treatment volume, or radiation dose.
- While many acute toxicities were not "significant", seizures, dysphasia, psychosis, acute hemiplegia and elevated intracranial pressure were temporary events that occurred, hours to days, after treatment in about 14%[9]. Toxicities following SRS could be significant.[10]

Fractionated Stereotactic Radiotherapy[11]

- Useful in patients with larger tumors or tumors in eloquent structures.
- Survival rate comparable to that of SRS, despite poorer pretreatment prognostic factors.
- Late complications also lower.

Hypofractionated Stereotactic Radiotherapy[12, 13]

- Noninvasive, well-tolerated outpatient method.
- Palliative, high-dose, focal irradiation.
- H-SRT used in previously irradiated patients with malignant glioma was not associated with the need for Reoperation.
- Doses ranging from 20 to 50 Gy initially on a dose escalation program.
- Three noncoplanar arcs or four to six noncoplanar fixed beams, at 5 Gy/fraction
- Neurological improvement in 45%.
- Decreased steroid requirements occurred in 60%.
- Further evaluation and dose escalation is justified.

Radiation Sensitisers:

- Preliminary trials in treating recurrent glioblastoma with fractionated stereotactic radiotherapy and radiation sensitizers like Taxol revealed a one-year survival of 50% with Taxol, compared to 11% for those without Taxol.[14]

- Paclitaxel[15] and Cis-platinum have also been used. In one study, CDDP (40 mg/m2) was given weekly, with SRT once or twice weekly, to 20 patients. Five patients required surgery for tumour progression or radiation necrosis.[11]

Effect of SRS on recurrent gliomas:

- Histological evaluation of recurrent astrocytic tumors revealed that the proliferative potential of malignant astrocytic tumors in the radiosurgically treated area is reduced after SRS. Enlargement of enhanced lesions on MR images is due to propagation of the residual tumor cells that were not covered by radiosurgical target volume or to radiation necrosis.[16]

CONCLUSIONS:

The role of Stereotactic Irradiation for recurrent gliomas is continually evolving. The last word has not yet been written. The pendulum may swing completely and perhaps in the future Stereotactic Irradiation may even be used pre operatively. At present the median interval from the time of initial diagnosis and therapy to radiosurgery for recurrent tumour is about 9.6 months. The median survival calculated from the time of diagnosis is 18 months and from the time of radiosurgery is 9 months. Today Stereotactic irradiation can be considered as a salvage option in selected patients with recurrent gliomas.[17] For recurrent gliomas, the most exciting area for Radiosurgery is the correct selection of patients and targeting of tissues based on metabolic information derived from readily available MRS data. Future directions should be based on the results of prospective clinical studies.

ACKNOWLEDGMENTS:

I wish to thank Mrs. Vijayalakshmi Ganapathy and Mr. G. Murali Krishnan for secretarial assistance.

REFERENCES:

1. Ganapathy K: Stereotactic Radiosurgery for malignant gliomas. *Indian Clinical Neurosurgery* Vol. 2 Editors Anil K Singh et al CBS Publishers & Distributors.
2. Nieder C, Grosu AL, Molls MA: comparison of treatment results for recurrent malignant gliomas. *Cancer Treat Rev.* 2000;26(6):397-409. Review.
3. Hall WA, Djalilian HR, Sperduto PW, Cho KH, Gerbi BJ, Gibbons JP, Rohr M, Clark HB: Streotactic *Radiosurgery.*
4. Van Kampen M, Engenhart-Cabillic R, Debus J, Fuss M, Rhein B, Wannenmacher M: The Radiosurgery of glioblastoma multimforme in cases of recurrence. *Strahlenther Onkol* 1998;174(1):19-24.
5. Essig M, Debus J, Schlemmer HP, Hawighorst H, Wannenmacher M, van Kaick G: Improved tumor contrast and delineation in the stereotactic radiotherapy planning of cerebral gliomas and metastases with contrast media-supported FLAIR imaging. *Strahlenther Onkol* 2000;176:84-94.
6. Graves EE, Nelson SJ, Vigneron DB, Chin C, Verhey L, McDermott M, Larson D, Sneed PK, Chang S, Prados MD, Lamborn K, Dillon WP: A preliminary study of the prognostic value of proton magnetic resonance spectroscopic imaging in gamma knife Radiosurgery of recurrent malignant gliomas, *Neurosurgery* 2000;46(2)319-26.

7. Graves EE, Nelson SJ, Vigneron DB, Verhey L, McDermott M, Larson D, Chang S, Prados MD, Dillon WP: Serial proton MR spectroscopic imaging of recurrent malignant gliomas after gamma knife Radiosurgery. *Am J Neuroradiology* 2001;22:613-24.
8. Kline JL, Noto RB, Glantz M: Single-photon emission CT in the evaluation of recurrent brain tumour in patients treated with gamma knife Radiosurgery or conventional radiation therapy. *AJNR Am J Neuroradiol* 1996;17:1681-86.
9. Hall WA, Djalilian HR, Sperduto PW, Cho KH, Gerbi BJ, Gibbons JP, Rohr M, Clark HB: Stereotactic Radiosurgery for recurrent malignant gliomas. *J Clin Oncol* 1995;13(7):1642-48.
10. Nieder C, Grosu AL, Molls M: A comparison of treatment results for recurrent malignant gliomas. *Cancer Treat Rev.* 2000; 26(6):397-409. Review.
11. Glass J, Silverman CL, Axelrod R, Corn BW, Andrews DW: Fractionated stereotactic radiotherapy with cis-platinum radiosensitisation in the treatment of recurrent, progressive or persistent malignant astrocytoma. *Am J Clin Oncol* 1997; 20(3): 226-29.
12. Shepherd SF, Laing RW, Cosgrove VP, Warrington AP, Hines F, Ashley SE, Brada M: Hypofractionated stereotactic radiotherapy in the management of recurrent glioma. *Int J Radiat Oncol Biol Phys* 1997 Jan 15;37(2):393-98.
13. Hudes RS, Corn BW, Werner-Wasik M, Andrews D, Rosenstock J, Thoron L, Downes B, Currant WJ Jr.: A phase I dose escalation study of hypofractionated stereotactic radiotherapy as salvage therapy for persistent or recurrent malignant gliomas. *Int J Radiat Oncol Biol Phys* 1999; 43(2):293-98.
14. Lederman G, Arbit E, Odaimi M, Wertheim S, Lombardi E: Recurrent glioblastoma multiforme: potential benefits using fractionated stereotactic radiotherapy and concurrent taxol. *Sterotact Funct Neurosurg* 1997;69(1-4 Pt 2):162-74.
15. Lederman G, Wronski M, Arbit E, Odaimi M, Wertheim S, Lombardi E, Wrzolek M: Treatment of recurrent glioblastoma multiforme using fractionated stereotactic Radiosurgery and concurrent paclitaxel. *Am J Clin Oncol* 2000;23(2):155-59.
16. Kodera T, Kubota T, Kabuto M, Nakagawa T, Takeuchi H, Arishima.H, Sato K, Kobayashi H, Kitabayashi M, Hirose S: Analysis of the proliferative potential of tumor cells after stereotactic radiosurgery for recurrent astrocytic tumors. *Neurol Res* 2000;22:802-08.
17. Van Kampen M, Engenhart-Cabillic R, Debus J, Fuss M, Rhein B, Wannenmacher M: The Radiosurgery of glioblastoma multimforme in cases of recurrence. *Strahlenther Onkol* 1998;174(1):19-24.

48

Gamma Knife Radiosurgery for Recurrent Gliomas and Metastasis

J Misra

INTRODUCTION

Recurrent glial tumors are difficult to control as presently there is no curative treatment option available. The clinical setting is different in patients with recurrent gliomas in contrast to those with newly diagnosed gliomas. At the time of recurrence the patient has probably exhausted all other treatment options. Primary cause of death in these patients is from local disease progression causing neurological deterioration related to the site of recurrence.[1] Radiosurgical booster radiation has been proposed as means of improving local tumor control and therefore survival in these patients.[2]

Brain metastasis develops in 10-30% patients with cancer.[13] Solitary brain metastasis is found in 40% cases, two or less metastasis in 54%, and four or fewer in 72% cases.[13] Recent developments in chemotherapy and immunotherapy regimens mandate brain screening to detect and treat even subclinical disease. Surgical resection is difficult to perform when more than metastasis is present in disparate locations. Role of radiosurgery as treatment option in metastasis centers on the fact that control of growth in imaging-detectable lesion does not stop growth of imaging-undetectable lesions and in solitary metastases fresh reseeding can occur after radiosurgery.

MATERIAL AND METHODS

Clinical Material: Glial Tumors

Gamma knife radiosurgery was performed by the standardized technique in both glial and metastatic tumors using Leksell stereotactic head frame. Anxious and uncooperative patients were given Inj.Propofol sedation, all others were treated with local anesthesia using Inj. 1% xylocaine at the pin sites.

Radiosurgical dose was determined on the basis of age of the patient, tumor volume, location of tumor, histological grade, and any prior fractionated irradiation. Pilocytic astrocytomas in optico hypothalamic region were delivered <10.0 Gy and the brainstem gliomas were delivered less <14.0 Gy. After radiosurgery follow-up imaging was performed at six-month intervals and reviewed. Between April 1998 and November 2001, 45 patients aged 8 to 78 years (mean 39.16 yr.) with newly diagnosed, residual or recurrent gliomas (33 male and 12 female) underwent radiosurgery with 201-source cobalt-60 Gamma Knife (Elekta Instruments, Atlanta, GA). Histological subtypes of gliomas are summarized in **(Table 1)**. Prior surgical resection was performed in 28/45 patients and in 3/27 patients more than one surgical resection was attempted for recurrent tumors. Newly diagnosed cases were determined by image morphology, MR-spectroscopy, open biopsy or stereotactic biopsy.

Table 1 : *Histological Subtypes of glial tumors*

Histological Subtypes	Number of patients
Xanthoastrocytoma	01
Oligoastrocytoma	05
Astrocytoma I-II	20
Astrocytoma III-IV	10
Glioblastoma	05
Juvenile Pilocytic	04
Subtotal	45

In locations like the brainstem we did not insist on tissue biopsy when the MRS finding was convincing. MRS alone prior to radiosurgery was performed in 11/45 patients in those patients who either refused stereotactic biopsy or those where we felt that the risk of the procedure-related morbidity was high. 15/44 patients had received prior fractionated XRT. Radiosurgical dosage between 7.0 and 30.0 Gy (mean 15.7 Gy) and 14.0 and 66.67 Gy (mean 38.52 Gy) prescription and maximum doses were administered, respectively.

CLINICAL MATERIAL: METASTATIC TUMORS

We have treated 20 patients aged 36 to 75 years (mean 57.65 yr.) with brain metastasis (8 male and 12 female) between April 1998 and November 2001. Primary site and histology of metastasis included ovarian carcinoma 2/20, adenocarcinoma lung 6/20, carcinoma breast 5/20, adenocarcinoma colon 1/20, renal cell carcinoma 3/20, adenocarcinoma uterus 1/20, chondrosarcoma 5th rib 1/20 and melanoma 1/20 **(Table 2)**. Second radiosurgery was performed in 2/20 patients for appearance of new deposit in 1/20 and local failure in 1/20 patient. None of the patients had had undergone prior fractionated XRT prior to radiosurgery. Radiosurgical dose was determined on the criteria of age of the patient, tumor volume, location of tumor, number of metastasis, and any prior fractionated irradiation.[15] Many patients had multiple metastasis and total 49 metastatic deposits were treated in these 20 patients. Between 10.0 and 25.0 Gy (mean 23.37 Gy) and 25.0 and 62.5 Gy (mean 44.44 Gy) prescription and maximum doses were administered, respectively.

Table 2: *Histological subtypes of metastatic tumors and the primary sites*

Primary Diagnosis	No. of Cases	Prior Surgery of intracranial metastasis	Prior Chemotherapy	Second Radiosurgery
Carcinoma Lung	6	3	1	1
Overian Carcinoma	2	1	2	0
Carcinoma Breast	5	0	3	0
Carcinoma Colon	1	0	0	0
Carcinoma Uterus	1	0	0	0
Carcinoma renal cell	3	1	0	1
Chondrocarcoma Rib	1	1	1	0
Melanoma	1	1	0	0
Subtotal	20	7	7	2

RESULTS AND DISCUSSION

Glial Tumors

Selection Criteria

Selection criterion for patients likely to benefit from radiosurgery has been the subject for much research and discussion in the past. Factors that influence increased survival in patients with glial tumors are younger age, lower pathologic grade; increased Karnofsky performance status, smaller tumor volume and unifocal tumor.[4] Survival is not significantly related to radiosurgical technical parameters (dose, number of isocenters, prescription isodose percent, inhomogeneity) or the extent of prior microsurgery.[4] In gliomas there is tumor cell infiltration beyond the contrast-enhanced tumor margin on imaging. Juvenile pilocytic astrocytomas is the only exception where the enhanced tumor margin on imaging precisely conforms to the extent of the tumor. The tumor cells outside the contrast enhanced margin however fall in the dose fall-off region of the prescription isodose of radiosurgery.

Radiotherapy

Conventional radiotherapy dose delivery is limited to 60 Gy to avoid unacceptable risk of diffuse brain necrosis.[3] Prospective trials at University of California at San Francisco (UCSF), Harvard JCRT and Brain Tumor Cooperative Group reported that dose escalation by 'Booster dose' in malignant glioma significantly improved survival.[2] Radosurgery is appealing because it is non-invasive single day procedure and achieves similar results and toxicity when compared with brachytherapy.

RECURRENT GLIOMAS

By the time clinical and imaging evidence of glial tumor recurrence is documented, the patient has already exhausted the treatment options available for local tumor control. The argument of radiosurgery in recurrent glioma will be different from the indication in newly diagnosed glioma where radiosurgery booster dose is delivered as part of initial management for tumor control. Local tumor control in recurrent gliomas can be achieved by any the focused procedures like microsurgery, brachytherapy or radiosurgery. McDermott et al[10] reported on use of radiosurgery in recurrent gliomas (Glioblastoma n = 34, anaplastic astrocytoma n = 12) achieving median survival for Glioblastoma 40.6 weeks and for anaplastic astrocytoma 61.6 weeks. Alexander and Loeffler[11] reported their experience on 118 recurrent glioma patients treated with brachytherapy (n = 32) and radiosurgery (n = 86). Median tumor volumes treated were 10.1 cubic cm and 29.0 cubic cms for radiosurgery and brachytherapy, respectively. Radiosurgery was reserved for smaller tumor volumes and non-implantable sites. 19/86 22% patients required repeat surgery in radiosurgery group compared to 44% in brachytherapy group. With more experience now, radiosurgery has became the preferred treatment option for dose escalation (except with larger and irregularly shaped lesions). In our material 15/45 tumors received primary radiosurgery for newly diagnosed gliomas (after MRS or stereotactic biopsy}. There were 15/45 residual tumors (residual following recent surgery or surgery plus XRT) where booster dose was delivered **(Table 2)**. The remaining 15/45 tumors were labeled as recurrent gliomas (recurrence after initial local control with surgery or surgery plus XRT in past). Second radiosurgery was performed in 2/45 patients for local disease failure two years after first radiosurgery. Chemotherapy (PCV and Temzolamide) was delivered in 5 patients (3/5 with oligoastrocytoma and 2/5 with astrocytoma grade-IV) with good disease remission **(Table 3)**.

Astrocytoma Grade-I (Juvenile Pilocytic)

These tumors are well delineated and amenable to microsurgical resection, even from eloquent areas and brain stem, and presently there is no role of radiotherapy.[5] Radiosurgery is indicated for precisely the same reason of good delineation on imaging as adjunct to microsurgery, or even as primary treatment option. Steiner et al[7] treated 11/12 pilocytic astrocytomas and 1/12 subependymal giant cell astrocytoma with mean maximum dose 32.7 Gy and mean prescription dose 13.3 Gy included in 30-50% isodose. Follow-up at 9-58 months revealed decreased tumor size in 7/12; complete resolution in 1/12; and increased tumor

Table 3: *Treatment options in radiosurgery groups of recurrent gliomas*

Radiosurgery group	Number of patients	Chemotherapy Postradiosurgery	Surgery Postradiosurgery	Second radiosurgery Postradiosurgery
Primary	15	1	1	1
Residual	15	2	2	0
Recurrent	15	2	0	1
Subtotal	45	5	3	2

size in 4/12 cases. Follow-up available from 12 to 24 months in 3/3 Juvenile pilocytic astrocytomas of optico-hypothalamic region treated at our Centre has revealed significant decrease in size in all cases with stable vision.

Astrocytoma Grade-II

There is lack of consensus on the optimal management of these tumors. Present management options include observation alone; low dose-rate XRT and high dose-rate XRT. Partial or complete response to radiosurgery was reported first by Pozza.[6] Steiner et al treated 14 grade-II astrocytomas (0.70-9.74 cc) with mean maximum dose 36.2 Gy and mean prescription dose 13.5 Gy included in 30-50% isodose. Tumor volume decreased in 6/14; resolved completely in 2/14; and increased in 6/14 cases during follow-up of 8-62 months. Similar experience with 7 tectal low-grade gliomas reported from Karolinska Institute with follow-up of 2-15 years demonstrated decreased size in 5/7; unchanged in 1/7; and continuous increase in partially treated 1/7 case.[7]

Malignant Glioma (Grades-III & IV)

Detailed analysis has not been performed at our centre while we await adequate followup data, but preliminary results revealed that all the patients with histological grade IV and glioblastoma died of intracranial disease within 12 months post radiosurgery. All the tumors initially decreased in size or resolved but they all recurred again due to local progression from tumor margin outside the prescription isodose margin. Steiner et al[7] reported 30 malignant gliomas, primarily with 20-40 Gy as maximum dose and 9-17 Gy prescription dose in 8/30 cases. Secondarily 22/30 cases received 'booster' with maximum doses 10-40 Gy post radiotherapy. Alexander[8] reported 11 malignant gliomas treated with gamma knife showing decreased tumor in 3/11; unchanged in 2/11; and increased tumor size in 6/11 cases. At follow-up of 3-24 months 6/11 cases died and 5/11 cases were stable. On the contrary at Karolinska Institute, the experience with gamma knife surgery on 30 malignant gliomas as an adjunct to surgery / radiation failed to demonstrate additional advantage.[9]

METASTATIC TUMORS

Solitary deposit is found in 40% intracranial metastasis and if untreated median survival of 1 month is reported. Surgery alone obtains mean survival of only 6 months,[18] surgery combined with XRT for 40 weeks with a recurrence in 20%, and XRT alone has survival of only 15 weeks with recurrence in 52%.[16] In autopsy series of 10,916 cancer patients Pickren et al[13] found incidence of brain metastasis in 8.7% cases (67% had systemic metastasis alone and 24% no metastasis either systemic or brain metastasis. This autopsy study also found solitary metastasis in 39%, two or less in 54%, four or fewer in 72%, and eight or more in 20% of the patients.[13] These findings suggest that metastasis to brain is in limited numbers and are amenable to control of local disease progression by surgery or radiosurgery.

Chemotherapy and Immunotherapy

Control of primary disease due to recent imrovements in systemic therapies like chemotherapy and immunotherapy mandates control of intracranial disease prior to initiation of these therapies. Systemic therapies do not cross the blood-brain barrier requiring local disease control first because they are known to increase the intracranial peritumoral edema making asymptomatic disease symptomatic. Screening to detect subclinical brain metastasis is now routinely performed.

Surgery and Radiotherapy

Surgery unfortunately is not an option with deeply located tumors or in patients with high medical risks. Mintz et al[17] failed to show any benefit for surgery followed by whole brain XRT in either duration or quality of survival (5.6 months for surgery plus XRT versus 6, 3 months for XRT alone). Rutigliano et al[19] reported higher treatment related morbidity and mortality for surgery (30% and 7%, respectively) versus radiosurgery (4% and 0%, respectively). In multiple metastatic deposits, unlike surgical resection, all the deposits can be treated in the same session limited only by the patience of the radiosurgical team. Patchel et al[16] reported an actuarial recurrence rate >80% for whole brain XRT (36 Gy in 12 fractions) of solitary metastasis, where it was compared unfavorably to surgery. Radiation Therapy Oncology Group (RTOG) investigation on five different fractionation regimen (20 Gy in 5 fractions, 30 Gy in 10 or 15 fractions, 40 Gy in 15 or 20 fractions) of whole brain XRT could not detect any difference in survival outcome on comparative analysis of various fractionation schedules.[14]

Radiosurgery in Radio-resistant Metastasis

Gamma Knife Users Group Study[24] has reported had better local tumor control in even relatively radio-resistant tumors like melanoma and renal cell carcinoma. We have treated with Gamma knife radiosurgery 1 patient with melanoma metastasis 3 with renal cell carcinoma. All the tumors continue to decrease in size with good local control in followup duration ranging 12 to 24 months. Uveal melanomas were treated with Gamma knife radiosurgery first time in Clinica Del Sol, in Buenos Aires.[25] Larger series with average prescription dose 40.0–45.0 Gy and maximum dose 75.0–80.0 Gy was reported from Vienna.[26] Even though the side effects were severe, local disease control was satisfactory.

CONCLUSIONS

Constant research effort is required to document additional survival benefit with Gamma Knife Radiosurgery. Hopefully with introduction of radiation sensitizers or radiation protectors there will be further improvement in the therapeutic index. Radiosurgery presently offers an additional treatment modality, compatible with and complementary to the other options in the management of recurrent gliomas. Radiosurgery when used for local dose escalation is effective in controlling local disease progression.

The best management brain metastasis has to defined individually based on disease and patient parameters. Amongst the available current options radiosurgery alone or combined with radiotherapy is an effective initial treament option by virtue of lower morbidity. Surgical resection should be reserved for large resectable solitary tumors causing neurologic deficit due to mass effect. Radiosurgery also attractive and effective initial treament option in multiple metastasis and relatively radio-resistant histologies.

REFERENCES:

1. Smith KA, Elder W, Rogers L, Fiedler J: Adjunctive gamma knife radiosurgical boost for patients with glioblastoma multiforme. *Abstract in proceedings* 10th International Meeting of the Leksell Gamma Knife Society, 2000;53.
2. Loeffler J, Alexander III E, Shea M, et al: Radiosurgery as part of the initial management of patients with malignant gliomas. *J Clin Oncol* 1992;10:1379-85.
3. Sheline GE, Wara WM, Smith V: Therapeutic irradiation and brain injury. *Int J Radiat Oncol Phys* 1980; 6:1215.
4. Larson DA, Gutin PH, McDermott M, et al: Gamma knife for glioma - selection factors and survival. *Int J Radi Onco Bio Phys* 1996;36(5):1045-53.
5. Shaw EG, Scheithauer BW, JR: Management of supratentorial low-grade gliomas. *Oncology* 1993; 7(7):97-104, 107; discussion 108-11.
6. Pozza F, Colombo F, Chierego G, et al: Low-grade astrocytomas: treatment with unconventionally fractionated external beam stereotactic radiation therapy. *Radiology* 1989; 171(2):565-69.
7. Steiner L, Prasad D, Lindquist C, Steiner M: Clinical aspects of gamma knife stereotactic radiosurgery. *Text Book of Stereotactic and Functional Neurosurgery / Gildenberg PL, Tasker RR eds.* McGraw Hill 1996;763-803.
8. Alexander III E, Coffey R, Loeffler J: Radiosurgery for gliomas, In: Alexander III E, Loeffler J, Lunsford L, eds. *Stereotactic radiosurgery*, New York: McGraw-Hill, 1993;207-19.
9. Lindquist C, Steiner L: Radiosurgery for tumors. In: Wilkins R, Rengachary S, eds, *Neurosurgery*. New York: McGraw-Hill, 1996;1887-1907.

10. McDermott MW, Sneed PK, Chang SM, Gutin P, et al: Results of Radiosurgery for Recurrent Gliomas. In Kondziolka D (eds): *Radiosurgery 1995*; Radiosurgery, Basel, Karger, Vol. 1 1996; 102-12.
11. Alexander III E, Loeffler JS: Clinical experience with Linac radiosurgery. *Text Book of Stereotactic and Functional Neurosugery / Gildenberg PL, Tasker RR eds*, McGraw Hill 1996;745-56.
12. Kondziolka D, Flickinger JC, Lunsford JC: Stereotactic radiosurgery for glial neoplasms. Gamma Knife Brain Surgery. *Prog Neurol Surg*, Basal, Karger, Vol. 14 1998;160-74.
13. Pickren JW, Lopez G, Tzukada Y, et al: Brain Metastasis. An autopsy study. *Cancer Treat Symp* 1983; 2;295-313.
14. Gelber R, Larson M, Borget BB, et al: Equivalence of radiation schedules for the palliative treatment of brain metastasis in patients with favourable prognosis. *Cancer* 1981;48:1749-53.
15. Flikinger JC, Lunsford LD, Konziolka D: Dose-volume considerations in radiosurgery. *Stereotact Funct Neurosurg* 1991;57:99-105.
16. Patchell RA, Tibbs PA, Walsh JW, et al: A randomized trial of surgery in the treatment of single metastases to the brain. *New England J Med* 1990;322(8):494-500.
17. Mintz AH, Kestle J, Rathbone MP, et al: A randomized trial to assess the efficacy of surgery in addition to radiotherapy in patients with a single cerebral metastasis. *Cancer* 1996;78:1470-76.
18. Salcman M: Metastatic brain tumors. *Contemp Neurosurg* 1990;26:1-5.
19. Rutigliano MJ, Lunsford LD, Kondziolka D, et al: The cost effectiveness of stereotactic radiosurgery versus surgical resection in the treatment of solitary metastatic brain tumors. *Neurosurgery* 1995;37: 445-55.
20. Auchter RM, Lamond JP, Alexander EA, Buatti JM, et al: A multi-institutional outcome and prognostic factor analysis of radiosurgery for resectable single brain metastasis. *Int J Radiat Oncol Biol Phys* 1996;35:27-36.
21. Noordijk EM, Vecht CJ, Haaxma-Reiche H, et al: The choice of treatment of single brain metastasis should be based on extracranial tumor activity and age. *Int J Radiat Oncol Biol Phys* 1994;29: 711-17.
22. Kihlstrom L, Karlsson B, Lindquist C: Gamma Knife surgery for cerebral metastases. implications for survival based on 16 years experience. *Stereotact Funct Neurosurg* 1993;61(Suppl 1):45-50.
23. Kihlstrom L, Karlsson B, Lindquist C, Noren G, Rahn T: Gamma knife surgery for cerebral metastasis. *Acta Neurochirurgica - Supplementum* 1991;52:87-89.114.
24. Flickinger JC, Kondziolka D, Lunsford LD, et al.: A multi-institutional experience with stereotactic radiosurgery for solitary brain metastasis. *Int J Radiat Oncol Biol Phys* 1994;28:797-802.
25. Chinela AB, Zambrano A, Bunge HY, et al: Gamma Knife radiosurgery in uveal melanomas; in Steiner L (ed): Radiosurgery: Baseline and Trends. New York, Raven Press, 1992; 161-69.
26. Marchini G, Gerosa M, Piovan E, et al: Gamma Knife stereotactic radiosurgery for uveal melanomas. *Stereotact Funct Neurosurg* 1996; 66(Suppl 1):208-13.

49
Epidemiology and Pathology of CNS Tumors in AIDS

V Santosh, A Mahadevan, V Ravi

INTRODUCTION

Since the original description of AIDS in 1981, HIV infection has become recognized worldwide and is now being reported from more than 175 countries. According to the recent UN AIDS/WHO estimates (end of year 2000), over 36 million persons are currently infected with HIV, and over 20 million have died from AIDS. The greatest impact of HIV infection to date has been felt by the developing countries of Sub Saharan Africa and more recently in South East Asia, especially Thailand India.

AIDS & NEOPLASIA

Epidemiological data from most developed and few developing countries have shown that upto 40% of patients with AIDS will ultimately develop some form of cancer.[1] Non Hodgkin's lymphoma (NHL), Kaposi Sarcoma (KS) and invasive cervical cancer have a higher incidence in persons with HIV infection and all three are now categorized as AIDS-defining illnesses. In addition, several reports suggest that a number of other malignancies may occur at an increased incidence in persons with HIV infection including squamous cell carcinoma in the head and neck region, plasmacytoma, melanoma etc.[2] The incidence of AIDS related malignancies is expected to further increase as more effective therapies for HIV and associated opportunistic infections allow patients to live long in an advanced stage of immunodeficiency.

Recent epidemiological data from Western Europe has shown that, among AIDS cases reported, 6.4% had Kaposi sarcoma and 5.3% had NHL as AIDS defining illnesses.[1] Another large study from the United States has reported NHL as a part of an AIDS defining illness in nearly 3.2% of all people with AIDS.[3] Population based cancer and AIDS registries linkage studies conducted in the United States, Italy and Australia and from a few cohort and case control studies have estimated that the relative risk of neoplasms in adults with HIV/AIDS was over

1000 fold for Kaposi sarcoma and over 300 fold for high grade NHL. For Hodgkin's disease, a consistent ten-fold higher relative risk was observed. For cervical and other anogenital tumors associated with human papilloma virus, relative risk increases were 2 and 12 fold, depending upon location. In Africa, the AIDS epidemic has led to Kaposi Sarcoma becoming the most common type of cancer in men in several areas. However, the relative risk of AIDS-associated tumors have been lower in Africa than those reported in the West.

INDIAN SCENARIO

HIV infection was recognised in India in the mid 1980's and WHO estimates that over 3.6 million people are already infected with HIV in this country during the past decade. **(Table 1)** shows the cumulative AIDS cases in India from May 1986 (NACO data). This however does not reflect the actual number of AIDS cases that have occurred in the country since there is gross under reporting of this disease due to various reasons. The clinical presentation of AIDS in India has been similar to that found in other developing countries, with majority of patients presenting with varied opportunistic infections and occasional ones with neoplastic diseases). Despite the marked increase in the number of AIDS cases in India, reports of AIDS associated lesions from different parts of the country document very few neoplasms. The first case of AIDS associated neoplasm reported from India was a case of NHL.[4] Another case of NHL associated with AIDS has been reported from Kerala.[5] Chacko et al from Vellore found only one case of immunoblastic lymphoma in a large series of 61 AIDS cases.[6] All the three AIDS associated NHLs reported from India were systemic lesions and there has been no cases of AIDS related primary CNS lymphoma (PCNSL) hitherto reported from India.

Table 1: *(Courtesy NAVO)*

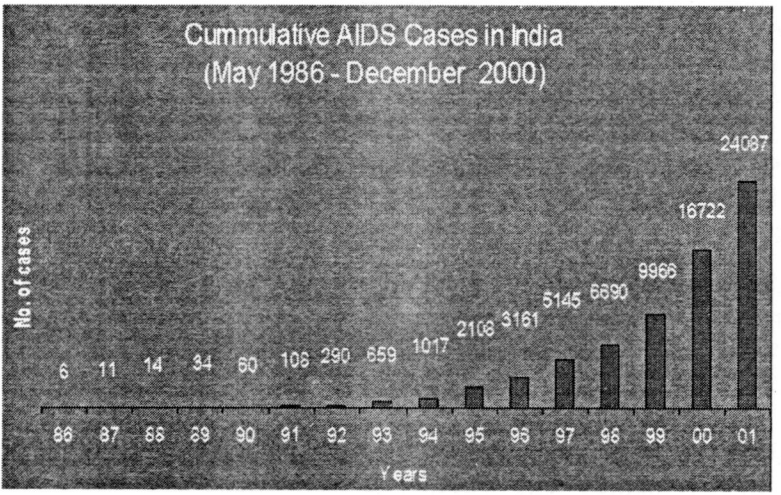

In another study from New Delhi, Sarkar et al on reviewing 80 cases of PCNSL did not find any case to be HIV seropositive. Majority of patients were young, immunocompetent and only one was renal transplant recepient. The authors remarked that despite the increasing numbers of AIDS cases in India there has been no increase or change in the trend of PCNSL.[7]

In a large autopsy study on 92 adult patients who succumbed to AIDS, Lanjewar and Co-workersfrom Mumbai noted that most patients had died of opportunistic infections. Among the neoplastic diseases, there was one case each of Kaposi sarcoma and hepatocellular carcinoma.[8] Following this, in another neuropathological study on 85 adult brains of patients who died of HIV/AIDS, the same authors noted CNS opportunistic infections were the commonest. Primary CNS tumors were conspicuously absent in their series.[9] Similarly, in an earlier study from NIMHANS, where 47 cases of HIV/AIDS were analysed on autopsy and surgical biopsy material, there were no neoplastic lesions involving the CNS.[10]

Gatpoh et al, in their study on 26 patients of AIDS from North East India, noted 4 to have AIDS associated malignancies. Kaposi sarcoma was noted in 2 cases and one case each had NHL and promyelocytic leukemia.[11] Very few cases of Kaposi sarcoma have been reported from India, none involving the nervous system.[8,12]

PATHOLOGY OF AIDS ASSOCIATED CNS NEOPLASIA

Unlike the data from India, in most series from the West, it has been noted that nearly 40% of AIDS patients develop some form of neoplastic lesions. CNS tumors alone comprise 5-10% of all neurological manifestations of AIDS. The most common neoplasms occurring in the CNS include Non-Hodgkin's lymphoma (NHL) followed by Kaposi sarcoma (KS), Hodgkin's disease (HD), gliomas, smooth muscle tumors, plasmacytoma while lymphomatoid granulomatosis occur less commonly in association with AIDS.

NON-HODGKIN'S LYMPHOMA (NHL)

These are the commonest AIDS associated neoplasms. The two types of CNS lymphomas recognised are:

1. Primary CNS lymphoma (PCNSL)
2. Systemic AIDS related lymphomas with secondary involvement
   Of the nervous system

PRIMARY CNS LYMPHOMA (PCNSL)

The incidence of PCNSL has increased markedly worldwide from 0.8-1.5% upto 6.6% of primary intracranial neoplasms in several neuropathological studies, as a consequence of the AIDS epidemic.[13] In immunocompetent patients, the incidence has increased in some but not in all series and populations. In patients with AIDS, the risk of developing this neoplasm is 1000 times greater than in the non-AIDS population.[1] It has been noted that 2-12% of AIDS patients develop PCNSL mainly during the late-stage of AIDS and occurs mostly in profoundly immunocompromised patients with CD_4 counts below 50 cells/Cu mm.[14] The finding of PCNSL in HIV positive patients is now considered as an AIDS-defining disease.[15] The peak incidence of PCNSL in immunocompetent individuals is during the sixth and seventh decades. In AIDS patients, the disease manifests mostly in young adult males in the third decade.

More than 75% of PCNSL are supratentorial and approximately 25% to 50% are multiple masses, particularly in AIDS patients. Preferential involvement of diencephalic structures and basal ganglia is seen even though other parts of cerebral hemispheres can be involved. Nearly 20% are periventricular in location. Brain stem and cerebellum are infrequently involved. Secondary meningeal spread is seen in 30-40% of PCNSL while primary leptomeningeal lymphoma is rare, accounting for about 8% of these tumors.[16] In contrast, secondary CNS malignant lymphomas prefer the dura and leptomeninges but parenchymal lesions may also occur. Occular disease, which may precede intracranial lesions is noted in 15-20% of cases.[17]

About 50-80% of PCNSL patients present with neurological deficits, 20-30% with neuropsychiatric symptoms, 10-30% with features of raised intracranial pressure and 5-20% with seizures. The duration of illness is much shorter in AIDS patients with PCNSL compared to the non-AIDS cases.

Cranial CT and MRI of PCNSL often show solitary or multiple, hyper or isodense diffusely enhancing lesions often solid and rarely cystic. FDG-PET scan or thallium-201-SPECT are helpful in the differential diagnosis of ring enhancing mass lesions that are frequently seen in AIDS related PCNSL and other non nonneoplastic conditions such as cerebral toxoplasmosis.[18,19]

Macroscopically, PCNSLs occur as single or multiple masses in the cerebral hemisphere often deep-seated and adjacent to the ventricular system. However superficial tumors can be encountered. They are often firm, friable and granular masses. Demarcation from the surrounding parenchyma is variable. Some tumors appear well-delineated, like a metastasis. When diffuse borders and effacement are present the lesions resemble gliomas. Diffusely infiltrating forms without a mass lesion are rarely encountered and have been referred to as 'lymphomatosis cerebri'.[20] AIDS related PCNSL tend to have more necrotic and haemorrhagic zones and often resemble cerebral toxoplasmosis.[21] The rare meningeal form can mimic meningioma or can appear normal macroscopically.

Histologically all cases of PCNL show a diffuse growth pattern, and unlike other nodal or extranodal NHLs, a follicular pattern is not described. The tumour is composed of lymphoid cells with a variable appearance and show a characteristic angiocentric infiltration, forming collars of tumor cells within concentric perivascular reticulin sheaths. From these perivascular cuffs, the tumor cells invade neural parenchyma, either as compact cellular aggregates or with single diffusely infiltrating tumor cells resembling encephalitis. Field necrosis may be seen with islands of viable tumor cells surrounded by large areas of coagulative necrosis.[22] There is no universally accepted classification scheme for PCNSL. The various classification systems currently used for nodal or extranodal NHLs[23] have limited value in the understanding of PCNSL.

On histology, most PCNSL are diffuse large cell (upto 50%) or diffuse large cell immunoblastic (20%) types. Small cleaved, non-cleaved, mixed small and large cell or atypical/unclassified types can also occur. Approximately 98% are B-cell lymphomas and express pan-B markers such as DC20 and CD79a. T-cell lymphomas constitute about 2% of all PCNSL an are often seen in immunocompetent individuals.[24, 25] A subset of Ki-1 lymphomas and lymphomatoid granulomatosis belong to the T-lymphoma category.[26]

In establishing the histological diagnosis of PCNSL, stereotactic biopsy (STB) is currently the method of choice, since the tumors are often deeply located and surgical resection is not possible. However, these biopsies have to be interpreted with caution particularly inpatients who have received steroids prior to the procedure. Steroid therapy can cause considerable shrinkage of the lymphoma cells and on histology can be misinterpreted as mature lymphocytes.

Abnormalities of CSF in cases of PCNSL included raised protein pleocytosis, elevated β_2-microglobulin and LDH. Elevated LDH isoenzyme is reported to be associated with leptomeningeal spread. Despite presence of tumor cells in the CSF, counts can be normal. CSF cytology is of diagnostic value in 5-30% of PCNSL cases as against metastatic malignant lymphomas where it is 70-90%, particularly if immunocytochemistry is used to determine monoclonality.[27]

The consistent association of EBV expression supports a pathogenic role for EBV in AIDS related PCNSL. The EBV genome has been demonstrated in the tumor cells in more than 95% of AIDS related PCNSL, but only in 0-20% of non-AIDS PCNSL[14, 28] and in less than 50% of AIDS associated systemic NHL. PCR assay for EBV DNA in the CSF is reported to be highly specific (100%) and sensitive (80%) for the detection of PCNSL in AIDS patients.[29] Genomic DNA of herpes virus (HHV-6 or HHV-8) has been identified in a few cases of PCNSL.[30, 31]

In general, PCNSL patients have a poor prognosis. Compared to non-AIDS PCNSL, patients with AIDS related PCNSL have a worse outcome, probably because of aggressive growth of tumour in an environment of impaired immunosurveillance and variable response to radiotherapy. Further, though improved survival has been reported with high dose radiation in AIDS related

PCNSL, there can be a risk of exacerbation of HIV-related encephalopathy. Hence these patients are often administered low-dose irradiation.[32]

Prognosis is still worse in AIDS patients who have an opportunistic infection co-existing with PCNSL. Some of the favourable prognostic factors include single intracranial lesions, absence of immunodeficiency, age below 60 yrs, a preoperative Karnofsky score of over 70 and absence of meningeal or periventricular tumors.[22]

In general, PCNSL is sensitive to radiation therapy and occasionally complete remissions have been observed.[32] A variety of chemotherapeutic agents have been used for both non-AIDS and AIDS related PCNSL.

HODGKIN'S DISEASE (HD)

Hodgkin's disease involving the CNS is rare, even in the absence of AIDS and classically involves the dura and leptomeninges. Parenchymal involvement is mainly by direct extension of the neoplastic process. In 1991, Hair and colleagues[33] observed that of the 42 intracerebral cases of HD reported till then, only 6 were intraparenchymal without contiguous meningeal or bony lesions. Till date, only two cases of intracerebral HD in HIV positive patients have been recorded in literature.[33, 34]

ANGIOTROPIC (INTRAVASCULAR) LYMPHOMA

This unusual form of lymphoma affects multiple organ systems, skin and nervous system being selectively affected. CNS is involved in more than 30% cases. In this form of lymphoma where large B lymphocytes get sequestered within lumina of blood vessels, plugging them and resulting in small infarcts in the brain and spinal cord.[35, 36] Absent or reduced expression of beta-2 integrins on tumor cells may contribute to an impaired capability for extravasation.[22] The association of this neoplasm with AIDS is uncommon.

LYMPHOMATOID GRANULOMATOSIS (LG)

This quasineoplastic entity was first described in 1972 as a disease that primarily affects the lung. The CNS is involved as part of the multisystem disease. It is frequently observed in immunocompromised patients. In AIDS, LG confined to the CNS has been rarely recorded.[37, 38]

Grossly the lesions are multifocal and scattered throughout the brain as necrotizing lesions. Microscopically, the lesions show a polymorphic infiltrate of lymphocytes, plasma cells, histiocytes and scattered atypical cells in an angiocentric distribution. This infitrate affects the medium and small sized arteries and veins causing necrosis of the walls and thrombosis with infarction, LG is considered as a reactive process probably stimulated by EBV.[39] It is supposed to begin as a non-neoplastic process with polymorphic infiltrates and T-cell predominance. However, eventually there is clonal expansion of T cells and some authors consider this to be a variant of T-cell lymphoma.[40]

KAPOSI SARCOMA (KS)

This is the most common neoplasm associated with AIDS and develops in about 30% of AIDS patients.[41] It preferentially involves homosexual males and manifests as multicentric tumors of the skin. Intracranial involvement is extremely rare. Only three cases have been reported and all these have been metastatic to the CNS.[42,43] Occasionally these tumors have been detected at autopsy suggesting that CNS involvement is an end-stage event in AIDS related KS.[44]

Grossly, CNS lesions present as multiple haemorrhagic nodules, most often in the cerebellum. Histology is identical to non-CNS counterparts characterized by atypical spindle cell proliferation with RBC's within slit-like vascular channels and hemosidderin pigment within cells. CNS tumors by themselves appear quite radiosensitive.

It is proposed that KS is not a true sarcoma but a proliferation of endothelial cells stimulated by the 'tat' protein of the HIV virus,[45] since KS cells are of putative endothelial lineage.

GLIOMAS

The incidence of glial tumors among focal mass lesions in AIDS is rare. However, in two series, it is reported that gliomas constitute approximately 6% of AIDS associated neoplasms.[46,47] The occurrence of gliomas in AIDS patients was generally considered to be coincidental. But many recent studies in literature suggest that HIV may have a strong influence on oncogenicity of glial cells though HIV-1 is by definition a non-transforming lentivirus.

Evidences from experimental studies suggest that the proteins encoded by the 'tat' and 'Nef' gene of the HIV-1 can be tumorigenic and promote malignant transformation of astrocytoma cells. Nef is now believed to be involved in the development of AIDS associated brain tumors.[48] Alternatively, co-infection by other oncogenic viruses can occur in immunosuppressed state. For instance, Papova viruses are oncogenic when inoculated into hamsters and produce CNS tumors, especially gliomas.[49]

OTHER TUMORS

These include intracranial and spinal leiomyomas and leiomyosarcoma which occur rarely. Occurrence in HIV positive patients has been occasionally reported.[50]

Plasmacytomas occur less frequently than lymphomas, but the incidence of these tumors is also being reported with increasing frequency in HIV positive patients. Intracranial plasmacytoma most often appears as a nodular or plaque-like dural mass, with variable infiltration of the underlying brain.[22] Although rare, primary intraparenchymal tumors have been described[51] and unlike their HIV-negative counterparts, these tumors occur at a younger age in HIV patients.

NIMHANS DATA

There has been a yearwise increase in the HIV seroprevalence amongst the patients seen at NIMHANS between 1989 to 2001. Amongst the total 31,786 samples tested between October 1989 to July 2001, 2110 were confirmed positive for HIV infection (6.6%) The HIV seroprevalence was 0.56% in 1989.[52] From 1995 onwards there has been a steep increase in the annual seroprevalence of HIV, from 3.32% in 1995 to 26.2% in 2001.

At NIMHANS, we have also analysed the various CNS mass lesions occurring in HIV seropositive patients between 1990 and August 2001. **(Table 2)** enlists the various mass lesions. Earlier to 1995, HIV testing was done mandatorily for all patients undergoing surgery and thus probably larger number of mass lesions associated with HIV were encountered, which included 7 gliomas, 5 miscellaneous tumors and other inflammatory lesions, many of which could be coincidental occurrence. From 1995 onwards, HIV testing was carried out only on clinical or radiological suspicion or if there was a history of high risk behaviour in the patient and hence the number of mass lesions were lower than in the first group. These probably represent the actual HIV associated mass lesions. Diagnosis was established either by histopathology or by serology. Inflammatory masses, such as tuberculoma, toxoplasmosis and abscesses were the predominating lesion. During this entire period, we have encountered only two cases of PCNSL diagnosed on biopsied tissue.

Table 2: *CNS mass lesions in HIV patients NIMHANS data (n=31)*

Lesions	Surgical/HP diag.		Clinical/radiological/ Serological diag.
	1990-95	1995-01	1995-01
Toxoplasmosis	-	2	1
Tuberculoma	-	1	4
PCNSL	1	1	-
Glioma	7	-	-
Abscess	2	1	1
Metastasis	1	-	-
HIV encephalitis	2	-	-
Miscellaneous	5	-	-
No diagnosis	-	-	2

Autopsy data on HIV/AIDS patients from NIMHANS from 1989 to August 2001 also showed that among the 89 patients who succumbed to several AIDS related diseases, majority had opportunistic CNS infection such as cryptococcal meningitis, Toxoplasma encephalitis, Neurotuberculosis. We have noted one case each of PCNSL and lymphomatoid granulomatosis. This observation is comparable to the other large autopsy series from Mumbai by Lanjewar and Co-workers.[8,9]

Among the 48 cases of histologically verified PCNSL, diagnosed over 10 years at NIMHANS, 29 were tested for HIV and only three cases were HIV seropositive and were categorized as AIDS related NHLs.

Unlike reports from the West, where the incidence of neoplasms in association with AIDS is documented to be on the rise, in India, opportunistic infections continue to be the most common lesions causing moribidity and mortality in AIDS patients. Only a few cases of AIDS associated neoplasms have been documented. In these cases, the cause and effect relationship has not been seriously considered. Further detailed systematic study and documentation of these cases will provide an opportunity to study the oncogenic potential of the HIV and evolution of CNS neoplasia in its multifarious cytological forms.

REFERENCES:

1. Dal Maso L, Serranio D, Franceschi S: Epidemiology of AIDS related tumors in developed and developing countries. *Eur J Cancer* 2001;37:1188-1201.
2. Smith C, Lilly S, Mann KP, Livingston E, et al: AIDS related malignancies. *Ann Med* 1998;30:323-24.
3. Cote TR, Biggar RJ, Rosenberg PS, Devesa SS, et al: Non-Hodgkin's lymphoma among people with AIDS: incidence, presentation and public health burden. AIDS/Cancer study group. *Int J Cancer* 1997;73:645-50.
4. Subramaniam S, Krishnan RK, Vijaysarathy K: Non-Hodgkin's lymphoma in AIDS – a case report. *JAPI* 1986;37:230-32.
5. Unnikrishna SS, Abhayambika K, Verghese R, Legor M, Sarojini: Clinical presentation of AIDS – A Kerala experience. *JAPI* 1993;41 (1):38-40.
6. Chacko S, John TL, Babu PG, et al: Clinical profile of AIDS in India. *JAPI* 1995;43;535-38.
7. Sarkar C, Sharma MC, Pandey RM, Mehta VS: Hospital based prevailence and clinicopathological features of primary CNS lymphoma: An Indian experience of 80 cases. *Brain Pathol* 2000;10:693.
8. Lanjewar DN, Shetty CR, Katdare G: Profile of AIDS pathology in India. An autopsy study. In: Wagholikar UL, Deodhar KP (eds). Recent advances in Pathology, New Delhi. Jaypee Brothers Medical Publishers 1998;pp 83-98.
9. Lanjewar DN, Jain PP, Shetty CR: Profile of central nervous system pathology in patients with AIDS;an autopsy study from India. *AIDS* 1998;12:309-13.
10. Santosh V, Yasha TC, Panda KM, Das S, Satishchandra P, Gourie-Devi M, Ravi V, et al: Pathology of AIDS study from a neuropsychiatric center from South India. *Ann Ind Acad Neurol* 1998;1:71-82.
11. Gatpoh ED, Zamzachin G, Babina Devi S, Punyabati P: AIDS related malignant disease at Regional Institute of Medical Sciences. *Indian J Pathol Microbiol* 2001;44 (1):1-4.
12. Shroff HJ, Dashatwar DR, Deshpande RP, et al: AIDS associated kaposi sarcoma in an Indian female. *JAPI* 1993;41:241-42.
13. Miller DC, Hochberg FH, Harris NL, et al: Pathology with clinical correlation of primary central

nervous system non-Hodgkin's lymphoma. The Masschusettes General Hospital experience 1958-1989. *Cancer* 1994;74:1383-97.
14. Camilleri-Broet S, Davi F, Feuillard J, et al: AIDS related primary brain lymphomas: histopathologic and immunohistochemical study of 51 cases. The French study group for HIV associated tumors. *Hum Pathol* 1997;28:367-74.
15. Revision of the case definitions of acquired immunodeficiency syndrome for national reporting in United States. *MMWR* 1985;34:373-75.
16. Grove A, Vyberg M: Primary leptomeningeal T-cell lymphoma of the central nervous system. *Clin Neuro pathol* 1993;12:7-12.
17. Brown MT, McClendon RE, Gockerman JP: Primary central nervous system lymphoma with systemic metastasis: case report and review. *J Neuro Neuro oncol* 1995;23:207-21.
18. Hoffman JM, Waskin HA, Schiefte T, et al: FDG-PET in differentiating lymphoma from non malignant central nervous system lesions in patients with AIDS. *J Nucl Med* 1993;34:567-75.
19. Antinori A, De Rossi G, Ammassuri A, et al: Value of combined approach with thallium – 201. single-photon emission computed tomography and Ebstein Barr Virus – DNA polymerase chain reaction in CSF for the diagnosis of AIDS related primary CNS lymphoma. *J Clin Oncol* 1999;17:554-60.
20. Bakshi R, Mazziotta JC, Mischel PS, et al: Lymphomatosis cerebri presenting as a rapidly progressive dementia: clinical, neuroimaging and pathologic findings. *Dement Geriatr Cogn Disorder* 1999;10:152-57.
21. Morgello S, Petito CK, Mouradian JA: Central nervous system lymphoma in the acquired immunodeficiency syndrome. *Clin neuropathol* 1990;9:205-15.
22. Paulus W, Jellinger K, Morgello S, Deckert – Schluter M: Malignant lymphomas. In: Pathology and Genetics. Tumors of the Nervous System (Ed) Kleihues P and Cavenee WK. International Agency for Research on Cancer, Lyon 1997;pp 198-203.
23. Mason DY, Harris NL: Human lymphoma. Clinical implications of the REAL classification. 1999. Springer: Berlin.
24. Morgello S, Maiese K, Petito CK: T-cell lymphoma in the CNS: Clinical and pathologic features. *Neurology* 1989;39:1190-96.
25. Nitta T, Uda K, Ebato M, et al: Primary peripheral postthymic T-cell lymphoma in the central nervous system: Immunological and molecular approaches to diagnosis. *J Neurosurg* 1995;82:77-82.
26. Hamilton MG, Demetrick DJ, Tranmer BI, Curry B: Isolated cerebellar lymphomatoid granulomatosis progressing to malignant lymphoma. Case report. *J Neurosurg* 1994;80:314-20.
27. Fine H, Mayer R: Primary central nervous system lymphoma. *Ann Intern Med* 1993;119:1093-1104.
28. Morgello S: Pathogenesis and classification of primary centralnervous system lymphoma: an update. *Brain Pathol* 1995;5:383-93.
29. Cingolani A, De Luca A, Larcocca LM, et al: Minimally invasive diagnosis of acquired immunodeficiency syndrome – related primary central nervous system lymphoma. *J Natl Cancer* 1998;90:364-69.
30. Morgello S, Tagliati M, Ewart MR: HH-8 and AIDS related CNS lymphoma. *Neurology* 1997;48:1333-35.
31. Paulus W, Jellinger K, Hallas C, et al: Human herpes virus – 6 and Ebstein-Barr virus genome in primary cerebral lymphomas. *Neurology* 1993;43:1591-93.
32. Baumgatner JE, Rachlin JR, Beckstead JH, et al: Primary central nervous system lymphoma: natural history and response to radiation therapy in 55 patients with acquired immunodeficiency syndrome. *J Neurosurg* 1990;73:206-11.
33. Hair LS, Rogers JD, Chadbum A, et al: Intracerebral Hodgkin's disease in human immunodeficiency virus-seropositive patient. *Cancer* 1991;67:2931-34.
34. Tirelli U, Vaccher E, Rezza G, et al: Hodgkin's disease and infection with the human immunodeficiency virus (HIV) in Italy. *Ann Int Med* 1988;108:309-10.
35. Liszka U, Drlicek M, Hitzenberger P, et al: Intravascular lymphomatosis: A clinicopathological study of three cases. *J Cancer Res Clin Oncol* 1994;120:164-68.

36. Vieren M, Sciot R, Robberecht W: Intravascular lymphomatosis of the brain: a diagnostic problem. *Clin Neurol Neurosurg* 1999;101:33-36.
37. Anders KH, Latta H, Chang BS, et al: Lymphomatoid granulomatosis and malignant lymphoma of the central nervous system in acquired immunodeficiency syndrome. *Hum Pathol* 1989;20:326-34.
38. Anders KH, Latta H, Vinters HV: Lymphomatoid granulomatosis and malignant lymphoma in the central nervous system of patients with AIDS. *J Neuropathol Exp Neurol* 1988;47:386.
39. Veltri RW, Raich PC, Mc Clung JE, et al: Lymphomatoid granulomatosis and Epstein-barr virus. *Cancer* 1982;50:1513-17.
40. Paulus W, Ott MM, Strik H, et al: Large cell anaplastic (K1-1) brain lymphoma of T-cell genotype. *Hum Pathol* 1994;25:1253-56.
41. Welch K, Fin kbeiner W, Apers CE: Autopsy findings in the acquired immunodeficiency syndrome. *JAMA.* 1984;252:1152-1155.
42. Ariza A, Kim JH: Kaposi sarcoma of dura mater. *Hum Pathol* 1988;19 (12):1461-62.
43. Gorin FA, Bale JF, et al: Kaposi sarcoma metastatic to the CNS. *Arch Neurol* 1985;42:162-65.
44. Buttner A, Marquart KH, Mehraein P, Weis S: Kaposi sarcoma in the cerebellum of a patient with AIDS. *Clin Neuropathol* 1997;16(4):185-87.
45. Ensoli B, Barillari G, et al: 'tat' protein of HIV-1 stimulates growth of cells derived from Kaposi sarcoma lesions of AIDS patients. *Nature* 1990;345:84-86.
46. Moulignier A, Mikol J, Pialoux G, et al: Cerebral glial tumor and HIV-1 infection. More than a co-incidental association. *Cancer* 1994;74:686-92.
47. Tacconi L, Stapleton S, Signorelli F, et al: AIDS and cerebral astrocytoma. *Clin Neurol N.Surg* 1996;98:149-51.
48. Kramer-Hemmerle S, Kohleisen B, Hohenadi C, et al: HIV type 1 Nef promotes neoplastic transformation of immortalized neural cells. *AIDS Res Hum Retroviruses* 2001;17(7);597-602.
49. Gulotta F, Masini T, Scariato G, et al: Progressive multifocal leukoencephalopathy and gliomas in HIV-negative patient. *Path Res Pract* 1992;188:964-72.
50. Brown HG, Burger PC, Olivi A, et al: Intracranial leiomyosarcoma in a patient with AIDS. *Neuroradiol* 1999;41:35-39.
51. Wisniewski T, Sisti M, Inhirami G, et al: Intracerebral solitary plasmacytoma. *Neurosurgery* 1990;27: 826-29.
52. Desai A, Puttaram S, Chandramuki A, Das S, Ravi V: Serosurveillance of HIV infection at a Neuropsychiatric center in South India. *J Acquired Immunodeficiency Syndromes* 1993;6 (5):534-36.

50

Management of Central Nervous System AIDS Tumours

PK Sethi, R Reddi, NK Sethi

INTRODUCTION

The disease now known as the Acquired Immuno deficiency syndrome (AIDS) was first reported 20 yrs. ago from the University and College of Los Angeles (UCLA) medical centre in the Morbidity and Mortality Weekly Report under the unassuming title of "Pneumocystis pneumonia-Los Angeles."[1] The disease however did not remain quiet for long and assumed pandemic proportions. It has been estimated that almost 22 million people have died of AIDS and that an even larger number are infected today; almost 36 million people, mostly in sub-saharan Africa and Asia are living today with human immunodeficiency virus (HIV) infection or AIDS.[2]

Over the last 20 years as the varied and widespread manifestations of this devastating disease have appeared, it has become clear that no organ nor system of the human body is spared by this apparently ubiqutous virus or the opportunistic infections as a result of infection by the virus. The protean manifestations of this disease can be compared to syphilis, the scourge of medieval times and the early part of the century. It would not be incorrect to say that the special place that syphillis had in Western medicine as the "great imitator" or "great imposter" can now be passed on to AIDS. It is ironic that after the appearance of the AIDS epidemic there has been a resurgence of syphillis.[3] This is no doubt because of concomitant HIV infection as the two diseases may enhance the acquisition and transmission of each other.[4] Since the 1990's the introduction of highly active antiretroviral therapy (HAART) fundamentally altered the epidemic in the United States and the AIDS death rate was shown to decrease. This however does not seem to be the final answer and the world still awaits a "magic bullet" for AIDS akin to penicillin for syphilis.

Dr. Julius Wagner Von Jauregg was awarded the Nobel Prize in Medicine in 1927 for the use of malaria infections for treating "paralytica dementia" (Neurosyphillis).[6] Earlier it was said that knowing syphillis is knowing medicine in its entirety; today the same can be said about AIDS. We use this analogy with syphillis and particularly neurosyphillis to coin and introduce the term "NeuroAIDS"- a term which encompasses the diverse manifestations of AIDS in the neurological system. During further discussion "NeuroAIDS" will be used where appropriate in this text instead of the cumbersome and rather laborious "Neurological manifestations of AIDS".

It has been shown that at least 10% cases of AIDS present as neuro AIDS and that during the course of the disease almost 30-65% patients develop neuro AIDS.[7-10] Impressively one prospective study found that neurologic findings were present in 90% patients if examined by a neurologist.[12] Postmortem examination of brains from patients who died of AIDS revealed abnormalities in almost 90% cases and often more than one disease process was present.[13,14] There seems little doubt that the nervous system, if not the primary target of HIV is certainly one of the most commonly involved parts of the human body.

TUMOURS IN NEURO AIDS

Primary central nervous system lymphoma (PCNSL)

Epidemology: Apart from the effect of direct and opportunistic infections of the nervous system due to HIV, tumours are another important manifestation of neuro AIDS. PCNSL is rare in the general population; the risk of developing PCNSL in AIDS is 100 times that of the general population.[16] Lymphomas account for 2-7% manifestations of neuroAIDS[7,10,11] and is second only to toxoplasmosis as a cause of mass lesions in neuroAIDS. The incidence of lymphoma rises with prolonged survival even with antiretroviral therapy.[16] The increased availability of HAART will possibly result in this manifestation of neuroAIDS becoming more and more common.

Etiology: The reasons for the increased risk of PCNSL in AIDS is not fully understood but is possibly related to defective internal surveillance mechanisms, oncogenic viruses or dysfunction of immune regulation.[17] DNA viruses, especially Epstein Barr Virus (EBV)[15] have been implicated primarily, apart from other DNA virus such as cytomegalovirus (CMV), herpes simplex virus (HSV), human papilloma virus (HPV) and RNA viruses including human T Lymphocytic virus-1 (HTLV-1) and HIV itself.[15,18]

Clinical Presentation and diagnosis: PCNSL presents mainly as a parenchymal mass lesion with signs that reflect its location and increased intracranial pressure. Aphasia, hemiparesis and altered mentation are common early findings but seizures that occur in a third of the patients usually occur later as the illness evolves. Meningeal involvement is infrequent in PCNSL as compared to systemic lymphoma[19] which explains why CSF examination in PCNSL is usually unhelpful.

Three diagnoses comprise a great majority of focal NeuroAIDS disorders- cerebral toxoplasmosis, primary CNS lymphoma and progressive multifocal leuco encephalopathy (PML).[20-22] If patients present with subacute onset of focal cerebral dysfunction and have less than 50 CD4+ lymphocytes/mm3 in blood then more than 90% will have one of the three conditions enumerated earlier.[23] All three conditions most often present with focal hemispherical dysfunction but less commonly present as a cerebrellar or brainstem lesion. Certain features **(Tables 1-2)** favour one of the conditions over the other–however it must be kept in mind that these features do not establish a diagnosis with absolute certainty and can be used only as a guideline until biopsy is done.[23, 24]

As seen in **Table I** and **Table 2** the clinical manifestations of toxoplasmosis, PCNSL and PML overlap significantly. Features used as a bedside guide which favour the diagnosis of PCNSL are:

Table 1 : *Comparison of common Focal Neuro AIDS disorders– Clinical and Diagnostic features.*

	Cerebral. Toxoplasmosis	PCNSL	PML
Clinical			
Onset and Progress	Rapid (Days)	Intemediate (1-2 weeks)	Indolent (Several weeks)
Fever, headache altered sensorium constitutional features	++++	±	±
Diagnosis			
Antitoxoplasma serology	95% positive	-ve	-ve
CSF	abnormal but unhelpful	CSF PCR for EBV	CSF PCR for JC Virus
Biopsy	+ve	+ve	+ve
Therapeutic trial			
antitoxoplasma agents	effective	-ve	-ve
Steroids	-ve	effective	-ve

Table 2: *Comparison of Neuro imaging in Common Focal Neuro AIDS disorders*

	Cerebral Toxoplasmosis	PCNSL	PML
CT/MRI			
Number of lesions	Multiple, occ. single	Multiple or solitary	small lesions, coalesce later
Mass effect and oedema	+++	+++	-ve
Location	grey matter (thalmus or basal ganglia), cortex	Paraventricular white matter, corpus callosum, subependymal extension along ventricular walls	white matter, predilection for occipital and parietal lobes
Contrast enhancement	Distinct, "eccentric" sign	Diffuse, less distinct, nodular, patchy, ring like.	-ve
SPECT/PET	"cold"	"hot"	

(i) History of systemic prophylaxis with trimethoprim/sulphamethoxazole for pneumocystics carinii pneumonia, as this drug inhibits toxoplasma also.[25]

(ii) Negative toxoplasma serology.[20]

(iii) Single lesion on neuroimaging.[26]

(iv) Failure of lesions to regress with empirical antitoxoplasma therapy for 10-14 days.[23]

As has been mentioned earlier, there are no pathognomic features that reliably distinguish a mass lesion due to lymphoma from that caused by toxoplasmosis. Also through the MRI and CT appearance of PCNSL in AIDS and non AIDS patients is similar, some differences have been described.[27] The potential utility of metabolic imaging of focal neuroAIDS disorders using positron emission tomography (PET) with labelled deoxyglucose to measure active tissue metabolism or single photon emission computed tomography (SPECT) to detect metabolically linked cerebral blood flow have been described but thse are not widely available.[28-30] Toxoplasma lesion have been described as "cold" by these methods.[30]

Since the outcome of PCNSL is likely to be influenced by how early the diagnosis is made and treatment started, an aggressive approach is often recommended. This is controversial in AIDS patients with PCNSL as the most common cause of death is opportunistic infection. Intensive chemotherapy after biopsy may result in greater mortality because of infections complicating therapy.[31] Brain Biopsy itself has its inherent complications which may occur more often in AIDS patients because of associated bleeding diasthesis due to thrombocytopenia, coagulopathy or DIC due to various causes. However some authorities still recommend that biopsy should be done immediately in patients with negative toxoplasma serology and neuro imaging abnormalities that suggest PCNSL. Biopsy is the gold standard for diagnosis and is facilitated by CT guided stereotactic approach.[32] However, as PCNSL are exquisitely sensitive to corticosteroids their use may be associated with a rapid and dramatic reduction in lesion size. Preoperative steroid administration may actually hinder attempts to locate the lesion.[33]

Algorithm for focal NeuroAIDS disorder

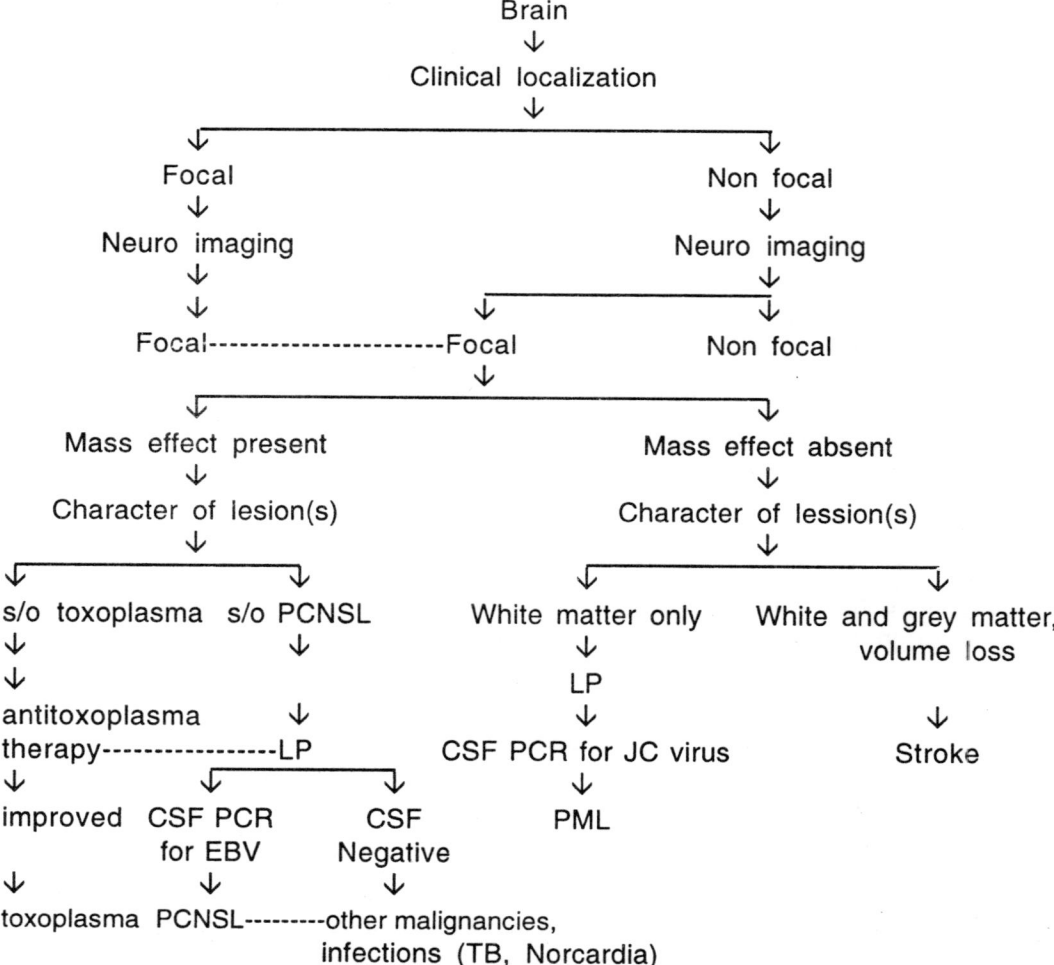

Treatment: PCNSL in non HIV patients is more responsive to chemotherapy and it has been seen that patients with AIDS are often too critically ill to tolerate intensive chemotherapy regimens. As mentioned earlier opportunistic infections result in the death of AIDS patients with PCNSL and this infection rate increases with chemotherapy.[31]

Radiation and corticosteroids result in an initial improvement and prolong survival but prognosis remains poor and death occurs in a few months.[17] Intensive chemotherapy is used after radiotherapy but the neurological and systemic toxicities are significant. Bermudez et al[34] reported a 52% response rate and median survival of 7 months using MACOP-B regimen **(Table 3)**. In their study patients without previous or concurrent opportunistic infections were found to survive longer. Gill et al[35] had a 54% response rate using M-BACOD regimen **(Table 3)** but only a 33% response rate using high dose cytosine arabinoside and methotrexate.

Table 3: *Common Chemotherapy Protocols for Lymphoma in AIDS*

CHOP	: cyclophosphamide, doxorubicin, vincristine prednisolone.
CEOP	: cyclophosphamide, epirubicin, vincristine, prednisolone.
PEN	: prednisolone, etoposide, mitoxantrone
MACOP-B	: methotrexate, bleomycin, doxorubicin, vincristine, prednisolone.
MBACOD	: methotrexate, bleomycin, doxorubicin, cyclophosphamide, vincristine, dexamethasone.

OTHER TUMOURS IN NEURO AIDS.

Metastatic lymphoma complicating systemic non Hodgkins lymphoma (NHL) involve the CNS in an epidural or leptomeningeal pattern.[15] Tumour deposits on spinal nerve roots cause radicular pain and segmental symptoms or signs.[33] Epidural metastases most often to the mid thoracic spine are common and may lead to spinal cord compression.[11] Diagnosis of meningeal lymphoma is made by examination of the CSF. Prophylactic treatment with intra thekeal intrathecal cytosine arabinoside has been successful in some cases.[36]

Other much rarer tumours in Neuroaids include metastases of.

(a) Kaposis sarcoma[7] to brain, parenchyma, dura, spinal cord and nerve roots.

(b) Immunoblastic sarcoma[37] to brain and epidural space.

(c) Extension of nasopharyngeal tumours[38] to brain

(d) Leiomyosarcoma[39] in children

(e) Rhabdomyosarcoma[40] to epidural space.

(f) Plasmacytoma[41] to epidural space.

(g) Progressive astrocytomas[42]

CONCLUSION:

As was mentioned earlier, the incidence of PCNSL in AIDS rises even with antiviral therapy.[16] It has also been seen that as patients with AIDS survive longer, more and more malignacies are diagnosed.[43] Thus we can assume that with the relative widespread availability and decreasing costs of HAART,[44] as well as more effective newer modalities of therapy, malignancies in neuroAIDS will possibly increase in incidence. Neurologists, neurosurgeons and neuropathologists have to be prepared to deal with this diverse and difficult neurological load even as the medical fraternity frantically continues the search for an elusive "magic bullet".

REFERENCES:

1. Pneumocystis pneumonia–Los Angeles–MMWR Morb Mortai Weekly Rep. 1981;30:250-52.
2. UNAIDS, WHO: AIDS epidemic update. December 2000. Geneva: Joint United Nations Programme on HIV/AIDS, 2000.
3. Ansell DA, Hu T, Straus M, et al: HIV and Syphilis seroprevalance among clients with sexually transmitted disease attending a walk in clinic at Cook County Hospital. Sex Transm Dis. 1993; 2:93-97.
4. Theus SA, Harrich DA, Gagnoi R, et al: Treponema pallidum, lipoproteins and synthetic lipoprotein analogues induce human immunodeficiency virus type I gene expression in monocytes via NFkB activation. *J Infect Dis* 1998;177:941-50.
5. Update: trends in AIDS incidence, deaths and prevalance, 1996: MMWR Morb Mortal Wkly Rep 1997;46:165-73.
6. Autin SC, Stolley PD, Lasky J: The history of malariotherapy for neurosyphillis. *JAMA* 1992;268: 516-19.
7. Levy RM, Bredesen DE, Rosenblum ML: Neurological manifestations of AIDS: experience at UCSF and review of the literature. *J Neurosurg* 1985;62:475-95.
8. McArthur JC: Neurologic manifestations of AIDS Medicine 1987;66:408-37.
9. Berger JR, Moskowitz L, Fischl M, Kelley RE: Neurologic disease as the presenting manifestation of acquired immunodeficiency syndrome. *South Med J* 1987;80:638-83.
10. Koppel BS, Wormser GP, Tuchman AJ, Maayan S, Hewlett D Jr., Daras M: Central nervous system involvement in patients with AIDS. *Acta Neurol Scand* 1985;71:337-53.
11. Snider WD, Simpson DM, Nielsen S, Gold JWM, Metroka CE, Posner JB: Neurological complication of acquired immunodeficiency syndrome: analysis of 50 patients. *Ann Neurol* 1983;14:403-18.
12. Malouf R, Jacquette G, Dobkin J, Brust JCM: Neurologic diseases in human immunodeficiency virus infected drug abusers. *Arch Neurol* 1990;47:1002-07.
13. Lantos PL, Mclaughlin JE, Scholtz CL, Berry CL, Tighe JR: Neuropathology of the brain in HIV infection. *Lancet* 1989;1:309-11.
14. Anders KH, Guerra WF, Tomiyasu U, Verity MA, Vinters HV: The Neuropathology of AIDS. UCLA experiences and review. *Am J Pathology* 1986;124:537-58.
15. So YT, Choucair A, Daris RL, et al: Neoplasms of the central nervous system in acquired immunodeficiency syndrome. In: Rosenblum ML, ed. AIDS and the nervous system. New York: Raven Press, 1988.

16. Conford ME, Holden JK, Boyd MC, Berry K. Vinters HV: Neuropathology of the Acquired Immunodeficiency syndrome. (AIDS). Report of 39 autopsies from Vancouver, British Columbia. *Can J Neurol Sci* 1992;84:516-29.
17. Levine AM: AIDS related malignancies: the emerging epidemic. *J Natl Cancer Inst* 1993;85: 1382-97.
18. Zeigler JL, Beckstead JH, Voldberding PA, et al: Non Hodgkins lymphoma in 90 homosexual men: Relation to generalized lymphadenopathy and the acquired immunodeficiency syndrome. *N Engl J Med* 1984;311:565-70.
19. Ioachim HL: Biopsy diagnosis in human immunodeficiency virus infection and acquired immunodeficiency syndrome. *Arch Pathol Lab Med* 1990;114:284-94.
20. Navia B, Petito C, Gold J, et al: Cerebral toxoplasmosis complicating the acquired immunodeficiency syndrome: clinical and neuropathological findings in 27 patients. *Ann Neurol* 1986;19:224-38.
21. Berger J, Kaszovitz B, Donovan Post M, Dickinson G: Progressive multifocal leucoencephalopathy in association with human immunodeficiency virus infection: a review of the literature with a report of sixteen cases. *Ann Intern Med* 1987;107:78-87.
22. So Y, Beckstead J, Davis R: Primary central nervous system lymphoma in acquired immune deficiency syndrome: a clinical and pathological study. *Ann Neurol* 1986;20:566-72.
23. Price RW. Neurologic Diseases In: Dolin R, Masur H, Saag MS, ed: Aids Therapy, New York.Churchill Livingstone, 1999.
24. Price RW: Neurological complications of HIV infection. *Lancet* 1996;348:445-52.
25. Antinori A, Ammassari A, De Luca A, et al: Diagnosis of AIDS related focal brain lesions: a decision-making analysis based on clinical and neuro radiologic characteristics combined with polymerare chain reaction assys in CSF. *Neurology* 1997;48:687.
26. Laissy JP, Soyer P, Tebboune J, et al: Contrast enhanced fast MRI in differentiating brain toxoplasmosis and lymphoma in AIDS patients. *Jr Comput Assist Tomogr* 1996;20:417.
27. Lee Y-Y, Bruner JM, Van Tassel P, Libshitz HI: Primary central nervous system lymphoma: CT and pathologic correlation. *Am J Rad* 1986;147:747-52.
28. Hoffman JM, Waskin HA, Schifter T, et al: FDG-PET in differtiating lymphoma from non malignant central nervous system lesions in patients with AIDS. *J Nuc Med* 1993;34:567.
29. Lorberboym M, Estok L, Machac J, et al: Rapid differential diagnosis of cerebral toxoplasmosis and primary central nervous system lymphoma by thallium 201 SPECT. *J Nuc Med* 1996;37:1150.
30. Catafau AM, Sola M, Lomena FJ, et al: Hyperperfusion and early technetium -99m HMPAO SPECT appearance of central nervous system toxoplasmosis. *J Nuc Med* 1994;35:1091.
31. Formanti SC, Gill Ps, Rarick M, et al: Primary central nervous system lymphoma in AIDS: Results of radiation therapy *Cancer* 1989;63:1101-07.
32. Feiden W, BiseK, Stende U, Pfister H-W, Moller AA: The stereotactic biopsy diagnosis of focal intracerebral lesions in AIDS patients. *Acta Neurol Scand* 1993;87:228-23, 37:1160-64.
33. De Angelis LM, Yahalom J, Heinemann MH: Cirrincione C, Thaler HT, Krol G: Primary CNS Lymphoma: Combined treatment with chemotherapy and radiotherapy. *Neurology* 1990;40:80-86.
34. Bermudez MA, Grant KM. Rodvein R, Mendes F: Non Hodgkins lymphoma in a population with or at risk for acquired immuno deficiency syndrome: indications for intensive chemotherapy. *Am JMed* 1989;86:71-76.
35. Gill PS, Levine AM, Kralis M, et al: AIDS related malignant lymphoma: results of prospective treatment trials. *J Chin oncol* 197;5:1322-28.
36. Levine AM, Wemz JC, Kaplan L, et al: Low close chemotherapy with central nervous system prophylaxis and zidovidine maintenance in AIDS related lymphoma. *JAMA* 1991;266:84-88.
37. Nielsen SL, Davis RL: Neuropathology of acquired immunodeficiency syndrome. In Rosenblum ML, ed. AIDS and the nervous system. New York: Raven Press, 1988;155-81.
38. Oksenhendler E, Lida H, D'Agay M-F, et al: Tumoral nasopharyngeal lymphoid hyperplasia in human immunodeficiency virus infected patients. *Arch Intern Med* 1989;149:2359-61.

39. Chadwick EG, Lconnor E J, Hancon CG, et al: Tumours of smooth muscle origin in HIV infected children. *JAMA* 1990;263:3182-84.
40. Case Records. Case 9-1986. *N Engl J Med* 1986;314:629-40.
41. Israel AM, Koziner B, Straus DJ: Plasmacytoma and the acquired immunodeficiency syndrome. *Ann Intern Med* 1983;99:635-36.
42. Gasnault J, Roux FX, Vedrenne C: Cerebral astrocytoma in association with HIV infection, *J Neurol Neurosurg Psych* 1988;51:422-24.
43. Tirelli U, Franchesi S, Carbone A: Malignant tumours in patients with HIV infection. BMJ 1994;308:1148-53.
44. Kent AS: AIDS—The first 20 years N. *Engl J Med* 2001;344:1764-71.